Discrete Hilbert-Type Inequalities

BY

Bicheng Yang

Department of Mathematics, Guangdong Education Institute, Guangdong Guangzhou, 510303,
People Republic of China
E-mail: bcyang@pub.guangzhou.gd.cn

CONTENTS

CHAPTERS

FOREWORD

In 1934 G. H. Hardy, J. E. Littlewood and G. Polya published the book entitled "Inequalities", in which a few theorems about Hilbert-type inequalities with the homogeneous kernels of degree -1 and some special cases were considered. In 1991, Lizhi Xu first raised the way of weight coefficients to make a strengthened version of Hilbert's inequality. In 1998, by applying the way of weight coefficients and introducing an independent parameter and the beta function, a best extension of Hilbert's integral inequality was published in Journal of Mathematical Analysis and Applications by the author. In 2004 the author added two pairs of conjugate exponents and provided an extended Hilbert's integral inequality. In 2009, the author published by Science Press (China) his book entitled "The Norm of Operator and Hilbert-Type Inequalities". In that book, two large classes of Hilbert-type inequalities with the homogeneous kernels of a negative number degree, including integrals and series were discussed. However the author did not consider it an important case of Hardy-type inequalities. In October 2009, the author published an e-book entitled "Hilbert-Type Integral Inequalities" by Bentham Science Publishers Ltd. In that book, the author studied the broader case of Hilbert-type integral inequalities with the homogeneous kernels of a real number degree, which provide the best recent extensions of corresponding results. A number of equivalent forms as well as their reverses with several extended multiple inequalities in a number of particular cases are considered.

The book is divided into six chapters. In Chapter 1, some preliminary materials on the theory and methods of Hilbert-type inequalities, including the classical Hilbert's inequality are discussed. Chapter 2 deals with an optimization of the methods of estimating the series and the weight coefficients. Some introductory theorems of improving the methods of Euler-Maclaurin summation formula are analyzed. In Chapter 3, by using the way of weight coefficients some fundamental theorems and corollaries on the discrete Hilbert-type inequalities with the homogeneous kernel of degree -1 are provided. The proofs regarding the best possible property of the constant factors are left to be studied in Chapter 4. In Chapter 4, some discrete Hilbert-type inequalities and their reverses with the general homogeneous kernel and the best constant factors are considered. These provide extensions of certain results of Chapter 3. By applying the improved Euler-Maclaurin summation formula and notions from Real Analysis, some particular examples are given. In Chapter 5 based upon some theorems of Chapter 4 and by applying techniques from Real Analysis, the author has explained how to use particular parameters to formulate some new Hilbert-type inequalities and their reverses with the best constant factors. In addition, a class of Hilbert-type inequalities with the general measurable kernels is considered. In Chapter 6, the author has decently formulated some lemmas and obtained two equivalent multiple Hilbert-type inequalities and their reverses with the homogeneous kernel of a real number degree. These inequalities are the best extensions of the corresponding inequalities in Chapter 4. Some special examples are also studied.

The author has succeeded to present in this book an extensive account of several Hilbert-type inequalities in a self contained and rigorous manner. The book will be very useful not only to graduate students who study inequalities in the broader domain of Mathematical Analysis but also to research mathematicians who need the latest information to refer in.

Themistocles M. Rassias

National Technical University of Athens
Athens
Greece

PREFACE

One Hundred years ago, in 1908, H. Wely published the well known Hilbert's inequality. In 1925, G. H. Hardy gave a best extension of it by introducing one pair of conjugate exponents (p, q), named as Hardy-Hilbert's inequality. The Hilbert-type inequalities are a more wide class of analysis inequalities which are with the bilinear kernels, including Hardy-Hilbert's inequality as the particular case. These inequalities are important in analysis and its applications. By making a great effort of mathematicians in the world at about one hundred years, the theory of Hilbert-type inequalities has now come into being. This book is a monograph about the theory of discrete Hilbert-type inequalities with the homogeneous kernels of real number-degree and its applications. Using the methods of series summation, Real Analysis, Functional Analysis and Operator Theory, and following the way of weight coefficients, the author introduces a few independent parameters to establish a number of discrete Hilbert-type inequalities with the homogeneous kernels of real number-degree and the best constant factors, including some multiple inequalities. The equivalent forms and the reverses with the best constant factors are also considered. As application, the author also considers some discrete Hilbert-type inequalities with the non-homogeneous kernels and a large number of particular examples.

For reading and understanding this book, readers should hold the basic knowledge of real analysis and functional analysis. This book is suited to the people who are interested in the fields of analysis inequalities and real analysis. The author expects this book to help many readers to make good progress in research for discrete Hilbert-type inequalities and their applications.

Bicheng Yang

Guangzhou, Guangdong,
P. R. China

CHAPTER 1

Introduction

Abstract: In this chapter, we introduce some evolvements for the theory and methods of Hilbert-type inequalities, including Hilbert's inequality. We must emphasize some excellent works on discrete Hilbert-type inequalities and Hilbert-type operators with multi-parameters in recent years, which have made more developments in this context. This chapter will enhance the understanding of the readers of the content of the following several chapters.

1.1. HILBERT'S INEQUALITIES AND HILBERT'S OPERATOR

1.1.1. RESEARCH BACKGROUND OF HILBERT'S INEQUALITIES AND HILBERT'S OPERATOR

In 1908, H. Weyl published the following well known Hilbert's inequality (Weyl ID 1908) [1]: If $\{a_n\}$ and $\{b_n\}$ are real sequences, satisfying $0<\sum_{n=1}^{\infty}a_n^2<\infty$ and $0<\sum_{n=1}^{\infty}b_n^2<\infty$, then we have

$$\sum_{n=1}^{\infty}\sum_{m=1}^{\infty}\frac{1}{m+n}a_m b_n<\pi(\sum_{n=1}^{\infty}a_n^2\sum_{n=1}^{\infty}b_n^2)^{\frac{1}{2}}, \tag{1.1.1}$$

where the constant factor π is the best possible. We named (1.1.1) as Hilbert's inequality. The best possible property of the constant factor π was proved by Schur in 1911 (Schur JM 1991) [2]. He also gave the following integral analogue of (1.1.1) at the same time: If $f(x)$ and $g(x)$ are measurable functions, $0<\int_0^{\infty}f^2(x)dx<\infty$ and $0<\int_0^{\infty}g^2(x)dx<\infty$, then we get

$$\int_0^{\infty}\int_0^{\infty}\frac{1}{x+y}f(x)g(y)dxdy$$
$$<\pi(\int_0^{\infty}f^2(x)dx\int_0^{\infty}g^2(x)dx)^{\frac{1}{2}}, \tag{1.1.2}$$

where the constant factor π is still the best possible. We call (1.1.2) as Hilbert's integral inequality. Inequalities (1.1.1) and (1.1.2) are important in analysis and applications. We can see a number of improvements and extensions in vast mathematics literature, especially in the books (Hardy CMP 1934) [3], (Mitrinovic KAP 1991) [4], (Kuang SSTP 2004) [5], (Hu WUP 2007) [6].

We may express inequality (1.1.1) by using the form of operator as follows: If l^2 is a space of real sequences, and $T:l^2\to l^2$ is a linear operator, for any $a=\{a_m\}_{m=1}^{\infty}\in l^2$, there exists a $c=\{c_n\}_{n=1}^{\infty}\in l^2$, satisfying

$$c_n:=(Ta)(n)=\sum_{m=1}^{\infty}\frac{a_m}{m+n},\ n\in\mathbf{N} \tag{1.1.3}$$

($\mathbf{N}$ is the set of positive integers). Hence, for any $b=\{b_n\}_{n=1}^{\infty}\in l^2$, we may indicate the inner product of Ta and b as follows:

$$(Ta,b)=(c,b)=\sum_{n=1}^{\infty}(\sum_{m=1}^{\infty}\frac{a_m}{m+n})b_n$$
$$=\sum_{n=1}^{\infty}\sum_{m=1}^{\infty}\frac{1}{m+n}a_m b_n. \tag{1.1.4}$$

Expressing the norm of a as $\|a\|_2=(\sum_{n=1}^{\infty}a_n^2)^{\frac{1}{2}}$, in view of (1.1.4), inequality (1.1.1) may be rewritten as

$$(Ta,b)<\pi\|a\|_2\|b\|_2, \tag{1.1.5}$$

where $\|a\|_2,\|b\|_2>0$. We may prove that T is a bounded operator and obtain the norm as $\|T\|=\pi$ (Wilhelm AJM 1950) [7]. We call T Hilbert's operator with the kernel $\frac{1}{m+n}$. For $\|a\|_2>0$, the equivalent form of (1.1.5) is given as $\|Ta\|_2<\pi\|a\|_2$, e.t.

$$\sum_{n=1}^{\infty}(\sum_{m=1}^{\infty}\frac{a_m}{m+n})^2<\pi^2\sum_{n=1}^{\infty}a_n^2, \tag{1.1.6}$$

where the constant factor π^2 is still the best possible. Obviously, inequalities (1.1.6) and (1.1.1) are equivalent (Hardy CUP 1934) [3].

Similarly, if $L^2(0,\infty)$ is a real function space, we may define Hilbert's integral operator as $\tilde{T}:L^2(0,\infty)\to L^2(0,\infty)$, for any $f\in L^2(0,\infty)$, there exists a $h=\tilde{T}f\in L^2(0,\infty)$, satisfying

$$(\tilde{T}f)(y)=h(y):=\int_0^{\infty}\frac{f(x)}{x+y}dx,\ y\in(0,\infty). \tag{1.1.7}$$

Hence for any $g\in L^2(0,\infty)$, we may still indicate the inner product of $\tilde{T}f$ and g as follows:

$$(\tilde{T}f,g)=\int_0^\infty(\int_0^\infty \frac{1}{x+y}f(x)dx)g(y)dy$$
$$=\int_0^\infty\int_0^\infty \frac{1}{x+y}f(x)g(y)dxdy. \quad (1.1.8)$$

Setting the norm of f as $\|f\|_2=(\int_0^\infty f^2(x)dx)^{\frac{1}{2}}$, if $\|f\|_2,\|g\|_2>0$, then (1.1.2) may be rewritten as

$$(\tilde{T}f,g)<\pi\|f\|_2\|g\|_2. \quad (1.1.9)$$

We have $\|\tilde{T}\|=\pi$ (Carleman U 1923) [8] and inequality $\|\tilde{T}f\|_2<\pi\|f\|_2$, which may be rewritten to the equivalent form of (1.1.2) as follows (Hardy CUP 1934) [3]:

$$\int_0^\infty(\int_0^\infty \frac{f(x)}{x+y}dx)^2dy<\pi^2\int_0^\infty f^2(x)dx, \quad (1.1.10)$$

where the constant factor π^2 is still the best possible. It is obvious that inequality (1.1.10) is the integral analogue of (1.1.6).

1.1.2. THE MORE ACCURATE HILBERT'S INEQUALITY

If we set the subscripts m, n of the double series from 0 to infinity, then we may rewrite inequality (1.1.1) equivalently in the following form:

$$\sum_{n=0}^\infty\sum_{m=0}^\infty \frac{a_m b_n}{m+n+2}<\pi(\sum_{n=0}^\infty a_n^2\sum_{n=0}^\infty b_n^2)^{\frac{1}{2}}, \quad (1.1.11)$$

where the constant factor π is still the best possible. Obviously, we may raise the following question: Is there a positive constant $\alpha(<2)$, that makes inequality (1.1.11) still valid as we replace 2 by α in the kernel $\frac{1}{m+n+2}$? The answer is positive. That is the following more accurate Hilbert's inequality (for short, Hilbert's inequality) (Hardy CUP 1934) [3]:

$$\sum_{n=0}^\infty\sum_{m=0}^\infty \frac{a_m b_n}{m+n+1}<\pi(\sum_{n=0}^\infty a_n^2\sum_{n=0}^\infty b_n^2)^{\frac{1}{2}}, \quad (1.1.12)$$

where the constant factor π is the best possible. Since for $a_m,b_n\ge 0,\alpha\ge 1$, we find

$$\sum_{n=0}^\infty\sum_{m=0}^\infty \frac{a_m b_n}{m+n+\alpha}\le\sum_{n=0}^\infty\sum_{m=0}^\infty \frac{a_m b_n}{m+n+1},$$

then by (1.1.12) and for $\alpha\ge 1$, we have

$$\sum_{n=0}^\infty\sum_{m=0}^\infty \frac{a_m b_n}{m+n+\alpha}<\pi(\sum_{n=0}^\infty a_n^2\sum_{n=0}^\infty b_n^2)^{\frac{1}{2}}. \quad (1.1.13)$$

For $1\le\alpha<2$, inequality (1.1.13) is a refinement of (1.1.11), which is equivalently a refinement of (1.1.1). Obviously, we have a refinement of (1.1.6), which is equivalent to (1.1.13) as follows:

$$\sum_{n=0}^\infty(\sum_{m=0}^\infty \frac{a_m}{m+n+\alpha})^2<\pi^2\sum_{n=0}^\infty a_n^2(1\le\alpha<2). (1.1.14)$$

For $0<\alpha<1$, in 1936, Ingham gave (Ingham JLMC 1936)[9]: If $\alpha\ge\frac{1}{2}$, then we have

$$\sum_{n=0}^\infty\sum_{m=0}^\infty \frac{a_m a_n}{m+n+\alpha}\le\pi\sum_{n=0}^\infty a_n^2; \quad (1.1.15)$$

if $0<\alpha<\frac{1}{2}$, then we get

$$\sum_{n=0}^\infty\sum_{m=0}^\infty \frac{a_m a_n}{m+n+\alpha}\le\frac{\pi}{\sin(\alpha\pi)}\sum_{n=0}^\infty a_n^2. \quad (1.1.16)$$

It is interesting that if we set $x=X+\frac{\alpha}{2}, y=Y+\frac{\alpha}{2}$, $F(X)=f(X+\frac{\alpha}{2})$ and $G(Y)=g(Y+\frac{\alpha}{2})$ ($\alpha\in\mathbf{R}$, **R** is the set of real numbers) in (1.1.2), then we obtain

$$\int_{-\frac{\alpha}{2}}^\infty\int_{-\frac{\alpha}{2}}^\infty \frac{1}{X+Y+\alpha}F(X)G(Y)dXdY$$
$$<\pi(\int_{-\frac{\alpha}{2}}^\infty F^2(X)dX\int_{-\frac{\alpha}{2}}^\infty G^2(X)dX)^{\frac{1}{2}}. \quad (1.1.17)$$

It is said that for $\alpha\ge\frac{1}{2}$, inequality (1.1.17) is an integral analogue of (1.1.5) (for $G=F$) and for $0<\alpha<\frac{1}{2}$, inequality (1.1.17) is not an integral analogue of (1.1.16), since two constant factors are different.

By using the improved Euler-Maclaurin summation formula and introducing one parameter, a few authors gave some more accurate Hilbert-type inequalities as (1.1.13) (see (Yang MIA 2004) [10], (Yang JXNU 2005)[11], (Yang CM 2005)[12], (Yang AMS 2006) [13], (Yang JM 2007)[14], (Yang AMS 2007)[15], (Yang IMF 2007)[16], (Zhong JZU 2008)[17]).

1.1.3. HILBERT'S INEQUALITY WITH ONE PAIR OF CONJUGATE EXPONENTS

In 1925, by introducing one pair of conjugate exponents $(p,q)(\frac{1}{p}+\frac{1}{q}=1)$, Hardy and Riesz gave an extension of (1.1.1) as follows (Hardy PLMS 1925) [18]: If $p>1$, $a_n,b_n\ge 0$, such that $0<\sum_{n=1}^\infty a_n^p<\infty$ and $0<\sum_{n=1}^\infty b_n^q<\infty$, then we obtain

$$\sum_{n=1}^\infty\sum_{m=1}^\infty \frac{a_m b_n}{m+n}<\frac{\pi}{\sin(\frac{\pi}{p})}(\sum_{n=1}^\infty a_n^p)^{\frac{1}{p}}(\sum_{n=1}^\infty b_n^q)^{\frac{1}{q}}, \quad (1.1.18)$$

where the constant factor $\frac{\pi}{\sin(\pi/p)}$ is the best possible. The equivalent form of (1.1.18) is as follows:

$$\sum_{n=1}^\infty(\sum_{m=1}^\infty \frac{a_m}{m+n})^p<[\frac{\pi}{\sin(\pi/p)}]^p\sum_{n=1}^\infty a_n^p, \quad (1.1.19)$$

where the constant factor $[\frac{\pi}{\sin(\pi/p)}]^p$ is still the best possible. In the same way, inequalities (1.1.12) and (1.1.14) (for $\alpha=1$) may be extended to the following equivalent forms (Hardy CUP 1934) [3]:

$$\sum_{n=0}^{\infty}\sum_{m=0}^{\infty}\frac{a_m b_n}{m+n+1}<\frac{\pi}{\sin(\frac{\pi}{p})}(\sum_{n=0}^{\infty}a_n^p)^{\frac{1}{p}}(\sum_{n=0}^{\infty}b_n^q)^{\frac{1}{q}}, \quad (1.1.20)$$

$$\sum_{n=0}^{\infty}(\sum_{m=0}^{\infty}\frac{a_m}{m+n+1})^p<[\frac{\pi}{\sin(\pi/p)}]^p\sum_{n=0}^{\infty}a_n^p, \quad (1.1.21)$$

where the constant factors $\frac{\pi}{\sin(\pi/p)}$ and $[\frac{\pi}{\sin(\pi/p)}]^p$ are the best possible. And the equivalent integral analogues of (1.1.18) and (1.1.19) are given as follows:

$$\int_0^{\infty}\int_0^{\infty}\frac{1}{x+y}f(x)g(y)dxdy$$
$$<\frac{\pi}{\sin(\frac{\pi}{p})}(\int_0^{\infty}f^p(x)dx)^{\frac{1}{p}}(\int_0^{\infty}g^q(x)dx)^{\frac{1}{q}}, \quad (1.1.22)$$

$$\int_0^{\infty}(\int_0^{\infty}\frac{f(x)}{x+y}dx)^p dy<[\frac{\pi}{\sin(\frac{\pi}{p})}]^p\int_0^{\infty}f^p(x)dx. \quad (1.1.23)$$

We call (1.1.18) and (1.1.20) as Hardy-Hilbert's inequality and call (1.1.22) as Hardy-Hilbert's integral inequality.

Inequality (1.1.20) may be expressed in the form of operator as follows: If l^p is a space of real sequences, $T_p: l^p\to l^p$ is a linear operator, such that for any non-negative sequence $a=\{a_m\}_{m=1}^{\infty}\in l^p$, there exists $T_p a=c=\{c_n\}_{n=1}^{\infty}\in l^p$, satisfying

$$c_n:=(T_p a)(n)=\sum_{m=0}^{\infty}\frac{a_m}{m+n+1},\ n\in\mathbf{N}_0 \quad (1.1.24)$$

($\mathbf{N}_0=\mathbf{N}\cup\{0\}$). And for any non-negative sequence $b=\{b_n\}_{n=1}^{\infty}\in l^q$, we can indicate the formal inner product of $T_p a$ and b as:

$$(T_p a,b)=\sum_{n=0}^{\infty}(\sum_{m=0}^{\infty}\frac{a_m}{m+n+1})b_n$$
$$=\sum_{n=0}^{\infty}\sum_{m=0}^{\infty}\frac{1}{m+n+1}a_m b_n. \quad (1.1.25)$$

Setting the norm of a as $\|a\|_p=(\sum_{n=0}^{\infty}a_n^p)^{\frac{1}{p}}$, then inequality (1.1.20) may be rewritten as follows

$$(T_p a,b)<\frac{\pi}{\sin(\pi/p)}\|a\|_p\|b\|_q, \quad (1.1.26)$$

where $\|a\|_p,\|b\|_q>0$. We call T_p Hardy-Hilbert's operator with the kernel $\frac{1}{m+n+1}$.

Similarly, if $L^p(0,\infty)$ is a real function space, we may define the following Hardy-Hilbert's integral operator $\tilde{T}_p: L^p(0,\infty)\to L^p(0,\infty)$ as: for any $f(\geq 0)\in L^p(0,\infty)$, there exists a $h=\tilde{T}_p f\in L^p(0,\infty)$, satisfying

$$(\tilde{T}_p f)(y)=h(y):=\int_0^{\infty}\frac{1}{x+y}f(x)dx,$$
$$y\in(0,\infty). \quad (1.1.27)$$

And for any $g(\geq 0)\in L^q(0,\infty)$, we can indicate the formal inner product of $\tilde{T}_p f$ and g as follows:

$$(\tilde{T}_p f,g)=\int_0^{\infty}\int_0^{\infty}\frac{1}{x+y}f(x)g(y)dxdy. \quad (1.1.28)$$

Setting the norm of f as $\|f\|_p=(\int_0^{\infty}f^p(x)dx)^{\frac{1}{p}}$, then inequality (1.1.22) may be rewritten as follows

$$(\tilde{T}_p f,g)<\frac{\pi}{\sin(\pi/p)}\|f\|_p\|g\|_q. \quad (1.1.29)$$

If (p,q) is not one pair of conjugate exponents, we get the following result (Hardy CUP 1934) [3]: If $p>1, q>1, \frac{1}{p}+\frac{1}{q}\geq 1, 0<\lambda=2-\frac{1}{p}+\frac{1}{q}\leq 1$, then

$$\sum_{n=1}^{\infty}\sum_{m=1}^{\infty}\frac{a_m b_n}{(m+n)^{\lambda}}\leq K(\sum_{n=1}^{\infty}a_n^p)^{\frac{1}{p}}(\sum_{n=1}^{\infty}b_n^q)^{\frac{1}{q}}, \quad (1.1.30)$$

where $K=K(p,q)$ relates to p,q; only for $\frac{1}{p}+\frac{1}{q}=1, \lambda=2-\frac{1}{p}+\frac{1}{q}=1$, the constant factor K is the best possible. The integral analogues of (1.1.30) are given as follows:

$$\int_0^{\infty}\int_0^{\infty}\frac{1}{(x+y)^{\lambda}}f(x)g(y)dxdy$$
$$\leq K(\int_0^{\infty}f^p(x)dx)^{\frac{1}{p}}(\int_0^{\infty}g^q(x)dx)^{\frac{1}{q}}. \quad (1.1.31)$$

We also find an extension of (1.1.31) as (Mitrinovic CAP 1991) [4]: If $p>1, q>1, \frac{1}{p}+\frac{1}{q}>1$, $0<\lambda=2-\frac{1}{p}+\frac{1}{q}<1$, then we get

$$\int_{-\infty}^{\infty}\int_{-\infty}^{\infty}\frac{f(x)g(y)}{|x+y|^{\lambda}}dxdy\leq k(p,q)$$
$$\times(\int_{-\infty}^{\infty}f^p(x)dx)^{\frac{1}{p}}(\int_{-\infty}^{\infty}g^q(x)dx)^{\frac{1}{q}}. \quad (1.1.32)$$

For $f(x),g(x)=0, x\in(-\infty,0]$, inequality (1.1.32) reduces to (1.1.31). Levin also studied the expression forms of the constant factors in (1.1.30) and (1.1.31) (Levin JIMS 1937) [19], but he could not prove their best possible property. In 1951, Bonsall considered the case of (1.1.31) in the general kernel (Bonsall JMOS 1951) [20].

1.1.4. A HILBERT-TYPE INEQUALITY WITH THE HOMOMGENEOUS KERNEL OF DEGREE -1 AND SOME PARTICULAR CASES

If $\alpha\in\mathbf{R}$, the function $k(x,y)$ is measurable in $(0,\infty)\times(0,\infty)$, satisfying for any $x,y,u>0$,

$k(ux,uy)=u^{\alpha}k(x,y)$, we name $k(x,y)$ as the homogeneous function of degree α.

Supposing that $(p,q)(\frac{1}{p}+\frac{1}{q}=1)$ is one pair of conjugate exponents with $p>1$, $k(x,y)\geq 0$ is a homogeneous function of degree -1 in $(0,\infty)\times(0,\infty)$, if $k=\int_0^\infty k(u,1)u^{-1/p}du$ is finite, then we have $k=\int_0^\infty k(1,v)v^{-\frac{1}{q}}dv$ and the following equivalent inequalities (Hardy CUP 1934) [3]:

$$\int_0^\infty\int_0^\infty k(x,y)f(x)g(y)dxdy \leq k(\int_0^\infty f^p(x)dx)^{\frac{1}{p}}(\int_0^\infty g^q(x)dx)^{\frac{1}{q}}, \quad (1.1.33)$$

$$\int_0^\infty(\int_0^\infty k(x,y)f(x)dx)^p dy \leq k^p\int_0^\infty f^p(x)dx, \quad (1.1.34)$$

where the constant factors k and k^p are the best possible.

If both $k(u,1)u^{-1/p}$ and $k(1,u)u^{-1/q}$ are decreasing functions in $(0,\infty)$, then we have the following equivalent inequalities:

$$\sum_{n=1}^\infty\sum_{m=1}^\infty k(m,n)a_m b_n \leq k(\sum_{n=1}^\infty a_n^p)^{\frac{1}{p}}(\sum_{n=1}^\infty b_n^q)^{\frac{1}{q}}, \quad (1.1.35)$$

$$\sum_{n=1}^\infty(\sum_{m=1}^\infty k(m,n)a_m)^p \leq k^p\sum_{n=1}^\infty a_n^p. \quad (1.1.36)$$

For $0<p<1$, if $k=\int_0^\infty k(u,1)u^{-1/p}du$ is finite, then we have the reverses of (1.1.33) and (1.1.34) (Note: we have not seen any proof of (1.1.33) - (1.1.36) and the reverse particular cases in (Hardy CUP 1934) [3]).

We name $k(x,y)$ the kernel of inequalities (1.1.33) - (1.1.36). If all the integrals and series in the right hand side of inequalities (1.1.33)-(1.1.36) are positive, we still obtain the following particular cases (Hardy CUP 1934) [3]: (1) For $k(x,y)=\frac{1}{x+y}$ in (1.1.33) -(1.1.36), they deduce to (1.1.22), (1.1.23), (1.1.18) and (1.1.19); (2) for $k(x,y)=\frac{1}{\max\{x,y\}}$ in (1.1.33)-(1.1.36), they deduce to the following two pairs of equivalent forms:

$$\int_0^\infty\int_0^\infty \frac{1}{\max\{x,y\}}f(x)g(y)dxdy < pq(\int_0^\infty f^p(x)dx)^{\frac{1}{p}}(\int_0^\infty g^q(x)dx)^{\frac{1}{q}}, \quad (1.1.37)$$

$$\int_0^\infty(\int_0^\infty \frac{f(x)}{\max\{x,y\}}dx)^p dy < (pq)^p\int_0^\infty f^p(x)dx; \quad (1.1.38)$$

$$\sum_{n=1}^\infty\sum_{m=1}^\infty \frac{a_m b_n}{\max\{m,n\}} < pq(\sum_{n=1}^\infty a_n^p)^{\frac{1}{p}}(\sum_{n=1}^\infty b_n^q)^{\frac{1}{q}}, \quad (1.1.39)$$

$$\sum_{n=1}^\infty(\sum_{m=1}^\infty \frac{a_m}{\max\{m,n\}})^p < (pq)^p\sum_{n=1}^\infty a_n^p; \quad (1.1.40)$$

(3) for $k(x,y)=\frac{\ln(x/y)}{x-y}$ in (1.1.33)-(1.1.36), they deduce to the following two pairs of equivalent forms:

$$\int_0^\infty\int_0^\infty \frac{\ln(x/y)}{x-y}f(x)g(y)dxdy < [\frac{\pi}{\sin(\frac{\pi}{p})}]^2(\int_0^\infty f^p(x)dx)^{\frac{1}{p}}(\int_0^\infty g^q(x)dx)^{\frac{1}{q}}, \quad (1.1.41)$$

$$\int_0^\infty[\int_0^\infty \frac{\ln(x/y)}{x-y}f(x)dx]^p dy < [\frac{\pi}{\sin(\frac{\pi}{p})}]^{2p}\int_0^\infty f^p(x)dx; \quad (1.1.42)$$

$$\sum_{n=1}^\infty\sum_{m=1}^\infty \frac{\ln(m/n)a_m b_n}{m-n} < [\frac{\pi}{\sin(\frac{\pi}{p})}]^2(\sum_{n=1}^\infty a_n^p)^{\frac{1}{p}}(\sum_{n=1}^\infty b_n^q)^{\frac{1}{q}}, \quad (1.1.43)$$

$$\sum_{n=1}^\infty[\sum_{m=1}^\infty \frac{\ln(m/n)a_m}{m-n}]^p < [\frac{\pi}{\sin(\frac{\pi}{p})}]^{2p}\sum_{n=1}^\infty a_n^p. \quad (1.1.44)$$

Note. The constant factors in the above inequalities are all the best possible. We name (1.1.39) and (1.1.43) as Hardy–Littlewood–Polya's inequalities, or for short, H-L-P inequalities, and call (1.1.37) and (1.1.41) H-L-P integral inequalities. We find that the kernels in the above inequalities are all decreasing, but this is not necessary. For example, we find the following two pairs of equivalent forms with the non-decreasing kernel (Yang JGEI 2006) [21], (Yang JJU 2007) [22]:

$$\int_0^\infty\int_0^\infty \frac{|\ln(x/y)|}{\max\{x,y\}}f(x)g(y)dxdy < (p^2+q^2)(\int_0^\infty f^p(x)dx)^{\frac{1}{p}}(\int_0^\infty g^q(x)dx)^{\frac{1}{q}}, \quad (1.1.45)$$

$$\int_0^\infty[\int_0^\infty \frac{|\ln(x/y)|}{\max\{x,y\}}f(x)dx]^p dy < (p^2+q^2)^p\int_0^\infty f^p(x)dx; \quad (1.1.46)$$

$$\sum_{n=1}^\infty\sum_{m=1}^\infty \frac{|\ln(m/n)|}{\max\{m,n\}}a_m b_n < (p^2+q^2)(\sum_{n=1}^\infty a_n^p)^{\frac{1}{p}}(\sum_{n=1}^\infty b_n^q)^{\frac{1}{q}}, \quad (1.1.47)$$

$$\sum_{n=1}^\infty(\sum_{m=1}^\infty \frac{|\ln(m/n)|}{\max\{m,n\}}a_m)^p < (p^2+q^2)^p\sum_{n=1}^\infty a_n^p, \quad (1.1.48)$$

where the constant factors p^2+q^2 and $(p^2+q^2)^p$ are the best possible. We call (1.1.47) Hilbert-Yang's inequality, or H-Y inequality, and call (1.1.45) as H-Y integral inequality.

1.1.5. TWO MULTIPLE INEQUALITIES WITH THE HOMOGENEOUS KERNELS OF DEGREE $-n+1$

Suppose $n \in \mathbf{N}\setminus\{1\}$, n numbers $p,q,\cdots,r$ satisfying $p,q,\cdots,r>1, \frac{1}{p}+\frac{1}{q}+\cdots+\frac{1}{r}=1$, $k(x,y,\cdots,z)\geq 0$ is a homogeneous function of degree $-n+1$. If

$$k=\int_0^\infty\int_0^\infty\cdots\int_0^\infty k(1,y,\cdots,z) \times y^{-\frac{1}{q}}\cdots z^{-\frac{1}{r}}dy\cdots dz \qquad (1.1.49)$$

is a finite number, then we have the following multiple integral inequality (Hardy CUP 1934) [3]:

$$\int_0^\infty\int_0^\infty\cdots\int_0^\infty k(x,y,\cdots,z) \times f(x)g(y)\cdots h(z)dxdy\cdots dz \leq k\left(\int_0^\infty f^p(x)dx\right)^{\frac{1}{p}} \times\left(\int_0^\infty g^q(y)dy\right)^{\frac{1}{q}}\cdots\left(\int_0^\infty h^r(z)dz\right)^{\frac{1}{r}}. \qquad (1.1.50)$$

If $k(1,y,\cdots,z)x^0y^{-\frac{1}{q}}\cdots z^{-\frac{1}{r}}$, $k(x,1,\cdots,z)x^{-\frac{1}{p}}y^0\cdots z^{-\frac{1}{r}},\cdots,$ $k(x,y,\cdots,1)x^{-\frac{1}{p}}y^{-\frac{1}{q}}\cdots z^0$ are all deceasing with respect to any single variable, then we have

$$\sum_{s=1}^\infty\cdots\sum_{n=1}^\infty\sum_{m=1}^\infty k(m,n,\cdots,s)a_m b_n\cdots c_s \leq k\left(\sum_{m=1}^\infty a_m^p\right)^{\frac{1}{p}}\left(\sum_{n=1}^\infty b_n^q\right)^{\frac{1}{q}}\cdots\left(\sum_{s=1}^\infty c_s^r\right)^{\frac{1}{r}}. \qquad (1.1.51)$$

For $n=2$, inequalities (1.1.50) and (1.1.51) reduce respectively to (1.1.33) and (1.1.35).

1.2. MODERN RESEARCH FOR HILBERT'S INEQUALITY

1.2.1. MODERN RESEARCH FOR HILBERT'S INTEGRAL INEQUALITY

(1) In 1979, based on an improvement of Hölder's inequality, Hu gave a refinement of (1.1.2) as follows (Hu JJTC 1979) [23]:

$$\int_0^\infty\int_0^\infty \frac{1}{x+y}f(x)f(y)dxdy < \pi\left[\left(\int_0^\infty f^2(x)dx\right)^2-\frac{1}{4}\left(\int_0^\infty f^2(x)\cos\sqrt{x}dx\right)^2\right]^{\frac{1}{2}}. \qquad (1.2.1)$$

Since then, he published many interesting research results (Hu WUP 2007) [6].

(2) In 1998, B. G. Pachpatte gave an inequality similar to (1.1.2) as follows (Pachpatte JMAA 1998) [24]:

$$\int_0^a\int_0^b \frac{1}{x+y}f(x)g(y)dxdy<\frac{1}{2}\sqrt{ab} \times\left[\int_0^a(a-x)f'^2(x)dx\int_0^b(b-x)g'^2(x)dx\right]^{\frac{1}{2}} \qquad (1.2.2)$$

($a,b>0$). Some improvements and extensions were made by (Zhao JMAA 2001)[25] (Lu TJM 2003)[26] (He JIPAM)[27]. We can see the other works of Pachpatte in (Pachpatte EBV 2005) [28].

(3) In 1998, by introducing a few independent parameters $\lambda\in(0,1]$ and $0<a<b<\infty$, Yang gave an extension of (1.1.2) as (Yang JMAA 1998) [29]:

$$\int_a^b\int_a^b \frac{f(x)g(y)}{(x+y)^\lambda}dxdy < B\left(\frac{\lambda}{2},\frac{\lambda}{2}\right)\left[1-\left(\frac{a}{b}\right)^{\frac{\lambda}{4}}\right] \times\left(\int_a^b x^{1-\lambda}f^2(x)dx\int_a^b x^{1-\lambda}g^2(x)dx\right)^{\frac{1}{2}}, \qquad (1.2.3)$$

where $B(u,v)$ is the beta function. In 1999, Kuang gave another extension of (1.1.2) as (Kuang JMAA 1999)[30]: For $\lambda\in(\frac{1}{2},1]$,

$$\int_0^\infty\int_0^\infty \frac{f(x)g(y)}{x^\lambda+y^\lambda}dxdy<\frac{\pi}{\lambda\sin(\frac{\pi}{2\lambda})} \times\left(\int_0^\infty x^{1-\lambda}f^2(x)dx\int_0^\infty x^{1-\lambda}g^2(x)dx\right)^{\frac{1}{2}}. \qquad (1.2.4)$$

We can see the other works of Kuang in (Kuang SSTP 2004) [5], (Kuang JBUU 2005) [31].

(4) In 1999, by using the methods of algebra and analysis, Gao gave an improvement of (1.1.2) as (Gao JMAA 1999) [32]:

$$\int_0^\infty\int_0^\infty \frac{f(x)g(y)}{x+y}dxdy<\pi\sqrt{1-R} \times\left(\int_0^\infty f^2(x)dx\int_0^\infty g^2(x)dx\right)^{\frac{1}{2}}, \qquad (1.2.5)$$

where $R=\frac{1}{\pi}\left(\frac{u}{\|g\|}-\frac{v}{\|f\|}\right)^2, u=\sqrt{\frac{2}{\pi}}(g,e),$ $v=\sqrt{2\pi}(f,e^{-x}), e(y)=\int_0^\infty\frac{e^x}{x+y}dx$. We can see the other works of Gao in (Gao JMRE 2005) [33].

(5) In 2002, by using the operator theory, K. Zhang gave an improvement of (1.1.2) as follows (Zhang JMAA 2002) [34]:

$$\int_0^\infty\int_0^\infty \frac{1}{x+y}f(x)g(y)dxdy \leq\frac{\pi}{\sqrt{2}}\left[\int_0^\infty f^2(x)dx\int_0^\infty g^2(x)dx +\left(\int_0^\infty f(x)g(x)dx\right)^2\right]^{\frac{1}{2}}. \qquad (1.2.6)$$

1.2.2. ON THE WAY OF WEIGHT COEFFICIENT FOR GIVING A STRENGTHENED VERSION OF HILBERT'S INEQUALITY

In 1991, for giving an improvement of (1.1.1), Hsu raised the way of weight coefficient as follows (Hsu JMRE 1991) [35]: First, using Cauchy's inequality in the left hand side of (1.1.1), it follows

$$\sum_{n=1}^{\infty}\sum_{m=1}^{\infty}\frac{a_m b_n}{m+n}=\sum_{n=1}^{\infty}\sum_{m=1}^{\infty}\Big[\frac{a_m}{(m+n)^{1/2}}\Big(\frac{m}{n}\Big)^{\frac{1}{4}}\Big]\Big[\frac{b_n}{(m+n)^{1/2}}\Big(\frac{n}{m}\Big)^{\frac{1}{4}}\Big]$$

$$\leq\Big\{\sum_{m=1}^{\infty}\Big[\sum_{n=1}^{\infty}\frac{1}{m+n}\Big(\frac{m}{n}\Big)^{\frac{1}{2}}\Big]a_m^2\sum_{n=1}^{\infty}\Big[\sum_{m=1}^{\infty}\frac{1}{m+n}\Big(\frac{n}{m}\Big)^{\frac{1}{2}}\Big]b_n^2\Big\}^{\frac{1}{2}}. \quad (1.2.7)$$

We then define the weight coefficient

$$\omega(n):=\sum_{m=1}^{\infty}\frac{1}{m+n}\Big(\frac{n}{m}\Big)^{\frac{1}{2}}, n\in\mathbf{N}, \quad (1.2.8)$$

and rewrite (1.2.7) as the following inequality :

$$\sum_{n=1}^{\infty}\sum_{m=1}^{\infty}\frac{a_m b_n}{m+n}\leq\Big\{\sum_{n=1}^{\infty}\omega(n)a_n^2\sum_{n=1}^{\infty}\omega(n)b_n^2\Big\}^{\frac{1}{2}}. \quad (1.2.9)$$

Afterwards, setting

$$\omega(n)=\pi-\frac{\theta(n)}{n^{1/2}}\ (n\in\mathbf{N}), \quad (1.2.10)$$

where, $\theta(n):=(\pi-\omega(n))n^{1/2}$, and estimating the series of $\theta(n)$, it follows

$$\theta(n)=\Big[\pi-\sum_{m=1}^{\infty}\frac{1}{m+n}\Big(\frac{n}{m}\Big)^{\frac{1}{2}}\Big]n^{\frac{1}{2}}>\theta:=1.1213^{+}. \quad (1.2.11)$$

Hence by (1.2.10), it yields

$$\omega(n)<\pi-\frac{\theta}{n^{1/2}}, n\in\mathbf{N}, \theta=1.1213^{+}. \quad (1.2.12)$$

And by (1.2.9), a strengthened version of (1.1.1) is given as follows:

$$\sum_{n=1}^{\infty}\sum_{m=1}^{\infty}\frac{a_m b_n}{m+n}\leq\Big\{\sum_{n=1}^{\infty}\Big[\pi-\frac{\theta}{n^{1/2}}\Big]a_n^2\sum_{n=1}^{\infty}\Big[\pi-\frac{\theta}{n^{1/2}}\Big]b_n^2\Big\}^{\frac{1}{2}}. \quad (1.2.13)$$

In this paper, Hsu raised an open problem to obtain the best value of θ in (1.2.13). In 1992, Gao obtained the best value of $\theta_0=1.281669^{+}$ (Gao HMA 1992) [36].

In the same year, by using the above way, a strengthened version of (1.1.8) was given as follows (Xu CQJM 1991) [37]:

$$\sum_{n=1}^{\infty}\sum_{m=1}^{\infty}\frac{a_m b_n}{m+n}<\Big\{\sum_{n=1}^{\infty}\Big[\frac{\pi}{\sin(\frac{\pi}{p})}-\frac{p-1}{n^{1/p}+n^{-1/q}}\Big]a_n^p\Big\}^{\frac{1}{p}}$$

$$\times\Big\{\sum_{n=1}^{\infty}\Big[\frac{\pi}{\sin(\frac{\pi}{p})}-\frac{q-1}{n^{1/q}+n^{-1/p}}\Big]b_n^q\Big\}^{\frac{1}{q}}. \quad (1.2.14)$$

In 1997, by using the way of weight coefficient and the improved Euler-Maclaurin's summation formula, Yang and Gao gave a new strengthened version of (1.1.18) as follows (Yang AM 1997) [38]:

$$\sum_{n=1}^{\infty}\sum_{m=1}^{\infty}\frac{a_m b_n}{m+n}<\Big\{\sum_{n=1}^{\infty}\Big[\frac{\pi}{\sin(\frac{\pi}{p})}-\frac{1-\gamma}{n^{1/q}}\Big]b_n^q\Big\}^{\frac{1}{q}}$$

$$\times\Big\{\sum_{n=1}^{\infty}\Big[\frac{\pi}{\sin(\frac{\pi}{p})}-\frac{1-\gamma}{n^{1/q}}\Big]b_n^q\Big\}^{\frac{1}{q}}, \quad (1.2.15)$$

where $1-\gamma=0.42278433^{+}$ (γ is Euler constant). We can see similar works in (Gao PAMS 1998) [39].

In 1998, Yang et.al. gave another strengthened version of (1.1.18), which is an improvement of (1.2.14) (Yang IJMMS 1998) [40]. We can see some strengthened versions of (1.1.12) and (1.1.20) in (Yang HJ 1997) [41], (Yang AMS 1999) [42], (Yang PJMS 2003) [43].

1.2.3. HILBERT'S INEQUALITY WITH INDEPENDENT PARAMETERS

In 1998, by using the optimized weight coefficient and introducing an independent parameter $0<\lambda\leq 1$, Yang gave an extension of (1.1.2) as follows (Yang JMAA 1998) [29]: If $0<\int_0^{\infty}x^{1-\lambda}f^2(x)dx<\infty$ and $0<\int_0^{\infty}x^{1-\lambda}g^2(x)dx<\infty$, then we have

$$\int_0^{\infty}\int_0^{\infty}\frac{1}{(x+y)^{\lambda}}f(x)g(y)dxdy<B\Big(\frac{\lambda}{2},\frac{\lambda}{2}\Big)$$

$$\times\Big(\int_0^{\infty}x^{1-\lambda}f^2(x)dx\int_0^{\infty}x^{1-\lambda}g^2(x)dx\Big)^{\frac{1}{2}}, \quad (1.2.16)$$

where the constant factor $B(\frac{\lambda}{2},\frac{\lambda}{2})$ is the best possible. The proof about the best possible property of the constant factor was given by (Yang CQJM 1998) [45], and the expressions of the beta function $B(u,v)$ are given as follows (Wang SP 1979) [46]:

$$B(u,v)=\int_0^{\infty}\frac{t^{u-1}}{(1+t)^{u+v}}dt=\int_0^1(1-t)^{u-1}t^{v-1}dt$$

$$=\int_1^{\infty}\frac{(t-1)^{u-1}}{t^{u+v}}dt\ (u,v>0). \quad (1.2.17)$$

Some extensions of (1.1.18), (1.1.20) and (1.1.22) were obtained by (Yang CAM 2000) [47], (Yang JMAA 2002) [48], (Yang JMAA 1999)[49] as follows: If $\lambda>2-\min\{p,q\}$, then

$$\int_0^{\infty}\int_0^{\infty}\frac{f(x)g(y)}{(x+y)^{\lambda}}dxdy<B\Big(\frac{p+\lambda-2}{p},\frac{q+\lambda-2}{q}\Big)$$

$$\times\Big(\int_0^{\infty}x^{1-\lambda}f^p(x)dx\Big)^{\frac{1}{p}}\Big(\int_0^{\infty}x^{1-\lambda}g^q(x)dx\Big)^{\frac{1}{q}}; \quad (1.2.18)$$

if $2-\min\{p,q\}<\lambda\leq 2$, then

$$\sum_{n=1}^{\infty}\sum_{m=1}^{\infty}\frac{a_m b_n}{(m+n)^{\lambda}} < B\left(\frac{p+\lambda-2}{p},\frac{q+\lambda-2}{q}\right) \times\left(\sum_{n=1}^{\infty} n^{1-\lambda}a_n^p\right)^{\frac{1}{p}}\left(\sum_{n=1}^{\infty} n^{1-\lambda}b_n^q\right)^{\frac{1}{q}}, \tag{1.2.19}$$

$$\sum_{n=0}^{\infty}\sum_{m=0}^{\infty}\frac{a_m b_n}{(m+n+1)^{\lambda}} < B\left(\frac{p+\lambda-2}{p},\frac{q+\lambda-2}{q}\right) \times\left\{\sum_{n=0}^{\infty} (n+\tfrac{1}{2})^{1-\lambda}a_n^p\right\}^{\frac{1}{p}}\left\{\sum_{n=0}^{\infty} (n+\tfrac{1}{2})^{1-\lambda}b_n^q\right\}^{\frac{1}{p}}, \tag{1.2.20}$$

where the constant factor $B(\frac{p+\lambda-2}{p},\frac{q+\lambda-2}{q})$ is the best possible (assuming that the right hand side of the above inequalities are all positive numbers). Yang also proved that (1.2.19) is valid for $p=q=2$ and $0<\lambda\le 4$ (Yang JNUMB 2001) [50]. Yang also gave other best extensions of (1.1.18) and (1.1.20) as (Yang CAM 2002)[51] (Yang AM 2006) [52]: If $0<\lambda\le\min\{p,q\}$, then

$$\sum_{n=1}^{\infty}\sum_{m=1}^{\infty}\frac{a_m b_n}{m^{\lambda}+n^{\lambda}} < \frac{\pi}{\lambda\sin(\pi/p)} \times\left\{\sum_{n=1}^{\infty} n^{(p-1)(1-\lambda)}a_n^p\right\}^{\frac{1}{p}}\left\{\sum_{n=1}^{\infty} n^{(q-1)(1-\lambda)}b_n^q\right\}^{\frac{1}{q}}; \tag{1.2.21}$$

if $0<\lambda\le 1$, then

$$\sum_{n=0}^{\infty}\sum_{m=0}^{\infty}\frac{a_m b_n}{(m+\frac{1}{2})^{\lambda}+(n+\frac{1}{2})^{\lambda}} < \frac{\pi}{\lambda\sin(\frac{\pi}{p})} \times\left\{\sum_{n=0}^{\infty} (n+\tfrac{1}{2})^{p-1-\lambda}a_n^p\right\}^{\frac{1}{p}}\left\{\sum_{n=0}^{\infty} (n+\tfrac{1}{2})^{q-1-\lambda}b_n^q\right\}^{\frac{1}{q}}. \tag{1.2.22}$$

In 2004, Yang discovered the following dual form of (1.1.18) (Yang ANH 2004) [53]:

$$\sum_{n=1}^{\infty}\sum_{m=1}^{\infty}\frac{a_m b_n}{m+n} < \frac{\pi}{\sin(\frac{\pi}{p})}\left(\sum_{n=1}^{\infty} n^{p-2}a_n^p\right)^{\frac{1}{p}}\left(\sum_{n=1}^{\infty} n^{p-2}b_n^q\right)^{\frac{1}{q}}. \tag{1.2.23}$$

Inequality (1.2.23) is similar to (1.1.18) but different, and for $p=q=2$, both of them reduce to (1.1.1). For $\lambda=1$, inequality (1.2.22) also reduces to the dual form of (1.1.20) as follows:

$$\sum_{n=0}^{\infty}\sum_{m=0}^{\infty}\frac{a_m b_n}{m+n+1} < \frac{\pi}{\sin(\frac{\pi}{p})} \times\left\{\sum_{n=0}^{\infty} (n+\tfrac{1}{2})^{p-2}a_n^p\right\}^{\frac{1}{p}}\left\{\sum_{n=0}^{\infty} (n+\tfrac{1}{2})^{q-2}b_n^q\right\}^{\frac{1}{q}}. \tag{1.2.24}$$

We can see some best extensions of the H-L-P inequalities such as (1.1.37)-(1.1.48) in (Yang MIA 2003)[54] , (Yang JJU 2004)[55], (Yang CJEM 2004) [56], (Yang JMRE 2005)[57] , (Yang SJM 2005)[58] , (Yang BBMS 2006)[59] , (Wang AJMAA 2006)[60], by introducing some independent parameters.

In 2001, by introducing some independent parameters, Hong gave a multiple integral inequality, which is an extension of (1.2.18) (Hong AMS 2001) [61]. He and Gao gave the similar work for particular conjugate exponents (He JSU 2002) [62]. For making an improvement of their works, Yang gave the following multiple integral inequality, which is a best extension of (1.2.18) (Yang CAM 2003) [63]: If $n\in\mathbf{N}\setminus\{1\}, p_i>1$, $\sum_{i=1}^{n}\frac{1}{p_i}=1$, $\lambda>n-\min_{1\le i\le n}\{p_i\}$, $f_i(t)\ge 0$ and $0<\int_0^{\infty} t^{n-1-\lambda}f_i^{p_i}(t)dt<\infty$ $(i=1,\cdots,n)$, then we have

$$\int_0^{\infty}\cdots\int_0^{\infty}\frac{\prod_{i=1}^{n}f_i(x_i)}{(\sum_{i=1}^{n}x_i)^{\lambda}}dx_1\cdots dx_n < \frac{1}{\Gamma(\lambda)}\prod_{i=1}^{n}\Gamma\left(\frac{p_i+\lambda-n}{p_i}\right)\left\{\int_0^{\infty} t^{n-1-\lambda}f_i^{p_i}(t)dt\right\}^{\frac{1}{p_i}}, \tag{1.2.25}$$

where the constant factor $\frac{1}{\Gamma(\lambda)}\prod_{i=1}^{n}\Gamma(\frac{p_i+\lambda-n}{p_i})$ is the best possible. In particular, for $\lambda=n-1$, it follows

$$\int_0^{\infty}\cdots\int_0^{\infty}\frac{\prod_{i=1}^{n}f_i(x_i)}{(\sum_{i=1}^{n}x_i)^{n-1}}dx_1\cdots dx_n < \frac{1}{(n-2)!}\prod_{i=1}^{n}\Gamma\left(\frac{p_i-1}{p_i}\right)\left\{\int_0^{\infty} f_i^{p_i}(t)dt\right\}^{\frac{1}{p_i}}. \tag{1.2.26}$$

In 2003, Yang and Rassias introduced the way of weight coefficient and considered its applications to Hilbert-type inequalities. They summarized how to use the way of weight coefficient to obtain some new improvements and generalizations of the Hilbert-type inequalities (Yang MIA 2003) [64]. Since then, a number of authors discussed this problem (see (Yang JGEI 2005) [65]-(Chen AMH 2007) [85]). But how to give a best extension of inequalities (1.2.23) and (1.1.18), this was solved in 2004 by introducing two pairs of conjugate exponents.

1.2.4. HILBERT-TYPE INEQUALITIES WITH MULTI-PARAMETERS

In 2004, by introducing an independent parameter $\lambda>0$ and two pairs of conjugate exponents (p,q) and (r,s) $(\frac{1}{p}+\frac{1}{q}=1,\frac{1}{r}+\frac{1}{s}=1)$, Yang gave an extension of (1.1.2) as follows (Yang AJMAA 2004) [86]: If $p,r>1$ and the integrals of the right hand side are positive, then we obtain

$$\int_0^\infty\int_0^\infty \frac{1}{x^\lambda+y^\lambda} f(x)g(y)dxdy < \frac{\pi}{\lambda\sin(\frac{\pi}{r})}$$
$$\times\{\int_0^\infty x^{p(1-\frac{\lambda}{r})-1} f^p(x)dx\}^{\frac{1}{p}}\{\int_0^\infty x^{q(1-\frac{\lambda}{s})-1} g^q(x)dx\}^{\frac{1}{q}}, \quad (1.2.27)$$

where the constant factor $\frac{\pi}{\lambda\sin(\pi/r)}$ is the best possible. For $\lambda=1, r=q, s=p$, inequality (1.2.27) reduces to (1.1.22); for $\lambda=1, r=p, s=q$, (1.2.27) reduces to the following dual form of (1.1.22):

$$\int_0^\infty\int_0^\infty \frac{1}{x+y} f(x)g(y)dxdy < \frac{\pi}{\sin(\frac{\pi}{p})}$$
$$\times\{\int_0^\infty x^{p-2} f^p(x)dx\}^{\frac{1}{p}}\{\int_0^\infty x^{q-2} g^q(x)dx\}^{\frac{1}{q}}. \quad (1.2.28)$$

In 2005, by introducing an independent parameter $\lambda>0$ and two pairs of generalized conjugate exponents $(p_1,\cdots,p_n)$, $(r_1,\cdots,r_n)$ $(p_i,r_i>1$, $\sum_{i=1}^n \frac{1}{p_i}=1$, $\sum_{i=1}^n \frac{1}{r_i}=1)$, Yang et al. gave a multiple inequality as follows (Yang MIA 2005)[87]:

$$\int_0^\infty\cdots\int_0^\infty \frac{\prod_{i=1}^n f_i(x_i)}{(\sum_{i=1}^n x_i)^\lambda}dx_1\cdots dx_n$$
$$<\frac{1}{\Gamma(\lambda)}\prod_{i=1}^n \Gamma(\frac{\lambda}{r_i})\{\int_0^\infty t^{p_i(1-\frac{\lambda}{r_i})-1} f_i^{p_i}(t)dt\}^{\frac{1}{p_i}}, \quad (1.2.29)$$

where the constant factor $\frac{1}{\Gamma(\lambda)}\prod_{i=1}^n\Gamma(\frac{\lambda}{r_i})$ is the best possible. For $r_i=\frac{p_i\lambda}{p_i-\lambda+n}$ $(i=1,2,\cdots,n)$, inequality (1.2.29) reduces to (1.2.25); for $n=2$, inequality (1.2.29) reduces to the following:

$$\int_0^\infty\int_0^\infty \frac{1}{(x+y)^\lambda} f(x)g(y)dxdy < B(\frac{\lambda}{r},\frac{\lambda}{s})$$
$$\times\{\int_0^\infty x^{p(1-\frac{\lambda}{r})-1} f^p(x)dx\}^{\frac{1}{p}}\{\int_0^\infty x^{q(1-\frac{\lambda}{s})-1} g^q(x)dx\}^{\frac{1}{q}}. \quad (1.2.30)$$

It is obvious that inequality (1.2.30) is another best extension of (1.1.22).

In 2006, Hong gave a multivariable integral inequalities as (Hong JIA 2006) [88]: If

$R_+^n=\{(x_1,x_2,\cdots,x_n); x_i>0, i=1,2,\cdots,n\}$,

$\|x\|_\alpha=\{\sum_{i=1}^n x_i^\alpha\}^{\frac{1}{\alpha}}, \alpha,\lambda,\beta>0, f,g\geq 0$,

$$0<\int_{R_+^n}\|x\|_\alpha^{p(n-\frac{\lambda\beta}{r})-n} f^p(x)dx$$

and $0<\int_{R_+^n}\|x\|_\alpha^{q(n-\frac{\lambda\beta}{s})-n} g^q(x)dx$, then we get

$$\int\int_{R_+^n\times R_+^n}\frac{f(x)g(y)}{(\|x\|_\alpha^\beta+\|y\|_\alpha^\beta)^\lambda}dxdy < \frac{\Gamma^n(\frac{1}{\alpha})}{\beta\alpha^{n-1}\Gamma(\frac{n}{\alpha})}B(\frac{\lambda}{r},\frac{\lambda}{s})$$
$$\times\{\int_0^\infty\|x\|_\alpha^{p(n-\frac{\lambda\beta}{r})-n} f^p(x)dx\}^{\frac{1}{p}}$$
$$\times\{\int_0^\infty\|x\|_\alpha^{q(n-\frac{\lambda\beta}{s})-n} g^q(x)dx\}^{\frac{1}{q}}, \quad (1.2.31)$$

where the constant factor $\frac{\Gamma^n(1/\alpha)}{\beta\alpha^{n-1}\Gamma(n/\alpha)}B(\frac{\lambda}{r},\frac{\lambda}{s})$ is the best possible. In particular, for $n=1$, (1.2.31) reduces to Hong's work in (Hong JIPAM 2005) [89]; for $n=1$, $\beta=1$, (1.2.31) reduces to (1.2.30). In 2007, Zhong and Yang generalized Hong's work to the general kernel and proposed the revision (Zhong JIA 2007) [90].

We can see another Hilbert's inequality with two parameters as follows (Yang JIPAM 2005) [91]:

$$\sum_{n=1}^\infty\sum_{m=1}^\infty \frac{a_m b_n}{(m^\alpha+n^\alpha)^\lambda} < \frac{1}{\alpha}B(\frac{\lambda}{r},\frac{\lambda}{s})$$
$$\times\{\sum_{n=1}^\infty n^{p(1-\frac{\alpha\lambda}{r})-1}a_n^p\}^{\frac{1}{p}}\{\sum_{n=1}^\infty n^{q(1-\frac{\alpha\lambda}{s})-1}b_n^q\}^{\frac{1}{q}}, \quad (1.2.32)$$

where $\alpha,\lambda>0, \alpha\lambda\leq\min\{r,s\}$. For $\alpha=1$, it follows $0<\lambda\leq\min\{r,s\}$ and

$$\sum_{n=1}^\infty\sum_{m=1}^\infty \frac{a_m b_n}{(m+n)^\lambda} < B(\frac{\lambda}{r},\frac{\lambda}{s})$$
$$\times\{\sum_{n=1}^\infty n^{p(1-\frac{\lambda}{r})-1}a_n^p\}^{\frac{1}{p}}\{\sum_{n=1}^\infty n^{q(1-\frac{\lambda}{s})-1}b_n^q\}^{\frac{1}{q}}. \quad (1.2.33)$$

For $\lambda=1, r=q$, inequality (1.2.33) reduces to (1.1.18), and for $\lambda=1, r=p$, (1.2.33) reduces to (1.2.23). Also we can see the reverse form as follows (Yang JSCNU 2005) [92]:

$$\sum_{n=0}^\infty\sum_{m=0}^\infty \frac{a_m b_n}{(m+n+1)^2}$$
$$>2\{\sum_{n=0}^\infty[1-\frac{1}{4(n+1)^2}]\frac{a_n^p}{2n+1}\}^{\frac{1}{p}}\{\sum_{n=0}^\infty\frac{b_n^q}{2n+1}\}^{\frac{1}{q}}, \quad (1.2.34)$$

where $0<p<1, \frac{1}{p}+\frac{1}{q}=1$. The other results on the reverse Hilbert-type inequalities are found in (Yang JJU 2004)[93], (Yang PAM 2006)[94], (Yang IMF 2006)[95], (Yang MPT 2006)[96], (Yang IJPAM 2006)[97], (Xi JIA 2007)[98], (Yang AMS 2006)[99].

In 2006, Xin gave a best extension of H-L-P integral inequality (1.1.41) as follows (Xin JIPAM 2006) [100]:

$$\int_0^\infty\int_0^\infty \frac{\ln(x/y)}{x^\lambda-y^\lambda} f(x)g(y)dxdy < [\frac{\pi}{\lambda\sin(\frac{\pi}{r})}]^2$$
$$\times\{\int_0^\infty x^{p(1-\frac{\lambda}{r})-1} f^p(x)dx\}^{\frac{1}{p}}\{\int_0^\infty x^{q(1-\frac{\lambda}{s})-1} g^q(x)dx\}^{\frac{1}{q}}; \quad (1.2.35)$$

Zhong et.al gave an extension of another H-L-P integral inequality (1.1.37) as (Zhong JJU 2007)[101]:

$$\int_0^\infty\int_0^\infty \frac{1}{\max\{x^\lambda,y^\lambda\}} f(x)g(y)dxdy < \frac{rs}{\lambda}$$

$$\times\{\int_0^\infty x^{p(1-\frac{\lambda}{r})-1}f^p(x)dx\}^{\frac{1}{p}}\{\int_0^\infty x^{q(1-\frac{\lambda}{s})-1}g^q(x)dx\}^{\frac{1}{q}};$$
(1.2.36)

Zhong et al. also gave the reverse form of (1.2.36) (Zhong PAM 2008)[102]. For some particular kernel and parameters, Yang gave (Yang JSU 2007) [103]

$$\sum_{n=1}^\infty\sum_{m=1}^\infty\frac{a_m b_n}{(\sqrt{m}+\sqrt{n})\sqrt{\max\{m,n\}}}$$

$$< 4\ln 2(\sum_{n=1}^\infty n^{\frac{p}{2}-1}a_n^p)^{\frac{1}{p}}(\sum_{n=1}^\infty n^{\frac{q}{2}-1}b_n^q)^{\frac{1}{q}}. \quad (1.2.37)$$

He also gave (Yang JXU 2006) [104]

$$\sum_{n=1}^\infty\sum_{m=1}^\infty\frac{a_m b_n}{(m+an)^2+n^2} < (\tfrac{\pi}{2}-\arctan a)$$

$$\times(\sum_{n=1}^\infty \tfrac{1}{n}a_n^p)^{\frac{1}{p}}(\sum_{n=1}^\infty \tfrac{1}{n}b_n^q)^{\frac{1}{q}} \ (a\ge 0). \quad (1.2.38)$$

By using the residue theory, Yang obtained (Yang JYU 2008) [105]

$$\int_0^\infty\int_0^\infty\frac{f(x)g(y)}{(x+ay)(x+by)(x+cy)}dxdy$$

$$< k(\int_0^\infty \tfrac{1}{t^{1+p/2}}f^p(t)dt)^{\frac{1}{p}}(\int_0^\infty \tfrac{1}{t^{1+q/2}}g^q(t)dt)^{\frac{1}{q}},$$
(1.2.39)

where $k=\frac{\pi}{(\sqrt{a}+\sqrt{b})(\sqrt{b}+\sqrt{c})(\sqrt{a}+\sqrt{c})}(a,b,c>0)$. The constant factors in the above new inequalities are all the best possible. We can see other new works in (Xie JJU 2007) [106], (Xie NSJXU 2007) [107], (Xie JMAA 2007)[108], (Xie SJM 2007)[109], (Li BAMS 2007) [110], (He CMA 2008)[111], (Yang JGEI 2007)[112].

1.3. OPERATOR EXPRESSIONS AND BASIC HILBERT-TYPE INEQUALITIES

1.3.1. MORDERN RESEARCH FOR HILBERT-TYPE OPERATORS

Supposing that H is a separable Hilbert space and $T:H\to H$ is a bounded self-adjoint semi -positive definite operator, in 2002, K. Zhang gave the following inequality (Zhang JMAA 2002) [32]:

$$(a,Tb)^2\le\tfrac{\|T\|^2}{2}(\|a\|^2\|b\|^2+(a,b)^2)(a,b\in H),$$
(1.3.1)

where (a,b) is the inner product of a and b, and $\|a\|=\sqrt{(a,a)}$ is the norm of a. Since the Hilbert integral operator $\tilde{T}$ is defined by (1.1.7) satisfying the conditions of (1.3.1), and $\|\tilde{T}\|=\pi$, then inequality (1.1.2) may be improved as (1.2.6). Since the operator T_p defined by (1.1.24) (for $p=2$) satisfies the conditions of (1.3.1) (Wilhelm AJM 1950) [7], we may improve (1.1.12) to the following form:

$$\sum_{n=0}^\infty\sum_{m=0}^\infty\frac{a_m b_n}{m+n+1}\le\tfrac{\pi}{\sqrt{2}}[(\sum_{n=0}^\infty a_n^2\sum_{n=0}^\infty b_n^2)+(\sum_{n=0}^\infty a_n b_n)^2]^{\frac{1}{2}}.$$
(1.3.2)

The key of applying (1.3.1) is to obtain the norm of the operator and the property of semi-definite. Now, we consider the concept and properties of Hilbert-type integral operator as follows.

Suppose $p>1,\frac{1}{p}+\frac{1}{q}=1, L^r(0,\infty)\ (r=p,q)$ are real normal linear spaces, $k(x,y)$ is a non-negative symmetric measurable function in $(0,\infty)\times(0,\infty)$ satisfying $k(x,y)=k(y,x)$ and

$$\int_0^\infty k(x,t)(\tfrac{x}{t})^{\frac{1}{r}}dt=k_0(p)\in\mathbf{R}(x>0). \quad (1.3.3)$$

We define an integral operator as

$$T:L^r(0,\infty)\to L^r(0,\infty)\ (r=p,q),$$

for any $f(\ge 0)\in L^p(0,\infty)$, there exists $h=Tf\in L^p(0,\infty)$, such that

$$(Tf)(y)=h(y):=\int_0^\infty k(x,y)f(x)dx(y>0);$$
(1.3.4)

or for any $g(\ge 0)\in L^q(0,\infty)$, there exists $\tilde{h}=Tg\in L^q(0,\infty)$, such that

$$(Tg)(x)=\tilde{h}(x):=\int_0^\infty k(x,y)g(y)dy(x>0).$$
(1.3.5)

Then we have the following expression :

$$(Tf,g)=\int_0^\infty\int_0^\infty k(x,y)f(x)g(y)dxdy.$$

In 2006, Yang prove that the operator T defined by (1.3.4) or (1.3.5) are bounded and $\|T\|\le k_0(p)$ (Yang JMAA 2006) [113]. The following are some results in this paper: If $\varepsilon>0$ is small enough, the integral $\int_0^\infty k(x,t)(\frac{x}{t})^{\frac{1+\varepsilon}{r}}dt\ (r=p,q;x>0)$ convergences to a constant $k_\varepsilon(p)$ independent of x, $k_\varepsilon(p)=k_0(p)+o(1)\ (\varepsilon\to 0^+)$, then we have $\|T\|=k_0(p)$; if $\|T\|>0$, and for $f,g\ge 0$, $f\in L^p(0,\infty), g\in L^q(0,\infty)$, $\|f\|_p$, $\|g\|_q>0$, we have the following equivalent inequalities:

$$(Tf,g)<\|T\|\cdot\|f\|_p\|g\|_q, \quad (1.3.6)$$

$$\|Tf\|_p<\|T\|\cdot\|f\|_p. \quad (1.3.7)$$

Some particular cases are considered in this paper.

Yang also considered some properties of Hilbert-type integral operator (for $p=q=2$) (Yang AMS 2007)

[114]; when the homogeneous kernel is degree -1, Yang considered some sufficient conditions to make $\| T \| = k_0(p) > 0$ (Yang TJM 2008) [115]. We can see some properties of the Hilbert-type operator in the disperse space in (Yang JMAA 2007) [116], (Yang BBMS 2006)[117], (Yang JIA 2007)[118], (Yang IJPAM 2008)[119]. A multiple integral operator is scored by (Arpad JIA 2006) [120]. In 2009, Yang summarized the above results in (Yang SP 2009) [121], (Yang BSP 2009) [121].

1.3.2. SOME BASIC HILBERT-TYPE INEQUALITIES

If the Hilbert-type integral inequality relates to a symmetric homogeneous kernel of degree -1 and the best constant factor, which is more brief and exhibits barbarism form and does not relate to any conjugate exponent (such as (1.1.2)), then we call it basic Hilbert-type integral inequality. Its series analogue (if exists) is called a basic Hilbert-type inequality. If the kernel of basic Hilbert-type (integral) inequality relates to a parameter, and the inequality can not be obtained by a simple transform to a basic Hilbert-type (integral) inequality, then we call it a basic Hilbert-type (integral) inequality with a parameter. For example, we call the following integral inequality (i.e. (1.1.2)):

$$\int_0^\infty \int_0^\infty \frac{1}{x+y} f(x)g(y)dxdy < \pi\left(\int_0^\infty f^2(x)dx \int_0^\infty g^2(x)dx\right)^{\frac{1}{2}} \quad (1.3.8)$$

and the following H-L-P integral inequalities (for $p = q = 2$ in (1.1.37) and (1.1.41))

$$\int_0^\infty \int_0^\infty \frac{1}{\max\{x,y\}} f(x)g(y)dxdy < 4\left(\int_0^\infty f^2(x)dx \int_0^\infty g^2(x)dx\right)^{\frac{1}{2}}, \quad (1.3.9)$$

$$\int_0^\infty \int_0^\infty \frac{\ln(x/y)}{x-y} f(x)g(y)dxdy < \pi^2\left(\int_0^\infty f^2(x)dx \int_0^\infty g^2(x)dx\right)^{\frac{1}{2}} \quad (1.3.10)$$

basic Hilbert-type integral inequalities.

In 2006, Yang gave the following H-Y integral inequality (Yang JGEI 2006) [21]:

$$\int_0^\infty \int_0^\infty \frac{|\ln(x/y)|}{\max\{x,y\}} f(x)g(y)dxdy < 8\left(\int_0^\infty f^2(x)dx \int_0^\infty g^2(x)dx\right)^{\frac{1}{2}}; \quad (1.3.11)$$

in 2008, Yang also gave the following H-Y integral inequalities (Yang CM 2008) [123], (Yang JGEI 2008) [124]:

$$\int_0^\infty \int_0^\infty \frac{|\ln(x/y)|}{x+y} f(x)g(y)dxdy < c_0\left(\int_0^\infty f^2(x)dx \int_0^\infty g^2(x)dx\right)^{\frac{1}{2}}, \quad (1.3.12)$$

$$\int_0^\infty \int_0^\infty \frac{\arctan\sqrt{x/y}}{x+y} f(x)g(y)dxdy < \frac{\pi^2}{4}\left(\int_0^\infty f^2(x)dx \int_0^\infty g^2(x)dx\right)^{\frac{1}{2}}, \quad (1.3.13)$$

where the constant factors $c_0 = 8\sum_{n=1}^\infty \frac{(-1)^n}{(2n-1)^2} = 7.3277^+$ and $\frac{\pi^2}{4}$ are the best possible. We still call (1.3.11)-(1.3.13) basic Hilbert-type integral inequalities.

In 2005, Yang gave the following H-Y integral inequality with a parameter (Yang JJU 2005) [125], (Yang JHU 2005) [126]:

$$\int_0^\infty \int_0^\infty \frac{1}{|x-y|^\lambda} f(x)g(y)dxdy < 2B(\tfrac{\lambda}{2}, 1-\lambda) \times\left(\int_0^\infty x^{1-\lambda} f^2(x)dx \int_0^\infty x^{1-\lambda} g^2(x)dx\right)^{\frac{1}{2}}, \quad (1.3.14)$$

where the constant factor $2B(\frac{\lambda}{2}, 1-\lambda)(0 < \lambda < 1)$ is the best possible. As in (1.2.16), i.e.

$$\int_0^\infty \int_0^\infty \frac{1}{(x+y)^\lambda} f(x)g(y)dxdy < B(\tfrac{\lambda}{2}, \tfrac{\lambda}{2}) \times\left(\int_0^\infty x^{1-\lambda} f^2(x)dx \int_0^\infty x^{1-\lambda} g^2(x)dx\right)^{\frac{1}{2}} \quad (1.3.15)$$

$(\lambda > 0)$, we call (1.3.14) and (1.3.15) basic Hilbert-type integral inequalities with a parameter.

It is noticed that the following inequality (for $p = r = 2$ in (1.2.27))

$$\int_0^\infty \int_0^\infty \frac{1}{x^\lambda + y^\lambda} f(x)g(y)dxdy < \frac{\pi}{\lambda} \times\left(\int_0^\infty x^{1-\lambda} f^2(x)dx \int_0^\infty x^{1-\lambda} g^2(x)dx\right)^{\frac{1}{2}} \ (\lambda > 0), \quad (1.3.16)$$

is not a basic Hilbert-type integral inequality with a parameter. By setting $x = X^\lambda, y = Y^\lambda$ in (1.3.8), we may get (1.3.16). Similarly, neither (1.2.35) nor (1.2.36) (for $p = r = 2$) are basic Hilbert-type integral inequalities with a parameter. Also we find the following basic Hilbert-type inequalities:

$$\sum_{n=1}^\infty \sum_{m=1}^\infty \frac{a_m b_n}{m+n} < \pi\left(\sum_{n=1}^\infty a_n^2 \sum_{n=1}^\infty b_n^2\right)^{\frac{1}{2}}, \quad (1.3.17)$$

$$\sum_{n=1}^\infty \sum_{m=1}^\infty \frac{a_m b_n}{\max\{m,n\}} < 4\left(\sum_{n=1}^\infty a_n^2 \sum_{n=1}^\infty b_n^2\right)^{\frac{1}{2}}, \quad (1.3.18)$$

$$\sum_{n=1}^\infty \sum_{m=1}^\infty \frac{\ln(m/n)}{m-n} a_m b_n < \pi^2\left(\sum_{n=1}^\infty a_n^2 \sum_{n=1}^\infty b_n^2\right)^{\frac{1}{2}}, \quad (1.3.19)$$

$$\sum_{n=1}^\infty \sum_{m=1}^\infty \frac{|\ln(m/n)| a_m b_n}{\max\{m,n\}} < 8\left(\sum_{n=1}^\infty a_n^2 \sum_{n=1}^\infty b_n^2\right)^{\frac{1}{2}}, \quad (1.3.20)$$

$$\sum_{n=1}^\infty \sum_{m=1}^\infty \frac{|\ln(m/n)| a_m b_n}{m+n} < c_0\left(\sum_{n=1}^\infty a_n^2 \sum_{n=1}^\infty b_n^2\right)^{\frac{1}{2}} \quad (1.3.21)$$

$$(c_0 = \sum_{k=0}^{\infty} \frac{8(-1)^k}{(2k+1)^2} = 7.3277^+ \text{ , (Xin JM 2010)[127]),}$$

$$\sum_{n=1}^{\infty}\sum_{m=1}^{\infty} \frac{a_m b_n}{(m+n)^{\lambda}} < B(\tfrac{\lambda}{2}, \tfrac{\lambda}{2})$$

$$\times(\sum_{n=1}^{\infty} n^{1-\lambda} a_n^2 \sum_{n=1}^{\infty} n^{1-\lambda} b_n^2)^{\frac{1}{2}} (0 < \lambda \leq 4). \qquad (1.3.22)$$

Among them, inequality (1.3.22) is called basic Hilbert-type inequality with a parameter.

By simple operation of the kernels in basic Hilbert-type inequalities, we may get some new Hilbert-type inequalities. For example, we found (Li IJMMS 2006) [128], (Xie JJU 2007) [129]

$$\int_0^{\infty}\int_0^{\infty} \frac{f(x)g(y)}{x+y+\max\{x,y\}} dxdy$$

$$< c(\int_0^{\infty} f^2(x)dx \int_0^{\infty} g^2(x)dx)^{\frac{1}{2}}, \qquad (1.3.23)$$

where the constant factor $c = \sqrt{2}(\pi - \arctan\sqrt{2})$ is the best possible. Still we found the following inequality (Yang JMAA 2006) [113]:

$$\int_0^{\infty}\int_0^{\infty} \frac{|x-y|}{(\max\{x,y\})^2} f(x)g(y)dxdy$$

$$< \tfrac{8}{3}(\int_0^{\infty} f^2(x)dx \int_0^{\infty} g^2(x)dx)^{\frac{1}{2}}, \qquad (1.3.24)$$

where the constant factor $\frac{8}{3}$ is the best possible. We can obtain some new Hilbert-type inequalities in the same way (Li IJMMS 2007) [130], (Li JIA 2007) [131], (Xie KMJ 2008) [132].

1.4. REFERENCES

1. Weyl H. Singulare integral gleichungen mit besonderer berucksichtigung des fourierschen integral theorems. Gottingen : Inaugeral-Dissertation, 1908.
2. Schur I. Bernerkungen sur Theorie der beschrankten Bilinearformen mit unendlich vielen veranderlichen. Journal of Math., 1911; 140: 1-28.
3. Hardy GH, Littlewood JE, Polya G. Inequalities. Cambridge : Cambridge University Press, 1934.
4. Mitrinovic J E, Pecaric J E, Fink A M. Inequalities involving functions and their integrals and derivatives. Boston: Kluwer Acaremic Publishers, 1991.
5. Kuang JC. Applied inequalities. Jinan: Shandong Science Technic Press, 2004.
6. Hu K. Some problems in analysis inequalities. Wuhan: Wuhan University Press, 2007.
7. Wilhelm M. On the spectrum of Hilbert's matrix. Amer J. Math., 1950; 72: 699-704.
8. Carleman T. Sur les equations integrals singulieres a noyau reel et symetrique. Uppsala, 1923.
9. Ingham AE. A note on Hilbert's inequality. J. London Math. Soc., 1936; 11: 237-240.
10. Yang BC. On a new Hardy-Hilbert's type inequality. Math. Inequalities & Applications, 2004; 7(3): 355-363.
11. Yang BC. A more accurate Hardy-Hilbert's type inequality. Journal of Xinyang Normal University, 2005; 18(2): 140-142.
12. Yang BC. A more accurate Hilbert-type inequality. College Mathematics, 2005; 21(5): 99-102.
13. Yang BC. On a more accurate Hardy-Hilbert's type inequality and its applications, Acta Mathematica Sinica, 2006; 49(3): 363-368.
14. Yang BC. A more accurate Hilbert type inequality. Journal of Mathematics, 2007; 27(6): 673-678.
15. Yang BC. On an extension of Hardy-Hilbert's type inequality and a reverse. Acta Mathematica Sinica, Chinese Series, 2007; 50(4): 861-868.
16. Yang BC. On a more accurate Hilbert's type inequality. International Mathematical Forum, 2007; 2(37): 1831~1837.
17. Zhong JH, Yang BC. On an extension of a more accurate Hilbert-type inequality. Journal of Zhejiang University (Science Edition), 2008; 35(2): 121-124.
18. Hardy GH. Note on a theorem of Hilbert concerning series of positive term. Proceedings of the London Mathematical Society, 1925; 23: 45-46.
19. Levin V. Two remarks on Hilbert's double series theorem. J. Indian Math. Soc., 1937; 11: 111~115.
20. Bonsall FF. Inequalities with non-conjugate parameter. J. Math. Oxford Ser. 1951; 2(2): 135-150.
21. Yang BC. On a basic Hilbert-type inequality. Journal of Guangdong Education Institute (Natural Science), 2006; 26(3): 1-5.
22. Yang BC. A Hilbert-type inequality with two pairs of conjugate exponents. Journal of Jilin University (Science Edition), 2007; 45(4): 524-528.
23. Hu K. A few important inequalities. Journal of Jianxi Teacher's College (Natural Science), 1979; 3(1): 1-4.
24. Pachpatte BG. On some new inequalities similar to Hilbert's inequality. J. Math. Anal. Appl., 1998; 226: 166-179.
25. Zhao CJ, Debnath L. Some new type Hilbert integral inequalities. J. Math. Anal. Appl., 2001; 262: 411~418.
26. Lu ZX. Some new inverse type Hilbert-Pachpatte inequalities. Tamkang Journal of Mathematics, 2003; 34(2): 155-161.
27. He B, Li YJ. On several new inequalities close to Hilbert-Pachpatte's inequality. J. Ineq. in Pure and Applied Math., 2006; 7(4), Art.154: 1-9.
28. Pachpatte BG. Mathematical inequalities. Netherland: Elsevier B. V., 2005.
29. Yang BC. On Hilbert's integral inequality. J. Math. Anal. Appl., 1998; 220: 778-785.
30. Kuang JC. On new extension of Hilbert's integral inequality. J. Math. Anal. Appl., 1999; 235: 608-614.
31. Kuang JC. New progress in inequality study in China. Journal of Beijing Union University (Natural Science), 2005; 19(1): 29- 37.
32. Gao MZ. On the Hilbert inequality . J. for Anal. Appl., 1999; 18 (4): 1117-1122.
33. Gao MZ, Hsu LC. A survey of various refinements and generalizations of Hilbert's inequalities. J. Math. Res. Exp., 2005; 25(2): 227-243.
34. Zhang KW . A bilinear inequality. J. Math. Anal. Appl., 2002; 271: 288~296.
35. Hsu LC, Wang YJ. A refinement of Hilbert's double series theorem. J. Math. Res. Exp., 1991; 11(1): 143~144.
36. Gao MZ. A note on Hilbert double series theorem. Hunan Mathematical Annal, 1992; 12(1-2): 143-147.
37. Xu LZ, Guo YK. Note on Hardy-Riesz's extension of Hilbert's inequality. Chin. Quart. J. Math., 1991; 6(1): 75-77.
38. Yang BC, Gao MZ. On a best value of Hardy-Hilbert's inequality. Advances in Math., 1997; 26(2): 159-164.
39. Gao MZ, Yang BC. On the extended Hilbert's inequality. Proc. Amer. Math. Soc., 1998; 126(3): 751-759.
40. Yang BC, Debnath L.On new strengthened Hardy-Hilbert's inequality. Internat. J. Math. & Math. Soc., 1998; 21(2): 403-408.
41. Yang BC. A refinement of Hilbert's inequality. Huanghuai Journal, 1997; 13(2): 47-51.
42. Yang BC. On a strengthened version of the more accurate Hardy-Hilbert's inequality. Acta Mathematica Sineca, 1999; 42(6): 1103-1110.

43. Yang BC, Debnath L. A strengthened Hardy-Hilbert's inequality. Proceedings of the Jangjeon Mathematical Society, 2003; 6(2): 119-124.
44. Yang BC. On generalization of Hardy-Hilbert's integral inequalities, Acta Math. Sineca, 1998; 41(4): 839-844.
45. Yang BC. A note on Hilbert's integral inequalities. Chinese Quarterly J. Math., 1998; 13(4): 83-86.
46. Wang ZQ, Guo DR. Introduction to special functions. Beijing: Science Press, 1979.
47. Yang BC. A general Hardy-Hilbert's integral inequality with a best value. Chinese Annal of Mathematics, 2000; 21A(4): 401-408.
48. Yang BC, Debnath L. On the extended Hardy-Hilbert's inequality. J. Math. Anal. Appl., 2002; 272: 187-199.
49. Yang BC, Debnath L. On a new generalization of Hardy-Hilbert's inequality. J. Math. Anal. Appl., 1999; 233: 484-497.
50. Yang BC. On a genealization of Hilbert's double series theorem. Journal of Nanjing University-Mathematical Biquarterly, 2001; 18(1): 145-151.
51. Yang BC. On a general Hardy-Hilbert's inequality. Chinese Annal of Mathematics, 2002; 23A(2): 247-254.
52. Yang BC. A dual Hardy-Hilbert's inequality and generalizations. Advances in Math., 2006; 35(1): 102-108.
53. Yang BC. On new extensions of Hilbert's inequality. Acta Math. Hungar., 2004; 104(4): 291-299.
54. Yang BC. On a new inequality similar to Hardy-Hilbert's inequality. Math. Ineq. Appl., 2003; 6(1): 37-44.
55. Yang BC. Best generalization of Hilbert's type of inequality. Journal of Jilin Univ. (Science Edition), 2004; 42(1): 30-34.
56. Yang BC. On a generalization of the Hilbert's type inequality and its applications. Chinese Journal of Engineering Mathematics, 2004; 21(5): 821-824.
57. Yang BC. Generalization of the Hilbert's type inequality with best constant factor and its applications. J. Math. Res. Exp., 2005; 25(2): 341-346.
58. Yang BC. On Mulholand's integral inequality. Soochow Journal of Mathematics, 2005; 31(4): 573-580.
59. Yang BC. A new Hilbert-type inequality. Bull. Belg. Math. Soc., 2006; 13: 479-487.
60. Wang WH, Yang BC. A strengthened Hardy-Hilbert's type inequality. The Australian Journal of Mathematical Analysis and Applications, 2006; 3(2), Art.17: 1-7.
61. Hong Y. All-side generalization about Hardy-Hilbert integral inequalities, Acta Mathematica Sinica, 2001; 44(4): 619-626.
62. He LP, Yu JM, Gao MZ. An extension of Hilbert's integral inequality. Journal of Shaoguan University (Natural Science) , 2002; 23(3): 25-30.
63. Yang BC. On a multiple Hardy-Hilbert's integral inequality. Chinese Annal of Mathematics, 2003, ; 24A(6): 743-750.
64. Yang BC, Rassias TM. On the way of weight coefficient and research for Hilber-type inequalities. Math. Ineq. Appl., 2003; 6(4): 625-658.
65. Yang BC. On the way of weight function and research for Hilbert's type integral inequalities. Journal of Guangdong Education Institute (Natural Science), 2005; 25(3): 1-6.
66. Sulaiman WT. On Hardy-Hilbert's integral inequality. J. Ineq. in Pure & Appl. Math., 2004; 5(2), Art.25: 1-9.
67. Brnetic I, Pecaric J. Generalization of Hilbert's integral inequality. Math. Ineq. & Appl., 2004; 7(2): 199-205.
68. Krnic M, Gao MZ, Pecaric J, Gao X M. On the best constant in Hilbert's inequality. Math. Ineq. & Appl., 2005; 8(2): 317-329.
69. Brnet I, Krnic M, Pecaric J. Multiple Hilbert and Hardy-Hilbert inequalities with non-conjugate parameters . Bull. Austral. Math. Soc., 2005; 71: 447-457.
70. Krnic M, Pecaric J. General Hilbert's and Hardy's inequalities. Math. Ineq. & Appl., 2005; 8(1): 29-51.
71. Sulaiman WT. New ideas on Hardy-Hilbert's integral inequality (I). Pan American Math. J., 2005; 15(2): 95-100.
72. Salem SR. Some new Hilbert type inequalities. Kyungpook Math. J., 2006; 46: 19-29.
73. Laith EA. On some extensions of Hardy-Hilbert's inequality and applications. Journal of Inequalities and Applications, volume 2008; Article ID 546828, 14 pages. Doi: 10.1155/2008/546829.
74. Jia WJ, Gao MZ, Debnath L. Some new improvement of the Hardy -Hilbert inequality with applications. International Journal of Pure and Applied Math., 2004; 11(1): 21-28.
75. Lu ZX. On new generalizations of Hilbert's inequalities. Tamkang Journal of Mathematics, 2004; 35(1): 77-86.
76. Xie H, Lu Z. Discrete Hardy-Hilbert's inequalities in $\mathbb{R}^n$. Northeast. Math., 2005; 21(1): 87-94.
77. Gao MZ. A new Hardy-Hilbert's type inequality for double series and its applications. The Australian Journal of Mathematical Analysis and Appl., 2005; 3(1), Art.13: 1-10.
78. He LP, Gao MZ, Jia WJ. On a new strengthened Hardy-Hilbert's inequality. J. Math. Res. Exp., 2006; 26(2): 276-282.
79. He LP, Jia WJ, Gao MZ. A Hardy-Hilbert's type inequality with gamma function and its applications. Integral Transforms and Special functions, 2006; 17(5): 355-363.
80. Jia WJ, Gao MZ, Gao XM. On an extension of the Hardy-Hilbert theorem. Studia Scientiarum Mathematicarum Hungarica, 2005; 42(1): 21-35.
81. Gao MZ, Jia WJ, Gao XM, On an improvement of Hardy-Hilbert's inequality. J. Math., 2006; 26(6): 647-651
82. Sun BJ. Best generalization of a Hilbert type inequality. J. Ineq. in Pure & Applied Math., 2006; 7(3), Art.113: 1-7.
83. Wang WH, Xin DM. On a new strengthened version of a Hardy-Hilbert type inequality and applications. J. Ineq. in Pure & Applied Math., 2006; 7(5), Art.180: 1-7.
84. Xu JS. Hardy-Hilbert's inequalities with two parameters. Advances in Mathematics, 2007; 36(2): 189-198.
85. Chen ZQ, Xu JS. New extensions of Hilbert's inequality with multiple parameters. Acta Math. Hungar., 2007; 117(4): 383-400.
86. Yang BC. On an extension of Hilbert's integral inequality with some parameters. The Australian Journal of Math. Analysis and Applications, 2004; 1(1), Art.11: 1-8.
87. Yang BC, Brnetc I, Krnic M, Pecaric J. Generalization of Hilbert and Hardy-Hilbert integral inequalities. Math. Ineq. & Appl., 2005; 8(2): 259-272.
88. Hong Y. On multiple Hardy-Hilbert integral inequalities with some parameters. Journal of Inequalities and Applications, Vol. 2006; Art.ID94960: 1-11.
89. Hong Y. On Hardy-Hilbert integral inequalities with some parameters. J. Ineq. in Pure & Applied Math., 2005; 6(4), Art. 92: 1-10.
90. Zhong WY, Yang BC. On Multiple's Hardy-Hilbert integral inequality with kernel. Journal of Inequalities and Applications, Vol.2007; Art.ID 27962, 17 pages, doi: 10.1155/ 2007/27962.
91. Yang BC. On best extensions of Hardy-Hilbert's inequality with two parameters. J. Ineq. in Pure & Applied Math., 2005; 6(3), Art. 81: 1-15.
92. Yang BC. A reverse of the Hardy-Hilbert's type inequality. Journal of Southwest China Normal University (Natural Science), 2005; 30(6): 1012-1015.
93. Yang BC. A reverse Hardy-Hilbert's integral inequality. Journal of Jilin University, 2004; 42(4): 489-493.
94. Yang BC. On a reverse of Hardy-Hilbert's integral inequality. Pure and Appl.Math., 2006; 22(3): 312-317.
95. Yang BC. On an extended Hardy-Hilbert's inequality and some reversed form. International Mathematical Forum, 2006; 1(39): 1905~1912.
96. Yang BC. A reverse of the Hardy-Hilbert's inequality. Math. in Practice and Theory, 2006; 36(11): 207-212.
97. Yang BC. On a reverse of a Hardy-Hilbert type inequality. J. Ineq. in Pure & Applied Math., 2006; 7(3), Art.115: 1-7.
98. Xi GW. A reverse Hardy-Hilbert-type inequality. Journal of Inequalities and Appl., Vol.2007; Art.ID79758: 1-7.
99. Yang BC. On a relation to Hardy-Hilbert's inequality and Mulholland's inequality. Acta Mathematica Sinica, 2006; 49 (3): 559-566.

100. Xin DM. Best generalization of Hardy-Hilbert's inequality with multi-parameters. J. Ineq. in Pure and Applied Math., 2006; 7(4), Art.153: 1-8.
101. Zhong WI, Yang BC. A best extension of Hilbert inequality involving several parameters. Journal of Jinan University (Natural Science), 2007; 28(1): 20-23.
102. Zhong WI, Yang BC. A reverse Hilbert's type integral inequality with some parameters and the equivalent forms. Pure and Applied Mathematics, 2008; 24(2): 401-407.
103. Yang BC. A new Hilbert-type inequality. Journal of Shanghai Univ. (Natural Science), 2007; 13(3): 274-278.
104. Yang BC. A bilinear inequality with the kernel of -2-order homogeneors. Journal of Xiamen University (Natural Science), 2006; 45(6): 752-755.
105. Yang BC. A Hilbert-type integral inequality with the kernel of -3-order homogeneous. Journal of Yunnam University, 2008; 30(4): 325-330.
106. Xie ZT. A new Hilbert-type inequality with the kernel of 3 μ -homogeneous. Journal of Jilin University (Science Edition), 2007; 45(3): 369-373.
107. Xie ZT, Zheng Z. A Hilbert-type inequality with parameters. J. Xiangtan Univ. (Natural Science), 2007;29(3): 24-28.
108. Xie ZT, Zheng Z. A Hilbert-type integral inequality whose kernel is a homogeneous form of degree -3 . J. Math. Anal. Appl., 2007; 339: 324-331.
109. Xie ZT, Zheng Z. A new Hilbert-type integral inequality and Its reverse. Soochow Journal of Math., 2007; 33(4): 751-759.
110. Li YJ, He B. On inequalities of Hilbert's type. Bull. Austral. Math. Soc., 2007; 76: 1-13.
111. He B, Qian Y, Li YJ. On analogues of the Hilbert's inequality. Comm. in Math. Anal., 2008; 4(2): 47-53.
112. Yang BC. On a Hilbert-type inequality with the homogeneous kernel of -3-order. Journal of Guangdong Education Institute (Natural Science), 2007; 27(5): 1-5.
113. Yang BC. On the norm of an integral operator and applications. J. Math. Anal. Appl., 2006; 321: 182-192.
114. Yang BC. On the norm of a self-adjoint operator and a new bilinear integral inequality. Acta Mathematica Sinica, English Series, 2007; 23(7): 1311-1316.
115. Yang BC. On the norm of a certain self-adjoint integral operator and applications to bilinear integral inequalities. Taiwan Journal of Mathematics, 2008; 12(2): 315-324.
116. Yang BC. On the norm of a Hilbert's type linear operator and applications. J. Math. Anal. Appl., 2007; 325: 529-541.
117. Yang BC. On the norm of a self-adjoint operator and applications to Hilbert's type inequalities. Bull. Belg. Math. Soc., 2006; 13: 577-584.
118. Yang BC. On a Hilbert-type operator with a symmetric homogeneous kernel of -1-order and applications. Journal of Inequalities and Applications, Volume 2007; Article ID 47812, 9 pages, doi: 10.1155/2007/47812.
119. Yang BC. On the norm of a linear operator and its applications. Indian Journal of Pure and Applied Mathematics, 2008; 39(3): 237-250.
120. Arpad B, Choonghong O. Best constants for certain multilinear integral operator. Journal of Inequalities and Applications, Vol.2006; Art. ID28582: 1-12.
121. Yang BC. On the norm of operator and Hilbert-type inequalities. Science Press, 2009.
122. Yang BC. Hilbrt-type integral inequalities, Bentham Science Publishers, 2009.
123. Yang BC. On a basic Hilbert-type integral inequality and extensions. College Mathematics, 2008; 24(1): 87-92.
124. Yang BC. A basic Hilbert-type integral inequality with the homogeneous kernel of -1-degree and extensions. Journal of Guangdong Education Institute (Natural Science), 2008; 28(3): 1-10.
125. Yang BC. A new Hilbert-type integral inequality and its generalization. Journal of Jilin University (Science Edition), 2005; 43(5): 580-584.
126. Yang BC, Liang HW. A new Hilbert-type integral inequality with a parameter. Journal of Henan University (Natural Science), 2005; 35(4): 4-8.
127. Xin DM, Yang BC. A basic Hilbert-type inequality. J. of Math., 2010,30(3): 552-560.
128 Li YJ, He B. A new Hilbert-type integral inequality and the equivalent form. Internat. J. Math. & Math. Soc., Vol. 2006; Art.ID45378: 1~6.
129. Xie CH. A best extension of a new Hilbert-type inequality. Journal of Jinan Univ. (Natural Science), 2007; 28(1): 24-27.
130. Li YJ, Qian Y, He B. On further analogs of Hilbert's inequality. Internat. J. Math. & Math. Soc., Vol. 2007; Art. ID76329: 1~6.
131. Li YJ, Wang Z, He B. Hilbert's type linear operator and some extensions of Hilbert's inequality. J. Ineq & Appl., Vol.2007; Art. ID82138: 1~10.
132. Xie ZT, Yang BC. A new Hilbert-type integral inequality with some parameters and its reverse . Kyungpook Math. J., 2008; 48: 93-100.

CHAPTER 2

Improvements of Euler –Maclaurin's Summation Formula: Preliminary Theorems

Abstract: In this chapter, for optimizing the methods of estimating the series and weight coefficients, we introduce some preliminary theorems of improving Euler-Maclaurin's summation formula. A few useful corollaries and inequalities are considered. By the way of applications, some new inequalities on Hurwitz ς-function restricted in the real axis are given.

2.1. SOME SPECIAL FUNCTIONS AND EULER -MACLAURIN'S SUMMATION FORMULA

2.1.1. BERNOULLI NUMBERS

We define a function $G(x)$ as

$$G(x) := \frac{x}{e^x - 1} \; (G(0) := \lim_{x\to 0} G(x) = 1).$$

It is obvious that the power series of $G(x)$ at $x = 0$ possesses a positive convergence radius. Assuming that the sequence $\{B_n\}_{n=0}^{\infty}$ is defined by the exponent creation function of $G(x)$, and $G(x) = \sum_{n=0}^{\infty} \frac{B_n}{n!} x^n$, we have

$$1 = \frac{e^x - 1}{x} \sum_{n=0}^{\infty} \frac{B_n}{n!} x^n = \sum_{n=0}^{\infty} \frac{1}{(n+1)!} x^n \sum_{n=0}^{\infty} \frac{B_n}{n!} x^n$$
$$= \sum_{n=0}^{\infty} \left[\sum_{k=0}^{n} \frac{B_k}{k!(n-k+1)!} \right] x^n.$$

In Comparison to the coefficients of x^n on two sides of the above equality, it follows that $B_0 = 1$, and $\sum_{k=0}^{n} \frac{B_k}{k!(n-k+1)!} = 0 \quad (n \in \mathbf{N})$. We can obtain the recursion formulas of $\{B_n\}_{n=0}^{\infty}$ as follows:

$$B_0 = 1, B_n = -n! \sum_{k=0}^{n-1} \frac{B_k}{k!(n-k+1)!}, \; n \in \mathbf{N}. \tag{2.1.1}$$

Since $G(-x) = \frac{-x}{e^{-x}-1} = \frac{xe^x}{e^x - 1} = x + G(x)$, then we find $\sum_{n=0}^{\infty} B_n[(-1)^n - 1] \frac{x^n}{n!} = x$. Comparing the coefficients of $\frac{x^n}{n!}$ on two sides of the above equality, we obtain

$$B_1 = -\frac{1}{2}, B_{2k+1} = 0 \; (k \in \mathbf{N}). \tag{2.1.2}$$

By (2.1.1) and (2.1.2), we can obtain the constants step by step as follows:

$$B_2 = \frac{1}{6}, B_4 = -\frac{1}{30}, B_6 = \frac{1}{42}, B_8 = -\frac{1}{30}, \cdots.$$

We call B_n $(n \in \mathbf{N}_0 = \mathbf{N} \cup \{0\})$ Bernoulli numbers. In a general way, we also have the following formula for the Bernoulli numbers (Xu HEP 1985)[1]:

$$B_{2k} = (-1)^{k+1} \frac{(2k)!}{2^{2k-1}\pi^{2k}} \sum_{n=1}^{\infty} \frac{1}{n^{2k}}, \; k \in \mathbf{N}. \tag{2.1.3}$$

2.1.2 BERNOULLI POLYNOMIALS

Suppose the function $B_n(t)$ is defined by the following exponent creation function:

$$e^{tx} G(x) = \sum_{n=0}^{\infty} B_n(t) \frac{x^n}{n!}. \tag{2.1.4}$$

In the convergence interval of (2.1.4), we have

$$e^{tx} G(x) = \sum_{n=0}^{\infty} \frac{t^n}{n!} x^n \sum_{n=0}^{\infty} \frac{B_n}{n!} x^n$$
$$= \sum_{n=0}^{\infty} \left[\sum_{k=0}^{n} \binom{n}{k} B_k t^{n-k} \right] \frac{x^n}{n!}, \tag{2.1.5}$$

In Comparison to the coefficients of the term $\frac{x^n}{n!}$ on two sides of (2.1.5) relating (2.1.4), we get the following formula:

$$B_n(t) = \sum_{k=0}^{n} \binom{n}{k} B_k t^{n-k} \; (n \in \mathbf{N}_0). \tag{2.1.6}$$

We call $B_n(t) (n \in \mathbf{N}_0)$ Bernoulli polynomials. In particular, we can obtain

$$B_0(t) = 1, B_1(t) = t - \frac{1}{2}, B_2(t) = t^2 - t + \frac{1}{6},$$
$$B_3(t) = t^3 - \frac{3}{2}t^2 + \frac{1}{2}t,$$
$$B_4(t) = t^4 - 2t^3 + t^2 - \frac{1}{30}, \cdots,$$

and prove the equations that (Xu HEP 1985)[1]

$$B_n'(t) = nB_{n-1}(t),$$
$$\int_0^1 B_n(t)dt = 0, B_{n-1}(0) = B_{n-1}, \; n \in \mathbf{N}.$$

On the other hand, in view of $B_0(t) = 1$,

$$B_n'(t) = nB_{n-1}(t)$$

and $B_n(0) = B_n$ $(n \in \mathbf{N})$, by integration, we find

$$B_n(t) = B_n + n\int_0^t B_{n-1}(t)dt, \quad n \in \mathbf{N}. \tag{2.1.7}$$

We may use (2.1.7) to define the Bernoulli polynomials in (2.1.6). In fact, by mathematical induction, for $n = 0$, we have $B_0(t) = 1$. Assuming that (2.1.6) is valid for n, then for $n+1$, it follows

$$B_{n+1}(t) = B_{n+1} + (n+1)\int_0^t B_n(t)dt$$

$$= B_{n+1} + \sum_{k=0}^{n}(n+1)\binom{n}{k}B_k\int_0^t t^{n-k}dt$$

$$= \sum_{k=0}^{n+1}\binom{n+1}{k}B_k t^{n+1-k}.$$

Hence the function (2.1.6) is the root of function equation (2.1.7).

2.1.3 BERNOULLI FUNCTIONS

For any $t \in \mathbf{R}$, we denote $[t]$ the maximal integer that dose not exceed t and $\{t\}$ as $\{t\} = t - [t]$. Then it follows that Bernoulli functions $P_k(t) := B_k(\{t\})$ $(k \in \mathbf{N})$ are periodic functions with the least positive periodic 1. It is obvious that $P_k(t)$ are of bounded variation in any finite interval; $P_1(t) = \{t\} - \frac{1}{2}$ is not continuous in the integers, but it is differentiable in the other points; $P_2(t)$ is continuous in **R** and differentiable in the set of all the non-integers; for $k \geq 3$, $P_k(t)$ are continuous and differentiable in **R.** We can prove that $P_{2k}(t)$ are even functions and $P_{2k-1}(t)$ are odd functions. In fact, setting $H(t,x) = e^{tx}G(x)$, and

$$e^{(1-t)x}\frac{x}{e^x-1} = e^{-tx}\frac{xe^x}{e^x-1} = e^{t(-x)}\frac{-x}{e^{-x}-1},$$

we obtain $H(1-t,x) = H(t,-x)$, and by (2.1.4), it follows

$$\sum_{n=0}^{\infty}\frac{B_n(1-t)}{n!}x^n = \sum_{n=0}^{\infty}\frac{B_n(t)}{n!}(-x)^n = \sum_{n=0}^{\infty}\frac{(-1)^n B_n(t)}{n!}x^n.$$

Hence in comparision to the coefficients of $\frac{x^n}{n!}$ on two sides of the above equality, we obtain

$$B_n(1-t) = (-1)^n B_n(t),$$

and then it follows

$$P_n(-t) = P_n(1-t) = (-1)^n P_n(t), \quad n \in \mathbf{N}.$$

Equivalently, we may define the Bernoulli functions $P_n(t)$ by the following recursional function equations:

$$P_1(t) = \{t\} - \tfrac{1}{2},$$

$$P_{n+1}(t) = B_{n+1} + (n+1)\int_0^t P_n(t)dt, n \in \mathbf{N}. \tag{2.1.8}$$

Since $P_1(t)$ is not continuous at any integer t, then in view of the integral properties, $P_2(t)$ is continuous at t, but it is not differentiable at t. And then $P_n(t)$ $(n \geq 3)$ are differentiable in the real axis.

Considering $P_2(t)$, at any integer k, it is obvious that $P_2(k)(= B_2 = \frac{1}{6})$ is the maximum value, and at $t = k + \frac{1}{2}$, since $P_2'(k + \frac{1}{2}) = 0$, then the minimum value of $P_2(t)$ is $P_2(k + \frac{1}{2})$ expressed by

$$P_2(k + \tfrac{1}{2}) = P_2(\tfrac{1}{2}) = B_2 + 2\int_0^{\frac{1}{2}}(t - \tfrac{1}{2})dt = -\tfrac{1}{12}.$$

The roots of $P_3(t)$ are divided in two classes. One class is the set of integers k, that make $P_3(k) = B_3 = 0$; the other class is the set of all $k + \frac{1}{2}$. In fact, by the definition of Bernoulli polynomial, it follows

$$P_3(k + \tfrac{1}{2}) = P_3(\tfrac{1}{2})$$

$$= 3\int_0^{\frac{1}{2}} P_2(t)dt = 3\int_0^{\frac{1}{2}}(t^2 - t + \tfrac{1}{6})dt = 0.$$

Since the points that make the values of $P_4(t)$ are maximum or minimum at the roots of $P_3(t)$, these points are divided into two classes. In a general way, by Rabbe formula (Xu HDP 1985)[1] (Proposition 98), we have

$$B_n(kt) = k^{n-1}\sum_{i=0}^{k-1}B_n(t + \tfrac{i}{k}), \quad k, n \in \mathbf{N}. \tag{2.1.9}$$

Setting $k = 2, t = 0$, we find

$$B_n(\tfrac{1}{2}) = (2^{1-n} - 1)B_n(0) = (2^{1-n} - 1)B_n. \tag{2.1.10}$$

Hence both $P_{2n+1}(t)$ and $P_3(t)$ possess the same two classes of roots (Xu HDP1985) [1], (Proposition 93). Since the positions of the points that make the value of $P_{2n+2}(t)$ are maximum or minimum positions of the points that make $P_{2n+2}'(t) = 0$, then both $P_{2n+2}(t)$ and $P_2(t)$ possess the same points that make the value of them as maximum or minimum. Since

$$P_{2n+2}(k) = P_{2n+2}(0) = B_{2n+2},$$

then by (2.1.10), we obtain

$$P_{2n+2}(k + \tfrac{1}{2}) = P_{2n+2}(\tfrac{1}{2})$$

$$= -(1 - \tfrac{1}{2^{2n+1}})B_{2n+2}, \tag{2.1.11}$$

$$|P_{2n+2}(k + \tfrac{1}{2})| < |B_{2n+2}|, \quad n \in \mathbf{N}_0. \tag{2.1.12}$$

In view of the above formulas, we discover that

$$(\max P_{2n+2}(t))(\min P_{2n+2}(t)) < 0,$$
$$\max P_{2n+2}(t) \neq -\min P_{2n+2}(t)(n \in \mathbf{N}_0).$$

2.1.4 EULER-MACLAURIN'S SUMMATION FORMULA

Assuming that $m,n \in \mathbf{N}_0$, $n > m$, $f(t)$ is a continuous differentiable function of 1-order in $[m,\infty)$, then we have the following formula:

$$\sum_{k=m+1}^{n} f(k) = \int_m^n f(t)dt + \tfrac{1}{2} f(t)|_m^n + \int_m^n P_1(t)f'(t)dt. \quad (2.1.13)$$

In fact, since it follows

$$P_1(t) = \{t\} - \tfrac{1}{2} = t - k - \tfrac{1}{2}, t \in [k,k+1),$$

integration by parts, we find

$$\int_m^n P_1(t)f'(t)dt = \sum_{k=m}^{n-1} \int_k^{k+1} P_1(t)f'(t)dt$$
$$= \sum_{k=m}^{n-1} \int_k^{k+1} (t-k-\tfrac{1}{2})df(t)$$
$$= \sum_{k=m}^{n-1} [(t-k-\tfrac{1}{2})f(t)|_k^{k+1} - \int_k^{k+1} f(t)dt]$$
$$= \tfrac{1}{2}\sum_{k=m}^{n-1} (f(k+1)+f(k)) - \int_m^n f(t)dt$$
$$= \tfrac{1}{2}(f(m)-f(n)) + \sum_{k=m+1}^{n} f(k) - \int_m^n f(t)dt.$$

Hence we obtain (2.1.13).

In a general way, we have the following Euler-Maclaurin summation formula (Knopp BSL 1928)[2]:

Theorem 2.1.1 Assuming that $m,n \in \mathbf{N}_0$, $n > m$, $f(t) \in C^q[m,\infty)$ $(q \in \mathbf{N})$, then we have

$$\sum_{k=m+1}^{n} f(k) = \int_m^n f(t)dt + \sum_{k=1}^{q} (-1)^k \tfrac{B_k}{k!} f^{(k-1)}(t)|_m^n + \tfrac{(-1)^{q+1}}{q!}\int_m^n P_q(t)f^{(q)}(t)dt. \quad (2.1.14)$$

Proof. We prove (2.1.14) by mathematical induction. For $q=1$ (indicating $f^{(0)}(t)=f(t)$),since $B_1 = -\tfrac{1}{2}$, by (2.1.13), we have (2.1.14). Supposing that (2.1.14) is valid for $q(\in \mathbf{N})$, then for $q+1$, since $P_{q+1}(m) = P_{q+1}(n) = B_{q+1}$ and for any non-integer t, $P'_{q+1}(t) = (q+1)P_q(t)$, integration by parts, we find

$$\int_m^n P_q(t)f^{(q)}(t)dt = \tfrac{1}{q+1}\int_m^n f^{(q)}(t)dP_{q+1}(t)$$
$$= \tfrac{B_{q+1}}{q+1} f^{(q)}(t)|_m^n - \tfrac{1}{q+1}\int_m^n P_{q+1}(t)f^{(q+1)}(t)dt. \quad (2.1.15)$$

Then substitution of (2.1.15) in (2.1.14), it follows that (2.1.14) is a value for $q+1$. □

Note. If $f^{(q)}(t)$ is a constant function, then it follows

$$\int_m^n P_q(t)f^{(q)}(t)dt = 0.$$

Since $B_{2q+1} = 0$ $(q \in \mathbf{N})$, we may reduce (2.1.14) in the following corollary:

Corollary 2.1.2 Assuming that $m,n,q \in \mathbf{N}_0$, $n > m$, $f(t) \in C^{2q+1}[m,\infty)$, then we have

$$\sum_{k=m}^{n} f(k) = \int_m^n f(t)dt + \tfrac{f(n)+f(m)}{2} + \sum_{k=1}^{q} \tfrac{B_{2k}}{(2k)!} f^{(2k-1)}(t)|_m^n + \delta_q(m,n), \quad (2.1.16)$$

$$\delta_q(m,n) := \tfrac{1}{(2q+1)!}\int_m^n P_{2q+1}(t)f^{(2q+1)}(t)dt, \quad (2.1.17)$$

where, $\delta_q(m,n)$ is called the residue term of $2q$-order in (2.1.16) (**Note.** for $q=0$, we define that the series in the right side of (2.1.16) is 0).

For any $r \in \mathbf{R}$ and $k \in \mathbf{N}_0$, we define the combination number $\binom{r}{k}$ as follows (Qu BUP 1989)[3]:

$$\binom{r}{k} := \begin{cases} 1, & k = 0, \\ \tfrac{r(r-1)\cdots(r-k+1)}{k!}, & k > 0. \end{cases}$$

Example 2.1.3 Suppose that

$$f(t) = (t+a)^l \quad (l \in \mathbf{N}, 0 < a \le 1; t \in (-a,\infty)).$$

Then for $i < l$, $f^{(i)}(t) = i!\binom{l}{i}(t+a)^{l-i}$; for $i \ge l$, $f^{(i)}(t) =$ constant. In (2.1.14), for $2q = l$ or $2q+1 = l$, we have $q = [\tfrac{l}{2}]$ and $\delta_q(m,n) = 0$. By (2.1.16) (setting $m=0$), we find

$$\sum_{k=0}^{n} (k+a)^l = \tfrac{1}{l+1}(n+a)^{l+1} + \tfrac{1}{2}(n+a)^l + \sum_{k=1}^{[\frac{l}{2}]} \tfrac{B_{2k}}{2k}\binom{l}{2k-1}(n+a)^{l-2k+1}$$

$$-\left(\frac{1}{l+1}a^{l+1}-\frac{1}{2}a^{l}+\sum_{k=1}^{[\frac{l}{2}]}\frac{B_{2k}}{2k}\binom{l}{2k-1}a^{l-2k+1}\right). \tag{2.1.18}$$

For $a=\frac{c}{b}$ $(b>c>0)$ in (2.1.18), we may deduce a summation formula of an arithmetical sequence with the power of non-negative integer (Yang JSCNU 1996)[4], and for $a=1$, replacing n by $n-1$ in (2.1.18), we still have (Yang NPC 1994)[5]

$$\sum_{k=1}^{n}k^{l}=\frac{1}{l+1}n^{l+1}+\frac{1}{2}n^{l}+\sum_{k=1}^{[\frac{l}{2}]}\frac{B_{2k}}{2k}\binom{l}{2k-1}n^{l-2k+1}. \tag{2.1.19}$$

2.2. ON ESTIMATIONS OF THE RESIDUE TERM ABOUT A CLASS SERIES

In (2.1.17), if $f^{(2q-1)}(t)$ is a bound variation function, then we find the following estimation (Cheng HEP 2003)[6]:

$$\begin{aligned}|\delta_q(m,n)| &= \frac{1}{(2q+1)!}\left|\int_m^n P_{2q+1}(t)df^{(2q)}(t)\right|\\ &= \frac{1}{(2q+1)!}\Big|[P_{2q+1}(t)f^{(2q)}(t)]_m^n\\ &\quad -(2q+1)\int_m^n P_{2q}(t)f^{(2q)}(t)dt\Big|\\ &= \frac{1}{(2q)!}\left|\int_m^n P_{2q}(t)df^{(2q-1)}(t)\right|\\ &\le \frac{1}{(2q)!}|B_{2q}|\bigvee_m^n(f^{(2q-1)}).\end{aligned} \tag{2.2.1}$$

We shall refine (2.2.1) in the following theorems by adding some conditions.

2.2.1 THE ESTIMATION UNDER THE MORE FORTIFIED CONDITIONS

Theorem 2.2.1 Assuming that $m,n,q\in \mathbf{N}_0$, $n>m$, $g(t)\in C^3[m,n]$, $(-1)^k g^{(k)}(t)\ge 0$ (≤ 0), $t\in[m,n]$ $(k=0,1,2,3)$, if there exist two intervals $I_k\subset[m,n]$, such that

$$g^{(k)}(t)<0\ (>0), t\in I_k\ (k=1,3),$$

then we have the following estimation (Yang ASNUS 1997)[7]

$$\begin{aligned}\tilde{\delta}_q(m,n) &:= \frac{1}{(2q+1)!}\int_m^n P_{2q+1}(t)g(t)dt\\ &= \varepsilon_q\frac{B_{2q+2}}{(2q+2)!}g(t)|_m^n,\ 0<\varepsilon_q<1.\end{aligned} \tag{2.2.2}$$

Setting $n=\infty$ and $g(\infty)=0$, then the integral $\int_m^\infty P_{2q+1}(t)g(t)dt$ is convergence and then we have

$$\begin{aligned}\tilde{\delta}_q(m,\infty) &= \frac{1}{(2q+1)!}\int_m^\infty P_{2q+1}(t)g(t)dt\\ &= -\varepsilon_q\frac{B_{2q+2}}{(2q+2)!}g(m),\ 0<\varepsilon_q<1.\end{aligned} \tag{2.2.3}$$

In particular, for $q=1$, it follows

$$\begin{aligned}\tilde{\delta}_1(m,\infty) &= \frac{1}{6}\int_m^\infty P_3(t)g(t)dt\\ &= \frac{\varepsilon_1}{720}g(m),\ 0<\varepsilon_1<1.\end{aligned} \tag{2.2.4}$$

Proof by (2.1.12), we have

$$\max_{t\in\mathbb{R}}|P_{2q+2}(t)|=|P_{2q+2}(m)|=|B_{2q+2}|.$$

Since $|P_{2q+2}(t)|$ is a non-constant continuous function, $|g'(t)|>0$, $t\in I_1\subset[m,n]$, and $g'(t)$ dose not change the sign in $[m,n]$, then we find

$$\begin{aligned}&\left|\int_m^n P_{2q+2}(t)g'(t)dt\right|\\ &\le\int_m^n|P_{2q+2}(t)|\cdot|g'(t)|dt\\ &<|B_{2q+2}|\int_m^n|g'(t)|dt\\ &=\left|B_{2q+2}\int_m^n g'(t)dt\right|=\left|B_{2q+2}g(t)|_m^n\right|,\end{aligned}$$

which is equivalent and there exists a $\varepsilon_q\in(0,2)$, such that the following expression is valid:

$$\begin{aligned}&\int_m^n P_{2q+2}(t)g'(t)dt\\ &=B_{2q+2}g(t)|_m^n(1-\varepsilon_q).\end{aligned} \tag{2.2.5}$$

Integration by parts and in view of (2.2.5), we find

$$\begin{aligned}\tilde{\delta}_q(m,n) &= \frac{1}{(2q+1)!}\int_m^n P_{2q+1}(t)g(t)dt\\ &= \frac{1}{(2q+2)!}\int_m^n g(t)dP_{2q+2}(t)\\ &= \frac{1}{(2q+2)!}[P_{2q+2}(t)g(t)|_m^n-\int_m^n P_{2q+2}(t)g'(t)dt]\\ &= \frac{1}{(2q+2)!}[B_{2q+2}g(t)|_m^n-B_{2q+2}g(t)|_m^n(1-\varepsilon_q)]\\ &= \varepsilon_q\frac{B_{2q+2}}{(2q+2)!}g(t)|_m^n,\ 0<\varepsilon_q<2.\end{aligned} \tag{2.2.6}$$

It is obvious that both the integrals $\int_m^n P_{2q+1}(t)g(t)dt$ and the term $B_{2q+2}g(t)|_m^n$ keep the same sign.

In the following, we show that $0<\varepsilon_q<1$ in (2.2.5). For this, we need to prove that both the integrals $\int_m^n P_{2q+2}(t)g'(t)dt$ and the term $B_{2q+2}g(t)|_m^n$ keep the same sign. In fact, integration by parts, we obtain

$$\int_m^n P_{2q+2}(t)g'(t)dt = \frac{1}{2q+3}\int_m^n g'(t)dP_{2q+3}(t)$$
$$= \frac{1}{2q+3}[g'(t)P_{2q+3}(t)|_m^n - \int_m^n P_{2q+3}(t)g''(t)dt]$$
$$= \frac{-1}{2q+3}\int_m^n P_{2q+3}(t)g''(t)dt. \qquad (2.2.7)$$

Since $g'''(t)$ and $g'(t)$ possess the same property of the sign, by the same way of obtaining (2.2.5) and (2.2.6), it follows that both $\int_m^n P_{2q+3}(t)g''(t)dt$ and $B_{2q+4}g''(t)|_m^n$ keep the same sign. Since by (2.1.3), we have $B_{2q+4}B_{2q+2} < 0$ ($q \in \mathbf{N}_0$), and we still have $g''(t)|_m^n \cdot g(t)|_m^n > 0$, then it follows that $B_{2q+4}g''(t)|_m^n$ and $B_{2q+2}g(t)|_m^n$ keep the different signs. Hence $B_{2q+2}g(t)|_m^n$ and $\int_m^n P_{2q+3}(t)g''(t)dt$ keep the different signs. But in (2.2.7), it follows that the integrals $\int_m^n P_{2q+3}(t)g''(t)dt$ and $\int_m^n P_{2q+2}(t)g'(t)dt$ keep the different signs, then both the integrals $\int_m^n P_{2q+2}(t)g'(t)dt$ and the term $B_{2q+2}g(t)|_m^n$ keep the same sign, and then (2.2.2) is valid.

In the above proof, if we set $n = \infty$, since $g(\infty) = 0$, then (2.2.5) is still valid. Under the case of both $\int_m^n P_{2q+2}(t)g'(t)dt$ and $B_{2q+2}g(t)|_m^n$ keep the same sign, setting $n = \infty$, it follows that both the integrals $\int_m^\infty P_{2q+2}(t)g'(t)dt$ and the term $-B_{2q+2}g(m)$ keeping the same sign (not zero). Then for $n = \infty$, there exists a constant $\varepsilon_q \in (0,1)$, such that (2.2.5) is valid , and then we have (2.2.3).W

Note. In view of the assumption of the theorem, it follows that $g^{(k-1)}(t)$ are strictly decreasing (or increasing) and do not change the sign in $I_k (k = 1,3)$. Hence for $g(t) \ge 0 (\le 0), t \in [m,n]$, we discover the fact that $g(m) > 0$ (< 0).

Example 2.2.2 We can show that the function
$$g(t) = \frac{\ln t}{t-1} \quad (g(1) := \lim_{t\to 1} g(t) = 1)$$
satisfies the conditions of
$$(-1)^k g^{(k)}(t) > 0, t \in (0,\infty) \ (k = 0,1,2,3).$$

In fact, we have the power series of $g(t)$ in $t_0 = 1$ as follows:
$$g(t) = \frac{\ln[1+(t-1)]}{t-1} = \sum_{k=0}^{\infty}(-1)^k \frac{(t-1)^k}{k+1}$$
$$= \sum_{k=0}^{\infty}\frac{(-1)^k k!}{k+1}\cdot\frac{(t-1)^k}{k!}, \ -1 < t-1 \le 1,$$
and then
$$g^{(k)}(1) = \frac{(-1)^k k!}{k+1} (k \in \mathbf{N}_0), g^{(0)}(1) = g(1) = 1,$$
$$g'(1) = \frac{-1}{2}, g''(1) = \frac{2}{3}, g'''(1) = \frac{-3}{2}, \cdots.$$
It is obvious that $g(t) > 0$ $(t > 0)$. We can find $g'(t) = \frac{h(t)}{t(t-1)^2}, h(t) := t-1-t\ln t$.

Since $h'(t) = -\ln t > 0, 0 < t < 1$; $h'(t) < 0$, $t > 1$, then $h(t)$ keeps the maximum value at $t = 1$, with $h(1) = 0$,and $h(t) < 0$, $t \in (0,\infty)\setminus\{1\}$ .Since $g'(1) = -\frac{1}{2} < 0$, then $g'(t) < 0$ $(t > 0)$. We find
$$g''(t) = \frac{J(t)}{t^2(t-1)^3},$$
$$J(t) := -(t-1)^2 - 2t(t-1) + 2t^2\ln t,$$
$$J'(t) = -4(t-1) + 4t\ln t.$$
Since $J''(t)(= 4\ln t)$ keeps different signs on two sides of $t = 1$, then it is obvious that $J'(t)$ keeps the minimum value at $t = 1$ with $J'(1) = 0$. Therefore
$$J'(t) > 0, t \in (0,\infty)\setminus\{1\}$$
and $J(t)$ is strictly increasing with $J(1) = 0$, and
$$J(t) < 0, 0 < t < 1; J(t) > 0, \ t > 1.$$
Since $g''(1) > 0$, then $g''(t) > 0$ $(t > 0)$. We find
$$g'''(t) = \frac{L(t)}{t^3(t-1)^4}$$
$$L(t) := 2(t-1)^3 + 3t(t-1)^2$$
$$+ 6t^2(t-1) - 6t^3\ln t,$$
$$L'(t) = 9(t-1)^2 + 18t(t-1) - 18t^2\ln t,$$
$$L''(t) = 36(t-1) - 36t\ln t, L'''(t) = -36\ln t.$$
Then $L'''(t)$ keeps different signs on two sides of $t = 1$, and $L''(t)$ keeps the maximum value at $t = 1$ with $L''(1) = 0$. Hence $L''(t) < 0, t \in (0,\infty)\setminus\{1\}$, $L'(t)$ is strictly decreasing with $L'(1) = 0$, and
$$L'(t) > 0, 0 < t < 1; \ L'(t) < 0, t > 1.$$
It is obvious that $L(t)$ keeps the maximum value at $t = 1$ with $L(1) = 0$ and $L(t) < 0, t \in (0,\infty)\setminus\{1\}$. Hence $g'''(1) < 0$, and then $g'''(t) < 0$ $(t > 0)$.

Therefore, by (2.2.2), we have the following estimation:

$$\tilde{\delta}_q(m,n) = \frac{1}{(2q+1)!}\int_m^n P_{2q+1}(t)\frac{\ln t}{t-1}dt$$
$$= \varepsilon_q \frac{B_{2q+2}}{(2q+2)!}(\frac{\ln n}{n-1} - \frac{\ln m}{m-1}),\ 0 < \varepsilon_q < 1. \quad (2.2.8)$$

Note. In the same way, for fixed $x > 0$ and $\lambda > 0$, the function

$$\tilde{g}_\lambda(t) = \frac{\ln(t/x)}{t^\lambda - x^\lambda} \quad (\varphi(x) := \lim_{t\to x}\tilde{g}_\lambda(t) = \frac{1}{\lambda x^\lambda})$$

keeps the same sign property as $g(t)$. In fact, setting $u = (\frac{t}{x})^\lambda$, we find

$$\tilde{g}_\lambda(t) = h(u) := \frac{1}{\lambda x^\lambda}\cdot\frac{\ln u}{u-1}.$$

Corollary 2.2.3 Assuming that $m_i, q \in \mathbf{N}_0$,

$$m_{i+1} > m_i,\ g(t) \in C^3[m_i, m_{i+1}],$$
$$(-1)^k g^{(k)}(t) \ge 0 \quad (\le 0),\ t \in [m_i, m_{i+1}]$$

$(k = 0,1,2,3; i = 1,\cdots,s)$ and $m = m_1, n = m_{s+1}$,

If there exist an i and two intervals $I_{k_i} \subset [m_i, m_{i+1}]$,

such that $g^{(k)}(t) < 0 \ (>0), t \in I_{k_i} \quad (k = 1,3)$,

then we still have (2.2.2), (2.2.3) and (2.2.4).

Proof Without any loss of generality, assuming that $(-1)^k g^{(k)}(t) \ge 0$, $t \in [m_i, m_{i+1}]$, and $B_{2q+2} < 0$, by (2.2.5), it follows

$$0 \le \int_{m_i}^{m_{i+1}} P_{2q+2}(t)g'(t)dt$$
$$\le B_{2q+2}g(t)\,|_{m_i}^{m_{i+1}} \ (i = 1,\cdots,s).$$

Since there exists an i, such that the corresponding above inequalities keep the signs of strict inequalities, then obtaining the sum of 1 to s, we find

$$0 < \int_m^n P_{2q+2}(t)g'(t)dt < B_{2q+2}g(t)\,|_m^n$$

and we have (2.2.2). In the same way, we have all the other results. □

2.2.2 THE ESTIMATIONS UNDER THE MORE IMPERFECT CONDITIONS

Corollary 2.2.4 Assuming that $m, n, q \in \mathbf{N}_0$, $n > m$, $g(t) \in C^1[m,n]$ with $(-1)^k g^{(k)}(t) \ge 0 \quad (\le 0)$, $t \in [m,n] \ (k = 0,1)$, if there exists an interval $I_1 \subset [m,n]$, such that $g'(t) < 0 \quad (>0)$, $t \in I_1$, then we have the following estimation:

$$\tilde{\delta}_q(m,n) := \frac{1}{(2q+1)!}\int_m^n P_{2q+1}(t)g(t)dt$$
$$= \tilde{\varepsilon}_q \frac{2B_{2q+2}}{(2q+2)!}g(t)\,|_m^n,\ 0 < \tilde{\varepsilon}_q < 1; \quad (2.2.9)$$

if $n = \infty$ and $g(\infty) = 0$, then we have

$$\tilde{\delta}_q(m,\infty) = \frac{1}{(2q+1)!}\int_m^\infty P_{2q+1}(t)g(t)dt$$
$$= -\tilde{\varepsilon}_q \frac{2B_{2q+2}}{(2q+2)!}g(m),\ 0 < \tilde{\varepsilon}_q < 1. \quad (2.2.10)$$

In particular, for $q = 1$, it follows

$$\tilde{\delta}_1(m,\infty) = \frac{1}{6}\int_m^\infty P_3(t)g(t)dt$$
$$= \frac{\tilde{\varepsilon}_1}{360}g(m),\ 0 < \tilde{\varepsilon}_1 < 1. \quad (2.2.11)$$

Proof It is obvious that in the conditions of the corollary, we have (2.2.6). Then (2.2.2) and (2.2.3) are valid for $0 < \varepsilon_q < 2$. Setting $\tilde{\varepsilon}_q = \frac{\varepsilon_q}{2}$, we have (2.2.9) and (2.2.10). □

We can refine Corollary 2.2.4 in the following theorem:

Theorem 2.2.5 As the assumption of Corollary 2.2.3, we still have

$$\tilde{\delta}_q(m,n) = \frac{1}{(2q+1)!}\int_m^n P_{2q+1}(t)g(t)dt$$
$$= \tilde{\varepsilon}_q \frac{2B_{2q+2}}{(2q+2)!}(1 - \frac{1}{2^{2q+2}})g(t)\,|_m^n\ (0 < \tilde{\varepsilon}_q < 1). \quad (2.2.12)$$

If $n = \infty$ and $g(\infty) = 0$, then we have $0 < \tilde{\varepsilon}_q < 1$ and

$$\tilde{\delta}_q(m,\infty) = \frac{1}{(2q+1)!}\int_m^\infty P_{2q+1}(t)g(t)dt$$
$$= \tilde{\varepsilon}_q \frac{2B_{2q+2}}{(2q+2)!}(\frac{1}{2^{2q+2}} - 1)g(m). \quad (2.2.13)$$

In particular, for $q = 1$, it follows

$$\tilde{\delta}_1(m,\infty) = \frac{1}{6}\int_m^\infty P_3(t)g(t)dt$$
$$= \frac{\tilde{\varepsilon}_1}{384}g(m),\ 0 < \tilde{\varepsilon}_1 < 1. \quad (2.2.14)$$

Proof We only prove the case in the condition that $(-1)^k g^{(k)}(t) \ge 0,\quad t \in [m,n] \ (k = 0,1)$ and there exists an interval $I_1 \subset [m,n]$, such that $g'(t) < 0$, $t \in I_1$. In this case, $g(t) \ge 0, g'(t) \le 0$ and $g(t)$ is a non-negative decreasing function with $g(t)\,|_m^n < 0$.

(1) If $B_{2q+2} < 0$, since it follows

$$\int_{m+k-1}^{m+k} P_{2q+1}(t)dt = 0 \ (k \in \mathbf{N}),$$

then by (2.2.6), we have

$$0 < \int_m^n P_{2q+1}(t)g(t)dt$$
$$= \sum_{k=1}^{n-m}\int_{m+k-1}^{m+k} P_{2q+1}(t)[g(t) - g(m+k)]dt$$

$$=\sum_{k=1}^{n-m}\{\int_{m+k-1}^{m+k-\frac{1}{2}}P_{2q+1}(t)[g(t)-g(m+k)]dt$$
$$+\int_{m+k-\frac{1}{2}}^{m+k}P_{2q+1}(t)[g(t)-g(m+k)]dt\}$$
$$=\sum_{k=1}^{n-m}\{[g(m+k-1)-g(m+k)]$$
$$\times\int_{m+k-1}^{m+k-\frac{1}{2}}P_{2q+1}(t)dt\}+\sum_{k=1}^{n-m}\alpha_k, \qquad (2.2.15)$$

where, α_k is defined by the following:

$$\alpha_k:=\int_{m+k-1}^{m+k-\frac{1}{2}}P_{2q+1}(t)[g(t)-g(m+k-1)]dt$$
$$+\int_{m+k-\frac{1}{2}}^{m+k}P_{2q+1}(t)[g(t)-g(m+k)]dt.$$

Since $g(t)$ is a decreasing function, then it follows

$$g(t)-g(m+k-1)\le 0,$$
$$t\in[m+k-1,m+k-\tfrac{1}{2}],$$
$$g(t)-g(m+k)\ge 0, t\in[m+k-\tfrac{1}{2},m+k].$$

In view of $P_{2q+2}(m+k-1)=B_{2q+2}<0$, by (2.1.9), it follows $P_{2q+2}(m+k-\frac{1}{2})>0$, $P_{2q+2}(t)$ is strictly increasing in $(m+k-1,m+k-\frac{1}{2})$, and

$$P_{2q+1}(t)=\tfrac{P'_{2q+2}(t)}{2q+2}>0.$$

In the same way, we can show that $P_{2q+2}(t)$ is strictly decreasing in $(m+k-\frac{1}{2},m+k)$ and

$$P_{2q+1}(t)=\tfrac{P'_{2q+2}(t)}{2q+2}<0.$$

Hence we find

$$P_{2q+1}(t)[g(t)-g(m+k-1)]\le 0,$$
$$t\in[m+k-1,m+k-\tfrac{1}{2}],$$
$$P_{2q+1}(t)[g(t)-g(m+k)]\le 0,$$
$$t\in[m+k-\tfrac{1}{2},m+k],$$

and then $\alpha_k\le 0$. Since $g(t)$ is strictly decreasing at least in a small interval, then there exists an integer k, such that $\alpha_k<0$ and then $\sum_{k=1}^{n-m}\alpha_k<0$. By (2.1.11), we have

$$\sum_{k=1}^{n-m}\{[g(m+k-1)-g(m+k)]$$
$$\times\int_{m+k-1}^{m+k-\frac{1}{2}}P_{2q+1}(t)dt\}$$
$$=-\int_{m-1}^{m-\frac{1}{2}}P_{2q+1}(t)dt$$
$$\times\sum_{k=1}^{n-m}[g(m+k)-g(m+k-1)]$$
$$=\tfrac{-1}{2q+2}[P_{2q+2}(m-\tfrac{1}{2})-P_{2q+2}(m-1)]g(t)|_m^n$$
$$=\tfrac{-1}{2q+2}(\tfrac{1}{2^{2q+1}}-2)B_{2q+2}g(t)|_m^n.$$

In view of (2.2.15), we find

$$0<\int_m^n P_{2q+1}(t)g(t)dt$$
$$<\tfrac{2}{2q+2}(1-\tfrac{1}{2^{2q+2}})B_{2q+2}g(t)|_m^n. \qquad (2.2.16)$$

Hence (2.2.12) is valid.

(2) For $B_{2q+2}>0$, in the same way, we have $\sum_{k=1}^{n-m}\alpha_k>0$, and the corresponding result of (2.2.16) is as follows:

$$0<-\int_m^n P_{2q+1}(t)g(t)dt$$
$$<\tfrac{-2}{2q+2}(1-\tfrac{1}{2^{2q+2}})B_{2q+2}g(t)|_m^n. \qquad (2.2.17)$$

Hence (2.2.12) is valid for $-g(t)$, and then dividing -1 on two sides, we obtain (2.2.12).W

Corollary 2.2.6 Assuming that $m_i,q\in\mathbf{N}_0$, $m_{i+1}>m_i$, $g(t)\in C^1[m_i,m_{i+1}]$, $(-1)^k g^{(k)}(t)\ge 0$ (≤ 0), $t\in[m_i,m_{i+1}]$ $(k=0,1;i=1,\cdots,s)$ and $m=m_1,n=m_{s+1}$, if there exist an integer i and an interval $I_1\subset[m_i,m_{i+1}]$,such that $g'(t)<0$ (>0), $t\in I_1$, then we still have (2.2.12), (2.2.13) and (2.2.14).

2.2.3 ESTIMATIONS OF $\delta_q(m,n)$ AND SOME INEQUALITIES

Setting $g(t)=f^{(2q+1)}(t)$ in Theorem 2.2.1 and Theorem 2.2.4, we have the following corollaries:

Corollary 2.2.7 Assuming that $m,n,q\in\mathbf{N}_0$, $n>m$, $f(t)\in C^{2q+4}[m,n]$ with

$$(-1)^k f^{(2q+1+k)}(t)\ge 0\ (\le 0),\ t\in[m,n]$$

$(k=0,1,2,3)$, and there exist two intervals $I_k\subset[m,n]$, such that

$$(-1)^k f^{(2q+1+k)}(t)>0\ (<0),\ t\in I_k$$

$(k=1,3)$, then we have

$$\delta_q(m,n)=\tfrac{1}{(2q+1)!}\int_m^n P_{2q+1}(t)f^{(2q+1)}(t)dt$$
$$=\tfrac{\varepsilon_q B_{2q+2}}{(2q+2)!}f^{(2q+1)}(t)|_m^n,\ 0<\varepsilon_q<1. \qquad (2.2.18)$$

Setting $n=\infty$ and $f^{(2q+1)}(\infty)=0$, then we have $0<\tilde{\varepsilon}_q<1$ and

$$\delta_q(m,\infty)=\tfrac{1}{(2q+1)!}\int_m^\infty P_{2q+1}(t)f^{(2q+1)}(t)dt$$
$$=\tfrac{-\tilde{\varepsilon}_q B_{2q+2}}{(2q+2)!}f^{(2q+1)}(m). \quad (2.2.19)$$

In particular, for $q=1$, we have

$$\delta_1(m,\infty)=\tfrac{1}{6}\int_m^\infty P_3(t)f'''(t)dt$$
$$=\tfrac{\tilde{\varepsilon}_1}{720}f'''(m),\ 0<\tilde{\varepsilon}_1<1. \quad (2.2.20)$$

Corollary 2.2.8 Assuming that $m,n,q\in \mathbf{N}_0$, $n>m$, $f(t)\in C^{2q+2}[m,n]$ with

$$(-1)^k f^{(2q+1+k)}(t)\ge 0\ (\le 0),\ t\in[m,n]$$

$(k=0,1)$, there exists an interval $I_1\subset[m,n]$, such that $f^{(2q+2)}(t)<0\ (>0)$, $t\in I_1$, then we have the following estimation:

$$\delta_q(m,n)=\tilde{\varepsilon}_q\tfrac{2B_{2q+2}}{(2q+2)!}(1-\tfrac{1}{2^{2q+2}})f^{(2q+1)}(t)|_m^n,$$
$$0<\tilde{\varepsilon}_q<1. \quad (2.2.21)$$

Setting $n=\infty$ and $f^{(2q+1)}(\infty)=0$, then we have

$$\delta_q(m,\infty)=\tilde{\varepsilon}_q\tfrac{2B_{2q+2}}{(2q+2)!}(\tfrac{1}{2^{2q+2}}-1)f^{(2q+1)}(m),$$
$$0<\tilde{\varepsilon}_q<1. \quad (2.2.22)$$

In particular, for $q=1$, we find

$$\delta_1(m,\infty)=\tfrac{1}{6}\int_m^\infty P_3(t)f'''(t)dt$$
$$=\tfrac{\tilde{\varepsilon}_1}{384}f'''(m)\ (0<\tilde{\varepsilon}_1<1). \quad (2.2.23)$$

For $q=0$ in Corollary 2.2.3, in view of (2.1.13), we have the following corollary:

Corollary 2.2.9 Assuming that $m,n\in\mathbf{N}_0$, $n>m$, $f(t)\in C^4[m,n]$ with

$$(-1)^k f^{(1+k)}(t)\ge 0\ (\le 0),\ t\in[m,n]$$

$(k=0,1,2,3)$, there exist two intervals $I_k\subset[m,n]$, such that $f^{(1+k)}(t)<0\ (>0)$, $t\in I_k\ (k=1,3)$, then we have

$$\sum_{k=m}^n f(k)=\int_m^n f(t)dt+\tfrac{f(n)+f(m)}{2}$$
$$+\tfrac{\varepsilon_0}{12}f'(t)|_m^n,\ 0<\varepsilon_0<1. \quad (2.2.24)$$

Setting $n=\infty$ and $f^{(k)}(\infty)=0\ (k=0,1)$, then both $\sum_{k=m}^\infty f(k)$ and $\int_m^\infty f(t)dt$ are convergence or divergence at the same time, and if both of them are convergence, then we have

$$\sum_{k=m}^\infty f(k)=\int_m^\infty f(t)dt+\tfrac{1}{2}f(m)$$
$$-\tfrac{\varepsilon_0}{12}f'(m),\ 0<\varepsilon_0<1. \quad (2.2.25)$$

If in (2.2.25), $f'(m)<0$, then we have the following inequalities:

$$\int_m^\infty f(t)dt+\tfrac{1}{2}f(m)<\sum_{k=m}^\infty f(k)$$
$$<\int_m^\infty f(t)dt+\tfrac{1}{2}f(m)-\tfrac{1}{12}f'(m). \quad (2.2.26)$$

For $q=0$ in Corollary 2.2.6, in view of (2.1.13), we have the following corollary:

Corollary 2.2.10 Assuming that $m,n\in\mathbf{N}_0$, $n>m$, $f(t)\in C^2[m,n]$ with

$$(-1)^k f^{(1+k)}(t)\ge 0\ (\le 0),\ t\in[m,n]$$

$(k=0,1)$, and there exists an interval $I_1\subset[m,n]$, such that $f''(t)<0\ (>0)$, $t\in I_1$, then we have

$$\sum_{k=m}^n f(k)=\int_m^n f(t)dt+\tfrac{f(n)+f(m)}{2}$$
$$+\tfrac{\varepsilon_0}{8}f'(t)|_m^n,\ 0<\varepsilon_0<1. \quad (2.2.27)$$

Setting $n=\infty$ and $f^{(k)}(\infty)=0\ (k=0,1)$, then both $\sum_{k=m}^\infty f(k)$ and $\int_m^\infty f(t)dt$ are convergence or divergence at the same time. If both of them are convergence, then we have

$$\sum_{k=m}^\infty f(k)=\int_m^\infty f(t)dt+\tfrac{1}{2}f(m)$$
$$-\tfrac{\varepsilon_0}{8}f'(m),\ 0<\varepsilon_0<1. \quad (2.2.28)$$

If in (2.2.28), $f'(m)<0$ (or $f''(t)>0$, $t\in I_1$), then we have the following inequalities:

$$\int_m^\infty f(t)dt+\tfrac{1}{2}f(m)<\sum_{k=m}^\infty f(k)$$
$$<\int_m^\infty f(t)dt+\tfrac{1}{2}f(m)-\tfrac{1}{8}f'(m). \quad (2.2.29)$$

Example 2.2.11 If $f(t)=\frac{1}{(1+t)t^{1/2}}\ (t\in(0,\infty))$, $m\in\mathbf{N}$, then we obtain $f(t)>0$,

$$f'(t) = -\frac{1}{(1+t)^2 t^{1/2}} - \frac{1}{2(1+t)t^{3/2}} < 0,$$

$f''(t) > 0, f'''(t) < 0$. Setting $u = t^{-1/2}$, we find

$$\int_m^\infty \frac{dt}{(1+t)t^{1/2}} = 2\int_0^{1/\sqrt{m}} \frac{du}{1+u^2} = 2\arctan\frac{1}{\sqrt{m}}.$$

By (2.2.26), for $m \in \mathbf{N}$, we have the following inequalities:

$$2\arctan\frac{1}{\sqrt{m}} + \frac{1}{2(1+m)\sqrt{m}} < \sum_{k=m}^\infty \frac{1}{(1+k)k^{1/2}}$$
$$< 2\arctan\frac{1}{\sqrt{m}} + \frac{1}{2(1+m)\sqrt{m}} + \frac{3m+1}{24(1+m)^2\sqrt{m^3}}. \quad (2.2.30)$$

Example 2.2.12 If $f(t) = \frac{1}{t^2+1}$ $(t \in (0,\infty)), m \in \mathbf{N}$, then we obtain $f(t) > 0$, $f'(t) = -\frac{2t}{(t^2+1)^2} < 0$,

$$\int_m^\infty f(t)dt = \int_m^\infty \frac{1}{t^2+1}dt = \frac{\pi}{2} - \arctan m.$$

By (2.2.29), for $m \in \mathbf{N}$, we have the following inequalities:

$$\frac{\pi}{2} - \arctan m + \frac{1}{2(m^2+1)} < \sum_{k=m}^\infty \frac{1}{k^2+1}$$
$$< \frac{\pi}{2} - \arctan m + \frac{1}{2(m^2+1)} + \frac{m}{4(m^2+1)^2}. \quad (2.2.31)$$

Since $\sum_{k=0}^\infty \frac{1}{k^2+1} = 1 + \sum_{k=1}^\infty \frac{1}{k^2+1}$, then by (2.2.31) (setting $m = 1$), we find the following inequalities:

$$\frac{\pi}{4} + \frac{5}{4} < \sum_{k=0}^\infty \frac{1}{k^2+1} < \frac{\pi}{4} + \frac{21}{16}.$$

2.3. ABOUT TWO CLASSES OF SERIES ESTIMATIONS

2.3.1 ONE CLASS OF CONVERGENCE SERIES ESTIMATIONS

For $n \to \infty$ in (2.1.16), by Corollary 2.2.5 and Corollary 2.2.6, we have the following theorem:

Theorem 2.3.1 Assuming that $m, n, q \in \mathbf{N}_0, n > m$, $f(t) \in C^{2q+1}[m,\infty)$, $f(\infty) = 0$, $f^{(2k-1)}(\infty) = 0$ $(k = 1,2,\cdots,q+1)$, $\delta_q(m,\infty)$ is convergence, then both $\sum_{k=m}^\infty f(k)$ and $\int_m^\infty f(t)dt$ are convergence or divergence at the same time. When both of them are convergence, we have the following estimation:

$$\sum_{k=m}^\infty f(k) = \int_m^\infty f(t)dt + \frac{1}{2}f(m)$$
$$-\sum_{k=1}^q \frac{B_{2k}}{(2k)!} f^{(2k-1)}(m) + \delta_q(m,\infty), \quad (2.3.1)$$

$$\delta_q(m,\infty) := \frac{1}{(2q+1)!}\int_m^\infty P_{2q+1}(t) f^{(2q+1)}(t)dt, \quad (2.3.2)$$

and $\delta_q(m,\infty)$ satisfies the following recursion formulas: $\delta_0(m,\infty) = \int_m^\infty P_1(t)f'(t)dt$,

$$\delta_q(m,\infty) = \frac{B_{2q}}{(2q)!} f^{(2q-1)}(m)$$
$$+\delta_{q-1}(m,\infty), q \in \mathbf{N}. \quad (2.3.3)$$

(1) If $(-1)^k f^{(2q+1+k)}(t) \ge 0$ (≤ 0), $t \in [m,n]$ $(k = 0,1,2,3)$, there exist two intervals $I_k \subset [m,n]$, such that $(-1)^k f^{(2q+1+k)}(t) > 0$ (< 0), $t \in I_k$ $(k = 1,3)$, then we have

$$\delta_q(m,\infty) = \frac{-\varepsilon_q B_{2q+2}}{(2q+2)!} f^{(2q+1)}(m), \ 0 < \varepsilon_q < 1; \quad (2.3.4)$$

(2) if $(-1)^k f^{(2q+1+k)}(t) \ge 0$ (≤ 0), $t \in [m,n]$ $(k = 0,1)$, there exists an interval $I_1 \subset [m,n]$, such that $f^{(2q+2)}(t) < 0$ (> 0), $t \in I_1$, then we have

$$\delta_q(m,\infty) = \tilde{\varepsilon}_q \frac{2B_{2q+2}}{(2q+2)!}\left(\frac{1}{2^{2q+2}} - 1\right) f^{(2q+1)}(m),$$
$$0 < \tilde{\varepsilon}_q < 1. \quad (2.3.5)$$

Example 2.3.2 If $0 < a \le 1, p > 1$,

$$f(t) = \frac{1}{(t+a)^p} \quad (t \in (0,\infty)),$$

then we find $f^{(k)}(t) = \binom{-p}{k}\frac{k!}{(t+a)^{p+k}}$. For $m, q \in \mathbf{N}_0$, by (2.3.1) and (2.3.4), we have

$$\sum_{k=m}^\infty \frac{1}{(k+a)^p} = \frac{1}{(p-1)(m+a)^{p-1}} + \frac{1}{2(m+a)^p}$$
$$-\sum_{k=1}^q \frac{B_{2k}}{2k}\binom{-p}{2k-1}\frac{1}{(m+a)^{p+2k-1}}$$
$$-\frac{\varepsilon_q B_{2q+2}}{2q+2}\binom{-p}{2q+1}\frac{1}{(m+a)^{p+2q+1}} \ (0 < \varepsilon_q < 1). \quad (2.3.6)$$

In particular, for $a = 1$, replacing $m+1$ by m in (2.3.6), we find the following estimation on the convergence p-series (Yang JGEI 1992)[8]:

$$\sum_{k=m}^\infty \frac{1}{k^p} = \frac{1}{(p-1)m^{p-1}} + \frac{1}{2m^p} - \sum_{k=1}^q \frac{B_{2k}}{2k}\binom{-p}{2k-1}\frac{1}{m^{p+2k-1}}$$
$$-\frac{\varepsilon_q B_{2q+2}}{2q+2}\binom{-p}{2q+1}\frac{1}{m^{p+2q+1}}, \ 0 < \varepsilon_q < 1, \quad (2.3.7)$$

2.3.2 ONE CLASS OF FINITE SUM ESTIMATION ON DIVERGENCE SERIES

Setting $\delta_q(m) := \delta_q(m,\infty), \int_m^n f(t)dt = F(t)|_m^n$, define the constant β_m as follows

$$\beta_m := -F(m) + \tfrac{1}{2}f(m) - \sum_{k=1}^{q} \tfrac{B_{2k}}{(2k)!} f^{(2k-1)}(m) + \delta_q(m). \quad (2.3.8)$$

By (2.1.16), we find

$$\sum_{k=m}^{n} f(k) = F(n) + \tfrac{1}{2}f(n) + \sum_{k=1}^{q} \tfrac{B_{2k}}{(2k)!} f^{(2k-1)}(n) + \beta_m - \delta_q(n), \quad (2.3.9)$$

$$\delta_q(n) = \tfrac{1}{(2q+1)!}\int_n^\infty P_{2q+1}(t) f^{(2q+1)}(t)dt. \quad (2.3.10)$$

Since $\delta_q(\infty) = 0$, then we still have

$$\beta_m = \lim_{n\to\infty}\Big[\sum_{k=m}^{n} f(k) - F(n) - \tfrac{1}{2}f(n) - \sum_{k=1}^{q} \tfrac{B_{2k}}{(2k)!} f^{(2k-1)}(n)\Big]. \quad (2.3.11)$$

Example 2.3.3 If $f(t) = \frac{\ln t}{t-1}$ $(t \in (0,\infty))$, in view of example 2.2.2, by (2.3.9) and (2.3.4) setting ($m=1, q=1$), we have the following estimation:

$$\sum_{k=1}^{n} f(k) = \int_1^n f(t)dt + \tilde{\gamma} + \tfrac{1}{2}f(n) + \tfrac{1}{12}f'(n) - \tfrac{\varepsilon}{720}f'''(n),$$

$$\tilde{\gamma} := \beta_1 + F(1) = \sum_{k=1}^{n} f(k) - \int_1^n f(t)dt - \tfrac{1}{2}f(n) - \tfrac{1}{12}f'(n) + \tfrac{\varepsilon}{720}f'''(n), \quad (2.3.12)$$

and

$$\tilde{\gamma} = \lim_{n\to\infty}\Big(\sum_{k=1}^{n} \tfrac{\ln k}{k-1} - \int_1^n \tfrac{\ln t}{t-1}dt\Big).$$

We estimate $\tilde{\gamma}$ in the following. Setting $u = \frac{1}{x}$, it follows

$$\int_1^n f(x)dx = \int_1^n \tfrac{\ln x}{x-1}dx = \int_{1/n}^1 \tfrac{\ln u}{(u-1)u}du$$

$$= \int_{1/n}^1 \tfrac{-\ln u}{u}du + \int_{1/n}^1 \tfrac{\ln u}{u-1}du$$

$$= \int_{1/n}^1 -\ln u\, d\ln u + \int_{1/n}^1 \sum_{k=1}^{\infty} \tfrac{(1-u)^{k-1}}{k}du$$

$$= \tfrac{1}{2}(\ln n)^2 + \sum_{k=1}^{\infty}\int_{1/n}^1 \tfrac{(1-u)^{k-1}}{k}du$$

$$= \tfrac{1}{2}(\ln n)^2 + \sum_{k=1}^{\infty} \tfrac{1}{k^2}(1-\tfrac{1}{n})^k.$$

By (2.3.12) and Example 2.2.2, we find

$$\tilde{\gamma} = \sum_{k=1}^{n} \tfrac{\ln k}{k-1} - \sum_{k=1}^{\infty} \tfrac{1}{k^2}(1-\tfrac{1}{n})^k - \tfrac{1}{2}(\ln n)^2 - \tfrac{\ln n}{2(n-1)} - \tfrac{1}{12n(n-1)} + \tfrac{\ln n}{12(n-1)^2}$$
$$+ \tfrac{\varepsilon}{720}\Big[\tfrac{2}{n^3(n-1)} + \tfrac{3}{n^2(n-1)^2} + \tfrac{6}{n(n-1)^3} - \tfrac{6\ln n}{(n-1)^4}\Big],$$
$$0 < \varepsilon < 1. \quad (2.3.13)$$

For $n = 3$ in (2.3.13), we obtain the following inequalities:

$$-\sum_{k=1}^{20} \tfrac{1}{k^2}(\tfrac{2}{3})^k - \tfrac{1}{20}(\tfrac{2}{3})^{20} < -\sum_{k=1}^{20} \tfrac{1}{k^2}(\tfrac{2}{3})^k - \int_{20}^\infty \tfrac{1}{x^2}(\tfrac{2}{3})^x dx$$
$$< -\sum_{k=1}^{\infty} \tfrac{1}{k^2}(1-\tfrac{1}{3})^k < -\sum_{k=1}^{20} \tfrac{1}{k^2}(\tfrac{2}{3})^k. \quad (2.3.14)$$

In view of (2.3.13) and (2.3.14), we find

$$0.539902 < \tilde{\gamma} < 0.539976,$$

and then it follows $\tilde{\gamma} = 0.5399^+$.

Example 2.3.4 If $f(t) = \frac{1}{t+a}$ $(0 < a \le 1, t \in [0,\infty))$, then $F(t) = \ln(t+a)$, $f^{(k)}(t) = \frac{(-1)^k k!}{(t+a)^{k+1}}$. By (2.3.11) (setting $m = q = 0$), we have (Yang NSJHTC 1997)[10], (Yang JMS)[11]

$$\beta_0 = \lim_{n\to\infty}\Big[\sum_{k=0}^{n} f(k) - F(n) - \tfrac{1}{2}f(n)\Big]$$
$$= \lim_{n\to\infty}\Big[\sum_{k=0}^{n} \tfrac{1}{k+a} - \ln(n+a)\Big] = \gamma_0(a). \quad (2.3.15)$$

We call $\gamma_0(a)$ Stieltjes constant (Pang SP 1990)[12]. By (2.3.9) and (2.2.19), we find (Yang NSJHTC 1997) [10]

$$\sum_{k=0}^{n} \tfrac{1}{k+a} = \gamma_0(a) + \ln(n+a) + \tfrac{1}{2(n+a)} - \sum_{k=1}^{q} \tfrac{B_{2k}}{2k}\tfrac{1}{(n+a)^{2k}}$$
$$- \tfrac{\varepsilon_q B_{2q+2}}{2(q+1)}\tfrac{1}{(n+a)^{2q+2}},\ 0 < \varepsilon_q < 1. \quad (2.3.16)$$

In particular, for $a = 1$, replacing $n+1$ by n in (2.3.16), we have the following estimation on the harmonic series (Yang JMS 1996)[11]:

$$\sum_{k=1}^{n}\frac{1}{k}=\gamma+\ln n+\frac{1}{2n}-\sum_{k=1}^{q}\frac{B_{2k}}{2k}\frac{1}{n^{2k}}$$

$$-\frac{\varepsilon_q B_{2q+2}}{2(q+1)}\frac{1}{n^{2q+2}},\ 0<\varepsilon_q<1,\qquad(2.3.17)$$

where, $\gamma=\gamma_0(1)=0.5772156649^+$ is called the Euler Constant.

Example 2.3.5 If $s\in\mathbf{R}\backslash\{1\}, 0<a\le 1$,

$$f(t)=\frac{1}{(t+a)^s}\ (t\in[0,\infty)),$$

then by (2.3.9), setting $\varsigma(s,a)=\beta_0$, for $q\ge\frac{1-s}{2}$, we have an estimation of Hurwitz ς -function $\varsigma(s,a)$ in the real axis as follows (Titchmarsh CP 1986)[13]:

$$\varsigma(s,a)=\sum_{k=0}^{n}\frac{1}{(k+a)^s}-\frac{1}{1-s}(n+a)^{1-s}$$

$$-\frac{1}{2(n+a)^s}-\sum_{k=1}^{q}\frac{\binom{-s}{2k-1}B_{2k}}{2k(n+a)^{s+2k-1}}$$

$$-\frac{\varepsilon_q\binom{-s}{2q+1}B_{2q+2}}{2(q+1)(n+a)^{s+2q+1}},\ 0<\varepsilon_q<1.\qquad(2.3.18)$$

In particular, for $a=1$, $q\ge\frac{1-s}{2}$ in (2.3.18), we have an estimation of Riemann ς - function $\varsigma(s)=\varsigma(s,1)$ in the real axis as follows (Zhu ASNUS 1997)[14]:

$$\varsigma(s)=\sum_{k=1}^{n}\frac{1}{k^s}-\frac{1}{1-s}n^{1-s}-\frac{1}{2n^s}-\sum_{k=1}^{q}\frac{\binom{-s}{2k-1}B_{2k}}{2kn^{s+2k-1}}$$

$$-\frac{\varepsilon_q\binom{-s}{2q+1}B_{2q+2}}{2(q+1)n^{s+2q+1}},\ 0<\varepsilon_q<1.\qquad(2.3.19)$$

Example 2.3.6 If $0<a\le 1$,

$$f(t)=\ln(t+a)\ (t\in[0,\infty)),$$

then we find $F(t)=(t+a)\ln(t+a)-t$,

$$f^{(k)}(t)=\frac{(-1)^{k-1}(k-1)!}{(t+a)^k},\ k\in\mathbf{N}.$$

By (2.3.8)-(2.3.11), we obtain

$$\beta_0(a)=\lim_{n\to\infty}[\sum_{k=0}^{n}\ln(k+a)$$

$$-(n+a+\tfrac{1}{2})\ln(n+a)+n],\qquad(2.3.20)$$

$$\ln\prod_{k=0}^{n}(k+a)=\sum_{k=0}^{n}\ln(k+a)$$

$$=\beta_0(a)+(n+a+\tfrac{1}{2})\ln(n+a)-n$$

$$+\sum_{k=1}^{q}\frac{B_{2k}}{2k(2k-1)}\frac{1}{(n+a)^{2k-1}}$$

$$+\frac{\varepsilon_q B_{2q+2}}{2(q+1)(2q+1)}\frac{1}{(n+a)^{2q+1}},0<\varepsilon_q<1,q\in\mathbf{N}.\qquad(2.3.21)$$

Equivalently, since $e^{\beta_0(a)+a}=\frac{\sqrt{2\pi}}{\Gamma(a)}$ (Xie MPC 2006)[15], we find

$$\prod_{k=0}^{n}(k+a)=\frac{\sqrt{2\pi}}{\Gamma(a)}\sqrt{n+a}\left(\frac{n+a}{e}\right)^{n+a}$$

$$\times\exp\{\sum_{k=1}^{q}\frac{B_{2k}}{2k(2k-1)}\frac{1}{(n+a)^{2k-1}}$$

$$+\frac{\varepsilon_q B_{2q+2}}{2(q+1)(2q+1)}\frac{1}{(n+a)^{2q+1}}\},0<\varepsilon_q<1,q\in\mathbf{N}.\qquad(2.3.22)$$

In particular, for $a=1$, replacing $n+1$ by n, we have the following extended Stirling formula (Knopp BSL 1928)[2], (Yang JGEI 2002)[16]:

$$n!=\sqrt{2\pi n}\left(\frac{n}{e}\right)^n\exp\{\sum_{k=1}^{q}\frac{B_{2k}}{2k(2k-1)}\frac{1}{n^{2k-1}}$$

$$+\frac{\varepsilon_q B_{2q+2}}{2(q+1)(2q+1)}\frac{1}{n^{2q+1}}\},\ 0<\varepsilon_q<1,q\in\mathbf{N}.\qquad(2.3.23)$$

For $q=1$, we obtain

$$\sqrt{2\pi n}\left(\frac{n}{e}\right)^n<\sqrt{2\pi n}\left(\frac{n}{e}\right)^n e^{\frac{1}{12n}}$$

$$<n!<\sqrt{2\pi n}\left(\frac{n}{e}\right)^n e^{\frac{1}{12n}(1-\frac{1}{30n^2})}.\qquad(2.3.24)$$

Example 2.3.7 About the Laurent series of Riemann ς -function $\varsigma(s)$ in the isolated singular point $s=1$, we have the following expression (Pang SP 1990)[12]:

$$\varsigma(s)=\frac{1}{s-1}+\sum_{n=0}^{\infty}(-1)^n\frac{\gamma_n}{n!}(s-1)^n,$$

$$0<|s-1|<\infty,\qquad(2.3.25)$$

where, $\{\gamma_n\}_{n=0}^{\infty}$ are called Stieltjes constants,

$$\gamma_0=\lim_{N\to\infty}\{\sum_{k=1}^{N}\frac{1}{k}-\ln N\}=\gamma$$

is called Euler constant, and

$$\gamma_n=\lim_{N\to\infty}\{\sum_{k=1}^{N}\frac{(\ln k)^n}{k}-\frac{(\ln N)^{n+1}}{n+1}\},n\in\mathbf{N}.\qquad(2.3.26)$$

(1) First, we estimate γ_1. Setting $f(t)=\frac{\ln t}{t}$, $m=1$, $n=30, q=0$ in (2.3.9), for $t\ge 20$, $g(t)=\frac{1-\ln t}{t^2}$ satisfies the condition of (2.2.3) ($g'''(t)=\frac{-50+24\ln t}{t^5}$). Then by (2.2.3) (for $q=0$), it follows that

$$\gamma_1=\beta_1=\sum_{k=1}^{30}f(k)-F(30)-\tfrac{1}{2}f(30)+\delta_0(30)$$

$$=\sum_{k=1}^{30}\frac{\ln k}{k}-\frac{1}{2}(\ln 30)^2-\frac{\ln 30}{60}+\int_{30}^{\infty}P_1(t)\left(\frac{1-\ln t}{t^2}\right)dt$$

$$=\sum_{k=1}^{30}\frac{\ln k}{k}-\frac{1}{2}(\ln 30)^2-\frac{\ln 30}{60}+\frac{\varepsilon}{12}\left(\frac{\ln 30-1}{30^2}\right).$$

We find $-0.07304 < \gamma_1 < -0.0728$, and then it follows $\gamma_1 = -0.073^+$.

(2) We estimate $|\gamma_n|$ $(n \in \mathbb{N} \setminus \{1\})$ (Wei JCNU 1996)[17], (Yang JSCNU 1996)[18]. Setting $m=1, n=N, q=n$ in (2.1.12), we have

$$\sum_{k=1}^{N} f(k) - \int_1^N f(t)dt$$

$$= f(1) + \sum_{k=1}^{n} (-1)^k \tfrac{B_k}{k!} f^{(k-1)}(t)|_1^N$$

$$+ \tfrac{(-1)^{n+1}}{n!} \int_1^N P_n(t) f^{(n)}(t)dt.$$

Putting $f(t) = \frac{(\ln t)^n}{t}, t \in (0,\infty)$, we find $f(1)=0$, $f^{(k-1)}(t)|_1^\infty = 0 \ (1 \le k \le n)$. In fact,

$$f^{(k)}(t) = \sum_{j=n-k}^{n} \frac{Q_{k,j}(n)(\ln t)^j}{t^{k+1}}, \quad 0 \le k \le n-1, \tag{2.3.27}$$

where $Q_{k,j}(n)$ are polynomials of n-order. Since

$$\gamma_n = \lim_{N\to\infty} [\sum_{k=1}^{N} f(k) - \int_1^N f(t)dt]$$

$$= \tfrac{(-1)^{n+1}}{n!} \int_1^\infty P_n(t) df^{(n-1)}(t), \tag{2.3.28}$$

then we obtain the following inequality:

$$|\gamma_n| \le \tfrac{1}{n!} \max_{1\le t<\infty} |P_n(t)| \lim_{N\to\infty} \overset{N}{\underset{1}{V}}(f^{(n-1)}). \tag{2.3.29}$$

By (2.3.27), it follows

$$\overset{N}{\underset{1}{V}}(f^{(n-1)}) \le \sum_{j=1}^{n} |Q_{n-1,j}(n)| \overset{N}{\underset{1}{V}} \tfrac{(\ln t)^j}{t^n}. \tag{2.3.30}$$

For $1 \le j \le n$, setting $g(t) = \frac{(\ln t)^j}{t^n}$, we find

$$g'(t) = \tfrac{(j - n\ln t)(\ln t)^{j-1}}{t^{n+1}}.$$

It yields that $t_0 = e^{j/n}$ is one and only single point satisfying $g(t_0) = \max g(t)$. For $N > e$, by (2.3.24), we have

$$\lim_{N\to\infty} \overset{N}{\underset{1}{V}} \tfrac{(\ln t)^j}{t^n} = [\tfrac{(\ln t_0)^j}{t_0^n} - \tfrac{(\ln 1)^j}{1^n}]$$

$$+ \lim_{N\to\infty} [\tfrac{(\ln t_0)^j}{t_0^n} - \tfrac{(\ln N)^j}{N^n}]$$

$$= (\tfrac{j}{e})^j \tfrac{2}{n^j} \le \tfrac{2 \cdot j!}{n^j \sqrt{2\pi j}}. \tag{2.3.31}$$

By Leibniz formula as follows (Wang CCP 1991)[19]:

$$(u_1 u_2 \cdots u_m)^{(k)}$$

$$= \sum_{\substack{0 \le k_i \le k \\ (\sum_{i=1}^m k_i = k)}} \tfrac{k!}{k_1! k_2! \cdots k_m!} u_1^{(k_1)} u_2^{(k_2)} \cdots u_m^{(k_m)}, \tag{2.3.32}$$

setting $m = n+1, k = n-1$,

$$u_i = \ln t \ (i = 1,2,\cdots,n), u_{n+1} = \tfrac{1}{t},$$

the absolute value of the coefficients of $\frac{(\ln t)^j}{x^n}$ in (2.3.32) $|P_{n-1,j}(n)|$ (viz. the term including $\frac{(\ln t)^j}{t^n}$) are as follows

$$\binom{n}{j} \sum \tfrac{(n-1)!}{(0!)^j k_{j+1}! k_{j+2}! \cdots k_{n+1}!}$$

$$\times u_1^{(0)} u_2^{(0)} \cdots u_j^{(0)} u_{j+1}^{(k_{j+1})} u_{j+2}^{(k_{j+2})} \cdots u_{n+1}^{(k_{n+1})}$$

$$= \binom{n}{j} \sum \tfrac{(n-1)!}{k_{j+1}! k_{j+2}! \cdots k_{n+1}!} (\ln t)^j$$

$$\times \tfrac{(-1)^{k_{j+1}} (k_{j+1}-1)!}{t^{k_{j+1}}} \cdots \tfrac{(-1)^{k_{n+1}} k_{n+1}!}{t^{k_{n+1}+1}}$$

$$= \binom{n}{j} \sum \tfrac{(-1)^{n-1}(n-1)!}{k_{j+1} k_{j+2} \cdots k_n} \tfrac{(\ln t)^j}{t^n} \quad (j = 1,2,\cdots,n),$$

where, $\sum$ is indicated making sum of all roots of integers in the following equation

$$\begin{cases} k_{j+1} + k_{j+2} + \cdots + k_n + k_{n+1} = n-1 \\ k_{j+i} > 0 \ (i = 1,2,\cdots,n-j), k_{n+1} \ge 0 \end{cases}.$$

Setting $\tilde{k}_{j+i} = k_{j+i} - 1 \ (i = 1,2,\cdots,n-j)$, then the above equation is changed as

$$\begin{cases} \tilde{k}_{j+1} + \tilde{k}_{j+2} + \cdots + \tilde{k}_n + k_{n+1} = j-1, \\ k_{n+1}, \tilde{k}_{j+i} \ge 0 \ (i = 1,2,\cdots,n-j). \end{cases}$$

The number of the different roots is

$$\binom{j-1+(n-j+1)-1}{j-1} = \binom{n-1}{j-1}$$

(Titchmarsh CP 1986)[13]. Then we find

$$|Q_{n-1,j}(n)| = \binom{n}{j} \sum \tfrac{(n-1)!}{k_{j+1} k_{j+2} \cdots k_{n+1}}$$

$$\le \binom{n}{j} (n-1)! \sum 1$$

$$= \binom{n}{j} (n-1)! \binom{n-1}{j-1}. \tag{2.3.33}$$

For $n \ge 2$, since it follows

$$\sum_{k=1}^{\infty} \tfrac{1}{k^n} = 1 + \sum_{k=2}^{\infty} \tfrac{1}{k^n}$$

$$< 1 + \int_1^\infty \tfrac{1}{t^n} dt = 1 + \tfrac{1}{n-1} \le 2,$$

then we find (Xu HDP 1985)[1]

$$|P_n(t)| \le \frac{2\cdot n!}{(2\pi)^n}\sum_{k=1}^{\infty}\frac{1}{k^n}$$

$$\le \frac{4\cdot n!}{(2\pi)^n},\ n \ge 2. \qquad (2.3.34)$$

In view of (2.3.29)-(2.3.31), (2.3.33) and (2.3.34), for $n \in \mathbf{N}\setminus\{1\}$, we obtain

$$|\gamma_n| \le \frac{1}{n!}\max_{1\le t<\infty}|P_n(t)|$$

$$\times\sum_{j=1}^{n}|Q_{n-1,j}(n)|\lim_{N\to\infty} V_1^N \frac{(\ln t)^j}{t^n}$$

$$< \frac{1}{n!}\frac{4\cdot n!}{(2\pi)^n}\sum_{j=1}^{n}\binom{n}{j}(n-1)!\binom{n-1}{j-1}\frac{2\cdot j!}{n^j\sqrt{2\pi j}}$$

$$< \frac{8\cdot(n-1)!}{(2\pi)^n\sqrt{2\pi}}\sum_{j=1}^{n}\binom{n-1}{j-1}\frac{n(n-1)\cdots(n-j+1)}{n^j}$$

$$< \frac{8\cdot(n-1)!}{(2\pi)^n\sqrt{2\pi}}\sum_{j=1}^{n}\binom{n-1}{j-1}$$

$$= \frac{8\cdot(n-1)!2^{n-1}}{(2\pi)^n\sqrt{2\pi}} = 2\sqrt{\frac{2}{\pi}}\frac{(n-1)!}{\pi^n}. \qquad (2.3.35)$$

By (2.3.24), for $n \ge 2$, we have

$$n! < \sqrt{2\pi n}\left(\frac{n}{e}\right)^n e^{\frac{1}{12n}\left(1-\frac{1}{30n^2}\right)}$$

$$< \sqrt{2\pi n}\left(\frac{n}{e}\right)^n e^{\frac{1}{24}}.$$

Then by (2.3.35), it follows that

$$|\gamma_n| < \frac{4e^{1/24}}{\sqrt{n}}\left(\frac{n}{e\pi}\right)^n,\ n \ge 2. \qquad (2.3.36)$$

2.4. REFERENCES

1. Xu LZ, WANG XH. Methods on mathematical analysis and examples. Beijing: Higher Education Press, 1985.
2. Knopp K.Theory and application of infinite series. Londen: Blackie & Son Limited,1928.
3. Qu WL. Combination mathematics. Beijing: Beijing University Press, 1989.
4. Yang BC. A new formula for evaluating the sum of d-th powers of the first n terms of an arithmetic sequence. Journal of South China Normal University, 1996(1):129-137.
5. Yang BC. The formula about the sum of powers of natural numbers relating Bernoulli numbers. Mathematical Practice and Cognition, 1994(4):52-56.
6. Cheng QX. Basic on real variable functions and functional analysis. Beijing: Higher Education Press, 2003.
7. Yang BC, ZHU YH. Inequalities on the Hurwitz Zeta-function restricted to the axis of positive reals. Acta Scientiarum Naturalium Universitis Sunyatseni, 1997,36(3):30-35.
8. Yang BC. The evaluating formulas on the convergence p-series relating Bernoulli numbers. Journal of Guangdong Education Institute, 1992 (3):19-27.
9. Zhu YH, Yang BC. Accurate inequalities of partial sums on a type of divergent series. Acta Scientiarum Naturalium Universitis Sunyatseni, 1998,37(4):33-37.
10. Yang BC, Li DC. Estimation of the sum for -1-th powers of the first n terms of an arithmetic sequence. Natural Science Journal of Hainan Teachers College, 1997,10(1): 19-24.
11. Yang BC, Wang GQ. Some inequalities on harmonic series. Journal of Mathematics Study, 1996,29(3):90-97.
12. Pang CD, Pang CB. Basic on analysis number theory. Beijing: Science Press, 1990.
13. Titchmarsh E C. The theory of the Riemann Zeta-function.Oxford: Clarendon Press,1986.
14. Zhu YH, Yang BC. Improvement on Euler's summation formula and some inequalities on sums of powers. Acta Scientiarum Naturalium Universitis Sunyatseni, 1997, 36(4): 21-26.
15. Xie ZT. A generalization of Stirling formula. Mathematical Practice and Cognition, 2006, 36 (6): 331-333.
16. Yang BC. Some new inequalities on step multiply. Journal of Guangdong Education Institute, 2002, 22(2):1-4.
17. Wei SR, Yang BC. An inequality of Stieltjes coefficients and estimation of their order. Journal of Central Nation University (Natural Science), 1996,5(2):149~152.
18. Yang BC, Wu K. Inequality on the Stieltjes coefficients, Journal of South China Normal University (Natural Science), 1996(2):17-20.
19. Wang ZK. Applied mathematical formulas. Chongqing: Chongqing Press, 1987.

CHAPTER 3

Hilbert-Type Inequalities with the Homogeneous Kernel of Degree -1

Abstract: In this chapter, by using the way of weight coefficients and the technique of real analysis, some basic theorems and corollaries on the discrete Hilbert-type inequalities with the homogeneous kernel of degree -1 are given. We apply some relating results mentioned in Chapter 2 to building some Hilbert-type inequalities with a particular homogeneous kernel of degree -1. The strengthened versions, the reverses and the more accurate Hilbert-type inequalities are considered. The proofs about the best possible property of the constant factors are left in Chapter 4. In some particular examples of this chapter, the improved Euler-Maclaurin summation formulas mentioned in Charter 2 and are used here.

3.1 SOME BASIC RESULTS

3.1.1 SOME THEOREMS AND COROLLARIES

Theorem 3.1.1 Assuming that $p>0(p\neq 1), r>1$, $\frac{1}{p}+\frac{1}{q}=1, \frac{1}{r}+\frac{1}{s}=1, k(x,y)(\geq 0)$ is a homogeneous function of degree -1 in $(0,\infty)\times(0,\infty)$, such that

$$k(r):=\int_0^\infty k(1,u)u^{\frac{-1}{r}}du$$

is a positive number, we define the weight coefficients $\varpi(r,m)$ and $\omega(s,n)$ as follows

$$\varpi(r,m):=\sum_{n=1}^\infty k(m,n)(\tfrac{m}{n})^{\frac{1}{r}},$$

$$\omega(s,n):=\sum_{m=1}^\infty k(m,n)(\tfrac{n}{m})^{\frac{1}{s}},\quad m,n\in\mathbf{N}. \qquad (3.1.1)$$

There are some non-negative measurable functions $\tilde{\mu}(x),\tilde{\kappa}(x)$ and $\kappa(x)$ in $(0,\infty)$ and constants $\tilde{l}(r)$ and $l(r)$, satisfying the following inequalities:

$$0<\tilde{l}(r)\leq\tilde{\mu}(m)<\varpi(r,m)<\tilde{\kappa}(m)\leq k(r),$$
$$0<l(r)\leq\omega(s,n)<\kappa(n)\leq k(r),$$
$$m,n\in\mathbf{N}. \qquad (3.1.2)$$

If $a_m,b_n\geq 0\ (m,n\in\mathbf{N})$, such that

$$0<\sum_{m=1}^\infty m^{\frac{p}{s}-1}a_m^p<\infty, 0<\sum_{n=1}^\infty n^{\frac{q}{r}-1}b_n^q<\infty, \qquad (3.1.3)$$

then (1) for $p>1$, we have the following equivalent inequalities:

$$I:=\sum_{n=1}^\infty\sum_{m=1}^\infty k(m,n)a_mb_n$$
$$<\{\sum_{m=1}^\infty\tilde{\kappa}(m)m^{\frac{p}{s}-1}a_m^p\}^{\frac{1}{p}}\{\sum_{n=1}^\infty\kappa(n)n^{\frac{q}{r}-1}b_n^q\}^{\frac{1}{q}}, \qquad (3.1.4)$$

$$J:=\sum_{n=1}^\infty\frac{n^{p/s-1}}{\kappa^{p-1}(n)}(\sum_{m=1}^\infty k(m,n)a_m)^p$$
$$<\sum_{m=1}^\infty\tilde{\kappa}(m)m^{\frac{p}{s}-1}a_m^p. \qquad (3.1.5)$$

(2) for $0<p<1$, we have the following reverse equivalent inequalities:

$$I=\sum_{n=1}^\infty\sum_{m=1}^\infty k(m,n)a_mb_n$$
$$>\{\sum_{m=1}^\infty\tilde{\mu}(m)m^{\frac{p}{s}-1}a_m^p\}^{\frac{1}{p}}\{\sum_{n=1}^\infty\kappa(n)n^{\frac{q}{r}-1}b_n^q\}^{\frac{1}{q}}, \qquad (3.1.6)$$

$$J=\sum_{n=1}^\infty\frac{n^{p/s-1}}{\kappa^{p-1}(n)}(\sum_{m=1}^\infty k(m,n)a_m)^p$$
$$>\sum_{m=1}^\infty\tilde{\mu}(m)m^{\frac{p}{s}-1}a_m^p, \qquad (3.1.7)$$

$$\tilde{J}:=\sum_{m=1}^\infty\frac{m^{q/r-1}}{\tilde{\mu}^{q-1}(m)}(\sum_{n=1}^\infty k(m,n)b_n)^q$$
$$<\sum_{n=1}^\infty\kappa(n)n^{\frac{q}{r}-1}b_n^q. \qquad (3.1.8)$$

Proof Under the condition of (3.1.2), it is obvious that inequalities (3.1.3) are equivalent to

$$0<\sum_{m=1}^\infty\tilde{\mu}(m)m^{\frac{p}{s}-1}a_m^p<\sum_{m=1}^\infty\tilde{\kappa}(m)m^{\frac{p}{s}-1}a_m^p<\infty,$$

$$0<\sum_{n=1}^\infty\kappa(n)n^{\frac{q}{r}-1}b_n^q<\infty. \qquad (3.1.9)$$

(1) For $p>1$, by Hölder's inequality (Kuang SSTP 2004)[1] and (3.1.1), we have

$$I=\sum_{n=1}^\infty\sum_{m=1}^\infty k(m,n)[\frac{m^{1/(sq)}}{n^{1/(rp)}}a_m][\frac{n^{1/(rp)}}{m^{1/(sq)}}b_n]$$

$$\leq \{\sum_{m=1}^{\infty}\sum_{n=1}^{\infty} k(m,n)\frac{m^{(p-1)/s}}{n^{1/r}} a_m^p\}^{\frac{1}{p}}$$

$$\times\{\sum_{n=1}^{\infty}\sum_{m=1}^{\infty} k(m,n)\frac{n^{(q-1)/r}}{m^{1/s}} b_n^q\}^{\frac{1}{q}}. \quad (3.1.10)$$

Then by (3.1.2), we have (3.1.4).

If $J=0$, then (3.1.5) is naturally valid; if $J>0$, there exists a $n_0 \in \mathbf{N}$, such that for any $N \geq n_0$, we have $\sum_{m=1}^{N} \tilde{\kappa}(m) m^{\frac{p}{s}-1} a_m^p > 0$ and $J(N) := \sum_{n=1}^{N} \frac{n^{\frac{p}{s}-1}}{\kappa^{p-1}(n)} (\sum_{m=1}^{N} k(m,n) a_m)^p > 0$. We set

$$b_n(N) := \frac{n^{\frac{p}{s}-1}}{\kappa^{p-1}(n)} (\sum_{m=1}^{N} k(m,n) a_m)^{p-1},$$

$$n \in \mathbf{N}, N \geq n_0.$$

Then by (3.1.4), we obtain

$$0 < \sum_{n=1}^{N} \kappa(n) n^{\frac{q}{r}-1} b_n^q(N) = J(N)$$

$$= \sum_{n=1}^{N}\sum_{m=1}^{N} k(m,n) a_m b_n(N)$$

$$< \{\sum_{m=1}^{N} \tilde{\kappa}(m) m^{\frac{p}{s}-1} a_m^p\}^{\frac{1}{p}}$$

$$\times\{\sum_{n=1}^{N} \kappa(n) n^{\frac{q}{r}-1} b_n^q(N)\}^{\frac{1}{q}} < \infty, \quad (3.1.11)$$

$$J(N) = \sum_{n=1}^{N} \kappa(n) n^{\frac{q}{r}-1} b_n^q(N)$$

$$< \sum_{m=1}^{\infty} \tilde{\kappa}(m) m^{\frac{p}{s}-1} a_m^p < \infty. \quad (3.1.12)$$

Hence we find

$$0 < J = J(\infty) = \sum_{n=1}^{\infty} \kappa(n) n^{\frac{q}{r}-1} b_n^q(\infty) < \infty,$$

and for $N \to \infty$, using (3.1.4), (3.1.11) keeps the form of strict sign-inequality; so does (3.1.12). Then we have (3.1.5).

On the other-hand, suppose that (3.1.5) is valid. By Hölder's inequality, we find

$$I = \sum_{n=1}^{\infty} [\frac{n^{1/q-1/r}}{\kappa^{1/q}(n)} \sum_{m=1}^{\infty} k(m,n) a_m][\kappa^{\frac{1}{q}}(n) n^{\frac{1}{r}-\frac{1}{q}} b_n]$$

$$\leq J^{\frac{1}{p}} \{\sum_{n=1}^{\infty} \kappa(n) n^{\frac{q}{r}-1} b_n^q\}^{\frac{1}{q}}. \quad (3.1.13)$$

Then by (3.1.5), we have (3.1.4), which is equivalent to (3.1.5).

(2) For $0<p<1$, similar to the proof of the case in $p>1$, by the reverse Hölder's inequality (Kuang SSTP 2004) [1], we have

$$I \geq \{\sum_{m=1}^{\infty} \varpi(r,m) m^{\frac{p}{s}-1} a_m^p\}^{\frac{1}{p}} \{\sum_{n=1}^{\infty} \omega(s,n) n^{\frac{q}{r}-1} b_n^q\}^{\frac{1}{q}}.$$

Then by (3.1.2), in view of (3.1.3) and $q<0$, we have (3.1.6).

Since $\sum_{m=1}^{\infty} m^{\frac{p}{s}-1} a_m^p > 0$, it follows $J>0$. If $J=\infty$, then (3.1.7) is naturally valid; if $0<J<\infty$, setting

$$b_n := \frac{n^{p/s-1}}{\kappa^{p-1}(n)} (\sum_{m=1}^{\infty} k(m,n) a_m)^{p-1}, n \in \mathbf{N},$$

by (3.1.6), we find

$$\sum_{n=1}^{\infty} \kappa(n) n^{\frac{q}{r}-1} b_n^q = J = I$$

$$> \{\sum_{m=1}^{\infty} \tilde{\kappa}(m) m^{\frac{p}{s}-1} a_m^p\}^{\frac{1}{p}} \{\sum_{n=1}^{\infty} \kappa(n) n^{\frac{q}{r}-1} b_n^q\}^{\frac{1}{q}}, \quad (3.1.14)$$

$$J = \sum_{n=1}^{\infty} \kappa(n) n^{\frac{q}{r}-1} b_n^q > \sum_{m=1}^{\infty} \tilde{\kappa}(m) m^{\frac{p}{s}-1} a_m^p. \quad (3.1.15)$$

Then we have (3.1.7).

On the other-hand, suppose that (3.1.7) is valid. By the reverse Hölder's inequality, we have the reverse (3.1.13). Then by (3.1.7), we have (3.1.6), which is equivalent to (3.1.7).

If $\tilde{J}=0$, then (3.1.8) is naturally valid; if $\tilde{J}>0.$, then there exists a $n_0 \in \mathbf{N}$, such that for any $N \geq n_0$, $\sum_{n=1}^{N} \kappa(n) n^{\frac{q}{r}-1} b_n^q > 0$ and

$$\tilde{J}(N) := \sum_{m=1}^{N} \frac{m^{\frac{q}{r}-1}}{\tilde{\mu}^{q-1}(m)} (\sum_{n=1}^{N} k(m,n) b_n)^q > 0.$$

Setting

$$a_m(N) = \frac{m^{q/r-1}}{\tilde{\mu}^{q-1}(m)} (\sum_{n=1}^{N} k(m,n) b_n)^{q-1},$$

$$m \in \mathbf{N}, N \geq n_0,$$

By (3.1.6), in view of $q<0$, we obtain

$$\infty > \sum_{m=1}^{N} \tilde{\mu}(m) m^{\frac{p}{s}-1} a_m^p(N) = \tilde{J}(N)$$

$$= \sum_{n=1}^{N}\sum_{m=1}^{N} k(m,n) a_m(N) b_n$$

$$> \{\sum_{m=1}^{N} \tilde{\mu}(m) m^{\frac{p}{s}-1} a_m^p(N)\}^{\frac{1}{p}}$$

$$\times\{\sum_{n=1}^{N}\kappa(n)n^{\frac{q}{r}-1}b_n^q\}^{\frac{1}{q}}>0,$$

$$\tilde{J}(N)=\sum_{m=1}^{N}\tilde{\mu}(m)m^{\frac{p}{s}-1}a_m^p(N)$$

$$<\sum_{n=1}^{\infty}\kappa(n)n^{\frac{q}{r}-1}b_n^q<\infty.$$

Hence we find

$$0<\tilde{J}=\sum_{m=1}^{\infty}\tilde{\mu}(m)m^{\frac{p}{s}-1}a_m^p(\infty)<\infty,$$

and for $N\to\infty$, using (3.1.6), both the above inequalities still keep the strict-sign inequalities. So we have (3.1.8).

On the other-hand, suppose that (3.1.8) is valid. By the reverse Hölder's inequality, we have

$$I=\sum_{m=1}^{\infty}[\tfrac{m^{1/p-1/s}}{\tilde{\mu}^{1/p}(m)}\sum_{n=1}^{\infty}k(m,n)b_n][\tilde{\mu}^{\frac{1}{p}}(m)m^{\frac{1}{s}-\frac{1}{p}}a_m]$$

$$\geq\tilde{J}^{\frac{1}{q}}\{\sum_{m=1}^{\infty}\tilde{\mu}(m)m^{\frac{p}{s}-1}a_m^p\}^{\frac{1}{p}}.$$

Then by (3.1.8), in view of $q<0$, we have (3.1.6), which is equivalent to (3.1.8). Hence (3.1.6)-(3.1.8) are all equivalent. □

Corollary 3.1.2 As the assumption of Theorem 3.1.1, if there exist constants $\alpha_1,\alpha_2,\alpha_3>0$, such that

$$\tilde{\kappa}(m)=k(r)-\tilde{O}(m^{-\alpha_1})\leq k(r),$$

$$\kappa(n)=k(r)-O(n^{-\alpha_2})\leq k(r),$$

$$0<\tilde{l}(r)\leq\tilde{\mu}(m)=k(r)-\tilde{O}(m^{-\alpha_3})\leq k(r),$$

then (1) For $p>1$, we have the following equivalent inequalities:

$$I<\{\sum_{m=1}^{\infty}[k(r)-\tilde{O}(m^{-\alpha_1})]m^{\frac{p}{s}-1}a_m^p\}^{\frac{1}{p}}$$

$$\times\{\sum_{n=1}^{\infty}[k(r)-O(n^{-\alpha_2})]n^{\frac{q}{r}-1}b_n^q\}^{\frac{1}{q}},\quad(3.1.16)$$

$$\sum_{n=1}^{\infty}\frac{n^{p/s-1}}{[k(r)-O(n^{-\alpha_2})]^{p-1}}(\sum_{m=1}^{\infty}k(m,n)a_m)^p$$

$$<\sum_{m=1}^{\infty}[k(r)-\tilde{O}(m^{-\alpha_1})]m^{\frac{p}{s}-1}a_m^p.\quad(3.1.17)$$

In particular (for $\kappa(n)=k(r)$), we have the following equivalent inequalities:

$$\sum_{n=1}^{\infty}\sum_{m=1}^{\infty}k(m,n)a_mb_n$$

$$<k(r)\{\sum_{m=1}^{\infty}[1-\tfrac{1}{k(r)}\tilde{O}(m^{-\alpha_1})]m^{\frac{p}{s}-1}a_m^p\}^{\frac{1}{p}}$$

$$\times\{\sum_{n=1}^{\infty}n^{\frac{q}{r}-1}b_n^q\}^{\frac{1}{q}},\quad(3.1.18)$$

$$\sum_{n=1}^{\infty}n^{\frac{p}{s}-1}(\sum_{m=1}^{\infty}k(m,n)a_m)^p$$

$$<k^p(r)\sum_{m=1}^{\infty}[1-\tfrac{1}{k(r)}\tilde{O}(m^{-\alpha_1})]m^{\frac{p}{s}-1}a_m^p.\quad(3.1.19)$$

(2) For $0<p<1$, we have the following equivalent inequalities:

$$I>\{\sum_{m=1}^{\infty}[k(r)-\tilde{O}(m^{-\alpha_3})]m^{\frac{p}{s}-1}a_m^p\}^{\frac{1}{p}}$$

$$\times\{\sum_{n=1}^{\infty}[k(r)-O(n^{-\alpha_2})]n^{\frac{q}{r}-1}b_n^q\}^{\frac{1}{q}},\quad(3.1.20)$$

$$\sum_{n=1}^{\infty}\frac{n^{p/s-1}}{[k(r)-O(n^{-\alpha_2})]^{p-1}}(\sum_{m=1}^{\infty}k(m,n)a_m)^p$$

$$>\sum_{m=1}^{\infty}[k(r)-\tilde{O}(m^{-\alpha_3})]m^{\frac{p}{s}-1}a_m^p,\quad(3.1.21)$$

$$\sum_{m=1}^{\infty}\frac{m^{q/r-1}}{[k(r)-\tilde{O}(m^{-\alpha_3})]^{q-1}}(\sum_{n=1}^{\infty}k(m,n)b_n)^q$$

$$<\sum_{n=1}^{\infty}[k(r)-O(n^{-\alpha_2})]n^{\frac{q}{r}-1}b_n^q.\quad(3.1.22)$$

In particular (for $\kappa(n)=k(r)$), we have the following equivalent inequalities:

$$\sum_{n=1}^{\infty}\sum_{m=1}^{\infty}k(m,n)a_mb_n$$

$$>k(r)\{\sum_{m=1}^{\infty}[1-\tfrac{1}{k(r)}\tilde{O}(m^{-\alpha_3})]m^{\frac{p}{s}-1}a_m^p\}^{\frac{1}{p}}$$

$$\times\{\sum_{n=1}^{\infty}n^{\frac{q}{r}-1}b_n^q\}^{\frac{1}{q}},\quad(3.1.23)$$

$$\sum_{n=1}^{\infty}n^{\frac{p}{s}-1}(\sum_{m=1}^{\infty}k(m,n)a_m)^p$$

$$>k^p(r)\sum_{m=1}^{\infty}[1-\tfrac{1}{k(r)}\tilde{O}(m^{-\alpha_3})]m^{\frac{p}{s}-1}a_m^p,\quad(3.1.24)$$

$$\sum_{m=1}^{\infty}\frac{m^{q/r-1}}{[1-\frac{1}{k(r)}\tilde{O}(m^{-\alpha_3})]^{q-1}}(\sum_{n=1}^{\infty}k(m,n)b_n)^q$$

$$<k^q(r)\sum_{n=1}^{\infty}n^{\frac{q}{r}-1}b_n^q.\quad(3.1.25)$$

Note 3.1.3 (1) By (3.1.18) and (3.1.19), we have the following equivalent inequalities: (for $\tilde{\kappa}(m)=\kappa(n)=k(r)$):

$$\sum_{n=1}^{\infty}\sum_{m=1}^{\infty}k(m,n)a_m b_n < k(r)\{\sum_{m=1}^{\infty}m^{\frac{p}{s}-1}a_m^p\}^{\frac{1}{p}}\{\sum_{n=1}^{\infty}n^{\frac{q}{r}-1}b_n^q\}^{\frac{1}{q}}, \qquad (3.1.26)$$

$$\sum_{n=1}^{\infty}n^{\frac{p}{s}-1}(\sum_{m=1}^{\infty}k(m,n)a_m)^p < k^p(r)\sum_{m=1}^{\infty}m^{\frac{p}{s}-1}a_m^p. \qquad (3.1.27)$$

It is obvious that (3.1.16) and (3.1.17) (or (3.1.18) and (3.1.19)) are the strengthened versions of (3.1.26) and (3.1.27).

(2) If the constant factor $k(r)$ is proved to be the best possible of (3.1.26), then by (3.1.11), we can prove that the constant factor $k^p(r)$ is the best value of (3.1.27). By the same way, we can show the best possible property of the constant factor in the case of the reverse equivalent inequalities.

3.1.2 SOME EXAMPLES

Example 3.1.4 If $k(x,y)=\frac{1}{x+y}$, then it follows

$$k(r)=\int_0^{\infty}\frac{1}{1+u}u^{\frac{-1}{r}}du=\frac{\pi}{\sin(\pi/r)}>0.$$

We find

$$\varpi(r,m)=\sum_{n=1}^{\infty}\frac{1}{m+n}(\frac{m}{n})^{\frac{1}{r}}$$

$$<\int_0^{\infty}\frac{1}{m+y}(\frac{m}{y})^{\frac{1}{r}}dy=\int_0^{\infty}\frac{1}{1+u}(\frac{1}{u})^{\frac{1}{r}}du=\frac{\pi}{\sin(\pi/r)},$$

$$\varpi(r,m)>\int_1^{\infty}\frac{1}{m+y}(\frac{m}{y})^{\frac{1}{r}}dy$$

$$=\int_{\frac{1}{m}}^{\infty}\frac{1}{1+u}(\frac{1}{u})^{\frac{1}{r}}du\ge\tilde{l}(r)$$

$$=\int_1^{\infty}\frac{1}{1+u}(\frac{1}{u})^{\frac{1}{r}}du>0,$$

$$0<l(r)=\int_1^{\infty}\frac{1}{1+u}(\frac{1}{u})^{\frac{1}{s}}du$$

$$<\omega(s,n)=\sum_{m=1}^{\infty}\frac{1}{m+n}(\frac{n}{m})^{\frac{1}{s}}<\frac{\pi}{\sin(\pi/r)}.$$

By (3.1.26) and (3.1.27), we have the following equivalent inequalities:

$$\sum_{n=1}^{\infty}\sum_{m=1}^{\infty}\frac{a_m b_n}{m+n}<\frac{\pi}{\sin(\pi/r)}\{\sum_{m=1}^{\infty}m^{\frac{p}{s}-1}a_m^p\}^{\frac{1}{p}}\{\sum_{n=1}^{\infty}n^{\frac{q}{r}-1}b_n^q\}^{\frac{1}{q}}, \qquad (3.1.28)$$

$$\sum_{n=1}^{\infty}n^{\frac{p}{s}-1}(\sum_{m=1}^{\infty}\frac{a_m}{m+n})^p<[\frac{\pi}{\sin(\pi/r)}]^p\sum_{m=1}^{\infty}m^{\frac{p}{s}-1}a_m^p. \qquad (3.1.29)$$

Example 3.1.5 If $k(x,y)=\frac{1}{\max\{x,y\}}$, then it follows

$$k(r)=\int_0^{\infty}\frac{1}{\max\{1,u\}}u^{\frac{-1}{r}}du=rs>0.$$

We find

$$\varpi(r,m)=\sum_{n=1}^{\infty}\frac{1}{\max\{m,n\}}(\frac{m}{n})^{\frac{1}{r}}$$

$$<\int_0^{\infty}\frac{1}{\max\{m,y\}}(\frac{m}{y})^{\frac{1}{r}}dy=\int_0^{\infty}\frac{1}{\max\{1,u\}}(\frac{1}{u})^{\frac{1}{r}}du=rs$$

,

$$\varpi(r,m)\ge\int_1^{\infty}\frac{1}{\max\{m,y\}}(\frac{m}{y})^{\frac{1}{r}}dy$$

$$=rs-\int_0^1\frac{1}{m}(\frac{m}{y})^{\frac{1}{r}}dy\ge\tilde{l}(r)=rs-s=r>0,$$

$$0<l(r)=s\le\omega(s,n)=\sum_{m=1}^{\infty}\frac{1}{\max\{m,n\}}(\frac{n}{m})^{\frac{1}{s}}<rs.$$

By (3.1.26) and (3.1.27), we have the following equivalent inequalities:

$$\sum_{n=1}^{\infty}\sum_{m=1}^{\infty}\frac{a_m b_n}{\max\{m,n\}}<rs\{\sum_{m=1}^{\infty}m^{\frac{p}{s}-1}a_m^p\}^{\frac{1}{p}}\{\sum_{n=1}^{\infty}n^{\frac{q}{r}-1}b_n^q\}^{\frac{1}{q}}, \qquad (3.1.30)$$

$$\sum_{n=1}^{\infty}n^{\frac{p}{s}-1}(\sum_{m=1}^{\infty}\frac{a_m}{\max\{m,n\}})^p<(rs)^p\sum_{m=1}^{\infty}m^{\frac{p}{s}-1}a_m^p. \qquad (3.1.31)$$

Example 3.1.6 If $k(x,y)=\frac{\ln(x/y)}{x-y}$, then it follows

$$k(r)=\int_0^{\infty}\frac{-\ln u}{1-u}u^{\frac{-1}{r}}du=[\frac{\pi}{\sin(\pi/r)}]^2>0.$$

In view of Example 2.2.2, we find

$$\varpi(r,m)=\sum_{n=1}^{\infty}\frac{\ln(m/n)}{m-n}(\frac{m}{n})^{\frac{1}{r}}$$

$$<\int_0^{\infty}\frac{\ln(m/y)}{m-y}(\frac{m}{y})^{\frac{1}{r}}dy$$

$$=\int_0^{\infty}\frac{-\ln u}{1-u}(\frac{1}{u})^{\frac{1}{r}}du=[\frac{\pi}{\sin(\pi/r)}]^2,$$

$$\varpi(r,m)\ge\int_1^{\infty}\frac{\ln(m/y)}{m-y}(\frac{m}{y})^{\frac{1}{r}}dy$$

$$\ge\tilde{l}(r)=\int_1^{\infty}\frac{\ln u}{u-1}(\frac{1}{u})^{\frac{1}{r}}du>0,$$

$$0<l(r)=\int_1^{\infty}\frac{\ln u}{u-1}(\frac{1}{u})^{\frac{1}{s}}du$$

$$\le\omega(s,n)=\sum_{m=1}^{\infty}\frac{\ln(m/n)}{m-n}(\frac{n}{m})^{\frac{1}{s}}<[\frac{\pi}{\sin(\pi/r)}]^2.$$

By (3.1.26) and (3.1.27), we have the following equivalent inequalities:

$$\sum_{n=1}^{\infty}\sum_{m=1}^{\infty}\frac{\ln(m/n)}{m-n}a_m b_n$$

$$<[\frac{\pi}{\sin(\pi/r)}]^2\{\sum_{m=1}^{\infty}m^{\frac{p}{s}-1}a_m^p\}^{\frac{1}{p}}\{\sum_{n=1}^{\infty}n^{\frac{q}{r}-1}b_n^q\}^{\frac{1}{q}}, \quad (3.1.32)$$

$$\sum_{n=1}^{\infty}n^{\frac{p}{s}-1}[\sum_{m=1}^{\infty}\frac{\ln(m/n)}{m-n}a_m]^p$$

$$<[\frac{\pi}{\sin(\pi/r)}]^{2p}\sum_{m=1}^{\infty}m^{\frac{p}{s}-1}a_m^p. \quad (3.1.33)$$

3.1.3 SOME EXTENDED RESULTS BY INTRODUCING VARIABLES

Theorem 3.1.7 Assuming that $p>0(p\neq 1)$, $r>1$, $\frac{1}{p}+\frac{1}{q}=1, \frac{1}{r}+\frac{1}{s}=1$, $k(x,y)(\geq 0)$ is a homogeneous function of degree -1 in $(0,\infty)\times(0,\infty)$, such that

$$k(r)=\int_0^{\infty}k(1,u)u^{\frac{-1}{r}}du$$

is a positive number. Suppose that $u(x)$ is differentiable in $[n_0,\infty)$ and $n_0\in\mathbf{N}$, $\{u(n)\}_{n_0}^{\infty}$ are strictly increasing sequences with $u(n_0)>0$ and $u(\infty)=\infty$. Setting $\tilde{k}(m,n):=k(u(m),u(n))$, we define the weight coefficients $\tilde{w}(r,m)$ and $w(s,n)$ as follows:

$$\tilde{w}(r,m):=\sum_{n=n_0}^{\infty}\tilde{k}(m,n)(\tfrac{u(m)}{u(n)})^{\frac{1}{r}}u'(n),$$

$$w(s,n):=\sum_{m=n_0}^{\infty}\tilde{k}(m,n)(\tfrac{u(n)}{u(m)})^{\frac{1}{s}}u'(m),$$

$$m,n(\geq n_0)\in\mathbf{N}_0. \quad (3.1.34)$$

There exist measurable functions $\tilde{\mu}(x),\tilde{\kappa}(x)$ and $\kappa(x)$ in $(0,\infty)$ and constants $\tilde{l}(r)$ and $l(r)$, satisfying

$$0<\tilde{l}(r)\leq\tilde{\mu}(u(m))<\tilde{w}(r,m)<\tilde{\kappa}(u(m))\leq k_\lambda(r),$$

$$0<l(r)\leq w(s,n)<\kappa(u(n))\leq k_\lambda(r),$$

$$m,n(\geq n_0)\in\mathbf{N}_0. \quad (3.1.35)$$

If $a_m,b_n\geq 0$, $0<\sum_{m=n_0}^{\infty}\frac{u^{(p/s)-1}(m)}{[u'(m)]^{p-1}}a_m^p<\infty$ and $0<\sum_{n=n_0}^{\infty}\frac{u^{(q/r)-1}(n)}{[u'(n)]^{q-1}}b_n^q<\infty$, then (1) for $p>1$, we have the following equivalent inequalities:

$$\sum_{n=n_0}^{\infty}\sum_{m=n_0}^{\infty}\tilde{k}(m,n)a_m b_n$$

$$<\{\sum_{m=n_0}^{\infty}\tilde{\kappa}(u(m))\frac{u^{p/s-1}(m)}{[u'(m)]^{p-1}}a_m^p\}^{\frac{1}{p}}\{\sum_{n=n_0}^{\infty}\kappa(u(n))\frac{u^{q/r-1}(n)}{[u'(n)]^{q-1}}b_n^q\}^{\frac{1}{q}}, \quad (3.1.36)$$

$$\sum_{n=n_0}^{\infty}\frac{u^{p/s-1}(n)u'(n)}{\kappa^{p-1}(u(n))}(\sum_{m=n_0}^{\infty}\tilde{k}(m,n)a_m)^p$$

$$<\sum_{m=n_0}^{\infty}\tilde{\kappa}(u(m))\frac{u^{p/s-1}(m)}{[u'(m)]^{p-1}}a_m^p; \quad (3.1.37)$$

(2) for $0<p<1$, we have the following equivalent inequalities:

$$\sum_{n=n_0}^{\infty}\sum_{m=n_0}^{\infty}\tilde{k}(m,n)a_m b_n$$

$$>\{\sum_{m=n_0}^{\infty}\tilde{\mu}(u(m))\frac{u^{p/s-1}(m)}{[u'(m)]^{p-1}}a_m^p\}^{\frac{1}{p}}\{\sum_{n=n_0}^{\infty}\kappa(u(n))\frac{u^{q/r-1}(n)}{[u'(n)]^{q-1}}b_n^q\}^{\frac{1}{q}}, \quad (3.1.38)$$

$$\sum_{n=n_0}^{\infty}\frac{u^{p/s-1}(n)u'(n)}{\kappa^{p-1}(u(n))}(\sum_{m=n_0}^{\infty}\tilde{k}(m,n)a_m)^p$$

$$>\sum_{m=n_0}^{\infty}\tilde{\mu}(u(m))\frac{u^{p/s-1}(m)}{[u'(m)]^{p-1}}a_m^p, \quad (3.1.39)$$

$$\sum_{m=n_0}^{\infty}\frac{u^{q/r-1}(m)u'(m)}{\tilde{\mu}^{q-1}(u(m))}(\sum_{n=n_0}^{\infty}\tilde{k}(m,n)b_n)^q$$

$$<\sum_{n=n_0}^{\infty}\kappa(u(n))\frac{u^{q/r-1}(n)}{[u'(n)]^{q-1}}b_n^q. \quad (3.1.40)$$

Proof (1) For $p>1$, by Hölder's inequality with weight and (3.1.34), we have

$$\sum_{n=n_0}^{\infty}\sum_{m=n_0}^{\infty}\tilde{k}(m,n)a_m b_n$$

$$=\sum_{n=n_0}^{\infty}\sum_{m=m_0}^{\infty}\tilde{k}(m,n)[\frac{u^{1/(sq)}(m)(u'(n))^{1/p}}{u^{1/(rp)}(n)(u'(m))^{1/q}}a_m]\times[\frac{u^{1/(rp)}(n)(u'(m))^{1/q}}{u^{1/(sq)}(m)(u'(n))^{1/p}}b_n]$$

$$\leq\{\sum_{m=n_0}^{\infty}\sum_{n=n_0}^{\infty}\tilde{k}(m,n)\frac{u^{(p-1)/s}(m)u'(n)}{u^{1/r}(n)(u'(m))^{p-1}}a_m^p\}^{\frac{1}{p}}\times\{\sum_{n=n_0}^{\infty}\sum_{m=n_0}^{\infty}\tilde{k}(m,n)\frac{u^{(q-1)/r}(n)u'(m)}{u^{1/s}(m)(u'(n))^{q-1}}b_n^q\}^{\frac{1}{q}}$$

$$=\{\sum_{m=n_0}^{\infty}\tilde{w}(r,m)\frac{u^{p/s-1}(m)}{(u'(m))^{p-1}}a_m^p\}^{\frac{1}{p}}\times\{\sum_{n=n_0}^{\infty}w(s,n)\frac{u^{q/r-1}(n)}{(u'(n))^{q-1}}b_n^q\}^{\frac{1}{q}}. \quad (3.1.41)$$

Then by (3.1.35), we have (3.1.36). By the same way of Theorem 3.1.1, we can show that (3.1.37) is valid, which is equivalent to (3.1.36).

(2) For $0 < p < 1$, by using the reverse Hölder's inequality with weight, (3.1.34) and (3.1.35), we can obtain the equivalent inequalities (3.1.38)-(3.1.40). □

Example 3.1.8 For $k(x,y) = \frac{1}{x+y}$ in Theorem 3.1.7, if $u(n) = n + \frac{1}{2}, n_0 = 0$, then we set $\tilde{k}(m,n) = \frac{1}{m+n+1}$, $m,n \in \mathbf{N}_0$; if $u(n) = \ln n, n_0 = 2$, then we set $\tilde{k}(m,n) = \frac{1}{\ln(mn)}$, $m,n \in \mathbf{N} \setminus \{1\}$; if $u(n) = n, n_0 = 1$, then we get Theorem 3.1.1, which follows that Theorem 3.1.7 is an extension of Theorem 3.1.1.

3.1.4 SOME LEMMAS

For estimating some inequalities of the weight coefficients, we introduce the following lemma:

Lemma 3.1.9 (1) If $f(x) \in C^1(0,\infty)$, $n > m$ $(m,n \in \mathbf{N})$, then we have (cf. (2.1.13))

$$\sum_{k=m}^{n} f(k) = \int_m^n f(x)dx + \frac{1}{2}(f(m)+f(n)) + \int_m^n P_1(t) f'(t)dt, \quad (3.1.42)$$

where, $P_1(t) = t - [t] - \frac{1}{2}$ is Bernoulli function of 1-order. If both $\sum_{k=m}^{\infty} f(k)$ and $\int_m^\infty f(x)dx$ are convergence (cf. (2.3.1) for $q = 0$), then

$$\sum_{k=m}^{\infty} f(k) = \int_m^\infty f(x)dx + \frac{1}{2} f(m) + \int_m^\infty P_1(t) f'(t)dt. \quad (3.1.43)$$

(2) If $G(x) \in C^3(0,\infty)$, $n > m$ $(m,n \in \mathbf{N})$, $(-1)^i G^{(i)}(x) > 0$, $G^{(i)}(\infty) = 0$ $(i = 0,1,2,3)$, then we have(cf. (2.2.2) and (2.2.3), for $q = 0,1$)

$$\int_m^\infty P_1(t) G(t)dt = -\frac{1}{12} G(m) + \frac{1}{6} \int_m^\infty P_3(t) G''(t)dt, \quad (3.1.44)$$

$$-\frac{1}{12}[G(m) - G(n)] < \int_m^n P_1(t) G(t)dt < 0, \quad (3.1.45)$$

$$-\frac{1}{12} G(m) < \int_m^\infty P_1(t) G(t)dt < 0. \quad (3.1.46)$$

We still have the following estimation (Boas MM 1978) [2]:

$$\frac{1}{2} G(m) < \sum_{k=m}^{\infty} (-1)^{k+m} G(k) < G(m); \quad (3.1.47)$$

if $G''(x) \in C^3(0,\infty)$,

$(-1)^i G^{(2+i)}(x) > 0, G^{(2+i)}(\infty) = 0$ $(i = 0,1,2,3)$,

then we have (cf. (2.2.4))

$$0 < \int_m^\infty P_3(t) G''(t)dt < \frac{1}{120} G''(m). \quad (3.1.48)$$

(3) If $f'(x) \in C^3(0,\infty)$, $n > m$ $(m,n \in \mathbf{N})$,

$(-1)^i f^{(i)}(x) > 0, f^{(i)}(\infty) = 0$ $(i = 1,2,3,4)$,

then we have (cf. (2.2.24) and (2.2.25))

$$0 < \sum_{k=m}^{n} f(k) - \left[\int_m^n f(x)dx + \frac{1}{2}(f(m)+f(n))\right] < \frac{1}{12} f'(x)\Big|_m^n, \quad (3.1.49)$$

$$0 < \sum_{k=m}^{\infty} f(k) - \left[\int_m^\infty f(x)dx + \frac{1}{2} f(m)\right] < -\frac{1}{12} f'(m). \quad (3.1.50)$$

3.2 SOME STRENGTHENED VERSIONS OF HILBERT-TYPE INEQUALITIES WITH THE HOMOGENEOUS KERNEL OF DEGREE -1

3.2.1 SOME STRENGTHENED VERSIONS OF HARDY-HILBERT'S INEQUALITY

In the following, we build two classes of strengthened versions of (3.1.28) and (3.1.29).

Lemma 3.2.1 If $F(t) \in C^2[1,\infty), F''(t) > 0$, $F(t) \downarrow 0 (t \to \infty)$, then we have (Zhao MPT 1993)[3]

$$-\frac{F(1)}{8} < \int_1^\infty P_1(t) F(t)dt < -\frac{F(3/2)}{12} < 0. \quad (3.2.1)$$

Setting $k(x,y) = \frac{1}{x+y}$ $(x,y > 0)$, we define the weight coefficients $\varpi(r,m)$ and $\omega(s,n)$ as follows:

$$\varpi(r,m) := \sum_{k=1}^{\infty} \frac{1}{m+k} \left(\frac{m}{k}\right)^{\frac{1}{r}},$$

$$\omega(s,n) := \sum_{k=1}^{\infty} \frac{1}{k+n} \left(\frac{n}{k}\right)^{\frac{1}{s}}, r,s > 1, m,n \in \mathbf{N}. \quad (3.2.2)$$

For fixed $m \in \mathbf{N}$, setting

$$f(x) = \frac{1}{m+x} \left(\frac{m}{x}\right)^{\frac{1}{r}} \quad (x \in (0,\infty)),$$

then $f(x)$ possesses the conditions of (3.1.50). Putting

$$\varpi(r,m)=\frac{\pi}{\sin(\pi/r)}-\frac{\theta(r,m)}{m^{1/s}},m\in\mathbf{N},\qquad(3.2.3)$$

we find $f(1)=\frac{m^{1/r}}{1+m}$,

$$f'(x)=\frac{-1}{(m+x)^2}(\frac{m}{x})^{\frac{1}{r}}-\frac{1}{r}\frac{m^{1/r}}{m+x}(\frac{1}{x})^{\frac{1}{r}+1},$$

and

$$\int_1^\infty f(x)dx=\int_1^\infty\frac{1}{m+x}(\frac{m}{x})^{\frac{1}{r}}dx$$

$$=\int_{1/m}^\infty\frac{u^{-\frac{1}{r}}}{1+u}du=\frac{\pi}{\sin(\pi/r)}-\int_0^{1/m}\frac{1}{1+u}u^{-\frac{1}{r}}du.$$

By (3.1.43), we obtain that

$$\theta(r,m)=A(m)+B(m)-\frac{m}{2(m+1)},$$

where define the functions

$$G(t,x):=\frac{(r+1)xt+x^2}{r(x+t)^2t^{1+1/r}},$$

$$A(x):=x^{1-\frac{1}{r}}\int_0^{1/x}\frac{1}{1+t}(\frac{1}{t})^{1/r}dt,\text{ and}$$

$$B(x):=\int_1^\infty P_1(t)G(t,x)dt,x\in[1,\infty).$$

Lemma 3.2.2 we have

$$\theta(r,m)\ge\theta(r,1)\ (m\in\mathbf{N},r>1).$$

Proof By (Gao MRE 1994)[4], Lemma 1, we have

$$\int_0^{1/x}\frac{1}{1+t}(\frac{1}{t})^{1/r}dt\ge\frac{r(2r-1)x^{1/r}}{(r-1)[(2r-1)x+r-1]}$$

$$>\frac{(2r-1)x^{1/r}}{(2r-1)x+r-1},x\in[1,\infty).$$

Then we find

$$A'(x)=(1-\frac{1}{r})x^{-\frac{1}{r}}\int_0^{1/x}\frac{1}{1+t}(\frac{1}{t})^{1/r}dt-\frac{1}{x+1}$$

$$>\frac{r}{(x+1)[(2r-1)x+r-1]}.$$

Setting $F_1(t):=\frac{1}{(x+t)^2t^{1/r}}$ and

$$F_2(t):=\frac{1}{(x+t)^3t^{1/r}},t\in[1,\infty)(r>1,x\ge1),$$

by Lemma 3.2.1, it follows

$$B'(x)=\int_1^\infty P_1(t)G_x'(t,x)dt$$

$$=\frac{r+1}{r}\int_1^\infty P_1(t)F_1(t)dt-2x\int_1^\infty P_1(t)F_2(t)dt$$

$$>\frac{r+1}{r}(-\frac{F_1(1)}{8})+\frac{2x}{12}F_2(\frac{3}{2})$$

$$=\frac{-(r+1)}{8r(1+x)^2}+\frac{x}{6(x+3/2)^3(3/2)^{1/r}}.$$

Then we find

$$\theta_x'(r,x)=A'(x)+B'(x)-\frac{1}{2(x+1)^2}$$

$$>\frac{r}{(x+1)[(2r-1)x+r-1]}-\frac{r+1}{8r(1+x)^2}$$

$$+\frac{x}{6(x+3/2)^3(3/2)^{1/r}}-\frac{1}{2(x+1)^2}$$

$$=\frac{(-2r^2+3r+1)x+(3r^2+4r+1)}{8r(x+1)^2[(2r-1)x+r-1]}+\frac{4x}{3(2x+3)^3}(\frac{2}{3})^{1/r}.$$

If $1<r<9.9$, $(2/3)^{1/r}>2/3$, we obtain $\theta_x'(r,x)>0$; if $r\ge9.9$, $(2/3)^{1/r}>0.9$, we still can obtain $\theta_x'(r,x)>0$. Therefore $\theta(r,x)$ is increasing in $[1,\infty)$ and $\theta(r,x)\ge\theta(r,1)$.□

Lemma 3.2.3 For $k\in\mathbf{N},k\ge5$, the function

$$I(r,k):=\int_0^k\frac{1}{1+t}(\frac{1}{t})^{1/r}dt$$

$$-\frac{1}{2(1+k)}(\frac{1}{k})^{1/r}-\sum_{m=1}^{k-1}\frac{1}{m+1}(\frac{1}{m})^{1/r}$$

is decreasing in $r\in(1,\infty)$.

Proof For $k\ge5$, we find

$$I_r'(r,k)=\frac{1}{r^2}\{\frac{-\ln k}{2(1+k)k^{1/r}}$$

$$+[\int_0^4\frac{(\ln t)dt}{(1+t)t^{1/r}}-\frac{\ln2}{3\cdot2^{1/r}}-\frac{\ln3}{4\cdot3^{1/r}}]$$

$$-[\sum_{m=4}^{k-1}\frac{\ln m}{(m+1)m^{1/r}}-\int_4^k\frac{(\ln t)dt}{(1+t)t^{1/r}}]\}.\qquad(3.2.4)$$

Setting $t=e^{-y}$, then we obtain

$$J(r):=\int_0^4\frac{(\ln t)}{(1+t)t^{1/r}}dt=-\int_{-\ln4}^\infty\frac{ye^{(-1+\frac{1}{r})y}}{1+e^{-y}}dy$$

$$<\frac{-1}{5}\int_{-\ln4}^\infty ye^{(-1+\frac{1}{r})y}dy=\frac{r4^{1-\frac{1}{r}}}{5(r-1)}(\ln4-\frac{r}{r-1}).$$

For $\ln4>\frac{r}{r-1}$, viz. $r>\frac{\ln4}{\ln4-1}\approx3.5887$, $J(r)>0$, then it follows

$$\int_0^4\frac{\ln t}{(1+t)t^{1/r}}dt-\frac{\ln2}{3\cdot2^{1/r}}-\frac{\ln3}{4\cdot3^{1/r}}$$

$$<\frac{4(\ln4-1)}{5}-\frac{\ln2}{3}\cdot2^{-1/3.5887}-\frac{\ln3}{4}\cdot2^{-1/3.5887}$$

$$<-0.083996<0;$$

for $r\le\frac{\ln4}{\ln4-1}$, it is obvious that $J(r)\le0$ and

$$\int_0^4\frac{(\ln t)dt}{(1+t)t^{1/r}}-\frac{\ln2}{3\cdot2^{1/r}}-\frac{\ln3}{4\cdot3^{1/r}}<0.$$

Since for $t\ge4$,

$$[\frac{\ln t}{(1+t)t^{1/r}}]_t'=\frac{1}{(1+t)t^{1/r}}(\frac{1}{t}-\frac{\ln t}{1+t}-\frac{\ln t}{rt})<0,$$

then $g(t)=\frac{\ln t}{(1+t)t^{1/r}}$ is decreasing in $t\in[4,\infty)$ and

$$\sum_{m=4}^{k-1}\frac{\ln m}{(m+1)m^{1/r}}-\int_4^k\frac{(\ln t)dt}{(1+t)t^{1/r}}>0\ (k\ge5).$$

Hence by (3.2.4), we have $I_r'(r,k)<0$ and $I(r,k)$ ($k\ge5$) is decreasing in $r\in(1,\infty)$.□

Lemma 3.2.4 If γ is Euler constant, then for $m,n\in\mathbf{N}$, we have the following inequalities:

$$\varpi(r,m)<\frac{\pi}{\sin(\pi/r)}-\frac{1-\gamma}{m^{1/s}},$$

$$\omega(s,n)<\frac{\pi}{\sin(\pi/r)}-\frac{1-\gamma}{n^{1/r}}.\qquad(3.2.5)$$

Proof Setting $f(t)=\frac{1}{(1+t)t^{1/r}}$ $(t\in(0,\infty))$, then we find $\int_0^\infty f(t)dt=\frac{\pi}{\sin(\pi/r)}$. By (3.2.1) and (3.2.2), we have

$$\theta(r,1)=\frac{\pi}{\sin(\pi/r)}-\varpi(r,1)$$
$$=\frac{\pi}{\sin(\pi/r)}-\sum_{m=1}^{\infty}\frac{1}{1+m}\left(\frac{1}{m}\right)^{\frac{1}{r}}$$
$$=\int_0^k\frac{dt}{(1+t)t^{1/r}}+\int_k^\infty\frac{dt}{(1+t)t^{1/r}}$$
$$-\sum_{m=1}^{k-1}\frac{1}{(m+1)m^{1/r}}-\sum_{m=k}^{\infty}\frac{1}{(m+1)m^{1/r}}.$$

Putting $g(t)=\frac{1}{(1+t)t^{1/r}}$, by (3.1.50), we have

$$\int_k^\infty\frac{dt}{(1+t)t^{1/r}}+\frac{1}{2(k+1)k^{1/r}}<\sum_{m=k}^{\infty}\frac{1}{(m+1)m^{1/r}}$$
$$<\int_k^\infty\frac{dt}{(1+t)t^{1/r}}+\frac{\ln k}{2(k+1)k^{1/r}}-\frac{1}{12}g'(k),$$
$$I(r,k)+\frac{1}{12}g'(k)<\theta(r,1)<I(r,k)$$

and then

$$\inf_{r>1}I(r,k)+\frac{1}{12}\inf_{r>1}g'(k)$$
$$\le\inf_{r>1}\theta(r,1)\le\inf_{r>1}I(r,k).$$

Setting $k\to\infty$, since

$$0\ge\inf_{r>1}g'(k)=-\sup_{r>1}\left[\frac{1}{r(1+k)k^{1+1/r}}+\frac{1}{(1+k)^2k^{1/r}}\right]$$
$$\ge-\left[\frac{1}{(1+k)k}+\frac{1}{(1+k)^2}\right]\to0\ (k\to\infty),$$

and for $k\ge5$, we have

$$\inf_{r>1}I(r,k)=\lim_{r\to\infty}I(r,k)$$
$$=1-\sum_{m=1}^{k+1}\frac{1}{m}+\ln(k+1)+\frac{1}{2(k+1)}$$
$$\to1-\gamma(k\to\infty).$$

Then it follows

$$1-\gamma\le\inf_{r>1}\theta(r,1)\le1-\gamma,$$

viz. $\inf_{r>1}\theta(r,1)=1-\gamma$ and $\theta(r,1)>1-\gamma$. By (3.2.2) and Lemma 3.2.2, we have (3.2.5).□

By Corollary 3.1.2, setting

$$\tilde{\kappa}(m)=\frac{\pi}{\sin(\pi/r)}-\frac{1-\gamma}{m^{1/s}}$$

and $\kappa(n)=\frac{\pi}{\sin(\pi/r)}-\frac{1-\gamma}{n^{1/r}}$, we have the following theorem:

Theorem 3.2.5 If $p,r>1$, $\frac{1}{p}+\frac{1}{q}=1,\frac{1}{r}+\frac{1}{s}=1$, $a_m,b_n\ge0$ $(m,n\in\mathbf{N})$, $0<\sum_{m=1}^{\infty}m^{\frac{p}{s}-1}a_m^p<\infty$ and $0<\sum_{n=1}^{\infty}n^{\frac{q}{r}-1}b_n^q<\infty$, then we have the following equivalent inequalities:

$$\sum_{n=1}^{\infty}\sum_{m=1}^{\infty}\frac{a_mb_n}{m+n}<\left\{\sum_{m=1}^{\infty}\left[\frac{\pi}{\sin(\frac{\pi}{r})}-\frac{1-\gamma}{m^{1/s}}\right]m^{\frac{p}{s}-1}a_m^p\right\}^{\frac{1}{p}}$$
$$\times\left\{\sum_{n=1}^{\infty}\left[\frac{\pi}{\sin(\frac{\pi}{r})}-\frac{1-\gamma}{n^{1/r}}\right]n^{\frac{q}{r}-1}b_n^q\right\}^{\frac{1}{q}},\qquad(3.2.6)$$

$$\sum_{n=1}^{\infty}\frac{n^{p/s-1}}{\left[\frac{\pi}{\sin(\pi/r)}-\frac{1-\gamma}{n^{1/r}}\right]^{p-1}}\left(\sum_{m=1}^{\infty}\frac{a_m}{m+n}\right)^p$$
$$<\sum_{m=1}^{\infty}\left[\frac{\pi}{\sin(\frac{\pi}{r})}-\frac{1-\gamma}{m^{1/s}}\right]m^{\frac{p}{s}-1}a_m^p,\qquad(3.2.7)$$

where, $1-\gamma=0.42278433^{+}$ (γ is Euler constant).

In particular (for $\kappa(n)=\frac{\pi}{\sin(\pi/r)}$), we may deduce the following equivalent inequalities:

$$\sum_{n=1}^{\infty}\sum_{m=1}^{\infty}\frac{a_mb_n}{m+n}<\left[\frac{\pi}{\sin(\frac{\pi}{r})}\right]^{\frac{1}{q}}$$
$$\times\left\{\sum_{m=1}^{\infty}\left[\frac{\pi}{\sin(\frac{\pi}{r})}-\frac{1-\gamma}{m^{1/s}}\right]m^{\frac{p}{s}-1}a_m^p\right\}^{\frac{1}{p}}\left\{\sum_{n=1}^{\infty}n^{\frac{q}{r}-1}b_n^q\right\}^{\frac{1}{q}},\qquad(3.2.8)$$

$$\sum_{n=1}^{\infty}n^{\frac{p}{s}-1}\left(\sum_{m=1}^{\infty}\frac{a_m}{m+n}\right)^p$$
$$<\left[\frac{\pi}{\sin(\frac{\pi}{r})}\right]^{p-1}\sum_{m=1}^{\infty}\left[\frac{\pi}{\sin(\frac{\pi}{r})}-\frac{1-\gamma}{m^{1/s}}\right]m^{\frac{p}{s}-1}a_m^p.\qquad(3.2.9)$$

In Lemma 3.2.2, setting $r=2$, we obtain

$$\theta(2,1)=\pi-\varpi(2,1)$$
$$=\pi-\sum_{m=1}^{k-1}\frac{1}{1+m}\left(\frac{1}{m}\right)^{\frac{1}{2}}-\sum_{m=k}^{\infty}\frac{1}{1+m}\left(\frac{1}{m}\right)^{\frac{1}{2}}.\qquad(3.2.10)$$

By (3.1.50), we have $0<\varepsilon<1$ and

$$\sum_{m=k}^{\infty}\frac{1}{(m+1)m^{1/2}}=2\arctan\sqrt{\frac{1}{k}}$$
$$+\frac{1}{2(k+1)k^{1/2}}+\frac{\varepsilon}{12}\left[\frac{1}{(k+1)^2k^{1/2}}+\frac{1}{2(k+1)k^{3/2}}\right].\qquad(3.2.11)$$

Setting more large integer k, by (3.2.10) and (3.2.11), we obtain (Gao HMA 1992)[5]

$$\theta_0:=\theta(2,1)=1.281669^{+}.$$

Then we have the following corollary:

Corollary 3.2.6 If $p>1$, $\frac{1}{p}+\frac{1}{q}=1$, $a_m,b_n\ge0$ $(m,n\in\mathbf{N})$, such that $0<\sum_{m=1}^{\infty}m^{\frac{p}{2}-1}a_m^p<\infty$ and $0<\sum_{n=1}^{\infty}n^{\frac{q}{2}-1}b_n^q<\infty$, then we have the following equivalent inequalities:

$$\sum_{n=1}^{\infty}\sum_{m=1}^{\infty}\frac{a_mb_n}{m+n}<\left\{\sum_{m=1}^{\infty}\left(\pi-\frac{\theta_0}{\sqrt{m}}\right)m^{\frac{p}{2}-1}a_m^p\right\}^{\frac{1}{p}}$$

$$\times\{\sum_{n=1}^{\infty}(\pi-\frac{\theta_0}{\sqrt{n}})n^{\frac{q}{2}-1}b_n^q\}^{\frac{1}{q}}, \qquad (3.2.12)$$

$$\sum_{n=1}^{\infty}\frac{n^{p/2-1}}{(\pi-\theta_0/\sqrt{n})^{p-1}}(\sum_{m=1}^{\infty}\frac{a_m}{m+n})^p$$

$$<\sum_{m=1}^{\infty}(\pi-\frac{\theta_0}{\sqrt{m}})m^{\frac{p}{2}-1}a_m^p, \qquad (3.2.13)$$

where, $\theta_0 = 1.281669^+$. In particular, we have the following equivalent inequalities:

$$\sum_{n=1}^{\infty}\sum_{m=1}^{\infty}\frac{a_m b_n}{m+n}<\pi\{\sum_{m=1}^{\infty}(1-\frac{\theta_0}{\pi\sqrt{m}})m^{\frac{p}{2}-1}a_m^p\}^{\frac{1}{p}}$$

$$\times\{\sum_{n=1}^{\infty}n^{\frac{q}{2}-1}b_n^q\}^{\frac{1}{q}}, \qquad (3.2.14)$$

$$\sum_{n=1}^{\infty}n^{\frac{p}{2}-1}(\sum_{m=1}^{\infty}\frac{a_m}{m+n})^p$$

$$<\pi^p\sum_{m=1}^{\infty}(1-\frac{\theta_0}{\pi\sqrt{m}})m^{\frac{p}{2}-1}a_m^p. \qquad (3.2.15)$$

3.2.2 ANOTHER STRENGTHENED VERSION OF HARDY-HILBERT'S INEQUALITY

Lemma 3.2.7 If $r>1, \frac{1}{r}+\frac{1}{s}=1, \theta(r,m)$ is define by (3.2.3), then we have

$$\theta(r,m)\geq f_m(s)+g_m(s),\ \ m\in\mathbf{N}, \qquad (3.2.16)$$

where, $f_m(s)$ and $g_m(s)$ are defined by

$$f_m(s):=s+\frac{1}{12s}+\frac{1}{(1+s)m}+\frac{1}{12sm^2}+\frac{1}{3(1+3s)m^3},$$

$$g_m(s):=-\frac{1}{12sm}-\frac{1}{2(1+2s)m^2}$$

$$-\frac{7}{12}-\frac{1}{2m}+\frac{1}{12m^2}-\frac{7}{12m^3}.$$

Proof By (3.2.3) and the relating expressions, we have

$$\theta(r,m)=A(m)+B(m)-\frac{m}{2(m+1)},$$

where we define

$$G(t,m)=\frac{m}{(m+t)^2t^{1/r}}+(1-\frac{1}{s})\frac{m}{(m+t)t^{1+1/r}},$$

$$A(m)=m^{1-\frac{1}{r}}\int_0^{1/m}\frac{1}{1+t}(\frac{1}{t})^{1/r}dt,$$

and $B(m)=\int_1^{\infty}P_1(t)G(t,m)dt$. By (3.1.46), we obtain

$$A(m)=m^{1-\frac{1}{r}}\int_0^{1/m}\frac{1}{1+t}(\frac{1}{t})^{1/r}dt$$

$$=m^{1-\frac{1}{r}}\int_0^{1/m}\sum_{k=0}^{\infty}(-1)^k t^{k-\frac{1}{r}}dt$$

$$=m^{1-\frac{1}{r}}\sum_{k=0}^{\infty}(-1)^k\int_0^{1/m}t^{k-\frac{1}{r}}dt$$

$$=\sum_{k=0}^{\infty}\frac{(-1)^k}{(k+\frac{1}{s})m^k}>s\sum_{k=0}^{3}\frac{(-1)^k}{(sk+1)m^k},$$

$$B(m)=\int_1^{\infty}P_1(t)G(t,m)dt$$

$$>-\frac{1}{12}[\frac{m}{(m+1)^2}+(1-\frac{1}{s})\frac{m}{m+1}].$$

For $m\geq 2$, we find

$$\frac{m}{1+m}=(1+\frac{1}{m})^{-1}<1-\frac{1}{m}+\frac{1}{m^2};$$

$$\frac{m}{(m+1)^2}=\frac{1}{m}(1+\frac{1}{m})^{-2}<\frac{1}{m}(1-\frac{2}{m}+\frac{3}{m^2}).$$

It is obvious that the above inequalities are valid for $m=1$. Then by simplification, we have (3.2.16). □

Lemma 3.2.8 For $m,n\in\mathbf{N}$, we have the following inequalities:

$$\varpi(r,m)<\frac{\pi}{\sin(\pi/r)}-\frac{1}{2m^{1/s}+n^{-1/r}},$$

$$\omega(s,n)<\frac{\pi}{\sin(\pi/r)}-\frac{1}{2n^{1/r}+n^{-1/s}}. \qquad (3.2.17)$$

Proof For $s>1, m\in\mathbf{N}$, we obtain

$$f_m'(s)=1-\frac{1}{12s^2}-\frac{1}{(1+s)^2m}-\frac{1}{12s^2m^2}-\frac{1}{(1+3s)^2m^3}$$

$$>1-\frac{1}{12}-\frac{1}{4m}-\frac{1}{12m^2}-\frac{1}{16m^3}>0,$$

$$g_m'(s)=\frac{1}{12s^2m}+\frac{1}{(1+2s)^2m^2}>0.$$

Then we find

$$f_m(s)+g_m(s)>\lim_{s\to 1^+}[f_m(s)+g_m(s)]$$

$$=\frac{1}{2}-\frac{1}{12m}-\frac{7}{12m^3}.$$

For $m\geq 3$, since

$$(\frac{1}{2}-\frac{1}{12m}-\frac{7}{12m^3})(1+\frac{1}{2m})$$

$$=\frac{1}{2}+\frac{1}{m}(\frac{1}{6}-\frac{1}{24m}-\frac{1}{2m^2}-\frac{1}{4m^3})>\frac{1}{2},$$

we have

$$\frac{1}{2}-\frac{1}{12m}-\frac{7}{12m^3}>\frac{1}{2(1+\frac{1}{2m})}=\frac{1}{2+m^{-1}},$$

and then by (3.2.2) and (3.2.16), we find

$$\varpi(r,m)<\frac{\pi}{\sin(\pi/r)}-\frac{1}{m^{1/s}}(\frac{1}{2}-\frac{1}{12m}-\frac{1}{2m^3})$$

$$<\frac{\pi}{\sin(\pi/r)}-\frac{1}{m^{1/s}}\cdot\frac{1}{2+m^{-1}}$$

$$=\frac{\pi}{\sin(\pi/r)}-\frac{1}{2m^{1/s}+n^{-1/r}},\ \ m\geq 3. \qquad (3.2.18)$$

Since $\gamma<0.6$, then for $m=1,2$, by (3.2.5), we have

$$\varpi(r,1)<\frac{\pi}{\sin(\pi/r)}-\frac{1-\gamma}{1^{1/s}}<\frac{\pi}{\sin(\pi/r)}-\frac{1}{2\cdot 1^{1/s}+1^{-1/r}}$$

and

$$\varpi(r,2)<\frac{\pi}{\sin(\pi/r)}-\frac{1-\gamma}{2^{1/s}}<\frac{\pi}{\sin(\pi/r)}-\frac{1}{2\cdot 2^{1/s}+2^{-1/r}}.$$

Hence (3.2.18) is valid for $m\in\mathbf{N}$.

By the same way, we can show the other inequalities and (3.2.17) are valid. □

Then by Corollary 3.1.2, setting

$$\tilde{\kappa}(m)=\frac{\pi}{\sin(\pi/r)}-\frac{1}{2m^{1/s}+n^{-1/r}}$$

and $\kappa(n)=\frac{\pi}{\sin(\pi/r)}-\frac{1}{2n^{1/r}+n^{-1/s}}$, we have the following theorem:

Theorem 3.2.9 If $p,\ r>1$, $\frac{1}{p}+\frac{1}{q}=1$, $\frac{1}{r}+\frac{1}{s}=1$, $a_m,b_n\geq 0$ $(m,n\in \mathbf{N})$, $0<\sum_{m=1}^{\infty} m^{\frac{p}{s}-1}a_m^p<\infty$ and $0<\sum_{n=1}^{\infty} n^{\frac{q}{r}-1}b_n^q<\infty$, then we have the following equivalent inequalities:

$$\sum_{n=1}^{\infty}\sum_{m=1}^{\infty}\frac{a_m b_n}{m+n}<\{\sum_{m=1}^{\infty}[\frac{\pi}{\sin(\frac{\pi}{r})}-\frac{1}{2m^{1/s}+m^{-1/r}}]m^{\frac{p}{s}-1}a_m^p\}^{\frac{1}{p}}$$
$$\times\{\sum_{n=1}^{\infty}[\frac{\pi}{\sin(\frac{\pi}{r})}-\frac{1}{2n^{1/r}+n^{-1/s}}]n^{\frac{q}{r}-1}b_n^q\}^{\frac{1}{q}}, \tag{3.2.19}$$

$$\sum_{n=1}^{\infty}\frac{n^{p/s-1}}{[\frac{\pi}{\sin(\pi/r)}-\frac{1}{2n^{1/r}+n^{-1/s}}]^{p-1}}(\sum_{m=1}^{\infty}\frac{a_m}{m+n})^p$$
$$<\sum_{m=1}^{\infty}[\frac{\pi}{\sin(\frac{\pi}{r})}-\frac{1}{2m^{1/s}+m^{-1/r}}]m^{\frac{p}{s}-1}a_m^p. \tag{3.2.20}$$

In particular (setting $\kappa(n)=\frac{\pi}{\sin(\pi/r)}$), we can deduce the following equivalent inequalities:

$$\sum_{n=1}^{\infty}\sum_{m=1}^{\infty}\frac{a_m b_n}{m+n}<[\frac{\pi}{\sin(\frac{\pi}{r})}]^{\frac{1}{q}}$$
$$\times\{\sum_{m=1}^{\infty}[\frac{\pi}{\sin(\frac{\pi}{r})}-\frac{1}{2m^{1/s}+m^{-1/r}}]m^{\frac{p}{s}-1}a_m^p\}^{\frac{1}{p}}\{\sum_{n=1}^{\infty}n^{\frac{q}{r}-1}b_n^q\}^{\frac{1}{q}}, \tag{3.2.21}$$

$$\sum_{n=1}^{\infty}n^{\frac{p}{s}-1}(\sum_{m=1}^{\infty}\frac{a_m}{m+n})^p$$
$$<[\frac{\pi}{\sin(\frac{\pi}{r})}]^{p-1}\sum_{m=1}^{\infty}[\frac{\pi}{\sin(\frac{\pi}{r})}-\frac{1}{2m^{1/s}+m^{-1/r}}]m^{\frac{p}{s}-1}a_m^p. \tag{3.2.22}$$

Note 3.2.10 It is obvious that (3.2.6) and (3.2.19) are different strengthened versions of (3.1.28). The above methods and results are also mentioned in (Yang AM 1997)[6]-(Yang HJ 1997)[9].

Setting $r=s=2$ in (3.2.16), we find

$$\theta(2,m)\geq f_m(2)+g_m(2)$$
$$>\frac{35}{24}-\frac{5}{24m}+\frac{1}{48m^2}-\frac{15}{24m^3}.$$

For $m\geq 2$, since

$$(\frac{35}{24}-\frac{5}{24m}+\frac{1}{48m^2}-\frac{15}{24m^3})(1+\frac{1}{m})$$
$$=\frac{35}{24}+\frac{1}{m}(\frac{60}{48}-\frac{9}{48m}-\frac{29}{48m^2}-\frac{30}{48m^3})>\frac{35}{24},$$

then we have

$$\frac{35}{24}-\frac{5}{24m}+\frac{1}{48m^2}-\frac{15}{24m^3}>\frac{35}{24(1+m^{-1})}$$

and

$$\varpi(2,m)<\pi-\frac{1}{m^{1/2}}\cdot\frac{35}{24(1+m^{-1})}$$
$$=\pi-\frac{35}{24(\sqrt{m}+\sqrt{m^{-1}})},\ m\geq 2; \tag{3.2.23}$$

for $m=1$, since $\theta_0=1.281669^+>35/48$, we have

$$\varpi(2,1)<\pi-\frac{\theta_0}{1^{1/2}}<\pi-\frac{35}{24(1^{1/2}+1^{-1/2})}.$$

Therefore (3.2.23) is valid for $m\in\mathbf{N}$. We have the following corollary:

Corollary 3.2.11 If $p>1$, $\frac{1}{p}+\frac{1}{q}=1$, $a_m,b_n\geq 0$ $(m,n\in\mathbf{N})$, such that $0<\sum_{m=1}^{\infty}m^{\frac{p}{2}-1}a_m^p<\infty$ and $0<\sum_{n=1}^{\infty}n^{\frac{q}{2}-1}b_n^q<\infty$, then we have the following equivalent inequalities:

$$\sum_{n=1}^{\infty}\sum_{m=1}^{\infty}\frac{a_m b_n}{m+n}<\{\sum_{m=1}^{\infty}[\pi-\frac{35}{24(\sqrt{m}+\sqrt{m^{-1}})}]m^{\frac{p}{2}-1}a_m^p\}^{\frac{1}{p}}$$
$$\times\{\sum_{n=1}^{\infty}[\pi-\frac{35}{24(\sqrt{n}+\sqrt{n^{-1}})}]n^{\frac{q}{2}-1}b_n^q\}^{\frac{1}{q}}, \tag{3.2.24}$$

$$\sum_{n=1}^{\infty}\frac{n^{p/2-1}}{[\pi-\frac{35}{24(\sqrt{n}+\sqrt{n^{-1}})}]^{p-1}}(\sum_{m=1}^{\infty}\frac{a_m}{m+n})^p$$
$$<\sum_{m=1}^{\infty}[\pi-\frac{35}{24(\sqrt{m}+\sqrt{m^{-1}})}]m^{\frac{p}{2}-1}a_m^p. \tag{3.2.25}$$

In particular (setting $\kappa(n)=\pi$), we may deduce the following equivalent inequalities:

$$\sum_{n=1}^{\infty}\sum_{m=1}^{\infty}\frac{a_m b_n}{m+n}<\pi\{\sum_{m=1}^{\infty}[1-\frac{35}{24\pi(\sqrt{m}+\sqrt{m^{-1}})}]m^{\frac{p}{2}-1}a_m^p\}^{\frac{1}{p}}$$
$$\times\{\sum_{n=1}^{\infty}n^{\frac{q}{2}-1}b_n^q\}^{\frac{1}{q}}, \tag{3.2.26}$$

$$\sum_{n=1}^{\infty}n^{\frac{p}{2}-1}(\sum_{m=1}^{\infty}\frac{a_m}{m+n})^p$$
$$<\pi^p\sum_{m=1}^{\infty}[1-\frac{35}{24\pi(\sqrt{m}+\sqrt{m^{-1}})}]m^{\frac{p}{2}-1}a_m^p. \tag{3.2.27}$$

3.2.3 A STRENGTHENED VERSION OF THE MORE ACCURATE HARDY-HILBERT'S INEQUALITY

If $r>1, \frac{1}{r}+\frac{1}{s}=1$, $k(x,y)=\frac{1}{x+y}$, $u(m)=m+\frac{1}{2}$ $(m\in\mathbf{N}_0)$, $n_0=0$, then we define the following weight coefficients

$$\tilde{w}(r,m):=(m+\tfrac{1}{2})^{\frac{1}{r}}\sum_{k=0}^{\infty}\frac{1}{m+k+1}(k+\tfrac{1}{2})^{\frac{-1}{r}},$$

$$w(s,n):=(n+\tfrac{1}{2})^{\frac{1}{s}}\sum_{k=0}^{\infty}\frac{1}{k+n+1}(k+\tfrac{1}{2})^{\frac{-1}{s}},$$

$$m,n\in\mathbf{N}_0, \qquad (3.2.28)$$

and set the following decompositions:

$$\tilde{w}(r,m)==\frac{\pi}{\sin(\pi/r)}-\frac{\tilde{\theta}(r,m)}{(2m+1)^{2-1/r}}, \qquad (3.2.29)$$

$$\tilde{\theta}(r,m):=\frac{\pi}{\sin(\pi/r)}(2m+1)^{2-\frac{1}{r}}$$
$$-(2m+1)^2\sum_{k=0}^{\infty}\frac{1}{m+k+1}(2k+1)^{\frac{-1}{r}}. \qquad (3.2.30)$$

We estimate the lower bound of $\tilde{\theta}(r,m)$ $(r>1,$ $m\in\mathbf{N}_0)$ in the following.

For fixed $m\in\mathbf{N}_0, r>1$, setting $f(x)$ as follows:

$$f(x):=\frac{1}{x+m+1}\left(\frac{1}{2x+1}\right)^{\frac{1}{r}}, x\in(-\tfrac{1}{2},\infty),$$

then $f(x)$ possesses the conditions of (3.1.43). We find $f(0)=\frac{1}{m+1}$, and

$$f'(x)=\frac{-1}{(x+m+1)^2}\left(\frac{1}{2x+1}\right)^{\frac{1}{r}}-\frac{2}{r(x+m+1)}\left(\frac{1}{2x+1}\right)^{\frac{1}{r}+1}.$$

Setting $u=\frac{2x+1}{2m+1}$, we obtain

$$\int_0^{\infty}f(x)dx=\frac{1}{(2m+1)^{1/r}}\int_{\frac{1}{2m+1}}^{\infty}\frac{1}{1+u}u^{-\frac{1}{r}}du$$
$$=\frac{1}{(2m+1)^{1/r}}\left[\frac{\pi}{\sin(\pi/r)}-\int_0^{\frac{1}{2m+1}}\frac{1}{1+u}u^{-\frac{1}{r}}du\right]$$
$$=\frac{1}{(2m+1)^{1/r}}\left[\frac{\pi}{\sin(\pi/r)}-\int_0^{1/(2m+1)}\sum_{k=0}^{\infty}(-1)^k u^{k-\frac{1}{r}}du\right]$$
$$=\frac{1}{(2m+1)^{\frac{1}{r}}}\left[\frac{\pi}{\sin(\frac{\pi}{r})}-\sum_{k=0}^{\infty}(-1)^k\int_0^{\frac{1}{2m+1}}u^{k-\frac{1}{r}}du\right]$$
$$=\frac{1}{(2m+1)^{1/r}}\left[\frac{\pi}{\sin(\frac{\pi}{r})}-\sum_{k=0}^{\infty}\frac{(-1)^k}{(k+1-\frac{1}{r})(2m+1)^{k+1-1/r}}\right].$$

By (3.1.43), we have

$$\sum_{k=0}^{\infty}\frac{1}{m+k+1}(2k+1)^{\frac{-1}{r}}$$
$$=\frac{1}{(2m+1)^{1/r}}\left[\frac{\pi}{\sin(\frac{\pi}{r})}-\sum_{k=0}^{\infty}\frac{(-1)^k}{(k+1-\frac{1}{r})(2m+1)^{k+1-1/r}}\right]+\frac{1}{2(m+1)}$$
$$-\int_0^{\infty}P_1(x)\left[\frac{1}{(x+m+1)^2}\left(\frac{1}{2x+1}\right)^{\frac{1}{r}}+\frac{2}{r(x+m+1)}\left(\frac{1}{2x+1}\right)^{\frac{1}{r}+1}\right]dx$$

.

Then by (3.2.30), we find

$$\tilde{\theta}(r,t)=-\frac{(2t+1)^2}{2(t+1)}+\sum_{k=0}^{\infty}\frac{(-1)^k}{(k+1-\frac{1}{r})(2t+1)^{k-1}}$$
$$+\int_0^{\infty}P_1(x)\left[\frac{(2t+1)^2}{(x+t+1)^2}\left(\frac{1}{2x+1}\right)^{\frac{1}{r}}+\frac{2(2t+1)^2}{r(x+t+1)}\left(\frac{1}{2x+1}\right)^{\frac{1}{r}+1}\right]dx,$$
$$t\geq 0. \qquad (3.2.31)$$

In view of (3.1.46), we have

$$\tilde{\theta}(r,t)>-\frac{(2t+1)^2}{2(t+1)}+\sum_{k=0}^{\infty}\frac{(-1)^k}{(k+1-\frac{1}{r})(2t+1)^{k-1}}$$
$$-\frac{(2t+1)^2}{12(t+1)^2}-\frac{(2t+1)^2}{6r(t+1)}=I(r,t)+J(r,t), \qquad (3.2.32)$$

where, $I(r,t):=\sum_{k=2}^{\infty}\frac{(-1)^k}{(k+1-\frac{1}{r})(2t+1)^{k-1}}$ and

$$J(r,t):=\frac{2t+1}{1-\frac{1}{r}}-\frac{1}{2-\frac{1}{r}}-\frac{(2t+1)^2}{2(t+1)}-\frac{(2t+1)^2}{12(t+1)^2}-\frac{(2t+1)^2}{6r(t+1)}.$$

We obtain $I_r'(r,t)=\sum_{k=2}^{\infty}\frac{(-1)^{k-1}}{r^2(k+1-\frac{1}{r})^2(2t+1)^{k-1}}<0$, and

$$I(r,t)>I(\infty,t)=\sum_{k=2}^{\infty}\frac{(-1)^k}{(k+1)(2t+1)^{k-1}}.$$ Since

$$J(r,t)=-\frac{(2t+1)^2}{2(t+1)}-\frac{(2t+1)^2}{12(t+1)^2}$$
$$+\frac{1}{r}\left[\frac{r^2(2t+1)}{r-1}-\frac{r^2}{2r-1}-\frac{(2t+1)^2}{6(t+1)}\right]$$
$$=-\frac{(2t+1)^2}{2(t+1)}-\frac{(2t+1)^2}{12(t+1)^2}$$
$$+\frac{1}{r}\left[\frac{(r^2-1)(2t+1)+(2t+1)}{r-1}-\frac{4r^2-1+1}{4(2r-1)}-\frac{(2t+2-1)^2}{6(t+1)}\right]$$
$$=-\frac{(2t+1)^2}{2(t+1)}-\frac{(2t+1)^2}{12(t+1)^2}$$
$$+\frac{1}{r}\left[(2tr+\frac{r}{2})+\frac{4t}{3}+\frac{3}{4}-\frac{1}{6(t+1)}-\frac{2t+1}{r-1}-\frac{1}{4(2r-1)}\right]$$
$$=-\frac{(2t+1)^2}{2(t+1)}-\frac{(2t+1)^2}{12(t+1)^2}+2t+\frac{1}{2}+g(r,t),$$

where,

$$g(r,t):=\frac{1}{r}\left\{\left[\frac{4t}{3}+\frac{1}{12}+\frac{4t+3}{6(t+1)}\right]+\left[\frac{2t+1}{r-1}-\frac{1}{4(2r-1)}\right]\right\}>0,$$

then we obtain

$$J(r,t)>-\frac{(2t+1)^2}{2(t+1)}-\frac{(2t+1)^2}{12(t+1)^2}+2t+\frac{1}{2}.$$

By (3.2.23), we have

$$\tilde{\theta}(r,t)>\sum_{k=2}^{\infty}\frac{(-1)^k}{(k+1)(2t+1)^{k-1}}$$
$$-\frac{(2t+1)^2}{2(t+1)}-\frac{(2t+1)^2}{12(t+1)^2}+2t+\frac{1}{2}$$
$$=\lambda(t):=\sum_{k=1}^{\infty}\frac{(-1)^k}{(k+1)(2t+1)^{k-1}}$$
$$+\frac{2}{3}-\frac{1}{6(t+1)}-\frac{1}{12(t+1)^2},\quad t\geq 0. \qquad (3.2.33)$$

For $t\geq 1$, we find

$$\lambda'(t)=\sum_{k=1}^{\infty}\frac{2(-1)^{k-1}(k-1)}{(k+1)(2t+1)^k}+\frac{1}{6(t+1)^2}+\frac{1}{6(t+1)^3}$$
$$=\sum_{k=2}^{\infty}\frac{2(-1)^{k-1}}{(2t+1)^k}+\sum_{k=2}^{\infty}\frac{4(-1)^k}{(k+1)(2t+1)^k}+\frac{1}{6(t+1)^2}+\frac{1}{6(t+1)^3}$$
$$>\left[\frac{-2}{(2t+1)^2}+\frac{2}{(2t+1)^3}-\frac{2}{(2t+1)^4}\right]+\left[\frac{4}{3(2t+1)^2}-\frac{4}{4(2t+1)^3}\right]$$
$$+\frac{1}{6(t+1)^2}+\frac{1}{6(t+1)^3}$$
$$>\frac{-2}{3(2t+1)^2}+\frac{1}{(2t+1)^3}-\frac{2}{(2t+1)^4}+\frac{1}{6(t+1)^2}+\frac{1}{6(t+1)^3}$$
$$>\frac{12t^4+18t^3-2t^2-17t-8}{6(2t+1)^4(t+1)^3}>0,$$

then we have

$$\lambda(t) \ge \lambda(1) > \sum_{k=1}^{5} \frac{(-1)^k}{(k+1)(2t+1)^{k-1}}$$
$$+\frac{2}{3} - \frac{1}{6(1+1)} - \frac{1}{12(1+1)^2} = 0.1518^+, \ t \ge 1.$$

Since

$$\lambda(0) = \ln 2 - \frac{7}{12} = 0.1098^+ < \lambda(1),$$

then we obtain

$$\tilde{\theta}(r,m) > \lambda(m) > \lambda(0) > \frac{1}{10}.$$

Lemma 3.2.12 For $r > 1$, we have the following inequalities:

$$\tilde{w}(r,m) < \tilde{\kappa}(u(m))$$
$$:= \frac{\pi}{\sin(\pi/r)} - \frac{1}{10(2m+1)^{2-1/r}}, \quad m \in \mathbf{N}_0. \tag{3.2.34}$$

Note 3.2.13 In view of the symmetric property, we have the following inequality:

$$w(s,n) < \kappa(u(n))$$
$$:= \frac{\pi}{\sin(\pi/r)} - \frac{1}{10(2n+1)^{2-1/s}}, \quad n \in \mathbf{N}_0. \tag{3.2.35}$$

By Theorem 3.1.7, setting $k(x,y) = \frac{1}{x+y}$, $u(m) = m + \frac{1}{2}$ $(m \in \mathbf{N}_0), n_0 = 0$, we have the following theorem:

Theorem 3.2.14 If $p > 1, r > 1$, $\frac{1}{p} + \frac{1}{q} = 1$, $\frac{1}{r} + \frac{1}{s} = 1$, $a_m, b_n \ge 0$ $(m,n \in \mathbf{N}_0)$, such that $0 < \sum_{m=0}^{\infty} (m+\frac{1}{2})^{\frac{p}{s}-1} a_m^p < \infty$ and $0 < \sum_{n=0}^{\infty} (n+\frac{1}{2})^{\frac{q}{r}-1} b_n^q < \infty$, then we have the following equivalent inequalities:

$$\tilde{I} := \sum_{n=0}^{\infty}\sum_{m=0}^{\infty} \frac{a_m b_n}{m+n+1}$$
$$< \{\sum_{m=0}^{\infty} [\frac{\pi}{\sin(\frac{\pi}{r})} - \frac{1}{10(2m+1)^{1+1/s}}](m+\frac{1}{2})^{\frac{p}{s}-1} a_m^p\}^{\frac{1}{p}}$$
$$\times \{\sum_{n=0}^{\infty} [\frac{\pi}{\sin(\frac{\pi}{r})} - \frac{1}{10(2n+1)^{1+1/r}}](n+\frac{1}{2})^{\frac{q}{r}-1} b_n^q\}^{\frac{1}{q}}, \tag{3.2.36}$$

$$\sum_{n=0}^{\infty} \frac{(n+\frac{1}{2})^{p/s-1}}{[\frac{\pi}{\sin(\pi/r)} - \frac{1}{10(2n+1)^{1+1/r}}]^{p-1}} (\sum_{m=0}^{\infty} \frac{a_m}{m+n+1})^p$$
$$< \sum_{m=0}^{\infty} [\frac{\pi}{\sin(\frac{\pi}{r})} - \frac{1}{10(2m+1)^{1+1/s}}](m+\frac{1}{2})^{\frac{p}{s}-1} a_m^p. \tag{3.2.37}$$

In particular (setting $\kappa(n) = \frac{\pi}{\sin(\pi/r)}$), we may deduce the following equivalent forms:

$$\tilde{I} < [\frac{\pi}{\sin(\frac{\pi}{r})}]^{\frac{1}{q}} \{\sum_{m=0}^{\infty} [\frac{\pi}{\sin(\frac{\pi}{r})} - \frac{1}{10(2m+1)^{1+1/s}}](m+\frac{1}{2})^{\frac{p}{s}-1} a_m^p\}^{\frac{1}{p}}$$
$$\times \{\sum_{n=0}^{\infty} (n+\frac{1}{2})^{\frac{q}{r}-1} b_n^q\}^{\frac{1}{q}}, \tag{3.2.38}$$

$$\sum_{n=0}^{\infty} (n+\frac{1}{2})^{\frac{p}{s}-1} (\sum_{m=0}^{\infty} \frac{a_m}{m+n+1})^p$$
$$< [\frac{\pi}{\sin(\frac{\pi}{r})}]^{p-1} \sum_{m=0}^{\infty} [\frac{\pi}{\sin(\frac{\pi}{r})} - \frac{1}{10(2m+1)^{1+1/s}}](m+\frac{1}{2})^{\frac{p}{s}-1} a_m^p. \tag{3.2.39}$$

3.2.4 ANOTHER STRENGTHENED VERSION OF THE MORE ACCURATE HARDY-HILBERT'S INEQUALITY

By (3.1.47), we have $\sum_{k=2}^{\infty} \frac{(-1)^k}{(k+1)(2t+1)^{k-1}} > \frac{1}{6(2t+1)}$. Then by (3.2.33), we find:

$$\tilde{\theta}(\infty,t) > \frac{1}{6} - \frac{1}{6(t+1)} - \frac{1}{12(t+1)^2} + \frac{1}{6(2t+1)},$$
$$t \ge 0. \tag{3.2.40}$$

Setting

$$g(t) := \frac{1}{12} - \frac{1}{6(2t+1)} + \frac{1}{12(t+1)} + \frac{1}{12(2t+1)^2} \quad (t \ge 0),$$

by (3.2.40), we obtain

$$\tilde{\theta}(\infty,m) > \frac{2m+1}{m+1} g(m) \quad (m \in \mathbf{N}_0).$$

Since

$$g'(t) := \frac{1}{3(2t+1)^2} - \frac{1}{12(t+1)^2} - \frac{1}{3(2t+1)^3}$$
$$= \frac{4t^2+2t-1}{12(t+1)^2(2t+1)^3} > 0, \ t \ge 1,$$

then $g(t)$ is strictly increasing in $[1,\infty)$ and

$$g(m) \ge g(1) = 0.0787 > \frac{1}{13} \quad (m \ge 1).$$

We have

$$\tilde{\theta}(\infty,m) > \frac{2m+1}{m+1} g(1) > \frac{2m+1}{13(m+1)}, \quad m \in \mathbf{N}. \tag{3.2.41}$$

Lemma 3.2.15 for $r > 1$, we have the following inequality:

$$\tilde{w}(r,m) < \tilde{\kappa}_1(u(m))$$
$$:= \frac{\pi}{\sin(\pi/r)} - \frac{1}{13(m+1)(2m+1)^{1-1/r}}, \quad m \in \mathbf{N}_0. \tag{3.2.42}$$

Proof By (3.2.29), (3.2.40) and (3.2.41), we find that (3.2.42) is valid for $m \in \mathbf{N}$. For $m = 0$, since by (3.2.34), we have

$$\tilde{w}(r,0) < \frac{\pi}{\sin(\pi/r)} - \frac{1}{10(2\times 0+1)^{2-1/r}}$$
$$< \frac{\pi}{\sin(\pi/r)} - \frac{1}{13(0+1)(2\times 0+1)^{1-1/r}},$$

then (3.2.42) is valid for $m \in \mathbf{N}_0$. □

In view of the symmetric property, we have

$$w(s,n) < \kappa_1(u(n)) := \frac{\pi}{\sin(\pi/r)} - \frac{1}{13(n+1)(2n+1)^{1-1/s}},$$
$$n \in \mathbf{N}_0. \tag{3.2.43}$$

Then by Theorem 3.1.7, we have the following theorem:

Theorem 3.2.16 If $p>1, r>1$, $\frac{1}{p}+\frac{1}{q}=1$, $\frac{1}{r}+\frac{1}{s}=1$, and $a_m, b_n \ge 0$ $(m,n \in \mathbf{N}_0)$, such that $0<\sum_{m=0}^{\infty}(m+\frac{1}{2})^{\frac{p}{s}-1}a_m^p<\infty$ and $0<\sum_{n=0}^{\infty}(n+\frac{1}{2})^{\frac{q}{r}-1}b_n^q<\infty$, then we have the following equivalent inequalities:

$$\sum_{n=0}^{\infty}\sum_{m=0}^{\infty}\frac{1}{m+n+1}a_m b_n$$
$$<\{\sum_{m=0}^{\infty}[\frac{\pi}{\sin(\frac{\pi}{r})}-\frac{1}{13(m+1)(2m+1)^{1/s}}](m+\frac{1}{2})^{\frac{p}{s}-1}a_m^p\}^{\frac{1}{p}}$$
$$\times\{\sum_{n=0}^{\infty}[\frac{\pi}{\sin(\frac{\pi}{r})}-\frac{1}{13(n+1)(2n+1)^{1/r}}](n+\frac{1}{2})^{\frac{q}{r}-1}b_n^q\}^{\frac{1}{q}}; \quad (3.2.44)$$

$$\sum_{n=0}^{\infty}\frac{(n+\frac{1}{2})^{p/s-1}}{[\frac{\pi}{\sin(\pi/r)}-\frac{1}{13(n+1)(2n+1)^{1/r}}]^{p-1}}(\sum_{m=0}^{\infty}\frac{a_m}{m+n+1})^p$$
$$<\sum_{m=0}^{\infty}[\frac{\pi}{\sin(\frac{\pi}{r})}-\frac{1}{13(m+1)(2m+1)^{1/s}}](m+\frac{1}{2})^{\frac{p}{s}-1}a_m^p. \quad (3.2.45)$$

In particular (setting $\kappa(n)=\frac{\pi}{\sin(\pi/r)}$), we may deduce the following equivalent inequalities:

$$\tilde{I}<[\frac{\pi}{\sin(\frac{\pi}{r})}]^{\frac{1}{q}}$$
$$\times\{\sum_{m=0}^{\infty}[\frac{\pi}{\sin(\frac{\pi}{r})}-\frac{1}{13(m+1)(2m+1)^{1/s}}](m+\frac{1}{2})^{\frac{p}{s}-1}a_m^p\}^{\frac{1}{p}}$$
$$\times\{\sum_{n=0}^{\infty}(n+\frac{1}{2})^{\frac{q}{r}-1}b_n^q\}^{\frac{1}{q}}, \quad (3.2.46)$$

$$\sum_{n=0}^{\infty}(n+\frac{1}{2})^{\frac{p}{s}-1}(\sum_{m=0}^{\infty}\frac{a_m}{m+n+1})^p<[\frac{\pi}{\sin(\frac{\pi}{r})}]^{p-1}$$
$$\times\sum_{m=0}^{\infty}[\frac{\pi}{\sin(\frac{\pi}{r})}-\frac{1}{13(m+1)(2m+1)^{1/s}}](m+\frac{1}{2})^{\frac{p}{s}-1}a_m^p. \quad (3.2.47)$$

Note 3.2.17 By (3.2.46) and (3.2.47), we may deduce the following equivalent forms, which are the best extensions of (1.1.20) and its equivalent form:

$$\sum_{n=0}^{\infty}\sum_{m=0}^{\infty}\frac{a_m b_n}{m+n+1}<\frac{\pi}{\sin(\frac{\pi}{r})}\{\sum_{m=0}^{\infty}(m+\frac{1}{2})^{\frac{p}{s}-1}a_m^p\}^{\frac{1}{p}}$$
$$\times\{\sum_{n=0}^{\infty}(n+\frac{1}{2})^{\frac{q}{r}-1}b_n^q\}^{\frac{1}{q}}, \quad (3.2.48)$$

$$\sum_{n=0}^{\infty}(n+\frac{1}{2})^{\frac{p}{s}-1}(\sum_{m=0}^{\infty}\frac{a_m}{m+n+1})^p$$
$$<[\frac{\pi}{\sin(\frac{\pi}{r})}]^p\sum_{m=0}^{\infty}(m+\frac{1}{2})^{\frac{p}{s}-1}a_m^p. \quad (3.2.49)$$

The above results and methods are also consulted in (Yang NMJ 2000)[10], (Yang PJMS 2003)[11]. In 2007, (Wang CM 2007)[12] gave the following another strengthened version of (1.1.20):

$$\sum_{n=0}^{\infty}\sum_{m=0}^{\infty}\frac{a_m b_n}{m+n+1}<\{\sum_{m=0}^{\infty}[\frac{\pi}{\sin(\frac{\pi}{p})}-\frac{2(p-1)}{3(2m+1)^{1/p}}]a_m^p\}^{\frac{1}{p}}$$
$$\times\{\sum_{n=0}^{\infty}[\frac{\pi}{\sin(\frac{\pi}{p})}-\frac{2(q-1)}{3(2n+1)^{1/q}}]b_n^q\}^{\frac{1}{q}}. \quad (3.2.50)$$

3.2.5 A STRENGTHENED VERSION OF A H-L-P INEQUALITY

Setting

$$k(x,y)=\frac{1}{\max\{x,y\}} \quad (x,y>0),$$

we define the following weight coefficients:

$$\varpi(r,m):=\sum_{k=1}^{\infty}\frac{1}{\max\{m,k\}}(\frac{m}{k})^{\frac{1}{r}}$$
$$=m^{\frac{1}{r}-1}\sum_{k=1}^{m}k^{\frac{-1}{r}}+m^{\frac{1}{r}}\sum_{k=m+1}^{\infty}k^{\frac{-1}{r}-1},$$
$$\omega(s,n):=\sum_{k=1}^{\infty}\frac{1}{\max\{k,n\}}(\frac{n}{k})^{\frac{1}{s}},$$
$$r,s>1, m,n\in\mathbf{N}. \quad (3.2.51)$$

For fixed $m\in\mathbf{N}$, putting

$$f(x)=\frac{1}{\max\{m,x\}}(\frac{m}{x})^{\frac{1}{r}} \quad (x\in(0,\infty)),$$

then we find

$$\int_0^{\infty}f(x)dx=\int_0^{\infty}\frac{1}{\max\{m,x\}}(\frac{m}{x})^{\frac{1}{r}}dx$$
$$=\int_0^m\frac{1}{m}(\frac{m}{x})^{\frac{1}{r}}dx+\int_m^{\infty}\frac{1}{x}(\frac{m}{x})^{\frac{1}{r}}dx=\frac{r^2}{r-1}.$$

We set the following decomposition:

$$\varpi(r,m)=\frac{r^2}{r-1}-\frac{\theta(r,m)}{m^{1/s}}. \quad (3.2.52)$$

Since by (3.1.42) and (3.1.43), we have

$$\sum_{k=1}^{m}k^{\frac{-1}{r}}=\frac{r}{r-1}m^{1-\frac{1}{r}}-\frac{r}{r-1}$$
$$+\frac{1}{2}+\frac{1}{2m^{1/r}}-\frac{1}{r}\int_1^m P_1(x)\frac{1}{x^{1+1/r}}dx,$$
$$\sum_{k=m}^{\infty}k^{\frac{-1}{r}-1}=\frac{r}{m^{1/r}}+\frac{1}{2m^{1+1/r}}$$
$$-(1+\frac{1}{r})\int_m^{\infty}P_1(x)\frac{1}{x^{2+1/r}}dx,$$

then by (3.2.51) and (3.2.52), we obtain

$$\theta(r,m)=(\frac{r^2}{r-1}-\varpi(r,m))m^{1/s}$$
$$=\frac{r^2}{r-1}m^{1-1/r}+\frac{1}{m^{1/r}}-\sum_{k=1}^{m}k^{\frac{-1}{r}}-m\sum_{k=m}^{\infty}k^{\frac{-1}{r}-1}$$

$$=\frac{1}{r-1}+\frac{3}{2}+\frac{1}{r}\int_1^m P_1(x)\frac{1}{x^{1+1/r}}dx$$
$$+(1+\frac{1}{r})m\int_m^\infty P_1(x)\frac{1}{x^{2+1/r}}dx. \quad (3.2.53)$$

Lemma 3.2.18 For $r>1$, we have
$$\inf_{m\in\mathbb{N}}\{\theta(r,m)\}=\theta(r,1).$$

Proof For fixed $r>1$, replacing m by t in (3.2.52), then $\theta(r,t)$ is continuous in $(0,\infty)$ and differentiable in $(0,\infty)\setminus\mathbf{N}$ and
$$\theta_t'(r,t)=\frac{1}{r}P_1(t)\frac{1}{t^{1+1/r}}$$
$$+(1+\frac{1}{r})\int_t^\infty P_1(x)\frac{1}{x^{2+1/r}}dx-(1+\frac{1}{r})P_1(t)\frac{1}{t^{1+1/r}}$$
$$=-P_1(t)\frac{1}{t^{1+1/r}}+(1+\frac{1}{r})\int_t^\infty P_1(x)\frac{1}{x^{2+1/r}}dx$$
$$=\int_t^\infty \frac{1}{x^{1+1/r}}dP_1(x)$$
$$=\int_t^{[t]+1}\frac{1}{x^{1+1/r}}d(x-[x]-\frac{1}{2})+\int_{[t]+1}^\infty \frac{1}{x^{1+1/r}}dP_1(x)$$
$$=\int_t^{[t]+1}\frac{1}{x^{1+1/r}}dx+\sum_{k=[t]+1}^\infty\int_k^{k+1}\frac{1}{x^{1+1/r}}d(x-k-\frac{1}{2})$$
$$=\int_t^{[t]+1}\frac{1}{x^{1+1/r}}dx+\sum_{k=[t]+1}^\infty\int_k^{k+1}\frac{1}{x^{1+1/r}}dx>0.$$
Hence $\theta(r,t)$ $(t>0)$ is strictly increasing and $\inf_{m\in\mathbb{N}}\{\theta(r,m)\}=\theta(r,1)$. □

Lemma 3.2.19 For $r>1$, we have
$$\theta(r,1)>\inf_{r\in(1,\infty)}\{\theta(r,1)\}$$
$$=\theta(\infty,1)=2-\gamma=1.42278433^+. \quad (3.2.54)$$

Proof By (3.2.53) and (3.1.43), we find
$$\theta(r,1)=\frac{r^2}{r-1}-1-\sum_{k=2}^\infty k^{\frac{-1}{r}-1}$$
$$=r+\frac{1}{r-1}-(r+\frac{1}{4})\frac{1}{2^{1/r}}$$
$$+(1+\frac{1}{r})\int_2^\infty P_1(x)\frac{1}{x^{2+1/r}}dx. \quad (3.2.55)$$
Since the point that makes $h(x)=\frac{\ln x}{x^{2+1/r}}$ keeping maximal value is $x=e^{1/(2+\frac{1}{r})}<2$, and
$$h(x)>0, h'(x)<0\ (x\geq 2),$$
then by (2.2.10) (setting $q=0$), we have
$$\int_2^\infty P_1(x)\frac{\ln x}{x^{2+1/r}}dx<0.$$
By (3.1.46), we still have
$$\int_2^\infty P_1(x)\frac{1}{x^{2+1/r}}dx>-\frac{1}{12\cdot 2^{2+1/r}}.$$
We find
$$\theta'(r,1)=1-\frac{1}{(r-1)^2}-\frac{1}{2^{1/r}}-(\frac{1}{r}+\frac{1}{4r^2})\frac{\ln 2}{2^{1/r}}$$
$$-\frac{1}{r^2}\int_2^\infty P_1(x)\frac{1}{x^{2+1/r}}dx+(\frac{1}{r^2}+\frac{1}{r^3})\int_2^\infty P_1(x)\frac{\ln x}{x^{2+1/r}}dx$$
$$<(1-\frac{1}{2^{1/r}})-\frac{1}{(r-1)^2}-\frac{\ln 2}{r\cdot 2^{1/r}}-(\ln 2-\frac{1}{12})\frac{1}{r^2\cdot 2^{2+1/r}}$$
$$<\tau(r):=(1-\frac{1}{2^{1/r}})-\frac{1}{(r-1)^2}-\frac{\ln 2}{r2^{1/r}}.$$
Since it follows
$$\tau'(r)=-\frac{\ln 2}{r^2 2^{1/r}}+\frac{1}{(r-1)^3}+\frac{\ln 2}{r^2 2^{1/r}}-\frac{(\ln 2)^2}{r^3 2^{1/r}}$$
$$=\frac{1}{(r-1)^3}-\frac{(\ln 2)^2}{r^3 2^{1/r}}>0,$$
then we find
$$\theta'(r,1)<\tau(r)\leq\tau(\infty)=0.$$
Hence $\theta(r,1)$ is strictly decreasing in $r>1$ and then
$$\theta(r,1)>\inf_{r\in(1,\infty)}\{\theta(r,1)\}=\theta(\infty,1)$$
$$=\frac{3}{2}+\int_1^\infty P_1(x)\frac{1}{x^2}dx$$
$$=\frac{3}{2}+[\frac{1}{2}-\lim_{n\to\infty}(\sum_{k=1}^n\frac{1}{k}-\int_1^n\frac{1}{t}dt)]=2-\gamma.$$
Inequality (3.2.54) is valid. □

By (3.2.54), (3.2.52) and Lemma 3.2.19, we have

Lemma 3.2.20 For $r>1, \frac{1}{r}+\frac{1}{s}=1$, we have the following inequalities:
$$\varpi(r,m)<\tilde{\kappa}(m):=rs-\frac{2-\gamma}{m^{1-1/r}},\ m\in\mathbf{N}, \quad (3.2.56)$$
$$\omega(s,n)<\kappa(n):=rs-\frac{2-\gamma}{n^{1-1/s}},\ n\in\mathbf{N}, \quad (3.2.57)$$
where, $2-\gamma=1.42278433^+$ (γ is Euler constant).

By Corollary 3.1.2, we have the following theorem:

Theorem 3.2.21 If $p>1, r>1$, $\frac{1}{p}+\frac{1}{q}=1$, $\frac{1}{r}+\frac{1}{s}=1$, $a_m,b_n\geq 0$ $(m,n\in\mathbf{N})$, such that $0<\sum_{m=1}^\infty m^{\frac{p}{s}-1}a_m^p<\infty$ and $0<\sum_{n=1}^\infty n^{\frac{q}{r}-1}b_n^q<\infty$, then we have the following equivalent inequalities:
$$\sum_{n=1}^\infty\sum_{m=1}^\infty\frac{a_m b_n}{\max\{m,n\}}<rs\{\sum_{m=1}^\infty[1-\frac{2-\gamma}{rs\cdot m^{1/s}}]m^{\frac{p}{s}-1}a_m^p\}^{\frac{1}{p}}$$
$$\times\{\sum_{n=1}^\infty[1-\frac{2-\gamma}{rs\cdot n^{1/r}}]n^{\frac{q}{r}-1}b_n^q\}^{\frac{1}{q}}, \quad (3.2.58)$$
$$\sum_{n=1}^\infty\frac{n^{p/s-1}}{(1-\frac{2-\gamma}{rs\cdot n^{1/r}})^{p-1}}(\sum_{m=1}^\infty\frac{a_m}{\max\{m,n\}})^p$$
$$<(rs)^p\sum_{m=1}^\infty[1-\frac{2-\gamma}{rs\cdot m^{1/s}}]m^{\frac{p}{s}-1}a_m^p, \quad (3.2.59)$$

where, $2-\gamma = 1.42278433^{+}$ (γ is Euler constant).

In particular (setting $\kappa(n) = \frac{\pi}{\sin(\pi/r)}$), we can deduce the following equivalent inequalities:

$$\sum_{n=1}^{\infty}\sum_{m=1}^{\infty}\frac{a_m b_n}{\max\{m,n\}} < rs\{\sum_{m=1}^{\infty}[1-\frac{2-\gamma}{rs\cdot m^{1/s}}]m^{\frac{p}{s}-1}a_m^p\}^{\frac{1}{p}}$$

$$\times\{\sum_{n=1}^{\infty}n^{\frac{q}{r}-1}b_n^q\}^{\frac{1}{q}}, \qquad (3.2.60)$$

$$\sum_{n=1}^{\infty}n^{\frac{p}{s}-1}(\sum_{m=1}^{\infty}\frac{a_m}{\max\{m,n\}})^p$$

$$<(rs)^p\sum_{m=1}^{\infty}[1-\frac{2-\gamma}{rs\cdot m^{1/s}}]m^{\frac{p}{s}-1}a_m^p. \qquad (3.2.61)$$

Note 3.2.22 By (3.2.60) and (3.2.61), we can deduce the H-L-P inequality (3.1.30) and its equivalent form (3.1.31). It follows that (3.2.58) and (3.2.60) are strengthened versions of (3.1.30). We can see the above methods and results in (Kuang JMAA 2000) [13]-(Yang JSU 2006)[16].

3.2.6 A STRENGTHENED VERSION OF ANOTHER H-L-P INEQUALITY

If $k(x,y) = \frac{\ln(x/y)}{x-y}$ $(x,y>0)$, then we define the following weight coefficients:

$$\varpi(r,m) := m^{\frac{1}{r}}\sum_{k=1}^{\infty}\frac{\ln(m/k)}{m-k}k^{\frac{-1}{r}},$$

$$\omega(s,n) := n^{\frac{1}{s}}\sum_{k=1}^{\infty}\frac{\ln(k/n)}{k-n}k^{\frac{-1}{s}},$$

$$r,s>1, m,n\in\mathbf{N}. \qquad (3.2.62)$$

Setting $g_m(x) := \frac{\ln(m/x)}{m-x}$, and

$$f_m(x) := \frac{\ln(m/x)}{m-x}(\frac{1}{x})^{\frac{1}{r}} = g_m(x)(\frac{1}{x})^{\frac{1}{r}},$$

we find $f_m(1) = g_m(1) = \frac{\ln m}{m-1}$,

$$g_m'(x) = \frac{-1}{x(m-x)} + \frac{\ln(m/x)}{(m-x)^2},$$

$$g_m'(1) = \frac{-1}{m-1} + \frac{\ln m}{(m-1)^2},$$

$$f_m'(x) = g_m'(x)(\frac{1}{x})^{\frac{1}{r}} - \frac{1}{r}g_m(x)(\frac{1}{x})^{\frac{1}{r}+1},$$

$$f_m'(1) = g_m'(1) - \frac{1}{r}g_m(1)$$

$$= \frac{-1}{m-1} + \frac{\ln m}{(m-1)^2} - \frac{\ln m}{r(m-1)}, \qquad (3.2.63)$$

$$\int_1^{\infty}f_m(x)dx = m^{\frac{-1}{r}}\int_{\frac{1}{m}}^{\infty}\frac{-\ln u}{1-u}(\frac{1}{u})^{\frac{1}{r}}du$$

$$= m^{\frac{-1}{r}}\{[\frac{\pi}{\sin(\pi/r)}]^2 + \int_0^{\frac{1}{m}}\frac{\ln u}{1-u}(\frac{1}{u})^{\frac{1}{r}}du\}.$$

By (3.1.43), we have

$$m^{\frac{1}{r}}\sum_{k=1}^{\infty}f_m(k) = m^{\frac{1}{r}}$$

$$\times[\int_1^{\infty}f_m(x)dx + \frac{1}{2}f_m(1) + \int_1^{\infty}P_1(x)f_m'(x)dx]$$

$$= [\frac{\pi}{\sin(\pi/r)}]^2 + \int_0^{\frac{1}{m}}\frac{\ln u}{1-u}(\frac{1}{u})^{\frac{1}{r}}du + \frac{m^{1/r}\ln m}{2(m-1)}$$

$$+m^{\frac{1}{r}}\int_1^{\infty}P_1(x)f_m'(x)dx. \qquad (3.2.64)$$

We set the following decompositions:

$$\varpi(r,m) = [\frac{\pi}{\sin(\pi/r)}]^2 - \frac{\theta(r,m)}{m^{1/s}}, \qquad (3.2.65)$$

$$\theta(r,m) := m^{\frac{1}{s}}\{[\frac{\pi}{\sin(\pi/r)}]^2 - m^{\frac{1}{r}}\sum_{k=1}^{\infty}f_m(k)\}.$$

$$(3.2.66)$$

Then by (3.2.64), we have

$$\theta(r,m) = m^{\frac{1}{s}}\int_0^{1/m}\frac{-\ln u}{1-u}(\frac{1}{u})^{\frac{1}{r}}du - \frac{m\ln m}{2(m-1)}$$

$$-m\int_1^{\infty}P_1(x)f_m'(x)dx. \qquad (3.2.67)$$

We obtain

$$m^{\frac{1}{s}}\int_0^{\frac{1}{m}}\frac{-\ln u}{1-u}(\frac{1}{u})^{\frac{1}{r}}du = m^{\frac{1}{s}}\int_0^{\frac{1}{m}}\sum_{k=0}^{\infty}(-\ln u)u^{k-\frac{1}{r}}du$$

$$= m^{\frac{1}{s}}\sum_{k=0}^{\infty}\int_0^{\frac{1}{m}}(-\ln u)u^{k-\frac{1}{r}}du$$

$$= m^{\frac{1}{s}}\sum_{k=0}^{\infty}\frac{1}{k-1/r+1}\int_0^{\frac{1}{m}}(-\ln u)du^{k-\frac{1}{r}+1}$$

$$= s(\ln m + s) + \sum_{k=1}^{\infty}\frac{1}{k-\frac{1}{r}+1}[\ln m + \frac{1}{k-\frac{1}{r}+1}]\frac{1}{m^k},$$

$$-m\int_1^{\infty}P_1(x)f_m'(x)dx > \frac{m}{12}f_m'(1)$$

$$= \frac{m}{12}[\frac{-1}{m-1} + \frac{\ln m}{(m-1)^2} - \frac{\ln m}{r(m-1)}],$$

and then we find

$$\theta(r,m) > \eta(r,m)$$

$$:= \sum_{k=1}^{\infty}\frac{1}{k-1/r+1}[\ln m + \frac{1}{k-1/r+1}]\frac{1}{m^k} - \frac{m\ln m}{2(m-1)}$$

$$-\frac{m}{12(m-1)} + \frac{m\ln m}{12(m-1)^2}$$

$$+[s\ln m + s^2 + \frac{m\ln m}{12s(m-1)} - \frac{m\ln m}{12(m-1)}].$$

It is obvious that $\sum_{k=1}^{\infty}\frac{1}{k-1/r+1}[\ln m + \frac{1}{k-1/r+1}]\frac{1}{m^k}$ is decreasing for $r>1$. We find

$$[s\ln m + s^2 + \frac{m\ln m}{12s(m-1)} - \frac{m\ln m}{12(m-1)}]_s'$$

$$= \ln m[1 - \frac{m}{12s^2(m-1)}] + 2s > 0,$$

and then $[s\ln m + s^2) + \frac{m\ln m}{12s(m-1)} - \frac{m\ln m}{12(m-1)}]$ is decreasing for $r>1$. Hence it follows that

$$\theta(r,m) > \eta(\infty,m)$$

$$= \sum_{k=0}^{\infty} \frac{1}{k+1}[\ln m + \frac{1}{k+1}]\frac{1}{m^k}$$

$$-\frac{m\ln m}{2(m-1)} - \frac{m}{12(m-1)} + \frac{m\ln m}{12(m-1)^2}. \qquad (3.2.68)$$

By (3.2.65) and the symmetric property , we have the following lemma:

Lemma 3.2.23 If $r > 1, \frac{1}{r}+\frac{1}{s}=1$, then we have the following inequalities:

$$\varpi(r,m) < [\frac{\pi}{\sin(\pi/r)}]^2 - \frac{\eta(\infty,m)}{m^{1/s}},$$

$$\omega(s,n) < [\frac{\pi}{\sin(\pi/r)}]^2 - \frac{\eta(\infty,n)}{n^{1/r}}, \quad m,n \in \mathbf{N}. \qquad (3.2.69)$$

We estimate $\eta(\infty,m)$ in the following. We find

$$\eta(\infty,1) = \sum_{k=0}^{\infty} \frac{1}{(k+1)^2} - \frac{1}{2} - \frac{1}{24}$$

$$= \frac{\pi^2}{6} - \frac{13}{24} = 1.1032^+, \qquad (3.2.70)$$

$$\eta(\infty,2) = \ln 2\sum_{k=1}^{\infty} \frac{1}{k2^{k-1}} + \sum_{k=1}^{\infty} \frac{1}{k^2 2^{k-1}} - \frac{5}{6}\ln 2 - \frac{1}{6}$$

$$> \ln 2\sum_{k=1}^{10} \frac{1}{k2^{k-1}} + \sum_{k=1}^{10} \frac{1}{k^2 2^{k-1}} - \frac{1}{6}(1+5\ln 2)$$

$$> 1.381212678. \qquad (3.2.71)$$

Since $g_1(u) = \frac{\ln u}{u-1}$ $(u \in (0,\frac{1}{m}], m \ge 3)$ is concave, the tangent of $g_1(u)$ at $(\frac{1}{m}, g_1(\frac{1}{m}))$ is under the curve $g_1(u)$ and the intercept of this tangent at y-axis is

$$b = g_1(\tfrac{1}{m}) - \tfrac{1}{m} g_1'(\tfrac{1}{m}) = m[\frac{1}{m-1} + \frac{\ln m}{m-1} - \frac{\ln m}{(m-1)^2}].$$

Hence we have the following inequality:

$$\frac{1}{2}[\frac{1}{m-1} + 2\frac{\ln m}{m-1} - \frac{\ln m}{(m-1)^2}] < \int_0^{1/m} \frac{\ln u}{u-1}du.$$

By (3.2.68), we find

$$\eta(\infty,m) = m\int_0^{1/m} \frac{\ln u}{u-1}du$$

$$-\frac{m\ln m}{2(m-1)} - \frac{m}{12(m-1)} + \frac{m\ln m}{12(m-1)^2}$$

$$> \frac{m}{2}[\frac{1}{m-1} + 2\frac{\ln m}{m-1} - \frac{\ln m}{(m-1)^2}]$$

$$-\frac{m\ln m}{2(m-1)} - \frac{m}{12(m-1)} + \frac{m\ln m}{12(m-1)^2}$$

$$= \frac{5}{12}[\frac{m(\ln m+1)}{m-1} - \frac{m\ln m}{(m-1)^2} + \frac{m\ln m}{5(m-1)}]$$

$$= \frac{5}{12}h(m), \quad m \ge 3, \qquad (3.2.72)$$

where,

$$h(x) := \frac{x(\ln x+1)}{x-1} - \frac{x\ln x}{(x-1)^2} + \frac{x\ln x}{5(x-1)} \quad (x \ge 3).$$

We obtain

$$h'(x) = \frac{x^2-4x+3+2\ln x}{(x-1)^3} + \frac{x-1-\ln x}{5(x-1)^2}$$

$$> \frac{x^2-4x+3+2\ln x}{(x-1)^3} = \frac{q(x)}{(x-1)^3},$$

where, $q(x) := x^2 - 4x + 3 + 2\ln x$. Since for $x \ge 3$,

$$q'(x) = 2x - 4 + \tfrac{2}{x} > 2\sqrt{2x\cdot\tfrac{2}{x}} - 4 = 0,$$

then we find $q(x) \ge q(3) = 2\ln 3 > 0$, and $h'(x) > 0$. Hence by (3.2.72), we have

$$\eta(\infty,m) > \tfrac{5}{12}h(m) \ge \tfrac{5}{12}h(3)$$

$$= 1.10564^+, \quad m \ge 3. \qquad (3.2.73)$$

In view of (3.2.70), (3.2.71) and (3.2.73), we find

$$\min_{m\in\mathbb{N}}\{\eta(\infty,m)\} = \eta(\infty,1).$$

Hence by (3.2.69), we have

Lemma 3.2.24 For $r > 1, \frac{1}{r}+\frac{1}{s}=1$,

$$\theta = \frac{\pi^2}{6} - \frac{13}{24} = 1.1032^+,$$

we have the following inequalities:

$$\varpi(r,m) < \tilde{\kappa}(m) := [\frac{\pi}{\sin(\pi/r)}]^2 - \frac{\theta}{m^{1/s}},$$

$$\omega(s,n) < \kappa(n) := [\frac{\pi}{\sin(\pi/r)}]^2 - \frac{\theta}{n^{1/r}},$$

$$m,n \in \mathbf{N}. \qquad (3.2.74)$$

By Corollary 3.1.2, we have the following theorem:

Theorem 3.2.25 If $p>1, r>1$, $\frac{1}{p}+\frac{1}{q}=1$, $\frac{1}{r}+\frac{1}{s}=1$, $\theta = \frac{\pi^2}{6} - \frac{13}{24} = 1.1032^+$, $a_m, b_n \ge 0$ ($m,n \in \mathbf{N}$), such that $0 < \sum_{m=1}^{\infty} m^{\frac{p}{s}-1}a_m^p < \infty$ and $0 < \sum_{n=1}^{\infty} n^{\frac{q}{r}-1}b_n^q < \infty$, then we have the following equivalent inequalities:

$$\sum_{n=1}^{\infty}\sum_{m=1}^{\infty} \frac{\ln(m/n)}{m-n}a_m b_n$$

$$< \{\sum_{m=1}^{\infty}[(\frac{\pi}{\sin(\pi/r)})^2 - \frac{\theta}{m^{1/s}}]m^{\frac{p}{s}-1}a_m^p\}^{\frac{1}{p}}$$

$$\times\{\sum_{n=1}^{\infty}[[\frac{\pi}{\sin(\pi/r)}]^2 - \frac{\theta}{n^{1/r}}]n^{\frac{q}{r}-1}b_n^q\}^{\frac{1}{q}}, \qquad (3.2.75)$$

$$\sum_{n=1}^{\infty} \frac{n^{p/s-1}}{\{[\frac{\pi}{\sin(\pi/r)}]^2 - \frac{\theta}{n^{1/r}}\}^{p-1}}[\sum_{m=1}^{\infty} \frac{\ln(m/n)a_m}{m-n}]^p$$

$$< \sum_{m=1}^{\infty}\{[\frac{\pi}{\sin(\pi/r)}]^2 - \frac{\theta}{m^{1/s}}\}m^{\frac{p}{s}-1}a_m^p. \qquad (3.2.76)$$

In particular (setting $\kappa(n) = [\frac{\pi}{\sin(\pi/r)}]^2$), we can deduce the following equivalent inequalities:

$$\sum_{n=1}^{\infty}\sum_{m=1}^{\infty} \frac{\ln(m/n)}{m-n}a_m b_n < [\frac{\pi}{\sin(\pi/r)}]^{\frac{2}{q}}$$

$$\times\{\sum_{m=1}^{\infty}[[\frac{\pi}{\sin(\pi/r)}]^2-\frac{\theta}{m^{1/s}}]m^{\frac{p}{s}-1}a_m^p\}^{\frac{1}{p}}$$

$$\times\{\sum_{n=1}^{\infty}n^{\frac{q}{r}-1}b_n^q\}^{\frac{1}{q}}, \quad (3.2.77)$$

$$\sum_{n=1}^{\infty}n^{\frac{p}{s}-1}[\sum_{m=1}^{\infty}\frac{\ln(m/n)}{m-n}a_m]^p$$

$$<[\frac{\pi}{\sin(\pi/r)}]^{2(p-1)}\sum_{m=1}^{\infty}\{[\frac{\pi}{\sin(\pi/r)}]^2-\frac{\theta}{m^{1/s}}\}m^{\frac{p}{s}-1}a_m^p. \quad (3.2.78)$$

Note 3.2.26 By (3.2.77) and (3.2.78), we can deduce another H-L-P inequality (3.1.32) and the equivalent form (3.1.33). We see (Huang JM 2008)[17] gave some midway results of (3.2.75) ($\theta=\frac{1}{6}(5+\ln 2)$ $=0.9488^{+}$).

3.3 SOME REVERSE HILBERT-TYPE INEQUALITIES WITH THE HOMOGENEOUS KERNEL OF DEGREE -1

3.3.1 A REVERSE HARDY-HILBERT'S INEQUALITY

If $k(x,y)=\frac{1}{x+y}$ $(x,y>0)$, then we define the following weight coefficients:

$$\varpi(r,m):=\sum_{n=1}^{\infty}\frac{1}{m+n}(\frac{m}{n})^{\frac{1}{r}},$$

$$\omega(s,n):=\sum_{m=1}^{\infty}\frac{1}{m+n}(\frac{n}{m})^{\frac{1}{s}}, r,s>1, m,n\in\mathbf{N}. \quad (3.3.1)$$

For fixed $m\in\mathbf{N}$, setting

$$f_m(x)=\frac{1}{m+x}(\frac{m}{x})^{\frac{1}{r}} \ (x\in(0,\infty)),$$

then $f_m(x)$ is strictly decreasing, and

$$\varpi(r,m)=m^{\frac{1}{r}}\sum_{n=1}^{\infty}\frac{1}{m+n}n^{\frac{-1}{r}}>\int_1^{\infty}f_m(x)dx$$

$$=\int_{\frac{1}{m}}^{\infty}\frac{1}{1+u}u^{-\frac{1}{r}}du=\frac{\pi}{\sin(\pi/r)}-\int_0^{\frac{1}{m}}\frac{1}{1+u}u^{-\frac{1}{r}}du$$

$$>\frac{\pi}{\sin(\pi/r)}-\int_0^{\frac{1}{m}}u^{-\frac{1}{r}}du$$

$$=\tilde{\mu}(m):=\frac{\pi}{\sin(\pi/r)}-\frac{s}{m^{1/s}}$$

$$\geq\tilde{l}(r)=\frac{\pi}{\sin(\pi/s)}-s>0.$$

By (3.2.5), setting

$$\omega(s,n)<\kappa(n):=\frac{\pi}{\sin(\pi/r)}-\frac{1-\gamma}{n^{1/r}}.$$

In view of Theorem 3.1.1, we have the following theorem:

Theorem 3.3.1 If $0<p<1, r>1$, $\frac{1}{p}+\frac{1}{q}=1$, $\frac{1}{r}+\frac{1}{s}=1$, $a_m,b_n\geq 0$ $(m,n\in\mathbf{N})$, such that $0<\sum_{m=1}^{\infty}m^{\frac{p}{s}-1}a_m^p<\infty$ and $0<\sum_{n=1}^{\infty}n^{\frac{q}{r}-1}b_n^q<\infty$, then we have the following equivalent inequalities:

$$\sum_{n=1}^{\infty}\sum_{m=1}^{\infty}\frac{a_mb_n}{m+n}>\{\sum_{m=1}^{\infty}[\frac{\pi}{\sin(\pi/r)}-\frac{s}{m^{1/s}}]m^{\frac{p}{s}-1}a_m^p\}^{\frac{1}{p}}$$

$$\times\{\sum_{n=1}^{\infty}[\frac{\pi}{\sin(\pi/r)}-\frac{1-\gamma}{n^{1/r}}]n^{\frac{q}{r}-1}b_n^q\}^{\frac{1}{q}}, \quad (3.3.2)$$

$$\sum_{n=1}^{\infty}\frac{n^{p/s-1}}{[\frac{\pi}{\sin(\pi/r)}-\frac{1-\gamma}{n^{1/r}}]^{p-1}}(\sum_{m=1}^{\infty}\frac{a_m}{m+n})^p$$

$$>\sum_{m=1}^{\infty}[\frac{\pi}{\sin(\pi/r)}-\frac{s}{m^{1/s}}]m^{\frac{p}{s}-1}a_m^p, \quad (3.3.3)$$

$$\sum_{m=1}^{\infty}\frac{m^{q/r-1}}{[\frac{\pi}{\sin(\pi/r)}-\frac{s}{m^{1/s}}]^{q-1}}(\sum_{n=1}^{\infty}\frac{b_n}{m+n})^q$$

$$<\sum_{n=1}^{\infty}[\frac{\pi}{\sin(\pi/r)}-\frac{1-\gamma}{n^{1/r}}]n^{\frac{q}{r}-1}b_n^q. \quad (3.3.4)$$

In particular (setting $\kappa(n)=\frac{\pi}{\sin(\pi/r)}$), we may deduce the following equivalent inequalities:

$$\sum_{n=1}^{\infty}\sum_{m=1}^{\infty}\frac{1}{m+n}a_mb_n>[\frac{\pi}{\sin(\pi/r)}]^{\frac{1}{q}}$$

$$\times\{\sum_{m=1}^{\infty}[\frac{\pi}{\sin(\pi/r)}-\frac{s}{m^{1/s}}]m^{\frac{p}{s}-1}a_m^p\}^{\frac{1}{p}}\{\sum_{n=1}^{\infty}n^{\frac{q}{r}-1}b_n^q\}^{\frac{1}{q}}, \quad (3.3.5)$$

$$\sum_{n=1}^{\infty}n^{\frac{p}{s}-1}(\sum_{m=1}^{\infty}\frac{1}{m+n}a_m)^p$$

$$>[\frac{\pi}{\sin(\pi/r)}]^{p-1}\sum_{m=1}^{\infty}[\frac{\pi}{\sin(\pi/r)}-\frac{s}{m^{1/s}}]m^{\frac{p}{s}-1}a_m^p; \quad (3.3.6)$$

$$\sum_{m=1}^{\infty}\frac{m^{q/r-1}}{[\frac{\pi}{\sin(\pi/r)}-\frac{s}{m^{1/s}}]^{q-1}}(\sum_{n=1}^{\infty}\frac{1}{m+n}b_n)^q$$

$$<\frac{\pi}{\sin(\pi/r)}\sum_{n=1}^{\infty}n^{\frac{q}{r}-1}b_n^q. \quad (3.3.7)$$

Note 3.3.2 Some particular reverse Hardy-Hilbert's inequalities are also consulted in (Yang IMF 1006)[18], (Yang SPT 2006)[19].

3.3.2 A REVERSE MORE ACCURATE HARDY -HILBERT'S INEQUALITY

If $r>1, \frac{1}{r}+\frac{1}{s}=1$, $k(x,y)=\frac{1}{x+y}$, $u(m)=m+\frac{1}{2}$ $(m\in\mathbf{N}_0), n_0=0$, then we define the following weight coefficients:

$$\tilde{w}(r,m) := (m+\tfrac{1}{2})^{\frac{1}{r}} \sum_{n=0}^{\infty} \tfrac{1}{m+n+1} (n+\tfrac{1}{2})^{\frac{-1}{r}},$$

$$w(s,n) := (n+\tfrac{1}{2})^{\frac{1}{s}} \sum_{m=0}^{\infty} \tfrac{1}{m+n+1} (m+\tfrac{1}{2})^{\frac{-1}{s}},$$

$$m,n \in \mathbf{N}_0. \qquad (3.3.8)$$

For fixed $m \in \mathbf{N}_0, r > 1$, setting

$$f_m(x) := \tfrac{1}{x+m+1} (\tfrac{2m+1}{2x+1})^{\frac{1}{r}}, x \in (-\tfrac{1}{2}, \infty),$$

then it follows that

$$\tilde{w}(r,m) > \int_0^{\infty} f_m(x)dx = \int_{\frac{1}{2m+1}}^{\infty} \tfrac{1}{1+u} u^{-\frac{1}{r}} du$$

$$= \tfrac{\pi}{\sin(\pi/r)} - \int_0^{\frac{1}{2m+1}} \tfrac{1}{1+u} u^{-\frac{1}{r}} du$$

$$> \tfrac{\pi}{\sin(\pi/r)} - \int_0^{\frac{1}{2m+1}} u^{-\frac{1}{r}} du$$

$$= \tilde{\mu}(u(m)) := \tfrac{\pi}{\sin(\pi/r)} - \tfrac{s}{(2m+1)^{1/s}}$$

$$\geq \tilde{l}(r) = \tfrac{\pi}{\sin(\pi/s)} - s > 0.$$

By (3.2.35), setting

$$w(s,n) < \kappa(n) := \tfrac{\pi}{\sin(\pi/r)} - \tfrac{1}{10(2n+1)^{1+1/r}}.$$

By Theorem 3.1.7, we have the following theorem:

Theorem 3.3.3 If $0 < p < 1, r > 1$, $\frac{1}{p}+\frac{1}{q}=1$, $\frac{1}{r}+\frac{1}{s}=1$, $a_m, b_n \geq 0$ $(m,n \in \mathbf{N}_0)$, such that $0 < \sum_{m=0}^{\infty} (m+\frac{1}{2})^{\frac{p}{s}-1} a_m^p < \infty$ and $0 < \sum_{n=0}^{\infty} (n+\frac{1}{2})^{\frac{q}{r}-1} b_n^q < \infty$, then we have the following equivalent inequalities:

$$\tilde{I} := \sum_{n=0}^{\infty} \sum_{m=0}^{\infty} \tfrac{a_m b_n}{m+n+1}$$

$$> \{\sum_{m=0}^{\infty} [\tfrac{\pi}{\sin(\pi/r)} - \tfrac{s}{(2m+1)^{1/s}}](m+\tfrac{1}{2})^{\frac{p}{s}-1} a_m^p\}^{\frac{1}{p}}$$

$$\times \{\sum_{n=0}^{\infty} [\tfrac{\pi}{\sin(\pi/r)} - \tfrac{1}{10(2n+1)^{1+1/r}}](n+\tfrac{1}{2})^{\frac{q}{r}-1} b_n^q\}^{\frac{1}{q}}, \qquad (3.3.9)$$

$$\sum_{n=0}^{\infty} \frac{(n+\frac{1}{2})^{p/s-1}}{[\frac{\pi}{\sin(\pi/r)} - \frac{1}{10(2n+1)^{1+1/r}}]^{p-1}} (\sum_{m=0}^{\infty} \tfrac{a_m}{m+n+1})^p$$

$$> \sum_{m=0}^{\infty} [\tfrac{\pi}{\sin(\pi/r)} - \tfrac{s}{(2m+1)^{1/s}}](m+\tfrac{1}{2})^{\frac{p}{s}-1} a_m^p, \qquad (3.3.10)$$

$$\sum_{n=0}^{\infty} \frac{(m+\frac{1}{2})^{q/r-1}}{[\frac{\pi}{\sin(\pi/r)} - \frac{s}{(2m+1)^{1/s}}]^{q-1}} (\sum_{m=0}^{\infty} \tfrac{b_n}{m+n+1})^q$$

$$< \sum_{n=0}^{\infty} [\tfrac{\pi}{\sin(\pi/r)} - \tfrac{1}{10(2n+1)^{1+1/r}}](n+\tfrac{1}{2})^{\frac{q}{r}-1} b_n^q. \qquad (3.3.11)$$

In particular (setting $\kappa(u(n)) = \frac{\pi}{\sin(\pi/r)}$), we may deduce the following equivalent inequalities:

$$\tilde{I} > \{\sum_{m=0}^{\infty} [\tfrac{\pi}{\sin(\pi/r)} - \tfrac{s}{(2m+1)^{1/s}}](m+\tfrac{1}{2})^{\frac{p}{s}-1} a_m^p\}^{\frac{1}{p}}$$

$$\times \{\tfrac{\pi}{\sin(\pi/r)} \sum_{n=0}^{\infty} (n+\tfrac{1}{2})^{\frac{q}{r}-1} b_n^q\}^{\frac{1}{q}}, \qquad (3.3.12)$$

$$\sum_{n=0}^{\infty} (n+\tfrac{1}{2})^{\frac{p}{s}-1} (\sum_{m=0}^{\infty} \tfrac{a_m}{m+n+1})^p$$

$$> [\tfrac{\pi}{\sin(\pi/r)}]^{p-1} \sum_{m=0}^{\infty} [\tfrac{\pi}{\sin(\pi/r)} - \tfrac{s}{(2m+1)^{1/s}}](m+\tfrac{1}{2})^{\frac{p}{s}-1} a_m^p, \qquad (3.3.13)$$

$$\sum_{n=0}^{\infty} \frac{(m+\frac{1}{2})^{q/r-1}}{[\frac{\pi}{\sin(\pi/r)} - \frac{s}{(2m+1)^{1/s}}]^{q-1}} (\sum_{m=0}^{\infty} \tfrac{b_n}{m+n+1})^q$$

$$< \tfrac{\pi}{\sin(\pi/r)} \sum_{n=0}^{\infty} (n+\tfrac{1}{2})^{\frac{q}{r}-1} b_n^q. \qquad (3.3.14)$$

Note 3.3.4 Some particular reverse results are also consulted in (Yang JWNU 2005)[20], (Yang JIPAM 2006)[21].

3.3.3 A REVERSE H-L-P INEQUALITY

If $k(x,y) = \frac{1}{\max\{x,y\}}$ $(x,y>0)$, then we define the following weight coefficients:

$$\varpi(r,m) := \sum_{n=1}^{\infty} \tfrac{1}{\max\{m,n\}} (\tfrac{m}{n})^{\frac{1}{r}},$$

$$\omega(s,n) := \sum_{m=1}^{\infty} \tfrac{1}{\max\{m,n\}} (\tfrac{n}{m})^{\frac{1}{s}},$$

$$r,s > 1, m,n \in \mathbf{N}. \qquad (3.3.15)$$

For fixed $m \in \mathbf{N}$, setting

$$f_m(x) = \tfrac{1}{\max\{m,x\}} (\tfrac{m}{x})^{\frac{1}{r}} \ (x > 0),$$

we find

$$\varpi(r,m) > \int_1^{\infty} f(x)dx$$

$$= \int_0^{\infty} \tfrac{1}{\max\{m,x\}} (\tfrac{m}{x})^{\frac{1}{r}} dx - \int_0^1 \tfrac{1}{m} (\tfrac{m}{x})^{\frac{1}{r}} dx$$

$$= \tilde{\mu}(m) := rs - \tfrac{s}{m^{1/s}} \geq r > 0.$$

By (3.2.57), setting

$$\omega(s,n) < \kappa(n) := rs - \tfrac{2-\gamma}{n^{1/r}},$$

in view of Theorem 3.1.1, we have the following theorem:

Theorem 3.3.5 If $0 < p < 1, r > 1$, $\frac{1}{p}+\frac{1}{q}=1$, $\frac{1}{r}+\frac{1}{s}=1$, $a_m, b_n \geq 0$ $(m,n \in \mathbf{N})$, such that $0 < \sum_{m=1}^{\infty} m^{\frac{p}{s}-1} a_m^p < \infty$ and $0 < \sum_{n=1}^{\infty} n^{\frac{q}{r}-1} b_n^q < \infty$, then we have the following equivalent inequalities:

$$\sum_{n=1}^{\infty}\sum_{m=1}^{\infty}\frac{a_m b_n}{\max\{m,n\}} > rs\{\sum_{m=1}^{\infty}[1-\frac{1}{r\cdot m^{1/s}}]m^{\frac{p}{s}-1}a_m^p\}^{\frac{1}{p}}$$

$$\times\{\sum_{n=1}^{\infty}[1-\frac{2-\gamma}{rs\cdot n^{1/r}}]n^{\frac{q}{r}-1}b_n^q\}^{\frac{1}{q}}, \quad (3.3.16)$$

$$\sum_{n=1}^{\infty}\frac{n^{p/s-1}}{[1-\frac{2-\gamma}{rs\cdot n^{1/r}}]^{p-1}}(\sum_{m=1}^{\infty}\frac{a_m}{\max\{m,n\}})^p$$

$$> (rs)^p\sum_{m=1}^{\infty}[1-\frac{1}{r\cdot m^{1/s}}]m^{\frac{p}{s}-1}a_m^p, \quad (3.3.17)$$

$$\sum_{n=1}^{\infty}\frac{m^{q/r-1}}{[1-\frac{1}{r\cdot m^{1/s}}]^{q-1}}(\sum_{m=1}^{\infty}\frac{b_n}{\max\{m,n\}})^q$$

$$< (rs)^q\sum_{n=1}^{\infty}[1-\frac{2-\gamma}{rs\cdot n^{1/r}}]n^{\frac{q}{r}-1}b_n^q. \quad (3.3.18)$$

In particular (setting $\kappa(n) = rs$), we may deduce the following equivalent inequalities:

$$\sum_{n=1}^{\infty}\sum_{m=1}^{\infty}\frac{1}{\max\{m,n\}}a_m b_n > rs$$

$$\times\{\sum_{m=1}^{\infty}[1-\frac{1}{r\cdot m^{1/s}}]m^{\frac{p}{s}-1}a_m^p\}^{\frac{1}{p}}\{\sum_{n=1}^{\infty}n^{\frac{q}{r}-1}b_n^q\}^{\frac{1}{q}}, \quad (3.3.19)$$

$$\sum_{n=1}^{\infty}n^{\frac{p}{s}-1}(\sum_{m=1}^{\infty}\frac{1}{\max\{m,n\}}a_m)^p$$

$$> (rs)^p\sum_{m=1}^{\infty}[1-\frac{1}{r\cdot m^{1/s}}]m^{\frac{p}{s}-1}a_m^p, \quad (3.3.20)$$

$$\sum_{n=1}^{\infty}\frac{m^{q/r-1}}{[1-\frac{1}{r\cdot m^{1/s}}]^{q-1}}(\sum_{m=1}^{\infty}\frac{1}{\max\{m,n\}}b_n)^q$$

$$< (rs)^q\sum_{n=1}^{\infty}n^{\frac{q}{r}-1}b_n^q. \quad (3.3.21)$$

3.3.4 ANOTHER REVERSE FORM OF H-L-P INEQUALITY

If $k(x,y)=\frac{\ln(x/y)}{x-y}$ $(x,y>0)$, then we define the following weight coefficients:

$$\varpi(r,m):=\sum_{n=1}^{\infty}\frac{\ln(m/n)}{m-n}(\frac{m}{n})^{\frac{1}{r}},$$

$$\omega(s,n):=\sum_{m=1}^{\infty}\frac{\ln(m/n)}{m-n}(\frac{n}{m})^{\frac{1}{s}},$$

$$r,s>1, m,n\in \mathbf{N}. \quad (3.3.22)$$

For fixed $m\in\mathbf{N}$, setting

$$f_m(x):=\frac{\ln(m/x)}{m-x}(\frac{m}{x})^{\frac{1}{r}} \ (x>0),$$

we find

$$\int_1^{\infty}f_m(x)dx=\int_{\frac{1}{m}}^{\infty}\frac{-\ln u}{1-u}(\frac{1}{u})^{\frac{1}{r}}du$$

$$=[\frac{\pi}{\sin(\pi/r)}]^2-\int_0^{\frac{1}{m}}\frac{-\ln u}{1-u}(\frac{1}{u})^{\frac{1}{r}}du$$

$$=[\frac{\pi}{\sin(\pi/r)}]^2-\int_0^{\frac{1}{m}}(\frac{-\ln u}{1-u}u^{\frac{1}{2s}})(\frac{1}{u})^{\frac{1}{r}+\frac{1}{2s}}du.$$

For $m\ge 2$, by integral median theorem, there exists a constant $\eta_m\in(0,\frac{1}{m}]$, such that $\tilde{\theta}_m(s):=\frac{-\ln\eta_m}{1-\eta_m}\eta_m^{\frac{1}{2s}}$ and

$$\int_0^{\frac{1}{m}}[\frac{(-\ln u)}{1-u}u^{\frac{1}{2s}}](\frac{1}{u})^{\frac{1}{r}+\frac{1}{2s}}du$$

$$=\tilde{\theta}_m(s)\int_0^{\frac{1}{m}}(\frac{1}{u})^{\frac{1}{r}+\frac{1}{2s}}du=\frac{2s\tilde{\theta}_m(s)}{m^{1/2s}}.$$

Setting $\tilde{\theta}_1(s):=\frac{1}{2s}\int_0^1\frac{-\ln u}{1-u}(\frac{1}{u})^{\frac{1}{r}}du$, then it follows that

$$\varpi(r,m)>\int_1^{\infty}f_m(x)dx$$

$$=\tilde{\mu}(m)=[\frac{\pi}{\sin(\pi/r)}]^2-\frac{2s\tilde{\theta}_m(s)}{m^{1/2s}}$$

$$\ge\int_1^{\infty}\frac{-\ln u}{1-u}(\frac{1}{u})^{\frac{1}{r}}du>0.$$

By (3.2.74), setting

$$\omega(s,n)<\kappa(n):=[\frac{\pi}{\sin(\pi/r)}]^2-\frac{\theta}{n^{1/r}},$$

$$\theta=\frac{\pi^2}{6}-\frac{13}{24}=1.1032^+,$$

in view of Theorem 3.1.1, we have

Theorem 3.3.6 If $0<p<1, r>1$, $\frac{1}{p}+\frac{1}{q}=1$, $\frac{1}{r}+\frac{1}{s}=1, \theta=1.1032^+, \tilde{\theta}_m(s):=\frac{-\ln\eta_m}{1-\eta_m}\eta_m^{\frac{1}{2s}}$ ($\eta_m\in(0,\frac{1}{m}]$), $a_m,b_n\ge 0$ ($m,n\in\mathbf{N}$), such that $0<\sum_{m=1}^{\infty}m^{\frac{p}{s}-1}a_m^p<\infty$ and $0<\sum_{n=1}^{\infty}n^{\frac{q}{r}-1}b_n^q<\infty$, then we have the following equivalent inequalities:

$$\sum_{n=1}^{\infty}\sum_{m=1}^{\infty}\frac{\ln(m/n)}{m-n}a_m b_n$$

$$>\{\sum_{m=1}^{\infty}[(\frac{\pi}{\sin(\pi/r)})^2-\frac{2s\tilde{\theta}_m(s)}{m^{1/2s}}]m^{\frac{p}{s}-1}a_m^p\}^{\frac{1}{p}}$$

$$\times\{\sum_{n=1}^{\infty}[(\frac{\pi}{\sin(\pi/r)})^2-\frac{\theta}{n^{1/r}}]n^{\frac{q}{r}-1}b_n^q\}^{\frac{1}{q}}, \quad (3.3.23)$$

$$\sum_{n=1}^{\infty}\frac{n^{p/s-1}}{\{[\frac{\pi}{\sin(\pi/r)}]^2-\frac{\theta}{n^{1/r}}\}^{p-1}}[\sum_{m=1}^{\infty}\frac{\ln(m/n)}{m-n}a_m]^p$$

$$>\sum_{m=1}^{\infty}\{[\frac{\pi}{\sin(\pi/r)}]^2-\frac{2s\tilde{\theta}_m(s)}{m^{1/2s}}\}m^{\frac{p}{s}-1}a_m^p, \quad (3.3.24)$$

$$\sum_{m=1}^{\infty}\frac{m^{q/r-1}}{\{[\frac{\pi}{\sin(\pi/r)}]^2-\frac{2s\tilde{\theta}_m(s)}{m^{1/2s}}\}^{q-1}}[\sum_{n=1}^{\infty}\frac{\ln(m/n)}{m-n}b_n]^q$$

$$<\sum_{n=1}^{\infty}\{[\frac{\pi}{\sin(\pi/r)}]^2-\frac{\theta}{n^{1/r}}\}n^{\frac{q}{r}-1}b_n^q. \quad (3.3.25)$$

In particular (setting $\kappa(n)=[\frac{\pi}{\sin(\pi/r)}]^2$), we may deduce the following equivalent inequalities:

$$\sum_{n=1}^{\infty}\sum_{m=1}^{\infty}\frac{\ln(m/n)}{m-n}a_m b_n$$

$$>\{\sum_{m=1}^{\infty}[(\frac{\pi}{\sin(\pi/r)})^2-\frac{2s\tilde{\theta}_m(s)}{m^{1/2s}}]m^{\frac{p}{s}-1}a_m^p\}^{\frac{1}{p}}$$

$$\times\{[\frac{\pi}{\sin(\pi/r)}]^2\sum_{n=1}^{\infty}n^{\frac{q}{r}-1}b_n^q\}^{\frac{1}{q}}, \quad (3.3.26)$$

$$\sum_{n=1}^{\infty}n^{\frac{p}{s}-1}[\sum_{m=1}^{\infty}\frac{\ln(m/n)a_m}{m-n}]^p>[\frac{\pi}{\sin(\pi/r)}]^{2(p-1)}$$

$$\times\sum_{m=1}^{\infty}\{[\frac{\pi}{\sin(\pi/r)}]^2-\frac{2s\tilde{\theta}_m(s)}{m^{1/2s}}\}m^{\frac{p}{s}-1}a_m^p, \quad (3.3.27)$$

$$\sum_{m=1}^{\infty}\frac{m^{q/r-1}}{\{[\frac{\pi}{\sin(\pi/r)}]^2-\frac{2s\tilde{\theta}_m(s)}{m^{1/2s}}\}^{q-1}}[\sum_{n=1}^{\infty}\frac{\ln(m/n)}{m-n}b_n]^q$$

$$<[\frac{\pi}{\sin(\pi/r)}]^2\sum_{n=1}^{\infty}n^{\frac{q}{r}-1}b_n^q. \quad (3.3.28)$$

3.4 SOME MORE ACCURATE HILBERT-TYPE INEQUALITY WITH THE HOMOGENEOUS KERNEL OF DEGREE -1

3.4.1 A MORE ACCURATE HILBERT-TYPE INEQUALITY

If $k(x,y)=\frac{1}{\max\{x,y\}}$, $u(m)=m+\alpha$ $(\alpha>0)$, $n_0=0$, then we define the following weight coefficients:

$$\tilde{w}(r,m):=\sum_{k=0}^{\infty}\frac{1}{\max\{m,k\}+\alpha}(\frac{m+\alpha}{k+\alpha})^{\frac{1}{r}}$$

$$=\sum_{k=0}^{m-1}\frac{1}{m+\alpha}(\frac{m+\alpha}{k+\alpha})^{\frac{1}{r}}+\sum_{k=m}^{\infty}\frac{1}{k+\alpha}(\frac{m+\alpha}{k+\alpha})^{\frac{1}{r}},$$

$$w(s,n):=\sum_{k=0}^{\infty}\frac{1}{\max\{k,n\}+\alpha}(\frac{n+\alpha}{k+\alpha})^{\frac{1}{s}},$$

$$r,s>1,m,n\in\mathbf{N}_0. \quad (3.4.1)$$

By (3.1. 49) and (3.1.50), we obtain

$$\sum_{k=0}^{m-1}\frac{1}{m+\alpha}(\frac{m+\alpha}{k+\alpha})^{\frac{1}{r}}$$

$$=\frac{1}{(m+\alpha)^{1-1/r}}[\sum_{k=0}^{m}(\frac{1}{k+\alpha})^{\frac{1}{r}}-(\frac{1}{m+\alpha})^{\frac{1}{r}}]$$

$$<\frac{1}{(m+\alpha)^{1-1/r}}\{\int_0^m\frac{1}{(x+\alpha)^{1/r}}dx+\frac{1}{2}[\frac{1}{(m+\alpha)^{1/r}}+\frac{1}{\alpha^{1/r}}]$$

$$-\frac{1}{12r}[\frac{1}{(m+\alpha)^{1+1/r}}-\frac{1}{\alpha^{1+1/r}}]-\frac{1}{(m+\alpha)^{1/r}}\}$$

$$=s-\frac{1}{2(m+\alpha)}-\frac{1}{12r(m+\alpha)^2}$$

$$+\frac{1}{(m+\alpha)^{1-1/r}}(-\frac{s}{\alpha^{-1+1/r}}+\frac{1}{2\alpha^{1/r}}+\frac{1}{12r\alpha^{1+1/r}}),$$

$$\sum_{k=m}^{\infty}\frac{1}{k+\alpha}(\frac{m+\alpha}{k+\alpha})^{\frac{1}{r}}=(m+\alpha)^{\frac{1}{r}}\sum_{k=m}^{\infty}\frac{1}{(k+\alpha)^{1+1/r}}$$

$$<(m+\alpha)^{\frac{1}{r}}[\int_m^{\infty}\frac{1}{(x+\alpha)^{1+1/r}}dx$$

$$+\frac{1}{2(m+\alpha)^{1+1/r}}+\frac{1}{12}(1+\frac{1}{r})\frac{1}{(m+\alpha)^{2+1/r}}]$$

$$=r+\frac{1}{2(m+\alpha)}+\frac{1}{12}(1+\frac{1}{r})\frac{1}{(m+\alpha)^2}.$$

Then we have $\tilde{w}(r,m)<rs-R(r,m)$, where,

$$R(r,m):=\frac{-1}{12(m+\alpha)^2}$$

$$-[-\frac{s}{\alpha^{-1+1/r}}+\frac{1}{2\alpha^{1/r}}+\frac{1}{12r\alpha^{1+1/r}}]\frac{1}{(m+\alpha)^{1-1/r}}.$$

Since for $m\in\mathbf{N}_0$, we have

$$\frac{1}{12(m+\alpha)^2}=\frac{1}{12(m+\alpha)^{1+1/r}(m+\alpha)^{1-1/r}}\le\frac{1}{12\alpha^{1+1/r}(m+\alpha)^{1-1/r}},$$

Then we find

$$R(r,m)\ge-[-\frac{s}{\alpha^{-1+1/r}}+\frac{1}{2\alpha^{1/r}}$$

$$+\frac{1}{12\alpha^{1+1/r}}+\frac{1}{12r\alpha^{1+1/r}}]\frac{1}{(m+\alpha)^{1-1/r}}$$

$$=[s\alpha^2-\frac{1}{2}\alpha-\frac{1}{6}+\frac{1}{12s}]\frac{1}{\alpha^{1+1/r}(m+\alpha)^{1-1/r}}$$

$$>[\alpha^2-\frac{1}{2}\alpha-\frac{1}{6}]\frac{1}{\alpha^{1+1/r}(m+\alpha)^{1-1/r}}. \quad (3.4.2)$$

It is obvious that for $\alpha\ge\frac{3}{4}$, $R(r,m)>0$, and

$$\tilde{w}(r,m)<\tilde{\kappa}(u(m)):=rs.$$

In view of the property of symmetric, we have

$$w(s,n)<\kappa(u(n)):=rs \ (n\in\mathbf{N}_0).$$

By Theorem 3.1.7, it follows:

Theorem 3.4.1 If $p>1,r>1$, $\frac{1}{p}+\frac{1}{q}=1,\frac{1}{r}+\frac{1}{s}=1$, $\alpha\ge\frac{3}{4}$, $a_m,b_n\ge0$ $(m,n\in\mathbf{N}_0)$, such that $0<\sum_{m=0}^{\infty}(m+\alpha)^{\frac{p}{s}-1}a_m^p<\infty$ and $0<\sum_{n=0}^{\infty}(n+\alpha)^{\frac{q}{r}-1}b_n^q<\infty$, then we have the following equivalent inequalities:

$$\sum_{n=0}^{\infty}\sum_{m=0}^{\infty}\frac{a_m b_n}{\max\{m,n\}+\alpha}<rs\{\sum_{m=0}^{\infty}(m+\alpha)^{\frac{p}{s}-1}a_m^p\}^{\frac{1}{p}}$$

$$\times\{\sum_{n=0}^{\infty}(n+\alpha)^{\frac{q}{r}-1}b_n^q\}^{\frac{1}{q}}, \quad (3.4.3)$$

$$\sum_{n=0}^{\infty}(n+\alpha)^{\frac{p}{s}-1}(\sum_{m=0}^{\infty}\frac{a_m}{\max\{m,n\}+\alpha})^p$$

$$<(rs)^p\sum_{m=0}^{\infty}(m+\alpha)^{\frac{p}{s}-1}a_m^p. \quad (3.4.4)$$

For $r=s=2$, since by (3.4.2), we have

$$R(2,m)\ge[\alpha^2-\tfrac{1}{4}\alpha-\tfrac{1}{16}]\frac{2}{\alpha^{1+1/r}(m+\alpha)^{1-1/r}},$$

Then for $\alpha\ge\frac{1+\sqrt{5}}{8}$, $R(2,m)\ge 0$. We have the following corresponding results:

Theorem 3.4.2 If $p>1$, , $\frac{1}{p}+\frac{1}{q}=1$, $\alpha\ge\frac{1+\sqrt{5}}{8}$, $a_m,b_n\ge 0$ $(m,n\in \mathbf{N}_0)$, such that $0<\sum_{m=0}^{\infty}(m+\alpha)^{\frac{p}{2}-1}a_m^p<\infty$ and $0<\sum_{n=0}^{\infty}(n+\alpha)^{\frac{q}{2}-1}b_n^q<\infty$, then we have the following equivalent inequalities:

$$\sum_{n=0}^{\infty}\sum_{m=0}^{\infty}\frac{a_mb_n}{\max\{m,n\}+\alpha}<4\{\sum_{m=0}^{\infty}(m+\alpha)^{\frac{p}{2}-1}a_m^p\}^{\frac{1}{p}}$$

$$\times\{\sum_{n=0}^{\infty}(n+\alpha)^{\frac{q}{2}-1}b_n^q\}^{\frac{1}{q}}, \quad (3.4.5)$$

$$\sum_{n=0}^{\infty}(n+\alpha)^{\frac{p}{2}-1}(\sum_{m=0}^{\infty}\frac{a_m}{\max\{m,n\}+\alpha})^p$$

$$<4^p\sum_{m=0}^{\infty}(m+\alpha)^{\frac{p}{2}-1}a_m^p. \quad (3.4.6)$$

By the same way, in view of (3.1.49) and (3.1.50), we have the following inequalities:

$$\sum_{k=0}^{m-1}\frac{1}{m+\alpha}(\frac{m+\alpha}{k+\alpha})^{\frac{1}{r}}$$

$$=\frac{1}{(m+\alpha)^{1-1/r}}[\sum_{k=0}^{m}(\frac{1}{k+\alpha})^{\frac{1}{r}}-(\frac{1}{m+\alpha})^{\frac{1}{r}}]$$

$$>\frac{1}{(m+\alpha)^{1-1/r}}\{\int_0^m\frac{1}{(x+\alpha)^{1/r}}dx$$

$$+\frac{1}{2}[\frac{1}{(m+\alpha)^{1/r}}+\frac{1}{\alpha^{1/r}}]-\frac{1}{(m+\alpha)^{1/r}}\}$$

$$=s-\frac{1}{2(m+\alpha)}+\frac{1}{(m+\alpha)^{1-1/r}}(-\frac{s}{\alpha^{-1+1/r}}+\frac{1}{2\alpha^{1/r}}),$$

$$\sum_{k=m}^{\infty}\frac{1}{k+\alpha}(\frac{m+\alpha}{k+\alpha})^{\frac{1}{r}}=(m+\alpha)^{\frac{1}{r}}\sum_{k=m}^{\infty}\frac{1}{(k+\alpha)^{1+1/r}}$$

$$>(m+\alpha)^{\frac{1}{r}}[\int_m^{\infty}\frac{1}{(x+\alpha)^{1+1/r}}dx+\frac{1}{2(m+\alpha)^{1+1/r}}]$$

$$=r+\frac{1}{2(m+\alpha)}.$$

Hence we find $\tilde{w}(r,m)>rs-\tilde{R}(r,m)$, where,

$$\tilde{R}(r,m):=[s\alpha^{\frac{1}{s}}-\frac{1}{2\alpha^{1/r}}]\frac{1}{(m+\alpha)^{1/s}}<\frac{s\alpha^{1/s}}{(m+\alpha)^{1/s}}.$$

Then it follows

$$\tilde{w}(r,m)>\tilde{\mu}(u(m)):=rs-\frac{s\alpha^{1/s}}{(m+\alpha)^{1/s}}.$$

Setting $w(s,n)<\kappa(u(n)):=rs$, by Theorem 3.1.7, we have the following theorem:

Theorem 3.4.3 If $0<p<1, r>1$, $\frac{1}{p}+\frac{1}{q}=1$, $\frac{1}{r}+\frac{1}{s}=1$, $\alpha\ge\frac{3}{4}$, $a_m,b_n\ge 0$ $(m,n\in \mathbf{N}_0)$, such that $0<\sum_{m=0}^{\infty}(m+\alpha)^{\frac{p}{s}-1}a_m^p<\infty$ and $0<\sum_{n=0}^{\infty}(n+\alpha)^{\frac{q}{r}-1}b_n^q<\infty$, then we have the following equivalent inequalities:

$$\sum_{n=0}^{\infty}\sum_{m=0}^{\infty}\frac{1}{\max\{m,n\}+\alpha}a_mb_n$$

$$>rs\{\sum_{m=0}^{\infty}[1-\frac{\alpha^{1/s}}{r\cdot(m+\alpha)^{1/s}}](m+\alpha)^{\frac{p}{s}-1}a_m^p\}^{\frac{1}{p}}$$

$$\times\{\sum_{n=0}^{\infty}(n+\alpha)^{\frac{q}{r}-1}b_n^q\}^{\frac{1}{q}}, \quad (3.4.7)$$

$$\sum_{n=0}^{\infty}(n+\alpha)^{\frac{p}{s}-1}(\sum_{m=0}^{\infty}\frac{1}{\max\{m,n\}+\alpha}a_m)^p$$

$$>(rs)^p\sum_{m=0}^{\infty}[1-\frac{\alpha^{1/s}}{r(m+\alpha)^{1/s}}](m+\alpha)^{\frac{p}{s}-1}a_m^p, \quad (3.4.8)$$

$$\sum_{n=0}^{\infty}\frac{(m+\alpha)^{q/r-1}}{[1-\frac{\alpha^{1/s}}{r(m+\alpha)^{1/s}}]^{q-1}}(\sum_{m=0}^{\infty}\frac{1}{\max\{m,n\}+\alpha}b_n)^q$$

$$<(rs)^q\sum_{n=0}^{\infty}(n+\alpha)^{\frac{q}{r}-1}b_n^q. \quad (3.4.9)$$

In particular, for $r=s=2$, $\alpha\ge\frac{1+\sqrt{5}}{8}$, we have the following equivalent inequalities:

$$\sum_{n=0}^{\infty}\sum_{m=0}^{\infty}\frac{1}{\max\{m,n\}+\alpha}a_mb_n$$

$$>4\{\sum_{m=0}^{\infty}[1-\frac{\alpha^{1/2}}{2(m+\alpha)^{1/2}}](m+\alpha)^{\frac{p}{2}-1}a_m^p\}^{\frac{1}{p}}$$

$$\times\{\sum_{n=0}^{\infty}(n+\alpha)^{\frac{q}{2}-1}b_n^q\}^{\frac{1}{q}}, \quad (3.4.10)$$

$$\sum_{n=0}^{\infty}(n+\alpha)^{\frac{p}{2}-1}(\sum_{m=0}^{\infty}\frac{1}{\max\{m,n\}+\alpha}a_m)^p$$

$$>4^p\sum_{m=0}^{\infty}[1-\frac{\alpha^{1/2}}{2(m+\alpha)^{1/2}}](m+\alpha)^{\frac{p}{2}-1}a_m^p, \quad (3.4.11)$$

$$\sum_{n=0}^{\infty}\frac{(m+\alpha)^{q/2-1}}{[1-\frac{\alpha^{1/2}}{2(m+\alpha)^{1/2}}]^{q-1}}(\sum_{m=0}^{\infty}\frac{1}{\max\{m,n\}+\alpha}b_n)^q$$

$$<4^q\sum_{n=0}^{\infty}(n+\alpha)^{\frac{q}{2}-1}b_n^q. \quad (3.4.12)$$

Note 3.4.4 We can see some forepart results of Theorem 3.4.1 and Theorem 3.4.2 in (Yang JXTC 2005)[22], (Yang CM 2005)[23].

3.4.2 ANOTHER MORE ACCURATE HILBERT-TYPE INEQUALITY

If $k(x,y)=\frac{\ln(x/y)}{x-y}$, $u(m)=m+\alpha$ $(\alpha>0)$, then we define the following weight coefficients:

$$\tilde{w}(r,m):=\sum_{k=0}^{\infty}\frac{\ln(\frac{m+\alpha}{k+\alpha})}{m-k}\left(\frac{m+\alpha}{k+\alpha}\right)^{\frac{1}{r}},$$

$$w(s,n):=\sum_{k=0}^{\infty}\frac{\ln(\frac{k+\alpha}{n+\alpha})}{k-n}\left(\frac{n+\alpha}{k+\alpha}\right)^{\frac{1}{s}},$$

$$r,s>1, m,n\in\mathbf{N}_0. \tag{3.4.13}$$

For fixed $m\in\mathbf{N}_0$, setting

$$f_\alpha(x):=\frac{\ln(\frac{m+\alpha}{x+\alpha})}{m-x}\left(\frac{m+\alpha}{x+\alpha}\right)^{\frac{1}{r}}, g_\alpha(x):=\frac{\ln(\frac{m+\alpha}{x+\alpha})}{m-x},$$

then we find $g'_\alpha(x)<0,$, $g''_\alpha(x)>0$,

$$f_\alpha(0)=g_\alpha(0)\left(\frac{m+\alpha}{\alpha}\right)^{\frac{1}{r}},$$

$$f_\alpha'(0)=g'_\alpha(0)\left(\frac{m+\alpha}{\alpha}\right)^{\frac{1}{r}}-\frac{1}{r\alpha}g_\alpha(0)\left(\frac{m+\alpha}{\alpha}\right)^{\frac{1}{r}},$$

$$\int_{-\alpha}^{\infty}f_\alpha(x)dx=\int_{-\alpha}^{\infty}\frac{\ln(\frac{m+\alpha}{x+\alpha})}{m-x}\left(\frac{m+\alpha}{x+\alpha}\right)^{\frac{1}{r}}dx$$

$$=\int_0^\infty\frac{\ln u}{u-1}u^{\frac{-1}{r}}du=\left[\frac{\pi}{\sin(\pi/r)}\right]^2,$$

$$\int_{-\alpha}^{0}f_\alpha(x)dx$$

$$=\frac{r}{r-1}(m+\alpha)^{\frac{1}{r}}\int_{-\alpha}^{0}g_\alpha(x)d(x+\alpha)^{1-\frac{1}{r}}$$

$$=\frac{r}{r-1}(m+\alpha)^{\frac{1}{r}}[g_\alpha(0)\alpha^{1-\frac{1}{r}}$$

$$-\int_{-\alpha}^{0}g'_\alpha(x)(x+\alpha)^{1-\frac{1}{r}}dx]$$

$$=\frac{r}{r-1}(m+\alpha)^{\frac{1}{r}}[g_\alpha(0)\alpha^{1-\frac{1}{r}}$$

$$-\frac{r}{2r-1}\int_{-\alpha}^{0}g'_\alpha(x)d(x+\alpha)^{2-\frac{1}{r}}]$$

$$=\frac{r}{r-1}(m+\alpha)^{\frac{1}{r}}[g_\alpha(0)\alpha^{1-\frac{1}{r}}-\frac{r}{2r-1}g'_\alpha(0)\alpha^{2-\frac{1}{r}}$$

$$+\frac{r}{2r-1}\int_{-\alpha}^{0}g''_\alpha(x)(x+\alpha)^{2-\frac{1}{r}}dx]$$

$$>\frac{r}{r-1}(m+\alpha)^{\frac{1}{r}}[g_\alpha(0)\alpha^{1-\frac{1}{r}}-\frac{r}{2r-1}g'_\alpha(0)\alpha^{2-\frac{1}{r}}].$$

By (3.1.50), we obtain

$$\tilde{w}(r,m)<\int_0^\infty f_\alpha(x)dx+\frac{1}{2}f_\alpha(0)-\frac{1}{12}f'_\alpha(0)$$

$$=\int_{-\alpha}^{\infty}f_\alpha(x)dx$$

$$-[\int_{-\alpha}^{0}f_\alpha(x)dx-\frac{1}{2}f_\alpha(0)+\frac{1}{12}f'_\alpha(0)]$$

$$=\left[\frac{\pi}{\sin(\pi/r)}\right]^2-\left(\frac{m+\alpha}{\alpha}\right)^{\frac{1}{r}}R(r,m),$$

$$R(r,m):=\left(\frac{r\alpha}{r-1}-\frac{1}{2}-\frac{1}{12r\alpha}\right)g(0)$$

$$+\left[\frac{r^2\alpha^2}{(r-1)(2r-1)}-\frac{1}{12}\right](-g'(0)). \tag{3.4.14}$$

For $\alpha\ge\frac{1}{2}$, we have

$$\frac{r\alpha}{r-1}-\frac{1}{2}-\frac{1}{12r\alpha}=\frac{6r^2\alpha(2\alpha-1)+(6\alpha-1)r+1}{12r(r-1)\alpha}>0,$$

$$\frac{r^2\alpha^2}{(r-1)(2r-1)}-\frac{1}{12}=\frac{2r^2(6\alpha^2-1)+3r-1}{12(r-1)(2r-1)}>0.$$

Hence we find $R(r,m)>0$ and

$$\tilde{w}(r,m)<\left[\frac{\pi}{\sin(\pi/r)}\right]^2.$$

In view of the symmetric property, we have $w(s,n)<\left[\frac{\pi}{\sin(\pi/r)}\right]^2$, then by Theorem 3.1.7, it follows:

Theorem 3.4.5 If $p>1, r>1, \frac{1}{p}+\frac{1}{q}=1, \frac{1}{r}+\frac{1}{s}=1$, $\alpha\ge\frac{1}{2}$, $a_m,b_n\ge 0$ $(m,n\in\mathbf{N}_0)$, such that $0<\sum_{m=0}^{\infty}(m+\alpha)^{\frac{p}{s}-1}a_m^p<\infty$ and $0<\sum_{n=0}^{\infty}(n+\alpha)^{\frac{q}{r}-1}b_n^q<\infty$, then we have the following equivalent inequalities:

$$\sum_{n=0}^{\infty}\sum_{m=0}^{\infty}\frac{\ln(\frac{m+\alpha}{n+\alpha})a_mb_n}{m-n}<\left[\frac{\pi}{\sin(\frac{\pi}{r})}\right]^2\{\sum_{m=0}^{\infty}(m+\alpha)^{\frac{p}{s}-1}a_m^p\}^{\frac{1}{p}}$$

$$\times\{\sum_{n=0}^{\infty}(n+\alpha)^{\frac{q}{r}-1}b_n^q\}^{\frac{1}{q}}, \tag{3.4.15}$$

$$\sum_{n=0}^{\infty}(n+\alpha)^{\frac{p}{s}-1}\left(\sum_{m=0}^{\infty}\frac{\ln(\frac{m+\alpha}{n+\alpha})a_m}{m-n}\right)^p$$

$$<\left[\frac{\pi}{\sin(\frac{\pi}{r})}\right]^{2p}\sum_{m=0}^{\infty}(m+\alpha)^{\frac{p}{s}-1}a_m^p. \tag{3.4.16}$$

For $r=s=2$, since by (3.4.12), we find

$$R(2,m)=\left(\frac{48\alpha^2-12\alpha-1}{24\alpha}\right)g(0)$$

$$+\left[\frac{4\alpha^2}{3}-\frac{1}{12}\right](-g'(0)),$$

For $\alpha\ge\frac{3+\sqrt{21}}{24}$, $R(2,m)>0$, and we have the following corresponding theorem:

Theorem 3.4.6 If $p>1$, $\frac{1}{p}+\frac{1}{q}=1$, $\alpha\ge\frac{3+\sqrt{21}}{24}=0.31594^+$, $a_m,b_n\ge 0$ $(m,n\in\mathbf{N}_0)$, such that $0<\sum_{m=0}^{\infty}(m+\alpha)^{\frac{p}{2}-1}a_m^p<\infty$ and $0<\sum_{n=0}^{\infty}(n+\alpha)^{\frac{q}{2}-1}b_n^q<\infty$, then we have the following equivalent inequalities:

$$\sum_{n=0}^{\infty}\sum_{m=0}^{\infty}\frac{\ln(\frac{m+\alpha}{n+\alpha})a_mb_n}{m-n}<\pi^2\{\sum_{m=0}^{\infty}(m+\alpha)^{\frac{p}{2}-1}a_m^p\}^{\frac{1}{p}}$$

$$\times\{\sum_{n=0}^{\infty}(n+\alpha)^{\frac{q}{2}-1}b_n^q\}^{\frac{1}{q}}, \quad (3.4.17)$$

$$\sum_{n=0}^{\infty}(n+\alpha)^{\frac{p}{2}-1}[\sum_{m=0}^{\infty}\frac{\ln(\frac{m+\alpha}{n+\alpha})a_m}{m-n}]^p$$

$$<\pi^{2p}\sum_{m=0}^{\infty}(m+\alpha)^{\frac{p}{2}-1}a_m^p. \quad (3.4.18)$$

In view of the integral median theorem, there exists a constant $\eta_m(\alpha)\in(0,\frac{\alpha}{m+\alpha}]$, such that

$$\tilde{\theta}_m(s,\alpha):=\frac{-\ln\eta_m(\alpha)}{1-\eta_m(\alpha)}\eta_m^{\frac{1}{2s}}(\alpha)>0, m\ge 1,$$

$$\tilde{\theta}_0(s,\alpha):=\frac{1}{2s}\int_0^1\frac{-\ln u}{1-u}(\frac{1}{u})^{\frac{1}{r}}du>0.$$

And we have the following inequality:

$$\tilde{w}(r,m)>\int_0^{\infty}f_\alpha(x)dx$$

$$=\int_{-\alpha}^{\infty}f_\alpha(x)dx-\int_{-\alpha}^{0}f_\alpha(x)dx$$

$$=[\frac{\pi}{\sin(\pi/r)}]^2-\int_0^{\alpha/(m+\alpha)}(\frac{-\ln u}{1-u}u^{\frac{1}{2s}})(\frac{1}{u})^{\frac{1}{r}+\frac{1}{2s}}du$$

$$=\tilde{\mu}(u(m)):=[\frac{\pi}{\sin(\pi/r)}]^2-2s\tilde{\theta}_m(s,\alpha)(\frac{\alpha}{m+\alpha})^{\frac{1}{2s}}$$

$$\ge\int_1^{\infty}\frac{-\ln u}{1-u}(\frac{1}{u})^{\frac{1}{r}}du.$$

Setting

$$w(s,n)<\kappa(u(n)):=[\frac{\pi}{\sin(\pi/r)}]^2,$$

by Theorem 3.1.7, we have the following theorem:

Theorem 3.4.7 f $0<p<1, r>1$, $\frac{1}{p}+\frac{1}{q}=1$, $\frac{1}{r}+\frac{1}{s}=1, \alpha\ge\frac{1}{2}, a_m, b_n\ge 0\ (m,n\in\mathbf{N}_0)$, such that $0<\sum_{m=0}^{\infty}(m+\alpha)^{\frac{p}{s}-1}a_m^p<\infty$ and $0<\sum_{n=0}^{\infty}(n+\alpha)^{\frac{q}{r}-1}b_n^q<\infty$, then we have the following equivalent inequalities:

$$\tilde{J}:=\sum_{n=0}^{\infty}\sum_{m=0}^{\infty}\frac{\ln(\frac{m+\alpha}{n+\alpha})}{m-n}a_m b_n$$

$$>\{\sum_{m=0}^{\infty}[(\frac{\pi}{\sin(\frac{\pi}{r})})^2-\frac{2s\tilde{\theta}_m(s,\alpha)\alpha^{1/(2s)}}{(m+\alpha)^{1/(2s)}}](m+\alpha)^{\frac{p}{s}-1}a_m^p\}^{\frac{1}{p}}$$

$$\times\{[\frac{\pi}{\sin(\frac{\pi}{r})}]^2\sum_{n=0}^{\infty}(n+\alpha)^{\frac{q}{r}-1}b_n^q\}^{\frac{1}{q}}, \quad (3.4.19)$$

$$\sum_{n=0}^{\infty}(n+\alpha)^{\frac{p}{s}-1}[\sum_{m=0}^{\infty}\frac{\ln(\frac{m+\alpha}{n+\alpha})a_m}{m-n}]^p>[\frac{\pi}{\sin(\frac{\pi}{r})}]^{2(p-1)}$$

$$\times\sum_{m=0}^{\infty}\{[\frac{\pi}{\sin(\frac{\pi}{r})}]^2-\frac{2s\tilde{\theta}_m(s,\alpha)\alpha^{1/(2s)}}{(m+\alpha)^{1/(2s)}}\}(m+\alpha)^{\frac{p}{s}-1}a_m^p, \quad (3.4.20)$$

$$\sum_{n=0}^{\infty}\frac{(m+\alpha)^{q/r-1}}{\{[\frac{\pi}{\sin(\pi/r)}]^2-\frac{2s\tilde{\theta}_m(s,\alpha)\alpha^{1/(2s)}}{(m+\alpha)^{1/(2s)}}\}^{q-1}}[\sum_{m=0}^{\infty}\frac{\ln(\frac{m+\alpha}{n+\alpha})b_n}{m-n}]^q$$

$$<[\frac{\pi}{\sin(\frac{\pi}{r})}]^2\sum_{n=0}^{\infty}(n+\alpha)^{\frac{q}{r}-1}b_n^q, \quad (3.4.21)$$

In particular, for $r=s=2$, we have the following equivalent inequalities:

$$\tilde{J}>\pi^2\{\sum_{m=0}^{\infty}[1-\frac{4\tilde{\theta}_m(2,\alpha)\alpha^{1/4}}{\pi^2(m+\alpha)^{1/5}}](m+\alpha)^{\frac{p}{2}-1}a_m^p\}^{\frac{1}{p}}$$

$$\times\{\sum_{n=0}^{\infty}(n+\alpha)^{\frac{q}{2}-1}b_n^q\}^{\frac{1}{q}}, \quad (3.4.22)$$

$$\sum_{n=0}^{\infty}(n+\alpha)^{\frac{p}{2}-1}[\sum_{m=0}^{\infty}\frac{\ln(\frac{m+\alpha}{n+\alpha})}{m-n}a_m]^p$$

$$>\pi^{2p}\sum_{m=0}^{\infty}\{1-\frac{4\tilde{\theta}_m(2,\alpha)\alpha^{1/4}}{\pi^2(m+\alpha)^{1/4}}\}(m+\alpha)^{\frac{p}{2}-1}a_m^p, \quad (3.4.23)$$

$$\sum_{n=0}^{\infty}\frac{(m+\alpha)^{q/2-1}}{\{1-\frac{4\tilde{\theta}_m(2,\alpha)\alpha^{1/4}}{\pi^2(m+\alpha)^{1/4}}\}^{q-1}}[\sum_{m=0}^{\infty}\frac{\ln(\frac{m+\alpha}{n+\alpha})}{m-n}b_n]^q$$

$$<\pi^{2q}\sum_{n=0}^{\infty}(n+\alpha)^{\frac{q}{2}-1}b_n^q, \quad (3.4.24)$$

where,

$$\tilde{\theta}_m(s,\alpha)=\frac{-\ln\eta_m(\alpha)}{1-\eta_m(\alpha)}\eta_m^{\frac{1}{2s}}(\alpha)>0$$

$$(\eta_m(\alpha)\in(0,\frac{\alpha}{m+\alpha}]), m\ge 1;$$

$$\tilde{\theta}_0(s,\alpha)=\frac{1}{2s}\int_0^1\frac{-\ln u}{1-u}(\frac{1}{u})^{\frac{1}{r}}du>0.$$

Note 3.4.8 Some results of Theorem 3.4.6 and Theorem 3.4.7 are consulted in (Yang AMS 2006)[24]-(Yang IJMA 2007)[28].

3.4.3 A MORE ACCURATE MULHOLLAND'S INEQUALITY

If $u(m)=\ln\sqrt{\alpha}m\ (\alpha>0), n_0=1$, then we define the following weight coefficients

$$\tilde{w}(r,m):=\sum_{k=1}^{\infty}\frac{1}{k\ln\alpha km}(\frac{\ln\sqrt{\alpha}m}{\ln\sqrt{\alpha}k})^{\frac{1}{r}},$$

$$w(s,n):=\sum_{k=1}^{\infty}\frac{1}{k\ln\alpha kn}(\frac{\ln\sqrt{\alpha}n}{\ln\sqrt{\alpha}k})^{\frac{1}{s}},$$

$$r,s>1, m,n\in\mathbf{N} \quad (3.4.25)$$

For fixed $m\in\mathbf{N}$, setting

$$f_\alpha(x):=\frac{1}{x\ln\alpha mx}(\frac{\ln\sqrt{\alpha}m}{\ln\sqrt{\alpha}x})^{\frac{1}{r}},\ x\in[1,\infty),$$

then we find $f_\alpha(1)=\frac{1}{\ln\alpha m}(\frac{\ln\sqrt{\alpha}m}{\ln\sqrt{\alpha}})^{\frac{1}{r}}$,

$$f_\alpha'(1)=-[(1+\frac{1}{r\ln\sqrt{\alpha}})\frac{1}{\ln\alpha m}+\frac{1}{\ln^2\alpha m}](\frac{\ln\sqrt{\alpha}m}{\ln\sqrt{\alpha}})^{\frac{1}{r}}.$$

Setting $u=(\ln\sqrt{\alpha}x)/(\ln\sqrt{\alpha}m)$, integration by parts, we obtain

$$\int_1^\infty f_\alpha(x)dx$$
$$=\int_1^\infty \frac{1}{x\ln\alpha m[1+(x\ln\sqrt{\alpha})/(\ln\sqrt{\alpha}m)]}\left(\frac{\ln\sqrt{\alpha}m}{\ln\sqrt{\alpha}x}\right)^{\frac{1}{r}}dx$$
$$=\int_{\frac{\ln\sqrt{\alpha}}{\ln\sqrt{\alpha}m}}^\infty \frac{1}{1+u}\left(\frac{1}{u}\right)^{\frac{1}{r}}du$$
$$=\frac{\pi}{\sin(\frac{\pi}{r})}-\int_0^{\frac{\ln\sqrt{\alpha}}{\ln\sqrt{\alpha}m}}\frac{1}{1+u}\left(\frac{1}{u}\right)^{\frac{1}{r}}du,$$
$$\int_0^{\frac{\ln\sqrt{\alpha}}{\ln\sqrt{\alpha}m}}\frac{1}{1+u}\left(\frac{1}{u}\right)^{\frac{1}{r}}du$$
$$=\frac{r}{r-1}\int_0^{\frac{\ln\sqrt{\alpha}}{\ln\sqrt{\alpha}m}}\frac{1}{1+u}du^{-\frac{1}{r}+1}$$
$$=\frac{r\ln\sqrt{\alpha}m}{(r-1)\ln\alpha m}\left(\frac{\ln\sqrt{\alpha}}{\ln\sqrt{\alpha}m}\right)^{1-\frac{1}{r}}$$
$$+\frac{r^2}{(r-1)(2r-1)}\int_0^{\frac{\ln\sqrt{\alpha}}{\ln\sqrt{\alpha}m}}\frac{du^{-\frac{1}{r}+2}}{(1+u)^2}$$
$$>\left[\frac{r\ln\sqrt{\alpha}}{(r-1)\ln\alpha m}+\frac{r^2}{(r-1)(2r-1)}\left(\frac{\ln\sqrt{\alpha}}{\ln\alpha m}\right)^2\right]\left(\frac{\ln\sqrt{\alpha}}{\ln\sqrt{\alpha}m}\right)^{-\frac{1}{r}}.$$

Then by (3.1.50), we have

$$\tilde{w}(r,m)<\int_1^\infty f_\alpha(x)dx+\frac{1}{2}f_\alpha(1)-\frac{1}{12}f_\alpha'(1)$$
$$=\frac{\pi}{\sin(\frac{\pi}{r})}-\left\{\int_0^{\frac{\ln\sqrt{\alpha}}{\ln\sqrt{\alpha}m}}\frac{1}{1+u}\left(\frac{1}{u}\right)^{\frac{1}{r}}du-\frac{1}{2\ln\alpha m}\left(\frac{\ln\sqrt{\alpha}m}{\ln\sqrt{\alpha}}\right)^{\frac{1}{r}}\right.$$
$$\left.-\left[\left(\frac{1}{12}+\frac{1}{12r\ln\sqrt{\alpha}}\right)\frac{1}{\ln\alpha m}+\frac{1}{12\ln^2\alpha m}\right]\left(\frac{\ln\sqrt{\alpha}m}{\ln\sqrt{\alpha}}\right)^{\frac{1}{r}}\right\}$$
$$<\frac{\pi}{\sin(\frac{\pi}{r})}-\left\{\left[\frac{r\ln\sqrt{\alpha}}{(r-1)\ln\alpha m}+\frac{r^2}{(r-1)(2r-1)}\left(\frac{\ln\sqrt{\alpha}}{\ln\alpha m}\right)^2\right]\left(\frac{\ln\sqrt{\alpha}}{\ln\sqrt{\alpha}m}\right)^{-\frac{1}{r}}\right.$$
$$-\frac{1}{2\ln\alpha m}\left(\frac{\ln\sqrt{\alpha}m}{\ln\sqrt{\alpha}}\right)^{\frac{1}{r}}$$
$$\left.-\left[\left(\frac{1}{12}+\frac{1}{12r\ln\sqrt{\alpha}}\right)\frac{1}{\ln\alpha m}+\frac{1}{12\ln^2\alpha m}\right]\left(\frac{\ln\sqrt{\alpha}m}{\ln\sqrt{\alpha}}\right)^{\frac{1}{r}}\right\}$$
$$=\frac{\pi}{\sin(\frac{\pi}{r})}-R(r,m)\left(\frac{\ln\sqrt{\alpha}m}{\ln\sqrt{\alpha}}\right)^{\frac{1}{r}},$$

$$R(r,m):=\left[\frac{r\ln\sqrt{\alpha}}{(r-1)}-\frac{7}{12}-\frac{1}{12r\ln\sqrt{\alpha}}\right]\frac{1}{\ln\alpha m}$$
$$+\left[\frac{r^2\ln^2\sqrt{\alpha}}{(r-1)(2r-1)}-\frac{1}{12}\right]\left(\frac{1}{\ln\alpha m}\right)^2. \qquad (3.4.26)$$

Since for $r>1,\alpha\geq e^{7/6}$, we have

$$\frac{r\ln\sqrt{\alpha}}{r-1}-\frac{7}{12}-\frac{1}{12r\ln\sqrt{\alpha}}$$
$$=\frac{r^2(12\ln\sqrt{\alpha}-7)\ln\sqrt{\alpha}+r(7\ln\sqrt{\alpha}-1)+1}{12r(r-1)\ln\sqrt{\alpha}}>0,$$
$$\frac{r^2\ln^2\sqrt{\alpha}}{(r-1)(2r-1)}-\frac{1}{12}\geq\frac{r^2}{4(r-1)(2r-1)}-\frac{1}{12}$$
$$>\frac{r}{4(2r)}-\frac{1}{12}>0.$$

Then we find $R(r,m)>0$,

$$\tilde{w}(r,m)<\tilde{\kappa}(u(m)):=\frac{\pi}{\sin(\pi/r)}$$

and

$$w(s,n)<\kappa(u(n)):=\frac{\pi}{\sin(\pi/r)}.$$

By Theorem 3.1.7, we have the following theorem:

Theorem 3.4.9 If $p>1,r>1,\frac{1}{p}+\frac{1}{q}=1,\ \frac{1}{r}+\frac{1}{s}=1$, $\alpha\geq e^{7/6}$, $a_m,b_n\geq 0$ $(m,n\in\mathbf{N})$, such that $0<\sum_{m=1}^\infty\frac{1}{m}(\ln\sqrt{\alpha}m)^{\frac{p}{s}-1}a_m^p<\infty$ and $0<\sum_{n=1}^\infty\frac{1}{n}(\ln\sqrt{\alpha}n)^{\frac{q}{r}-1}b_n^q<\infty$, then we have the following equivalent inequalities:

$$\sum_{n=1}^\infty\sum_{m=1}^\infty\frac{a_mb_n}{mn\ln\alpha mn}<\frac{\pi}{\sin(\pi/r)}\left\{\sum_{m=1}^\infty\frac{1}{m}(\ln\sqrt{\alpha}m)^{\frac{p}{s}-1}a_m^p\right\}^{\frac{1}{p}}$$
$$\times\left\{\sum_{n=1}^\infty\frac{1}{n}(\ln\sqrt{\alpha}n)^{\frac{q}{r}-1}b_n^q\right\}^{\frac{1}{q}}, \qquad (3.4.27)$$

$$\sum_{n=1}^\infty\frac{1}{n}(\ln\sqrt{\alpha}n)^{\frac{p}{s}-1}\left(\sum_{m=1}^\infty\frac{1}{m\ln\alpha mn}a_m\right)^p$$
$$<\left[\frac{\pi}{\sin(\pi/r)}\right]^p\sum_{m=1}^\infty\frac{1}{n}(\ln\sqrt{\alpha}m)^{\frac{p}{s}-1}a_m^p. \qquad (3.4.28)$$

For $r=s=2,\alpha\geq e^{\frac{7+\sqrt{97}}{24}}$, by (3.4.26), we have

$$R(2,m)=\left[2\ln\sqrt{\alpha}-\frac{7}{12}-\frac{1}{24\ln\sqrt{\alpha}}\right]\frac{1}{\ln\alpha m}$$
$$+\left[\frac{4\ln^2\sqrt{\alpha}}{3}-\frac{1}{12}\right]\left(\frac{1}{\ln\alpha m}\right)^2>0.$$

The corresponding theorem is as follows.

Theorem 3.4.10 If $p>1$, $\frac{1}{p}+\frac{1}{q}=1$, $\alpha\geq e^{\frac{7+\sqrt{97}}{24}}$, $a_m,b_n\geq 0$ $(m,n\in\mathbf{N})$, such that $0<\sum_{m=1}^\infty\frac{1}{m}(\ln\sqrt{\alpha}m)^{\frac{p}{2}-1}a_m^p<\infty$ and $0<\sum_{n=1}^\infty\frac{1}{n}(\ln\sqrt{\alpha}n)^{\frac{q}{2}-1}b_n^q<\infty$, then we have the following equivalent inequalities:

$$\sum_{n=1}^\infty\sum_{m=1}^\infty\frac{a_mb_n}{mn\ln\alpha mn}<\pi\left\{\sum_{m=1}^\infty\frac{1}{m}(\ln\sqrt{\alpha}m)^{\frac{p}{2}-1}a_m^p\right\}^{\frac{1}{p}}$$
$$\times\left\{\sum_{n=1}^\infty\frac{1}{n}(\ln\sqrt{\alpha}n)^{\frac{q}{2}-1}b_n^q\right\}^{\frac{1}{q}}, \qquad (3.4.29)$$

$$\sum_{n=1}^\infty\frac{1}{n}(\ln\sqrt{\alpha}n)^{\frac{p}{2}-1}\left(\sum_{m=1}^\infty\frac{a_m}{m\ln\alpha mn}\right)^p$$
$$<\pi^p\sum_{m=1}^\infty\frac{1}{m}(\ln\sqrt{\alpha}m)^{\frac{p}{2}-1}a_m^p. \qquad (3.4.30)$$

Since we have

$$\tilde{w}(r,m)>\int_1^\infty f_\alpha(x)dx$$
$$=\frac{\pi}{\sin(\frac{\pi}{r})}-\int_0^{\frac{\ln\sqrt{\alpha}}{\ln\sqrt{\alpha}m}}\frac{1}{1+u}\left(\frac{1}{u}\right)^{\frac{1}{r}}du$$
$$>\frac{\pi}{\sin(\frac{\pi}{r})}-\int_0^{\frac{\ln\sqrt{\alpha}}{\ln\sqrt{\alpha}m}}\left(\frac{1}{u}\right)^{\frac{1}{r}}du$$
$$=\tilde{\mu}(u(m)):=\frac{\pi}{\sin(\frac{\pi}{r})}-s\left(\frac{\ln\sqrt{\alpha}}{\ln\sqrt{\alpha}m}\right)^{1/s}$$

$$\geq \frac{\pi}{\sin(\frac{\pi}{r})} - s > 0,$$

then in view of $w(s,n) < \kappa(u(n)) := \frac{\pi}{\sin(\pi/r)}$ and Theorem 3.1.7, we have the following theorem:

Theorem 3. 4.11 If $0 < p < 1, r > 1$, $\frac{1}{p} + \frac{1}{q} = 1$, $\frac{1}{r} + \frac{1}{s} = 1$, $\alpha \geq e^{7/6}$, $a_m, b_n \geq 0$, such that $0 < \sum_{m=1}^{\infty} \frac{1}{m} (\ln \sqrt{\alpha m})^{\frac{p}{s}-1} a_m^p < \infty$ and $0 < \sum_{n=1}^{\infty} \frac{1}{n} (\ln \sqrt{\alpha n})^{\frac{q}{r}-1} b_n^q < \infty$, then we have the following equivalent inequalities:

$$\tilde{H} := \sum_{n=1}^{\infty} \sum_{m=1}^{\infty} \frac{1}{mn \ln \alpha mn} a_m b_n$$

$$> \{\sum_{m=1}^{\infty} [\frac{\pi}{\sin(\pi/r)} - s(\frac{\ln\sqrt{\alpha}}{\ln\sqrt{\alpha m}})^{\frac{1}{s}}] \frac{(\ln\sqrt{\alpha m})^{p/s-1}}{m} a_m^p\}^{\frac{1}{p}}$$

$$\times \{\frac{\pi}{\sin(\pi/r)} \sum_{n=1}^{\infty} \frac{1}{n} (\ln \sqrt{\alpha n})^{q/r-1} b_n^q\}^{\frac{1}{q}}, \tag{3.4.31}$$

$$\sum_{n=1}^{\infty} \frac{1}{n} (\ln \sqrt{\alpha n})^{\frac{p}{s}-1} (\sum_{m=1}^{\infty} \frac{1}{m \ln \alpha mn} a_m)^p$$

$$> [\frac{\pi}{\sin(\pi/r)}]^{p-1}$$

$$\times \sum_{m=1}^{\infty} [\frac{\pi}{\sin(\pi/r)} - s(\frac{\ln\sqrt{\alpha}}{\ln\sqrt{\alpha m}})^{\frac{1}{s}}] \frac{(\ln\sqrt{\alpha m})^{p/s-1}}{m} a_m^p, \tag{3.4.32}$$

$$\sum_{m=1}^{\infty} \frac{(\ln\sqrt{\alpha m})^{q/r-1}}{m[\frac{\pi}{\sin(\pi/r)} - s(\frac{\ln\sqrt{\alpha}}{\ln\sqrt{\alpha m}})^{1/s}]^{q-1}} (\sum_{n=1}^{\infty} \frac{1}{n \ln \alpha mn} b_n)^q$$

$$< [\frac{\pi}{\sin(\pi/r)}]^{q-1} \sum_{n=1}^{\infty} \frac{1}{n} (\ln \sqrt{\alpha n})^{\frac{q}{r}-1} b_n^q. \tag{3.4.33}$$

In particular, for $r = s = 2$, $\alpha \geq e^{\frac{7+\sqrt{97}}{24}}$, we have the following equivalent inequalities:

$$\tilde{H} > \pi\{\sum_{m=1}^{\infty} [1 - \frac{2}{\pi}(\frac{\ln\sqrt{\alpha}}{\ln\sqrt{\alpha m}})^{\frac{1}{2}}] \frac{(\ln\sqrt{\alpha m})^{p/2-1}}{m} a_m^p\}^{\frac{1}{p}}$$

$$\times \{\sum_{n=1}^{\infty} \frac{1}{n} (\ln \sqrt{\alpha n})^{\frac{q}{2}-1} b_n^q\}^{\frac{1}{q}}, \tag{3.4.34}$$

$$\sum_{n=1}^{\infty} \frac{1}{n} (\ln \sqrt{\alpha n})^{\frac{p}{2}-1} (\sum_{m=1}^{\infty} \frac{1}{m \ln \alpha mn} a_m)^p > \pi^p$$

$$\times \sum_{m=1}^{\infty} [1 - \frac{2}{\pi}(\frac{\ln\sqrt{\alpha}}{\ln\sqrt{\alpha m}})^{\frac{1}{2}}] \frac{(\ln\sqrt{\alpha m})^{p/2-1}}{m} a_m^p, \tag{3.4.35}$$

$$\sum_{m=1}^{\infty} \frac{(\ln\sqrt{\alpha m})^{q/2-1}}{m[1 - \frac{2}{\pi}(\frac{\ln\sqrt{\alpha}}{\ln\sqrt{\alpha m}})^{1/2}]^{q-1}} (\sum_{n=1}^{\infty} \frac{1}{n \ln \alpha mn} b_n)^q$$

$$< \pi^q \sum_{n=1}^{\infty} \frac{1}{n} (\ln \sqrt{\alpha n})^{\frac{q}{2}-1} b_n^q. \tag{3.4.36}$$

Note 3.4.12 Some results of Theorem 3.4.10 and Theorem 3.4.11 are consulted in (Yang MIA 2004)[29]. The well known Mulholland's inequality is as follows (Hardy CUP 1934)[30]:

$$\sum_{n=2}^{\infty} \sum_{m=2}^{\infty} \frac{1}{mn \ln mn} a_m b_n$$

$$< \frac{\pi}{\sin(\frac{\pi}{p})} \{\sum_{m=2}^{\infty} \frac{1}{m} a_m^p\}^{\frac{1}{p}} \{\sum_{n=2}^{\infty} \frac{1}{n} b_n^q\}^{\frac{1}{q}}. \tag{3.4.37}$$

Replacing a_m, b_n by ma_m, nb_n in (3.4.7), equivalently, we obtain the following inequality:

$$\sum_{n=1}^{\infty} \sum_{m=1}^{\infty} \frac{a_m b_n}{\ln(m+1)(n+1)} < \frac{\pi}{\sin(\frac{\pi}{p})} \{\sum_{m=1}^{\infty} (m+1)^{p-1} a_m^p\}^{\frac{1}{p}}$$

$$\times \{\sum_{n=1}^{\infty} (n+1)^{q-1} b_n^q\}^{\frac{1}{q}}. \tag{3.4.38}$$

We still call (3.4.38) Mulholland's inequality.

3.4.4 ANOTHER MORE ACCURATE EXTENDED MULHOLLAND'S INEQUALITY

If $u(m) = \ln(m+\alpha)$ $(\alpha \geq \frac{1}{2}), n_0 = 1$, then we define the following weight coefficients

$$\tilde{w}(r,m) := \sum_{k=1}^{\infty} \frac{1}{\ln(k+\alpha)(m+\alpha)} [\frac{\ln(m+\alpha)}{\ln(k+\alpha)}]^{\frac{1}{r}} \frac{1}{k+\alpha},$$

$$w(s,n) := \sum_{k=1}^{\infty} \frac{1}{\ln(k+\alpha)(n+\alpha)} [\frac{\ln(n+\alpha)}{\ln(k+\alpha)}]^{\frac{1}{s}} \frac{1}{k+\alpha},$$

$$r, s > 1, m, n \in \mathbf{N} \tag{3.4.39}$$

For fixed $m \in \mathbf{N}$, setting

$$f_\alpha(x) := \frac{1}{\ln(x+\alpha)(m+\alpha)} [\frac{\ln(m+\alpha)}{\ln(x+\alpha)}]^{\frac{1}{r}} \frac{1}{x+\alpha},$$
$$x \in [1, \infty),$$

since $f_\alpha'(x) < 0, f_\alpha''(x) > 0$, we have Hadamard's inequality as follows (Kuang SSTP 2004)[1]:

$$f_\alpha(k) < \int_{k-\frac{1}{2}}^{k+\frac{1}{2}} f_\alpha(x)dx, k \in \mathbf{N}. \tag{3.4.40}$$

Then setting $u = \frac{\ln(x+\alpha)}{\ln(m+\alpha)}$, we find

$$\tilde{w}(r,m) = \sum_{k=1}^{\infty} f_\alpha(k) < \int_{\frac{1}{2}}^{\infty} f_\alpha(x)dx$$

$$= \int_{\frac{\ln(\frac{1}{2}+\alpha)}{\ln(m+\alpha)}}^{\infty} \frac{1}{1+u} u^{-\frac{1}{r}} du \leq \int_0^\infty \frac{1}{1+u} u^{-\frac{1}{r}} du = \frac{\pi}{\sin(\pi/r)}.$$

By the same way, we obtain $w(s,n) < \frac{\pi}{\sin(\pi/r)}$.

By Theorem 3.1.7, we have

Theorem 3.4.13 If $p > 1, r > 1$, $\frac{1}{p} + \frac{1}{q} = 1$, $\frac{1}{r} + \frac{1}{s} = 1$, $\alpha \geq \frac{1}{2}$, $a_m, b_n \geq 0$ $(m, n \in \mathbf{N})$, such that

$0<\sum_{m=1}^{\infty}\frac{1}{(m+\alpha)^{1-p}}[\ln(m+\alpha)]^{\frac{p}{s}-1}a_m^p<\infty$ and $0<\sum_{n=1}^{\infty}\frac{1}{(n+\alpha)^{1-p}}[\ln(n+\alpha)]^{\frac{q}{r}-1}b_n^q<\infty$, then we have the following equivalent inequalities:

$$\sum_{n=1}^{\infty}\sum_{m=1}^{\infty}\frac{1}{\ln(m+\alpha)(n+\alpha)}a_m b_n$$

$$<\frac{\pi}{\sin(\pi/r)}\{\sum_{m=1}^{\infty}\frac{[\ln(m+\alpha)]^{\frac{p}{s}-1}}{(m+\alpha)^{1-p}}a_m^p\}^{\frac{1}{p}}\{\sum_{n=1}^{\infty}\frac{[\ln(n+\alpha)]^{\frac{q}{r}-1}}{(n+\alpha)^{1-p}}b_n^q\}^{\frac{1}{q}}, \tag{3.4.41}$$

$$\sum_{n=1}^{\infty}\frac{[\ln(n+\alpha)]^{\frac{p}{s}-1}}{(n+\alpha)}[\sum_{m=1}^{\infty}\frac{1}{\ln(m+\alpha)(n+\alpha)}a_m]^p$$

$$<[\frac{\pi}{\sin(\pi/r)}]^p\sum_{m=1}^{\infty}\frac{[\ln(m+\alpha)]^{\frac{p}{s}-1}}{(m+\alpha)^{1-p}}a_m^p. \tag{3.4.42}$$

In view of the decreasing property of $f_\alpha(x)$ and $\sin(\pi/r)=\sin(\pi/s)<\pi/s$, we find

$$\tilde{w}(r,m)=\sum_{k=1}^{\infty}f_\alpha(k)>\int_1^{\infty}f_\alpha(x)dx$$

$$=\int_{\frac{\ln(1+\alpha)}{\ln(m+\alpha)}}^{\infty}\frac{u^{-\frac{1}{r}}}{1+u}du=\frac{\pi}{\sin(\pi/r)}-\int_0^{\frac{\ln(1+\alpha)}{\ln(m+\alpha)}}\frac{u^{-\frac{1}{r}}}{1+u}du$$

$$\geq\frac{\pi}{\sin(\pi/r)}-\int_0^{\frac{\ln(1+\alpha)}{\ln(m+\alpha)}}u^{-\frac{1}{r}}du$$

$$=\frac{\pi}{\sin(\pi/r)}-s[\frac{\ln(1+\alpha)}{\ln(m+\alpha)}]^{\frac{1}{s}}$$

$$=\frac{\pi}{\sin(\pi/r)}\{1-\frac{s}{\pi}\sin(\frac{\pi}{r})[\frac{\ln(1+\alpha)}{\ln(m+\alpha)}]^{\frac{1}{s}}\}$$

$$\geq\frac{\pi}{\sin(\pi/r)}[1-\frac{s}{\pi}\sin(\frac{\pi}{r})]>0.$$

Then by Theorem 3.1.7, it follows

Theorem 3. 4.14 If $0<p<1, r>1$, $\frac{1}{p}+\frac{1}{q}=1$, $\frac{1}{r}+\frac{1}{s}=1, \alpha\geq\frac{1}{2}, a_m, b_n\geq 0$, such that $0<\sum_{m=1}^{\infty}\frac{1}{(m+\alpha)^{1-p}}[\ln(m+\alpha)]^{\frac{p}{s}-1}a_m^p<\infty$ and $0<\sum_{n=1}^{\infty}\frac{1}{(n+\alpha)^{1-p}}[\ln(n+\alpha)]^{\frac{q}{r}-1}b_n^q<\infty$, then we have the following equivalent inequalities:

$$\sum_{n=1}^{\infty}\sum_{m=1}^{\infty}\frac{1}{\ln(m+\alpha)(n+\alpha)}a_m b_n>\frac{\pi}{\sin(\pi/r)}$$

$$\times\{\sum_{m=1}^{\infty}\left[1-\frac{s}{\pi}\sin(\frac{\pi}{r})[\frac{\ln(1+\alpha)}{\ln(m+\alpha)}]^{\frac{1}{s}}\right]\frac{[\ln(m+\alpha)]^{\frac{p}{s}-1}}{(m+\alpha)^{1-p}}a_m^p\}^{\frac{1}{p}}$$

$$\times\{\sum_{n=1}^{\infty}\frac{[\ln(n+\alpha)]^{\frac{q}{r}-1}}{(n+\alpha)^{1-p}}b_n^q\}^{\frac{1}{q}}, \tag{3.4.43}$$

$$\sum_{n=1}^{\infty}\frac{[\ln(n+\alpha)]^{\frac{p}{s}-1}}{(n+\alpha)}[\sum_{m=1}^{\infty}\frac{a_m}{\ln(m+\alpha)(n+\alpha)}]^p>[\frac{\pi}{\sin(\pi/r)}]^p$$

$$\times\sum_{m=1}^{\infty}\left[1-\frac{s}{\pi}\sin(\frac{\pi}{r})[\frac{\ln(1+\alpha)}{\ln(m+\alpha)}]^{\frac{1}{s}}\right]\frac{[\ln(m+\alpha)]^{\frac{p}{s}-1}}{(m+\alpha)^{1-p}}a_m^p. \tag{3.4.44}$$

$$\sum_{m=1}^{\infty}\frac{[\ln(m+\alpha)]^{q/r-1}}{(m+\alpha)\left[1-\frac{s}{\pi}\sin(\frac{\pi}{r})[\frac{\ln(1+\alpha)}{\ln(m+\alpha)}]^{\frac{1}{s}}\right]^{q-1}}[\sum_{n=1}^{\infty}\frac{b_n}{\ln(m+\alpha)(n+\alpha)}]^q$$

$$<[\frac{\pi}{\sin(\pi/r)}]^q\sum_{n=1}^{\infty}\frac{[\ln(n+\alpha)]^{\frac{q}{r}-1}}{(n+\alpha)^{1-p}}b_n^q. \tag{3.4.45}$$

Note 3.4.15 For $\alpha=1, r=q, s=p$ in (3.4.41), we have (3.4.38). It follows that (3.4.41) is a more accurate extended Mulholland's inequality.

3.5 REFERENCES

1. Kuang JC. Applied inequalities. Jinan: Shangdong Science Technology Press, 2004.
2. Boas R P. Estimating remainders. Mathematics Magazine,1978,51(2):83-89.
3. Zhao DJ. On a refinement of Hilbert double series theorem. Mathematics in Practice and Theory, 1993,(1): 85-90.
4. Gao MZ. On an improvement of Hilbert's inequality extended by Hardy-Riesz. Journal of Mathematical Research and Exposition, 1994, 14(2): 255-259.
5. Gao MJ. A note on Hilbert double theorem. Hunan Math. Annual, 1992,11(1-2): 142-147.
6. Yang BC, Gao MJ. On a best value of Hardy-Hilbert's inequality. Advances in Mathematices, 1997,26(2): 159-164.
7. Gao MZ, Yang BC. On the extended Hilbert's inequality. Proc. Amer. Math. Soc.,1998, 126(3): 751-759.
8. Yang BC, Debnath L. On new strengthened Hardy-Hilbert's inequality. Internat. J. Math. & Math. Soc.,1998,21(2):403-408.
9. Yang BC. A refinement of Hilbert's inequality. Huanghuai Journal, 1997,13(2):47~51.
10. Yang BC. On a refinement of Hardy-Hilbert's inequality and its applications. Northeast. Math. J., 2000,16(3):279-286.
11. Yang BC, Debnath L. A strengthened Hardy-Hilbert's inequality. Proceedings of the Jangjeon Mathematical Society, 2003,6(2):119-124.
12. Wang WH, Yang BC. An improvement of Hardy-Hilbert inequality. College Mathematics, 2007,23(6):92-95.
13. Kuang JC, Debnath L. On new generalizations of Hilbert's inequality and their applications. J. Math. Anal. Appl.,2000,245:248-265.
14. Huang QL. A strengthened Hilbert-type inequality. Journal of Guangdong Education Institute, 2006,26(3):15-18.
15. Yang BC. On a strengthened Hilbert-type inequality. Journal of Guangdong Education Institute, 2006,26(5):1-4.

16. Yang BC. A strengthened Hardy-Hilbert's inequality. Journal of Shanghai University (Natural Science), 2006,12(3): 256-259.
17. Huang QL. A strengthened version of a Hilbert-type inequality and the equivalent form. Journal of Mathematics, 2010,30(3):503-508.
18. Yang BC. On an extended Hardy-Hilbert's inequality and some reversed form. International Mathematical Forum,2006,1(39):1905-1912.
19. Yang BC. A reverse of the Hardy-Hilbert's type inequality. Mathematics in Practice and Theory, 2006,36(11): 207-212.
20. Yang BC. A reverse of the Hardy-Hilbert's type inequality. Journal of Southwest China Normal University (Science), 2005,30(6):1012-1015.
21. Yang BC. On a reverse of a Hardy-Hilbert type inequality. Journal of Inequalities in Pure and Applied Mathematics, 2006,7(3): Art.115.
22. Yang BC. A more accurate Hardy-Hilbert type inequality. Journal of Xinyang Normal University (Natural Science), 2005,18(2):140-142.
23. Yang BC. A more accurate Hilbert's type inequality, College Math., 2005,21(5):99-102.
24. Yang BC. On a more accurate Hardy-Hilbert's type inequality and its applications. Acta Mathematica Sinica, Chinese Series, 2006,49(2): 363-368.
25. Yang BC. A more accurate Hilbert's type inequality. Journal of Mathematics, 2007,27(6): 673-678.
26. Yang BC. On a more accurate Hilbert's type inequality. International Mathematical Forum, 2007,2(37):1831-1837.
27. Wang WH, Yang BC. A strengthened Hardy-Hilbert's type inequality. The Australian Journal of Mathematical Analysis and Applications, 2006,3(2),Art.17:1-7.
28. Yang BC. On a new Hardy-Hilbert's type inequality with a parameter. International Journal of Mathematical Analysis, 2007,1(1-4):123-131.
29. Yang BC. On a new Hardy-Hilbert's type inequality. Math. Ineq. Appl., 2004,7(3): 355-363.
30 Hardy GH, Littlewood JE, Polya G. Inequalities. Cambridge : Cambridge University Press, 1934.

CHAPTER 4

Hilbert-Type Inequalities with the General Homogeneous Kernel

Abstract: In this chapter, Hilbert-type inequalities and the reverses with the general homogeneous kernel of the real number degree and the best constant factors are considered, which are some extensions of the results in Chapter 3. By using the improved Euler-Maclaurin summation formula and the technique of real analysis, some particular examples with the best constant factors are given.

4.1 HILBERT-TYPE INEQUALITIES WITH AN INDEPENDENT PARAMETER

4.1.1 THE NORM OF OPERATOR AND HILBERT-TYPE INEQUALITIES

If $\lambda \in \mathbf{R}, \omega(x) := x^{1-\lambda}$ $(x \in (0,\infty))$, then we define a real sequences space as follows

$$l_\omega^2 = \{a; a = \{a_n\}_{n=1}^\infty,$$
$$\| a \|_{2,\omega} = \{\sum_{n=1}^\infty \omega(n) a_n^2\}^{\frac{1}{2}} < \infty\}.$$

Assuming that $k_\lambda(x,y) (\geq 0)$ is a finite homogeneous function of degree $-\lambda$ in $\mathbf{R}_+^2$, we define the following linear operator as: $T : l_\omega^2 \to l_{\omega^{-1}}^2$, for $a = \{a_m\}_{m=1}^\infty \in l_\omega^2$, there exists $c = \{c_n\}_{n=1}^\infty \in l_{\omega^{-1}}^2$, satisfying

$$c_n = (Ta)(n) = \sum_{m=1}^\infty k_\lambda(m,n) a_m, \ n \in \mathbf{N}. \quad (4.1.1)$$

If there exists a constant $K > 0$, such that for any $a \in l_\omega^2$, $\| Ta \|_{2,\omega^{-1}} \leq K \| a \|_{2,\omega}$, then it follows that the operator is bounded and the norm of T satisfies the following inequality:

$$\| T \|_{2,\omega^{-1}} := \sup_{a(\neq\theta)\in l_\omega^2} \frac{\|Ta\|_{2,\omega^{-1}}}{\|a\|_{2,\omega}} \leq K. \quad (4.1.2)$$

We define the weight coefficients $\varpi_\lambda(m)$ and $\vartheta_\lambda(n)$ as follows:

$$\varpi_\lambda(m) := \sum_{n=1}^\infty k_\lambda(m,n) \frac{m^{\frac{\lambda}{2}}}{n^{1-\frac{\lambda}{2}}},$$

$$\vartheta_\lambda(n) := \sum_{m=1}^\infty k_\lambda(m,n) \frac{n^{\frac{\lambda}{2}}}{m^{1-\frac{\lambda}{2}}}, \ m,n \in \mathbf{N}. \quad (4.1.3)$$

Hence we have the following theorem:

Theorem 4.1.1 If there exist constants $\tilde{k}_\lambda, k'_\lambda > 0$, satisfying the following inequalities:

$$\varpi_\lambda(m) < \tilde{k}_\lambda, \vartheta_\lambda(n) < k'_\lambda, \ m,n \in \mathbf{N}, \quad (4.1.4)$$

then the linear operator T defined by (3.1.1) exists, which is bounded and $\| T \|_{2,\omega^{-1}} \leq \sqrt{\tilde{k}_\lambda k'_\lambda}$.

Proof By Cauchy's inequality with weight and (4.1.3)-(4.1.4), we have

$$(\sum_{m=1}^\infty k_\lambda(m,n) a_m)^2$$
$$= \{\sum_{m=1}^\infty k_\lambda(m,n) [\frac{m^{(1-\frac{\lambda}{2})/2}}{n^{\lambda/4}} a_m][\frac{n^{\lambda/4}}{m^{(1-\frac{\lambda}{2})/2}}]\}^2$$
$$\leq (\sum_{m=1}^\infty k_\lambda(m,n) \frac{m^{1-\frac{\lambda}{2}}}{n^{\lambda/2}} a_m^2)(\sum_{m=1}^\infty k_\lambda(m,n) \frac{n^{\lambda/2}}{m^{1-\frac{\lambda}{2}}})$$
$$= \vartheta_\lambda(n) \sum_{m=1}^\infty k_\lambda(m,n) \frac{m^{1-\frac{\lambda}{2}}}{n^{\lambda/2}} a_m^2$$
$$\leq k'_\lambda \sum_{m=1}^\infty k_\lambda(m,n) \frac{m^{1-\frac{\lambda}{2}}}{n^{\lambda/2}} a_m^2. \quad (4.1.5)$$

Then by (4.1.5), (4.1.3) and (4.1.4), it follows

$$\| Ta \|^2_{2,\omega^{-1}} = \| c \|^2_{2,\omega^{-1}}$$
$$= \sum_{n=1}^\infty n^{\lambda-1} (\sum_{m=1}^\infty k_\lambda(m,n) a_m)^2$$
$$\leq k'_\lambda \sum_{n=1}^\infty n^{\lambda-1} \sum_{m=1}^\infty k_\lambda(m,n) \frac{m^{1-\frac{\lambda}{2}}}{n^{\lambda/2}} a_m^2$$
$$= k'_\lambda \sum_{m=1}^\infty (\sum_{n=1}^\infty k_\lambda(m,n) \frac{m^{\frac{\lambda}{2}}}{n^{1-\lambda/2}}) m^{1-\lambda} a_m^2$$
$$= k'_\lambda \sum_{m=1}^\infty \varpi_\lambda(m) \, m^{1-\lambda} a_m^2$$
$$\leq k'_\lambda \tilde{k}_\lambda \sum_{m=1}^\infty m^{1-\lambda} \, a_m^2 = k'_\lambda \tilde{k}_\lambda \| a \|^2_{2,\omega}. \quad (4.1.6)$$

Bicheng Yang

Hence $c = Ta \in l^2_{\omega^{-1}}$, T exists, which is also bounded. By (4.1.6), we have $\|T\|_{2,\omega^{-1}} \le \sqrt{k'_\lambda \tilde{k}_\lambda}$. □

Theorem 4.1.2 Assuming that $\lambda \in \mathbf{R}$, $k_\lambda(x,y)$ (≥ 0) is a finite homogeneous function of degree $-\lambda$ in $\mathbf{R}^2_+$, satisfying

$$0 < k_\lambda := \int_0^\infty k_\lambda(u,1)u^{\frac{\lambda}{2}-1}du < \infty \tag{4.1.7}$$

and $k_\lambda(1,u) = O(\frac{1}{u^\delta})(\frac{\lambda}{2} < \delta < \frac{\lambda}{2}+1; u \to \infty)$, if

$$\varpi_\lambda(m) < k_\lambda, \vartheta_\lambda(n) < k_\lambda, \ m,n \in \mathbf{N}, \tag{4.1.8}$$

there exists a constant $\lambda' > 0$, such that for any $m \in \mathbf{N}$,

$$k_\lambda - O(\tfrac{1}{m^{\lambda'}}) \le \varpi_\lambda(m) \ (m \to \infty), \tag{4.1.9}$$

then the bounded linear operator T defined by (4.1.1) exists with $\|T\|_{2,\omega^{-1}} = k_\lambda$.

Proof Setting $\tilde{k}_\lambda = k'_\lambda = k_\lambda$, by (4.1.8), we have (4.1.4). By Theorem 4.1.1, the bounded linear operator T exists satisfying $\|T\|_{2,\omega^{-1}} \le k_\lambda$.

For any $a,b \in l^2_\omega$, by Cauchy's inequality and (4.1.6) (setting $\tilde{k}_\lambda = k'_\lambda = k_\lambda$), we have

$$\sum_{n=1}^\infty \sum_{m=1}^\infty k_\lambda(m,n)a_m b_n = (Ta,b)$$

$$\le \|Ta\|_{2,\omega^{-1}} \|b\|_{2,\omega} \le k_\lambda \|a\|_{2,\omega} \|b\|_{2,\omega}. \tag{4.1.10}$$

Assuming that there exists a constant $0 < k \le k_\lambda$, such that (4.1.10) is valid as we replace k_λ by k, then for large enough $N \in \mathbf{N}$, such that

$$k_\lambda(1,u) \le L(\tfrac{1}{u^\delta})(u \ge N; L > 0),$$

setting $\tilde{b} = \tilde{a} = \{\tilde{a}_n\}_{n=1}^\infty$, $\tilde{a}_n = n^{\frac{\lambda}{2}-1}, n \le N$; $\tilde{a}_n = 0$, $n > N$, it follows

$$\tilde{I} := \sum_{n=1}^\infty \sum_{m=1}^\infty k_\lambda(m,n)\tilde{a}_m \tilde{a}_n$$

$$\le k\|\tilde{a}\|^2_{2,\omega} = k\sum_{m=1}^N \tfrac{1}{m}. \tag{4.1.11}$$

Since we find

$$\sum_{m=1}^N \sum_{n=N+1}^\infty k_\lambda(m,n)m^{\frac{\lambda}{2}-1}n^{\frac{\lambda}{2}-1}$$

$$= \sum_{m=1}^N \sum_{n=N+1}^\infty k_\lambda(1,\tfrac{n}{m})m^{\frac{-\lambda}{2}-1}n^{\frac{\lambda}{2}-1}$$

$$\le L\sum_{m=1}^N \sum_{n=N+1}^\infty (\tfrac{n}{m})^{-\delta} m^{\frac{-\lambda}{2}-1}n^{\frac{\lambda}{2}-1}$$

$$= L\sum_{m=1}^N \frac{1}{m^{1-\delta+\lambda/2}} \sum_{n=N+1}^\infty \frac{1}{n^{1+\delta-\lambda/2}}$$

$$< L\int_0^N \frac{dx}{x^{1-\delta+\lambda/2}} \int_N^\infty \frac{dy}{y^{1+\delta-\lambda/2}} = \frac{L}{(\delta-\lambda/2)^2},$$

then by (4.1.11) and (4.1.9), we obtain

$$k\sum_{m=1}^N \tfrac{1}{m} \ge \tilde{I} = \sum_{m=1}^N \sum_{n=1}^N k_\lambda(m,n)m^{\frac{\lambda}{2}-1}n^{\frac{\lambda}{2}-1}$$

$$= \sum_{m=1}^N \tfrac{1}{m} \sum_{n=1}^\infty k_\lambda(m,n)m^{\frac{\lambda}{2}}n^{\frac{\lambda}{2}-1}$$

$$- \sum_{m=1}^N \sum_{n=N+1}^\infty k_\lambda(m,n)m^{\frac{\lambda}{2}-1}n^{\frac{\lambda}{2}-1}$$

$$\ge \sum_{m=1}^N \tfrac{1}{m}[k_\lambda - O(\tfrac{1}{m^{\lambda'}})] - \frac{L}{(\delta-\lambda/2)^2}$$

$$= k_\lambda \sum_{m=1}^N \tfrac{1}{m} - \sum_{m=1}^N \tfrac{1}{m}O(\tfrac{1}{m^{\lambda'}}) - \frac{L}{(\delta-\lambda/2)^2}$$

$$= \sum_{m=1}^N \tfrac{1}{m}\{k_\lambda - (\sum_{m=1}^N \tfrac{1}{m})^{-1}[\sum_{m=1}^N O(\tfrac{1}{m^{1-\lambda'}}) + \frac{L}{(\delta-\lambda/2)^2}]\},$$

$$k \ge k_\lambda - (\sum_{m=1}^N \tfrac{1}{m})^{-1}[\sum_{m=1}^N O(\tfrac{1}{m^{1+\lambda'}}) + \frac{L}{(\delta-\lambda/2)^2}].$$

For $N \to \infty$, we have $k \ge k_\lambda$. Therefore $k = k_\lambda$ is the best value of (4.1.10). By (4.1.10), dividing out $\|b\|_{2,\omega} (> 0)$, we have

$$\|Ta\|_{2,\omega^{-1}} \le k_\lambda \|a\|_{2,\omega}. \tag{4.1.12}$$

It is obvious that $k = k_\lambda$ is still the best value of (4.1.12) and $\|T\|_{2,\omega^{-1}} = k_\lambda$. □

Theorem 4.1.3 Assuming that $\lambda \in \mathbf{R}$, $k_\lambda(x,y)(\ge 0)$ is a finite homogeneous function of degree $-\lambda$ in $\mathbf{R}^2_+$, such that $0 < k_\lambda = \int_0^\infty k_\lambda(u,1)u^{\frac{\lambda}{2}-1}du < \infty$, if $k_\lambda(x,y)\frac{1}{x^{1-\lambda/2}}$ is decreasing for $x > 0$ and $k_\lambda(x,y)\frac{1}{y^{1-\lambda/2}}$ is decreasing for $y > 0$, and respectively strictly decreasing in a subinterval, then the bounded linear operator T defined by (4.1.1) exists with $\|T\|_{2,\omega^{-1}} = k_\lambda$.

Proof By the decreasing property, we find

$$\varpi_\lambda(m)=\sum_{n=1}^{\infty}k_\lambda(m,n)\frac{m^{\lambda/2}}{n^{1-\lambda/2}}$$
$$<\int_0^\infty k_\lambda(m,y)\frac{m^{\lambda/2}}{y^{1-\lambda/2}}dy.$$

Setting $u=\frac{m}{y}$, we have $\varpi_\lambda(m)<k_\lambda$. Since

$$\vartheta_\lambda(n)=\sum_{m=1}^{\infty}k_\lambda(m,n)\frac{n^{\lambda/2}}{m^{1-\lambda/2}}$$
$$<\int_0^\infty k_\lambda(x,n)\frac{n^{\lambda/2}}{x^{1-\lambda/2}}dx,$$

setting $u=\frac{x}{n}$, we still have $\vartheta_\lambda(n)<k_\lambda$. Hence (4.1.4) is valid for $\tilde{k}_\lambda=k'_\lambda=k_\lambda$, and by Theorem 4.1.1, T exists satisfying $\|T\|_{2,\omega^{-1}}\le k_\lambda$ and then (4.1.10) is valid.

If there exists a constant $0<k\le k_\lambda$, such that (4.1.10) is valid as we replace k_λ by k, then for $\varepsilon>0$, setting $\tilde{b}=\tilde{a}=\{\tilde{a}_n\}_{n=1}^\infty=\{n^{\frac{\lambda-\varepsilon}{2}-1}\}_{n=1}^\infty$, we find

$$\varepsilon\tilde{I}\le\varepsilon k\sum_{n=1}^{\infty}\frac{1}{n^{1+\varepsilon}}=\varepsilon k(1+\sum_{n=2}^{\infty}\frac{1}{n^{1+\varepsilon}})$$
$$<\varepsilon k(1+\int_1^\infty\frac{1}{y^{1+\varepsilon}}dy)=k(\varepsilon+1), \tag{4.1.13}$$
$$\varepsilon\tilde{I}=\varepsilon\sum_{n=1}^{\infty}\sum_{m=1}^{\infty}k_\lambda(m,n)(mn)^{\frac{\lambda-\varepsilon}{2}-1}$$
$$\ge\varepsilon\int_1^\infty x^{\frac{\lambda-\varepsilon}{2}-1}(\int_1^\infty k_\lambda(x,y)y^{\frac{\lambda-\varepsilon}{2}-1}dy)dx.$$

Setting $u=\frac{x}{y}$, by Fubini theorem, we have

$$\varepsilon\tilde{I}\ge\varepsilon\int_1^\infty x^{-1-\varepsilon}(\int_0^x k_\lambda(u,1)u^{\frac{\lambda+\varepsilon}{2}-1}du)dx$$
$$=\int_0^1 k_\lambda(u,1)u^{\frac{\lambda+\varepsilon}{2}-1}du$$
$$+\varepsilon\int_1^\infty x^{-1-\varepsilon}(\int_1^x k_\lambda(u,1)u^{\frac{\lambda+\varepsilon}{2}-1}du)dx$$
$$=\int_0^1 k_\lambda(u,1)u^{\frac{\lambda+\varepsilon}{2}-1}du$$
$$+\varepsilon\int_1^\infty(\int_u^\infty x^{-1-\varepsilon}dx)k_\lambda(u,1)u^{\frac{\lambda+\varepsilon}{2}-1}du$$
$$=\int_0^1 k_\lambda(u,1)u^{\frac{\lambda+\varepsilon}{2}-1}du+\int_1^\infty k_\lambda(u,1)u^{\frac{\lambda-\varepsilon}{2}-1}du.$$

In view of (4.1.13), we find

$$\int_0^1 k_\lambda(u,1)u^{\frac{\lambda+\varepsilon}{2}-1}du+\int_1^\infty k_\lambda(u,1)u^{\frac{\lambda-\varepsilon}{2}-1}du$$
$$<k(\varepsilon+1).$$

By Fatou lemma, we find

$$k_\lambda=\int_0^1\lim_{\varepsilon\to0^+}k_\lambda(u,1)u^{\frac{\lambda+\varepsilon}{2}-1}du$$
$$+\int_1^\infty\lim_{\varepsilon\to0^+}k_\lambda(u,1)u^{\frac{\lambda-\varepsilon}{2}-1}du$$
$$\le\underline{\lim}_{\varepsilon\to0^+}[\int_0^1 k_\lambda(u,1)u^{\frac{\lambda+\varepsilon}{2}-1}du$$
$$+\int_1^\infty k_\lambda(u,1)u^{\frac{\lambda-\varepsilon}{2}-1}du]\le k.$$

Hence $k=k_\lambda$ is the best value of (4.1.10) and (4.1.12), and then $\|T\|_{2,\omega^{-1}}=k_\lambda$. □

Note 4.1.4 If $0<\lambda\le2$, then the decreasing property condition of $k_\lambda(x,y)\frac{1}{x^{1-\lambda/2}}$ and $k_\lambda(x,y)\frac{1}{y^{1-\lambda/2}}$ in Theorem 4.1.2 may change for " $k_\lambda(x,y)$ is decreasing for $x(y)$ and strictly decreasing for a subinterval".

Theorem 4.1.5 Let the assumptions of Theorem 4.1.2 (or Theorem 4.1.3) be fulfilled and additionally, $\omega(x)=x^{1-\lambda}$, $a,b\in l_\omega^2$, such that $0<\|a\|_{2,\omega}=\{\sum_{n=1}^\infty n^{1-\lambda}a_n^2\}^{\frac{1}{2}}<\infty$ and $0<\|b\|_{2,\omega}=\{\sum_{n=1}^\infty n^{1-\lambda}b_n^2\}^{\frac{1}{2}}<\infty$, then we have the following equivalent inequalities:

$$(Ta,b)=\sum_{n=1}^{\infty}\sum_{m=1}^{\infty}k(m,n)a_mb_n$$
$$<k_\lambda\|a\|_{2,\omega}\|b\|_{2,\omega}, \tag{1.1.14}$$
$$\|Ta\|_{2,\omega^{-1}}^2=\sum_{n=1}^{\infty}n^{\lambda-1}(\sum_{m=1}^{\infty}k(m,n)a_m)^2$$
$$<k_\lambda^2\|a\|_{2,\omega}^2, \tag{4.1.15}$$

where the constant factors k_λ and k_λ^2 are the best possible.

Proof Since $0<\|a\|_{2,\omega}<\infty$, then the last inequality of (4.1.5) (setting $\tilde{k}_\lambda=k'_\lambda=k_\lambda$) takes the form of strict-sign inequality, and we have (4.1.15). By (4.1.15), it follows that the last inequality of (4.1.10) takes the form of strict-sign inequality, and then (4.1.14) yields.

On the other-hand, assuming that (4.1.14) is valid, sine $\|a\|_{2,\omega}>0$, there exists a $n_0\in\mathbf{N}$, such that $\sum_{n=1}^{n_0}n^{1-\lambda}a_n^2>0$. For $N\ge n_0$, setting

$$b_n(N)=n^{\lambda-1}\sum_{m=1}^{N}k(m,n)a_m,n\in\mathbf{N},$$

then by (4.1.14), we have

$$0<\sum_{n=1}^{N} n^{1-\lambda} b_n^2(N) = \sum_{n=1}^{N} n^{\lambda-1}(\sum_{m=1}^{N} k(m,n)a_m)^2$$
$$=\sum_{n=1}^{N}\sum_{m=1}^{N} k(m,n)a_m b_n(N)$$
$$<k_\lambda\{\sum_{n=1}^{N} n^{1-\lambda}a_n^2 \sum_{n=1}^{N} n^{1-\lambda} b_n^2(N)\}^{\frac{1}{2}}<\infty, \quad (4.1.16)$$
$$0<\sum_{n=1}^{N} n^{1-\lambda} b_n^2(N) = \sum_{n=1}^{N} n^{\lambda-1}(\sum_{m=1}^{N} k(m,n)a_m)^2$$
$$<k_\lambda(\sum_{n=1}^{\infty} n^{1-\lambda}a_n^2)^{\frac{1}{2}}. \quad (4.1.17)$$

It follows $0<\sum_{n=1}^{\infty} n^{1-\lambda} b_n^2(\infty)<\infty$. For $N\to\infty$, using (4.1.14), both (4.1.16) and (4.1.17) still take the strict-sign inequalities. Hence we have (4.1.15), which is equivalent to (4.1.14).

By the proof of Theorem 4.1.2 and Theorem 4.1.3, it follows that the constant factors in (4.1.14) and (4.1.15) are the best possible. □

4.1.2 SOME EXAMPLES FOR APPLYING THEOREM 4.1.3 AND THEOREM 4.1.5

In the following examples, we set $\omega(x)=x^{1-\lambda}$, $a,b\in l_\omega^2$, $0<\|a\|_{2,\omega}=\{\sum_{n=1}^{\infty} n^{1-\lambda}a_n^2\}^{\frac{1}{2}}<\infty$ and $0<\|b\|_{2,\omega}=\{\sum_{n=1}^{\infty} n^{1-\lambda}b_n^2\}^{\frac{1}{2}}<\infty$. The words that the constant factors are the best possible are omitted.

Example 4.1.6 If $\alpha>0, 0<\lambda\le 2$, $k_\lambda(x,y) =\frac{1}{(x^\alpha+y^\alpha)^{\lambda/\alpha}}$, setting $v=u^\alpha$, then we have

$$k_\lambda=\int_0^\infty k_\lambda(u,1)u^{\frac{\lambda}{2}-1}du$$
$$=\int_0^\infty \frac{1}{(1+u^\alpha)^{\lambda/\alpha}}u^{\frac{\lambda}{2}-1}du$$
$$=\frac{1}{\alpha}\int_0^\infty \frac{1}{(1+v)^{\lambda/\alpha}}v^{\frac{\lambda}{2\alpha}-1}dv=\frac{1}{\alpha}B(\frac{\lambda}{2\alpha},\frac{\lambda}{2\alpha}).$$

Since $k_\lambda(x,y)$ satisfies the assumption of Theorem 4.1.3, for $\alpha>0, 0<\lambda\le 2$, by Theorem 4.1.5, we have the following equivalent inequalities:

$$\sum_{n=1}^{\infty}\sum_{m=1}^{\infty}\frac{a_m b_n}{(m^\alpha+n^\alpha)^{\lambda/\alpha}}$$
$$<\frac{1}{\alpha}B(\frac{\lambda}{2\alpha},\frac{\lambda}{2\alpha})\|a\|_{2,\omega}\|b\|_{2,\omega}, \quad (4.1.18)$$
$$\sum_{n=1}^{\infty} n^{\lambda-1}[\sum_{m=1}^{\infty}\frac{a_m}{(m^\alpha+n^\alpha)^{\lambda/\alpha}}]^2$$
$$<[\frac{1}{\alpha}B(\frac{\lambda}{2\alpha},\frac{\lambda}{2\alpha})]^2\|a\|_{2,\omega}^2. \quad (4.1.19)$$

In particular, (1) for $0<\alpha=\lambda\le 2$, we have the following equivalent inequalities:

$$\sum_{n=1}^{\infty}\sum_{m=1}^{\infty}\frac{a_m b_n}{m^\lambda+n^\lambda}<\frac{\pi}{\lambda}\|a\|_{2,\omega}\|b\|_{2,\omega}, \quad (4.1.20)$$
$$\sum_{n=1}^{\infty} n^{\lambda-1}(\sum_{m=1}^{\infty}\frac{a_m}{m^\lambda+n^\lambda})^2<(\frac{\pi}{\lambda})^2\|a\|_{2,\omega}^2; \quad (4.1.21)$$

(2) for $\alpha=1, 0<\lambda\le 2$, we have the following equivalent inequalities:

$$\sum_{n=1}^{\infty}\sum_{m=1}^{\infty}\frac{a_m b_n}{(m+n)^\lambda}<B(\frac{\lambda}{2},\frac{\lambda}{2})\|a\|_{2,\omega}\|b\|_{2,\omega}, \quad (4.1.22)$$
$$\sum_{n=1}^{\infty} n^{\lambda-1}[\sum_{m=1}^{\infty}\frac{a_m}{(m+n)^\lambda}]^2<[B(\frac{\lambda}{2},\frac{\lambda}{2})]^2\|a\|_{2,\omega}^2. \quad (4.1.23)$$

Example 4.1.7 If $0<\lambda\le 2, k_\lambda(x,y)=\frac{1}{(\max\{x,y\})^\lambda}$, then we have

$$k_\lambda=\int_0^\infty k_\lambda(u,1)u^{\frac{\lambda}{2}-1}du$$
$$=\int_0^\infty \frac{1}{(\max\{u,1\})^\lambda}u^{\frac{\lambda}{2}-1}du=\frac{4}{\lambda}.$$

Since $k_\lambda(x,y)$ satisfies the condition of Theorem 4.1.3 for $0<\lambda\le 2$, by Theorem 4.1.5, we have the following equivalent inequalities

$$\sum_{n=1}^{\infty}\sum_{m=1}^{\infty}\frac{a_m b_n}{(\max\{m,n\})^\lambda}<\frac{4}{\lambda}\|a\|_{2,\omega}\|b\|_{2,\omega}, \quad (4.1.24)$$
$$\sum_{n=1}^{\infty} n^{\lambda-1}[\sum_{m=1}^{\infty}\frac{a_m}{(\max\{m,n\})^\lambda}]^2<(\frac{4}{\lambda})^2\|a\|_{2,\omega}^2. \quad (4.1.25)$$

Example 4.1.8 If $0<\lambda\le 2, k_\lambda(x,y)=\frac{\ln(x/y)}{x^\lambda-y^\lambda}$, setting $v=u^\lambda$, then we have

$$k_\lambda=\int_0^\infty k_\lambda(u,1)u^{\frac{\lambda}{2}-1}du=\int_0^\infty \frac{\ln u}{u^\lambda-1}u^{\frac{\lambda}{2}-1}du$$
$$=\frac{1}{\lambda^2}\int_0^\infty \frac{\ln v}{v-1}v^{\frac{-1}{2}}dv=(\frac{\pi}{\lambda})^2.$$

It is obvious that $k_\lambda(x,y)$ satisfies the condition of Theorem 4.1.3 for $0<\lambda\le 2$. By Theorem 4.1.5, we have the following equivalent inequalities:

$$\sum_{n=1}^{\infty}\sum_{m=1}^{\infty}\frac{\ln(m/n)a_m b_n}{m^\lambda-n^\lambda}<(\frac{\pi}{\lambda})^2\|a\|_{2,\omega}\|b\|_{2,\omega}, \quad (4.1.26)$$
$$\sum_{n=1}^{\infty} n^{\lambda-1}[\sum_{m=1}^{\infty}\frac{\ln(m/n)a_m}{m^\lambda-n^\lambda}]^2<(\frac{\pi}{\lambda})^4\|a\|_{2,\omega}^2. \quad (4.1.27)$$

4.1.3 SOME EXAMPLES FOR APPLYING THEOREM 4.1.2 AND THEOREM 4.1.5

Example 4.1.9 If $0<\lambda\le 4, k_\lambda(x,y)=\frac{1}{(x+y)^\lambda}$ $(x,y>0)$, then we define the following weight coefficients:

$$\varpi_\lambda(m):=\sum_{n=1}^{\infty}\frac{1}{(m+n)^\lambda}\frac{m^{\frac{\lambda}{2}}}{n^{1-\frac{\lambda}{2}}},$$

$$\vartheta_\lambda(n):=\sum_{m=1}^{\infty}\frac{1}{(m+n)^\lambda}\frac{n^{\frac{\lambda}{2}}}{m^{1-\frac{\lambda}{2}}},\quad m,n\in\mathbf{N}. \tag{4.1.28}$$

Setting $f_{\lambda,m}(y)=\frac{1}{(m+y)^\lambda y^{1-\lambda/2}}$, by (3.1.43), we have

$$\varpi_\lambda(m)=m^{\frac{\lambda}{2}}\sum_{n=1}^{\infty}f_{\lambda,m}(n)$$

$$=m^{\frac{\lambda}{2}}[\int_1^\infty f_{\lambda,m}(y)dy +\tfrac{1}{2}f_{\lambda,m}(1)+\int_1^\infty P_1(y)f'_{\lambda,m}(y)dy]$$

$$=m^{\frac{\lambda}{2}}\int_0^\infty f_{\lambda,m}(y)dy-m^{\frac{\lambda}{2}}\theta_\lambda(m), \tag{4.1.29}$$

$$\theta_\lambda(m)=\int_0^1 f_{\lambda,m}(y)dy -\tfrac{1}{2}f_{\lambda,m}(1)-\int_1^\infty P_1(y)f'_{\lambda,m}(y)dy. \tag{4.1.30}$$

We find $\frac{-1}{2}f_{\lambda,m}(1)=\frac{-1}{2(m+1)^\lambda}$. Setting $u=\frac{y}{m}$, we obtain

$$m^{\frac{\lambda}{2}}\int_0^\infty f_{\lambda,m}(y)dy=\int_0^\infty\frac{m^{\lambda/2}}{(m+y)^\lambda y^{1-\lambda/2}}dy$$

$$=\int_0^\infty\frac{1}{(u+1)^\lambda}u^{\frac{\lambda}{2}-1}du=B(\tfrac{\lambda}{2},\tfrac{\lambda}{2}), \tag{4.1.31}$$

$$\int_0^1 f_{\lambda,m}(y)dy=\frac{2}{\lambda}\int_0^1\frac{1}{(m+y)^\lambda}dy^{\frac{\lambda}{2}}$$

$$=\frac{2}{\lambda(m+1)^\lambda}+2\int_0^1\frac{1}{(m+y)^{\lambda+1}}y^{\frac{\lambda}{2}}dy$$

$$=\frac{2}{\lambda(m+1)^\lambda}+\frac{4}{2+\lambda}\int_0^1\frac{1}{(m+y)^{\lambda+1}}dy^{\frac{\lambda}{2}+1}$$

$$=\frac{2}{\lambda(m+1)^\lambda}+\frac{4}{(2+\lambda)(m+1)^{\lambda+1}}+\frac{4(\lambda+1)}{(2+\lambda)}\int_0^1\frac{y^{\lambda/2+1}dy}{(m+y)^{\lambda+2}}$$

$$=\frac{2}{\lambda(m+1)^\lambda}+\frac{4}{(2+\lambda)(m+1)^{\lambda+1}}$$

$$+\frac{8(\lambda+1)}{(2+\lambda)(4+\lambda)(m+1)^{\lambda+2}}+\frac{8(\lambda+1)(\lambda+2)}{(2+\lambda)(4+\lambda)}\int_0^1\frac{y^{\lambda/2+2}dy}{(m+y)^{\lambda+3}}$$

$$>\frac{2}{\lambda(m+1)^\lambda}+\frac{4}{(2+\lambda)(m+1)^{\lambda+1}}+\frac{8(\lambda+1)}{(2+\lambda)(4+\lambda)(m+1)^{\lambda+2}}$$

$$>\frac{2}{\lambda(m+1)^\lambda}+\frac{2}{3(m+1)^{\lambda+1}}+\frac{1}{6(m+1)^{\lambda+2}},\quad 0<\lambda\le 4,$$

$$f'_{\lambda,m}(y)=-\frac{\lambda}{(m+y)^{\lambda+1}y^{1-\lambda/2}}-\frac{(1-\lambda/2)}{(m+y)^\lambda y^{2-\lambda/2}}$$

$$=-\frac{(1+\lambda/2)}{(m+y)^\lambda y^{2-\lambda/2}}+\frac{m\lambda}{(m+y)^{\lambda+1}y^{2-\lambda/2}},$$

$$-\int_1^\infty P_1(y)f'_{\lambda,m}(y)dy$$

$$=\int_1^\infty P_1(y)\frac{(1+\lambda/2)dy}{(m+y)^\lambda y^{2-\lambda/2}}$$

$$-m\lambda\int_1^\infty P_1(y)\frac{dy}{(m+y)^{\lambda+1}y^{2-\lambda/2}}. \tag{4.1.32}$$

By (3.1.46), (3.1.44) and (3.1.48), since $0<\lambda\le 4$, it follows

$$-\int_1^\infty P_1(y)f'_{\lambda,m}(y)dy$$

$$>-\frac{(1+\lambda/2)}{12(m+1)^\lambda}+\frac{m\lambda}{12(m+1)^{\lambda+1}}$$

$$-\frac{m\lambda}{720}[\frac{(\lambda+1)(\lambda+2)}{(m+1)^{\lambda+3}}+\frac{(\lambda+1)(4-\lambda)}{(m+1)^{\lambda+2}}+\frac{(4-\lambda)(6-\lambda)}{4(m+1)^{\lambda+1}}]$$

$$>-\frac{(1+\lambda/2)}{12(m+1)^\lambda}+\frac{(m+1)\lambda-\lambda}{12(m+1)^{\lambda+1}}$$

$$-\frac{(m+1)4}{720}[\frac{30}{(m+1)^{\lambda+3}}+\frac{20}{(m+1)^{\lambda+2}}+\frac{24}{4(m+1)^{\lambda+1}}]$$

$$>-\frac{(1+\lambda/2)}{12(m+1)^\lambda}+\frac{\lambda}{12(m+1)^\lambda}-\frac{4}{12(m+1)^{\lambda+1}}$$

$$-\frac{1}{6(m+1)^{\lambda+2}}-\frac{1}{9(m+1)^{\lambda+1}}-\frac{1}{30(m+1)^\lambda}$$

$$>(-\tfrac{7}{60}+\tfrac{\lambda}{24})\frac{1}{(m+1)^\lambda}-\frac{4}{9(m+1)^{\lambda+1}}-\frac{1}{6(m+1)^{\lambda+2}}.$$

Then by (4.1.30), we find

$$\theta_\lambda(m)>(\tfrac{2}{\lambda}-\tfrac{37}{60}+\tfrac{\lambda}{24})\frac{1}{(m+1)^\lambda}+(\tfrac{2}{3}-\tfrac{4}{9})\frac{1}{(m+1)^{\lambda+1}}+(\tfrac{1}{6}-\tfrac{1}{6})\frac{1}{(m+1)^{\lambda+2}}$$

$$>(5\lambda^2-74\lambda+240)\frac{1}{120\lambda(m+1)^\lambda}$$

$$\ge\frac{1}{5\lambda(m+1)^\lambda}>0,\quad 0<\lambda\le 4. \tag{4.1.33}$$

Hence by (4.1.29), we have $\varpi_\lambda(m)<B(\frac{\lambda}{2},\frac{\lambda}{2})$. By the same way, we have $\vartheta_\lambda(n)<B(\frac{\lambda}{2},\frac{\lambda}{2})$, and (4.1.8) is valid.

Since we find

$$\int_0^1 f_{\lambda,m}(y)dy<\int_0^1\frac{1}{m^\lambda}y^{\frac{\lambda}{2}-1}dy=\frac{2}{\lambda m^\lambda},$$

then by (4.1.32) and (3.1.50), we have

$$-\int_1^\infty P_1(y)f'_{\lambda,m}(y)dy<\frac{m\lambda}{12(m+1)^{\lambda+1}}.$$

Hence by (4.1.30), it follows

$$0<m^{\frac{\lambda}{2}}\theta_\lambda(m)<m^{\frac{\lambda}{2}}[\frac{2}{\lambda m^\lambda}-\frac{1}{2(m+1)^\lambda}+\frac{m\lambda}{12(m+1)^{\lambda+1}}]$$

$$<(\tfrac{2}{\lambda}+\tfrac{\lambda}{12})\frac{1}{m^{\lambda/2}},$$

and $m^{\frac{\lambda}{2}}\theta_\lambda(m)=O(\frac{1}{m^{\lambda/2}})(m\to\infty)$. Then by (4.1.29) and (4.1.31), we obtain (4.1.9). Since

$$k_\lambda(1,u)=\frac{1}{(1+u)^\lambda}=O(\frac{1}{u^{2\lambda/3}})$$

$$(\tfrac{\lambda}{2}<\delta=\tfrac{2\lambda}{3}<\tfrac{\lambda}{2}+1;u\to\infty),$$

then by Theorem 4.1.5, for $0<\lambda\le 4$, we have the following equivalent inequalities:

$$\sum_{n=1}^{\infty}\sum_{m=1}^{\infty}\frac{a_m b_n}{(m+n)^\lambda}<B(\tfrac{\lambda}{2},\tfrac{\lambda}{2})\,||\,a\,||_{2,\omega}||\,b\,||_{2,\omega}, \tag{4.1.34}$$

$$\sum_{n=1}^{\infty}n^{\lambda-1}[\sum_{m=1}^{\infty}\frac{a_m}{(m+n)^\lambda}]^2<[B(\tfrac{\lambda}{2},\tfrac{\lambda}{2})]^2\,||\,a\,||_{2,\omega}^2. \tag{4.1.35}$$

Note 4.1.10 We can see the above methods and results in (Yang JNUMB 2001)[1].

Example 4.1.11 Assuming that $0<\lambda\le 2$, $k_\lambda(x,y)=\frac{|\ln(x/y)|}{(\max\{x,y\})^\lambda}$, we define the following weight coefficients:

$$\varpi_\lambda(m):=\sum_{n=1}^{\infty}\frac{|\ln m/n)|}{(\max\{m,n\})^\lambda}\frac{m^{\lambda/2}}{n^{1-\lambda/2}},$$

$$\vartheta_\lambda(n):=\sum_{m=1}^{\infty}\frac{|\ln(m/n)|}{(\max\{m,n\})^\lambda}\frac{n^{\lambda/2}}{m^{1-\lambda/2}},m,n\in\mathbf{N}. \quad (4.1.36)$$

Setting $f_{\lambda,m}(y)=\frac{|\ln(m/y)|}{(\max\{m,y\})^\lambda y^{1-\lambda/2}}$, by (3.1.43), it follows

$$\varpi_\lambda(m)=m^{\frac{\lambda}{2}}\sum_{n=1}^{\infty}f_{\lambda,m}(n)=m^{\frac{\lambda}{2}}[\int_1^\infty f_{\lambda,m}(y)dy$$
$$+\tfrac{1}{2}f_{\lambda,m}(1)+\int_1^\infty P_1(y)f'_{\lambda,m}(y)dy]$$
$$=m^{\frac{\lambda}{2}}\int_0^\infty f_{\lambda,m}(y)dy-m^{\frac{\lambda}{2}}\theta_\lambda(m); \quad (4.1.37)$$
$$\theta_\lambda(m)=\int_0^1 f_{\lambda,m}(y)dy$$
$$-\tfrac{1}{2}f_{\lambda,m}(1)-\int_1^\infty P_1(y)f'_{\lambda,m}(y)dy. \quad (4.1.38)$$

We find $-\frac{1}{2}f_{\lambda,m}(1)=-\frac{\ln m}{2m^\lambda}$. Setting $u=\frac{y}{m}$, we obtain

$$m^{\frac{\lambda}{2}}\int_0^\infty f_{\lambda,m}(y)dy=\int_0^\infty\frac{|\ln(m/y)|m^{\lambda/2}}{(\max\{m,y\})^\lambda y^{1-\lambda/2}}dy$$
$$=\int_0^\infty\frac{|\ln u|du}{(\max\{u,1\})^\lambda u^{1-\lambda/2}}=\int_0^1\frac{-\ln udu}{u^{1-\lambda/2}}+\int_1^\infty\frac{\ln udu}{u^{1+\lambda/2}}$$
$$=\tfrac{2}{\lambda}\int_0^1(-\ln u)du^{\frac{\lambda}{2}}-\tfrac{2}{\lambda}\int_1^\infty\ln udu^{\frac{-\lambda}{2}}=\tfrac{8}{\lambda^2}, \quad (4.1.39)$$
$$\int_0^1 f_{\lambda,m}(y)dy=\frac{1}{m^{\lambda/2}}\int_0^{\frac{1}{m}}\frac{-\ln udu}{u^{1-\lambda/2}}$$
$$=\frac{2}{\lambda m^{\lambda/2}}\int_0^{\frac{1}{m}}(-\ln u)du^{\frac{\lambda}{2}}=\frac{2}{\lambda m^\lambda}(\ln m+\tfrac{2}{\lambda}).$$

For $1\le y\le m$, since $f_{\lambda,m}(y)=\frac{-\ln(y/m)}{m^\lambda y^{1-\lambda/2}}$, we have

$$f'_{\lambda,m}(y)=[\frac{-1}{y^{2-\lambda/2}}+(1-\tfrac{\lambda}{2})\frac{\ln(y/m)}{y^{2-\lambda/2}}]\frac{1}{m^\lambda}$$
$$=[\frac{-1}{y^{2-\lambda/2}}-(1-\tfrac{\lambda}{2})\frac{m\ln(y/m)}{(y-m)y^{2-\lambda/2}}$$
$$+(1-\tfrac{\lambda}{2})\frac{\ln(y/m)}{(y-m)y^{1-\lambda/2}}]\frac{1}{m^\lambda};$$

for $y\ge m$, since $f_{\lambda,m}(y)=\frac{\ln(y/m)}{y^{1+\lambda/2}}$, we find

$$f'_{\lambda,m}(y)=\frac{1}{y^{2+\lambda/2}}-(1+\tfrac{\lambda}{2})\frac{\ln(\frac{y}{m})}{y^{2+\lambda/2}}$$
$$=\frac{1}{y^{2+\lambda/2}}+(1+\tfrac{\lambda}{2})\frac{m\ln(y/m)}{(y-m)y^{2+\lambda/2}}-(1+\tfrac{\lambda}{2})\frac{\ln(y/m)}{(y-m)y^{1+\lambda/2}}.$$

Then by (3.1.45) and (3.1.46), it follows

$$\int_1^\infty P_1(y)f'_{\lambda,m}(y)dy$$
$$=\int_1^m P_1(y)f'_{\lambda,m}(y)dy+\int_m^\infty P_1(y)f'_{\lambda,m}(y)dy$$
$$<\frac{1}{12m^\lambda}[\frac{-1}{y^{2-\lambda/2}}-(1-\tfrac{\lambda}{2})\frac{m\ln(y/m)}{(y-m)y^{2-\lambda/2}}]_1^m$$
$$+\tfrac{1}{12}(1+\tfrac{\lambda}{2})\frac{\ln(y/m)}{(y-m)y^{1+\lambda/2}}\Big|_{y=m}$$
$$=\frac{1}{12m^\lambda}-\tfrac{1}{12}(2-\tfrac{\lambda}{2})\frac{1}{m^{2+\lambda/2}}+\tfrac{1}{12}(1-\tfrac{\lambda}{2})\frac{m\ln m}{(m-1)m^\lambda}$$
$$+\tfrac{1}{12}(1+\tfrac{\lambda}{2})\frac{1}{m^{2+\lambda/2}}.$$

Hence we have

$$m^{\frac{\lambda}{2}}\theta_\lambda(m)>\tilde{\theta}_\lambda(m):=(\tfrac{2}{\lambda}-\tfrac{1}{2})\frac{\ln m}{m^{\lambda/2}}$$
$$+(\tfrac{4}{\lambda^2}-\tfrac{1}{12})\frac{1}{m^{\lambda/2}}+\tfrac{1}{12}(1-\lambda)\frac{1}{m^2}-\tfrac{1}{12}(1-\tfrac{\lambda}{2})\frac{m\ln m}{(m-1)m^{\lambda/2}}$$

.

For $m=1$, we find

$$\tilde{\theta}_\lambda(1)=(\tfrac{4}{\lambda^2}-\tfrac{1}{12})+\tfrac{1}{12}(1-\lambda)-\tfrac{1}{12}(1-\tfrac{\lambda}{2})$$
$$=\tfrac{4}{\lambda^2}-\tfrac{1}{12}-\tfrac{\lambda}{24}>0,\ 0<\lambda\le 2;$$

For $m\ge 2$, since

$$0<\lambda\le 2,-\tfrac{1}{m-1}\ge-\tfrac{2}{m}\text{ and}\tfrac{1}{m^{\lambda/2}}\ge\tfrac{1}{m^2},$$

we find

$$\tilde{\theta}_\lambda(m)\ge(\tfrac{2}{\lambda}-\tfrac{2}{3}+\tfrac{\lambda}{12})\frac{\ln m}{m^{\lambda/2}}+(\tfrac{4}{\lambda^2}-\tfrac{\lambda}{12})\frac{1}{m^2}>0.$$

Hence by (4.1.39) and (4.1.37), we have $\varpi_\lambda(m)<\frac{8}{\lambda^2}$. By the same way, it follows $\vartheta_\lambda(n)<\frac{8}{\lambda^2}$, and (4.1.8) is valid.

Since by (3.1.45) and (3.1.46), we have

$$\int_1^\infty P_1(x)f'_{\lambda,m}(y)dy$$
$$>\frac{1}{12m^\lambda}[(1-\tfrac{\lambda}{2})\frac{\ln(y/m)}{(y-m)y^{1-\lambda/2}}]_1^m$$
$$-\tfrac{1}{12}[\frac{1}{y^{2+\lambda/2}}+(1+\tfrac{\lambda}{2})\frac{m\ln(y/m)}{(y-m)y^{2+\lambda/2}}]_{y=m}$$
$$=-\tfrac{1}{12}(1+\lambda)\frac{1}{m^{2+\lambda/2}}-\tfrac{1}{12}(1-\tfrac{\lambda}{2})\frac{\ln m}{(m-1)m^\lambda},$$

then by (4.1.38) and the above results, we obtain

$$0<m^{\frac{\lambda}{2}}\theta_\lambda(m)<\frac{2}{\lambda m^{\lambda/2}}(\ln m+\tfrac{2}{\lambda})-\frac{\ln m}{2m^{\lambda/2}}$$
$$+\tfrac{1}{12}(1+\lambda)\frac{1}{m^2}+\tfrac{1}{12}(1+\tfrac{\lambda}{2})\frac{\ln m}{(m-1)m^{\lambda/2}}$$
$$<[\tfrac{2}{\lambda}\ln m+\tfrac{4}{\lambda^2}+\tfrac{1}{12}(1+\lambda)+\tfrac{1}{12}(1+\tfrac{\lambda}{2})\tfrac{\ln m}{m-1}]\frac{1}{m^{\lambda/2}},$$

and $m^{\frac{\lambda}{2}}\theta_\lambda(m)=O(\frac{1}{m^{\lambda/4}})$ $(m\to\infty)$. By (4.1.37) and (4.1.39), we have (4.1.9). Since

$$k_\lambda(1,u)=\frac{|\ln u|}{(1+u)^\lambda}=O(\frac{1}{u^{2\lambda/3}})$$
$$(\tfrac{\lambda}{2}<\delta=\tfrac{2\lambda}{3}<\tfrac{\lambda}{2}+1;u\to\infty),$$

then by Theorem 4.1.5, for $0<\lambda\le 2$, we have the following equivalent inequalities:

$$\sum_{n=1}^{\infty}\sum_{m=1}^{\infty}\frac{|\ln(m/n)|a_m b_n}{(\max\{m,n\})^{\lambda}}<\frac{8}{\lambda^2}\|a\|_{2,\omega}\|b\|_{2,\omega}, \quad (4.1.40)$$

$$\sum_{n=1}^{\infty}n^{\lambda-1}[\sum_{m=1}^{\infty}\frac{|\ln(m/n)|a_m}{(\max\{m,n\})^{\lambda}}]^2<\frac{64}{\lambda^4}\|a\|_{2,\omega}^2. \quad (4.1.41)$$

4.1.4 SOME IMPROVEMENTS OF THE BASIC HILBERT-TYPE INEQUALITIES

For $\lambda=1$ in (4.1.22), (4.1.24), (4.1.26) and (4.1.40), we obtain the following basic Hilbert-type inequalities:

$$\sum_{n=1}^{\infty}\sum_{m=1}^{\infty}\frac{a_m b_n}{m+n}<\pi\|a\|_2\|b\|_2, \quad (4.1.42)$$

$$\sum_{n=1}^{\infty}\sum_{m=1}^{\infty}\frac{a_m b_n}{\max\{m,n\}}<4\|a\|_2\|b\|_2, \quad (4.1.43)$$

$$\sum_{n=1}^{\infty}\sum_{m=1}^{\infty}\frac{\ln(m/n)a_m b_n}{m-n}<\pi^2\|a\|_2\|b\|_2, \quad (4.1.44)$$

$$\sum_{n=1}^{\infty}\sum_{m=1}^{\infty}\frac{|\ln(m/n)|a_m b_n}{\max\{m,n\}}<8\|a\|_2\|b\|_2, \quad (4.1.45)$$

where the constant factors are all the best possible.

In 2002, (Zhang JMAA 2002)[2] gave

$$\sum_{n=1}^{\infty}\sum_{m=1}^{\infty}\frac{a_m b_n}{m+n-1}\le\frac{\pi}{\sqrt{2}}[\|a\|_2^2\|b\|_2^2+(a,b)^2]^{\frac{1}{2}}. \quad (4.1.46)$$

By Cauchy' inequality $(a,b)\le\|a\|_2^2\|b\|_2^2$ and (4.1.46), we deduce the following Hilbert's inequality:

$$\sum_{n=1}^{\infty}\sum_{m=1}^{\infty}\frac{a_m b_n}{m+n-1}<\pi\|a\|_2\|b\|_2, \quad (4.1.47)$$

where the constant factor π is the best possible.

(4.1.47) is a more accurate form of (4.1.42). We can proof that the constant factor $\frac{\pi}{\sqrt{2}}$ in (4.1.46) is still the best possible. If $a_m,b_n\ge 0$, then we have

$$\sum_{n=1}^{\infty}\sum_{m=1}^{\infty}\frac{a_m b_n}{m+n}\le\sum_{n=1}^{\infty}\sum_{m=1}^{\infty}\frac{a_m b_n}{m+n-1},$$

by (4.1.46), we deduce the following inequality:

$$\sum_{n=1}^{\infty}\sum_{m=1}^{\infty}\frac{a_m b_n}{m+n}\le\frac{\pi}{\sqrt{2}}[\|a\|_2^2\|b\|_2^2+(a,b)^2]^{\frac{1}{2}}. \quad (4.1.48)$$

For general signs of a_m,b_n, (4.1.48) is still valid, which is an improvement of (4.1.42) and the constant factor $\frac{\pi}{\sqrt{2}}$ is still the best possible.

Since for $a_m,b_n\ge 0$, we have the following inequality (Hardy CUP 1934)[3]:

$$\sum_{n=1}^{\infty}\sum_{m=1}^{\infty}\frac{\ln(m/n)a_m b_n}{m-n}\le\pi\sum_{n=1}^{\infty}\sum_{m=1}^{\infty}\frac{a_m b_n}{m+n},$$

then by (4.1.48), we obtain a refinement of (4.1.42) as follows:

$$\sum_{n=1}^{\infty}\sum_{m=1}^{\infty}\frac{\ln(m/n)a_m b_n}{m-n}\le\frac{\pi^2}{\sqrt{2}}[\|a\|_2^2\|b\|_2^2+(a,b)^2]^{\frac{1}{2}}, \quad (4.1.49)$$

where the constant factor $\frac{\pi^2}{\sqrt{2}}$ is the best possible, and (4.1.49) is valid for general signs of a_m,b_n.

In the following, we give some improvements of (4.1.43) and (4.1.45). For this, we introduce some results of (Zhang JMAA 2002)[2] as follows:

Assuming that H is a real separable Hilbert space and $T:H\to H$ is a bounded self-adjoint semi-positive definite operator, we have

$$(Ta,b)\le\frac{\|T\|}{\sqrt{2}}[\|a\|^2\|b\|^2+(a,b)^2]^{\frac{1}{2}},\quad a,b\in H, \quad (4.1.50)$$

where, (a,b) is indicated the inner product of a and b, $\|a\|=\sqrt{(a,a)}$ is indicated the norm of a. It is obvious that (4.1.50) is an improvement of

$$(Ta,b)\le\|T\|\cdot\|a\|\cdot\|b\|\ (a,b\in H). \quad (4.1.51)$$

Since $\|T\|$ is the best value of (4.1.51), then we can conclude that $\frac{\|T\|}{\sqrt{2}}$ is the best value of (4.1.50).

Setting $H=l^2$, we define the operator $T:l^2\to l^2$ as: for $a=\{a_m\}_{m=1}^{\infty}\in l^2$, $(Ta)(n)=c_n=\sum_{m=1}^{\infty}\frac{a_m}{m+n-1}$. (Wilhelm AJM 1950)[4] proved that T is a bounded self-adjoint semi-positive definite operator and $\|T\|=\pi$. Then by (4.1.50), we have (4.1.46).

Example 4.1.12 For $k\in\mathbf{N}\setminus\{1\}$, we define a linear operator $T_k:\mathbb{R}^k\to\mathbb{R}^k$ as: for $a^{(k)}=\{a_m\}_{m=1}^{k}\in\mathbf{R}^k$, it follows

$$(T_k a^{(k)})(n)=c_n=\sum_{m=1}^{k}\frac{a_m}{\max\{m,n\}}\ (n=1,\cdots,k).$$

Equivalently we have

$$T_k\begin{pmatrix}a_1\\a_2\\a_3\\\vdots\\a_k\end{pmatrix}=A_{k\times k}\begin{pmatrix}a_1\\a_2\\a_3\\\vdots\\a_k\end{pmatrix},$$

where, $A_{k\times k}$ is called the matrix of T_k. By simplification, it follows

$$A_{k\times k}:=\begin{pmatrix} 1 & \frac{1}{2} & \frac{1}{3} & \cdots & \frac{1}{k} \\ \frac{1}{2} & \frac{1}{2} & \frac{1}{3} & \cdots & \frac{1}{k} \\ \vdots & \vdots & \vdots & \vdots & \vdots \\ \frac{1}{k-1} & \frac{1}{k-1} & \frac{1}{k-1} & \cdots & \frac{1}{k} \\ \frac{1}{k} & \frac{1}{k} & \frac{1}{k} & \cdots & \frac{1}{k} \end{pmatrix}.$$

In the following, we prove that $A_{k\times k}$ is definite positive, and then T_k is also definite positive.

By making primary row transforms in $A_{k\times k}$, we finally find

$$B_{k\times k}:=\begin{pmatrix} \frac{1}{2} & 0 & 0 & \cdots & 0 \\ \frac{1}{6} & \frac{1}{6} & 0 & \cdots & 0 \\ \vdots & \vdots & \vdots & \vdots & \vdots \\ \frac{1}{k(k-1)} & \frac{1}{k(k-1)} & \frac{1}{k(k-1)} & \cdots & 0 \\ \frac{1}{k} & \frac{1}{k} & \frac{1}{k} & \cdots & \frac{1}{k} \end{pmatrix}.$$

Obviously $B_{k\times k}$ is definite positive; so are $A_{k\times k}$ and T_k. Then by (4.1.43), we have

$$\| T_k a^{(k)} \| = \sum_{n=1}^{k}\left(\sum_{m=1}^{k}\frac{a_m}{\max\{m,n\}}\right)^2 \le 4\| a^{(k)} \|.$$

It is obvious that $\| T_k \| \le 4$ and T_k is bounded. Since

$$(T_k a^{(k)}, b^{(k)}) = (a^{(k)}, T_k b^{(k)})$$
$$=\sum_{n=1}^{k}\sum_{m=1}^{k}\frac{a_m b_n}{\max\{m,n\}},$$

T_k is self-adjoint, by (4.1.50), we have

$$\sum_{n=1}^{k}\sum_{m=1}^{k}\frac{a_m b_n}{\max\{m,n\}}$$
$$\le \frac{4}{\sqrt{2}}[\| a^{(k)} \|_2^2 \| b^{(k)} \|_2^2 + (a^{(k)}, b^{(k)})^2]^{\frac{1}{2}},$$

For $k\to\infty$ in the above inequality, we obtain

$$\sum_{n=1}^{\infty}\sum_{m=1}^{\infty}\frac{a_m b_n}{\max\{m,n\}}$$
$$\le 2\sqrt{2}[\| a \|_2^2 \| b \|_2^2 + (a,b)^2]^{\frac{1}{2}}. \qquad (4.1.52)$$

It is obvious that (4.1.52) is an improvement of (4.1.43) and the constant factor is the best possible.

For $a_m, b_n \ge 0$, we have the following inequality (Hardy CUP 1934)[3]:

$$\sum_{n=1}^{\infty}\sum_{m=1}^{\infty}\frac{|\ln(m/n)| a_m b_n}{\max\{m,n\}} \le 2\sum_{n=1}^{\infty}\sum_{m=1}^{\infty}\frac{a_m b_n}{\max\{m,n\}},$$

then by (4.1.52), we obtain the following improvement of (4.1.45):

$$\sum_{n=1}^{\infty}\sum_{m=1}^{\infty}\frac{|\ln(m/n)| a_m b_n}{\max\{m,n\}}$$
$$\le 4\sqrt{2}[\| a \|_2^2 \| b \|_2^2 + (a,b)^2]^{\frac{1}{2}}, \qquad (4.1.53)$$

where, (4.1.53) is valid for general signs of a_m, b_n and the constant factor is the best possible.

Thus we give some improvements of four basic Hilbert-type inequalities (4.1.42)-(4.1.45).

4.2 HILBERT-TYPE INEQUALITIES WITH MULTI-PARAMETERS

4.2.1 THE NORM OF OPERATOR AND HILBRT-TYPE INEQUALITIES WITH PARAMETERS

Suppose that $p>1, \frac{1}{p}+\frac{1}{q}=1$, $\lambda_1, \lambda_2, \lambda \in \mathbf{R}$, and $\lambda_1+\lambda_2=\lambda$. $k_\lambda(x,y)(\ge 0)$ is a finite homogeneous function of degree $-\lambda$ in $\mathbf{R}_+^2$,

$$\phi(x)=x^{p(1-\lambda_1)-1}, \psi(x)=x^{q(1-\lambda_2)-1},$$

and $[\psi(x)]^{1-p}=x^{p\lambda_2-1} (x\in(0,\infty))$. We define the following real sequences space:

$$l_\phi^p=\{a=\{a_n\}_{n=1}^{\infty};$$
$$\| a \|_{p,\phi}=\{\sum_{n=1}^{\infty}\phi(n)|a_n|^p\}^{\frac{1}{p}}<\infty\}.$$

By the same way, we may define the real spaces l_ψ^q and $l_{\psi^{1-p}}^p$.

Assuming that $a=\{a_m\}_{m=1}^{\infty}\in l_\phi^p$,

$$c_n=\sum_{m=1}^{\infty}k_\lambda(m,n)a_m \ (n\in\mathbf{N}),$$

such that $c=\{c_n\}_{n=1}^{\infty}\in l_{\psi^{1-p}}^p$, we define a linear operator as: $T: l_\phi^p \to l_{\psi^{1-p}}^p$, for any $a=\{a_m\}_{m=1}^{\infty}\in l_\phi^p$,

$$(Ta)(n):=c_n=\sum_{m=1}^{\infty}k_\lambda(m,n)a_m,\ n\in\mathbf{N}. \qquad (4.2.1)$$

If there exists a constant $K>0$, such that for any $a\in l_\phi^p$, $\| Ta \|_{p,\psi^{1-p}} \le K\| a \|_{p,\phi}$, then the operator T is bounded and the norm of T satisfies

$$\|T\|_{p,\psi^{1-p}}=\sup_{a\in l_\phi^p\,(a\neq\theta)}\frac{\|Ta\|_{p,\psi^{1-p}}}{\|a\|_{p,\phi}}\le K. \qquad (4.2.2)$$

We define the weight coefficients $\varpi(\lambda_2,m)$ and $\vartheta(\lambda_1,n)$ as follows

$$\varpi(\lambda_2,m):=\sum_{n=1}^{\infty}k_\lambda(m,n)\frac{m^{\lambda_1}}{n^{1-\lambda_2}},$$

$$\vartheta(\lambda_1,n):=\sum_{m=1}^{\infty}k_\lambda(m,n)\frac{n^{\lambda_2}}{m^{1-\lambda_1}}\quad m,n\in\mathbf{N}. \qquad (4.2.3)$$

Hence it follows:

Theorem 4.2.1 If there exist constants $\tilde{k}(\lambda_2)$ and $k'(\lambda_1)>0$, satisfying the following inequalities

$$\varpi(\lambda_2,m)<\tilde{k}(\lambda_2),\vartheta(\lambda_1,n)<k'(\lambda_1),$$
$$m,n\in\mathbf{N}, \qquad (4.2.4)$$

then the bounded linear operator T defined by (4.2.1) exists and

$$\|T\|_{p,\psi^{1-p}}\le\tilde{k}^{\frac{1}{p}}(\lambda_2)k'^{\frac{1}{q}}(\lambda_1).$$

Proof For $a_m\ge 0$, $a=\{a_m\}_{m=1}^{\infty}\in l_\phi^p$. By Hölder's inequality with weight and (4.2.3)-(4.2.4), we have

$$[\sum_{m=1}^{\infty}k_\lambda(m,n)a_m]^p$$
$$=\{\sum_{m=1}^{\infty}k_\lambda(m,n)[\frac{m^{(1-\lambda_1)/q}}{n^{(1-\lambda_2)/p}}a_m][\frac{n^{(1-\lambda_2)/p}}{m^{(1-\lambda_1)/q}}]\}^p$$
$$\le[\sum_{m=1}^{\infty}k_\lambda(m,n)\frac{m^{(1-\lambda_1)(p-1)}}{n^{1-\lambda_2}}a_m^p]$$
$$\times[\sum_{m=1}^{\infty}k_\lambda(m,n)\frac{n^{(1-\lambda_2)(q-1)}}{m^{1-\lambda_1}}]^{p-1}$$
$$=[\vartheta(\lambda_1,n)]^{p-1}n^{1-p\lambda_2}$$
$$\times\sum_{m=1}^{\infty}k_\lambda(m,n)\frac{m^{(1-\lambda_1)(p-1)}}{n^{1-\lambda_2}}a_m^p. \qquad (4.2.5)$$

Then by (4.2.5) and (4.2.4), we obtain

$$\|Ta\|_{p,\psi^{1-p}}=\|c\|_{p,\psi^{1-p}}$$
$$=\{\sum_{n=1}^{\infty}n^{p\lambda_2-1}[\sum_{m=1}^{\infty}k_\lambda(m,n)a_m]^p\}^{\frac{1}{p}}$$
$$\le k'^{\frac{1}{q}}(\lambda_1)\{\sum_{n=1}^{\infty}\sum_{m=1}^{\infty}k_\lambda(m,n)\frac{m^{(1-\lambda_1)(p-1)}}{n^{1-\lambda_2}}a_m^p\}^{\frac{1}{p}}$$
$$=k'^{\frac{1}{q}}(\lambda_1)\{\sum_{m=1}^{\infty}[\sum_{n=1}^{\infty}k_\lambda(m,n)\frac{m^{\lambda_1}}{n^{1-\lambda_2}}]m^{p(1-\lambda_1)-1}a_m^p\}^{\frac{1}{p}}$$
$$=k'^{\frac{1}{q}}(\lambda_1)\{\sum_{m=1}^{\infty}\varpi(\lambda_2,m)m^{p(1-\lambda_1)-1}a_m^p\}^{\frac{1}{p}}$$
$$\le\tilde{k}^{\frac{1}{p}}(\lambda_2)k'^{\frac{1}{q}}(\lambda_1)\{\sum_{m=1}^{\infty}m^{p(1-\lambda_1)-1}a_m^p\}^{\frac{1}{p}}$$
$$=\tilde{k}^{\frac{1}{p}}(\lambda_2)k'^{\frac{1}{q}}(\lambda_1)\|a\|_{p,\phi}, \qquad (4.2.6)$$

Hence $c=Ta\in l_{\psi^{1-p}}^p$, and the bounded linear operator T exists, and by (4.2.6), we still have

$$\|T\|_{p,\psi^{1-p}}\le\tilde{k}^{\frac{1}{p}}(\lambda_2)k'^{\frac{1}{q}}(\lambda_1).$$

□

Theorem 4.2.2 Assuming that $\lambda_1,\lambda_2,\lambda\in\mathbf{R}$, $\lambda_1+\lambda_2=\lambda$, $k_\lambda(x,y)(\ge 0)$ is a finite homogeneous function of degree $-\lambda$ in $\mathbf{R}_+^2$,

$$0<k(\lambda_1)=\int_0^\infty k_\lambda(u,1)u^{\lambda_1-1}du<\infty \qquad (4.2.7)$$

and $k_\lambda(1,u)=O(\frac{1}{u^\delta})(\lambda_2<\delta<\lambda_2+1;u\to\infty)$, if

$$\varpi(\lambda_2,m)<k(\lambda_1),\vartheta(\lambda_1,n)<k(\lambda_1),\quad m,n\in\mathbf{N}, \qquad (4.2.8)$$

there exists a $\lambda'>0$, such that for any $m\in\mathbf{N}$,

$$k(\lambda_1)-O(\frac{1}{m^{\lambda'}})\le\varpi(\lambda_2,m)\ (m\to\infty), \qquad (4.2.9)$$

then the bounded linear operator T defined by (4.2.1) exists, and $\|T\|_{p,\psi^{1-p}}=k(\lambda_1)$.

Proof Setting $\tilde{k}(\lambda_2)=k'(\lambda_1)=k(\lambda_1)$, then (4.2.4) is valid. By Theorem 4.2.1, the bounded linear operator T exists with $\|T\|_{p,\psi^{1-p}}\le k(\lambda_1)$. Setting $a_m,b_n\ge 0$, $a=\{a_m\}_{m=1}^{\infty}\in l_\phi^p$, $b=\{b_n\}_{n=1}^{\infty}\in l_\psi^q$, by Hölder's inequality and (4.2.6) (for $\tilde{k}(\lambda_2)=k'(\lambda_1)=k(\lambda_1)$), we have

$$\sum_{n=1}^{\infty}\sum_{m=1}^{\infty}k_\lambda(m,n)a_m b_n$$
$$=\sum_{n=1}^{\infty}[n^{\lambda_2-\frac{1}{p}}\sum_{m=1}^{\infty}k_\lambda(m,n)a_m][n^{\frac{1}{p}-\lambda_2}b_n]$$
$$\le\|Ta\|_{p,\psi^{1-p}}\|b\|_{q,\psi}$$
$$\le k(\lambda_1)\|a\|_{p,\phi}\|b\|_{q,\psi}. \qquad (4.2.10)$$

If there exists a constant $0<k\le k(\lambda_1)$, such that (4.2.10) is valid as we replace $k(\lambda_1)$ by k, then for large enough $N\in\mathbf{N}$, such that

$$k_\lambda(1,u) \le L(\tfrac{1}{u^\delta})(u \ge N; L > 0),$$

setting $\tilde{a} = \{\tilde{a}_n\}_{n=1}^\infty, \tilde{b} = \{\tilde{b}_n\}_{n=1}^\infty$ as $\tilde{a}_n = n^{\lambda_1 - 1}$, $\tilde{b}_n = n^{\lambda_2 - 1} n \le N;\ \tilde{a}_n = \tilde{b}_n = 0, n > N$, it follows

$$\tilde{I} := \sum_{n=1}^\infty \sum_{m=1}^\infty k_\lambda(m,n)\tilde{a}_m \tilde{b}_n$$

$$\le k \|\tilde{a}\|_{p,\phi} \|\tilde{b}\|_{q,\psi} = k \sum_{m=1}^N \tfrac{1}{m}. \tag{4.2.11}$$

Since we find

$$\sum_{m=1}^N \sum_{n=N+1}^\infty k_\lambda(m,n) m^{\lambda_1 - 1} n^{\lambda_2 - 1}$$

$$= \sum_{m=1}^N \sum_{n=N+1}^\infty k_\lambda(1, \tfrac{n}{m}) m^{-\lambda_2 - 1} n^{\lambda_2 - 1}$$

$$\le L \sum_{m=1}^N \sum_{n=N+1}^\infty (\tfrac{n}{m})^{-\delta} m^{-\lambda_2 - 1} n^{\lambda_2 - 1}$$

$$= L \sum_{m=1}^N \tfrac{1}{m^{1-\delta+\lambda_2}} \sum_{n=N+1}^\infty \tfrac{1}{n^{1+\delta-\lambda_2}}$$

$$< L \int_0^N \frac{dx}{x^{1-\delta+\lambda_2}} \int_N^\infty \frac{dy}{y^{1+\delta-\lambda_2}} = \frac{L}{(\delta - \lambda_2)^2},$$

then by (4.2.11) and (4.2.9), we obtain

$$k \sum_{m=1}^N \tfrac{1}{m} \ge \tilde{I} = \sum_{m=1}^N \sum_{n=1}^N k_\lambda(m,n) m^{\lambda_1 - 1} n^{\lambda_2 - 1}$$

$$= \sum_{m=1}^N \tfrac{1}{m} \sum_{n=1}^\infty k_\lambda(m,n) m^{\lambda_1} n^{\lambda_2 - 1}$$

$$- \sum_{m=1}^N \sum_{n=N+1}^\infty k_\lambda(m,n) m^{\lambda_1 - 1} n^{\lambda_2 - 1}$$

$$\ge \sum_{m=1}^N \tfrac{1}{m}[k(\lambda_1) - O(\tfrac{1}{m^{\lambda'}})] - \tfrac{L}{(\delta - \lambda_2)^2}$$

$$= k(\lambda_1) \sum_{m=1}^N \tfrac{1}{m} - \sum_{m=1}^N \tfrac{1}{m} O(\tfrac{1}{m^{\lambda'}}) - \tfrac{L}{(\delta - \lambda_2)^2}$$

$$= \sum_{m=1}^N \tfrac{1}{m} \{k(\lambda_1) - (\sum_{m=1}^N \tfrac{1}{m})^{-1} [\sum_{m=1}^N O(\tfrac{1}{m^{1+\lambda'}}) + \tfrac{L}{(\delta - \lambda_2)^2}]\},$$

$$k \ge k(\lambda_1) - (\sum_{m=1}^N \tfrac{1}{m})^{-1} [\sum_{m=1}^N O(\tfrac{1}{m^{1+\lambda'}}) + \tfrac{L}{(\delta - \lambda_2)^2}].$$

For $N \to \infty$, we have $k \ge k(\lambda_1)$. Therefore $k = k(\lambda_1)$ is the best value of (4.2.10). Dividing out $\|b\|_{q,\psi}$ (> 0) in (4.2.10), we have

$$\|Ta\|_{p,\psi^{1-p}} \le k(\lambda_1) \|a\|_{p,\phi}, \tag{4.2.12}$$

Hence $k = k(\lambda_1)$ is still the best value of (4.2.12) and $\|T\|_{p,\psi^{1-p}} = k(\lambda_1)$. □

Theorem 4.2.3 Assuming that $\lambda_1, \lambda_2, \lambda \in \mathbf{R}$, with $\lambda_1 + \lambda_2 = \lambda$, $k_\lambda(x,y)(\ge 0)$ is a finite homogeneous function of degree $-\lambda$ in $\mathbf{R}_+^2$,

$$0 < k(\lambda_1) = \int_0^\infty k_\lambda(u,1) u^{\lambda_1 - 1} du < \infty,$$

if $k_\lambda(x,y)\frac{1}{x^{1-\lambda_2}}$ is decreasing for $x > 0$, $k_\lambda(x,y)\frac{1}{y^{1-\lambda_1}}$ is decreasing for $y > 0$ and they are strictly decreasing in a subinterval respectively, then the bounded linear operator T defined by (4.2.1) exists with the norm

$$\|T\|_{p,\psi^{1-p}} = k(\lambda_1).$$

Proof We find

$$\varpi(\lambda_2, m) = \sum_{n=1}^\infty k_\lambda(m,n) \tfrac{m^{\lambda_1}}{n^{1-\lambda_2}}$$

$$< \int_0^\infty k_\lambda(m,y) \tfrac{m^{\lambda_1}}{y^{1-\lambda_2}} dy,$$

and setting $u = \frac{m}{y}$, it follows

$$\varpi(\lambda_2, m) < \int_0^\infty k_\lambda(1,u) u^{\lambda_2 - 1} du = k(\lambda_1).$$

By the same way, we have

$$\vartheta(\lambda_1, n) = \sum_{m=1}^\infty k_\lambda(m,n) \tfrac{n^{\lambda_2}}{m^{1-\lambda_1}}$$

$$< \int_0^\infty k_\lambda(x,n) \tfrac{n^{\lambda_2}}{x^{1-\lambda_1}} dx,$$

and setting $u = \frac{x}{n}$, it follows $\vartheta(\lambda_1, n) < k(\lambda_1)$. Setting $\tilde{k}(\lambda_2) = k'(\lambda_1) = k(\lambda_1)$, we have (4.2.4). By Theorem 4.2.1, the bounded linear operator T exists, $\|T\|_{p,\psi^{p-1}} \le k(\lambda_1)$ and (4.2.10) is valid.

If there exists a constant $0 < k \le k(\lambda_1)$, such that (4.2.10) is still valid as we replace $k(\lambda_1)$ by k, then in particular, for $\varepsilon > 0$, setting $\tilde{a} = \{\tilde{a}_m\}_{m=1}^\infty$ and $\tilde{b} = \{\tilde{b}_n\}_{n=1}^\infty$ as: $\tilde{a}_m = m^{\lambda_1 - \frac{\varepsilon}{p} - 1}, \tilde{b}_n = n^{\lambda_2 - \frac{\varepsilon}{q} - 1}$, we find

$$\varepsilon \tilde{I} \le \varepsilon k \sum_{n=1}^\infty \tfrac{1}{n^{1+\varepsilon}} = \varepsilon k (1 + \sum_{n=2}^\infty \tfrac{1}{n^{1+\varepsilon}})$$

$$< \varepsilon k (1 + \int_1^\infty \tfrac{1}{y^{1+\varepsilon}} dy) = k(\varepsilon + 1), \tag{4.2.13}$$

$$\varepsilon \tilde{I} = \varepsilon \sum_{n=1}^\infty \sum_{m=1}^\infty k_\lambda(m,n) m^{\lambda_1 - \frac{\varepsilon}{p} - 1} n^{\lambda_2 - \frac{\varepsilon}{q} - 1}$$

$$\ge \varepsilon \int_1^\infty x^{\lambda_1 - \frac{\varepsilon}{p} - 1} (\int_1^\infty k_\lambda(x,y) y^{\lambda_2 - \frac{\varepsilon}{q} - 1} dy) dx.$$

Setting $u = \frac{x}{y}$, it follows

$$k(\varepsilon + 1) \ge \varepsilon \tilde{I}$$

$$\geq \varepsilon\int_1^\infty x^{-1-\varepsilon}(\int_0^x k_\lambda(u,1)u^{\lambda_1+\frac{\varepsilon}{q}-1}du)dx$$

$$= \int_0^1 k_\lambda(u,1)u^{\lambda_1+\frac{\varepsilon}{q}-1}du$$

$$+\varepsilon\int_1^\infty x^{-1-\varepsilon}(\int_1^x k_\lambda(u,1)u^{\lambda_1+\frac{\varepsilon}{q}-1}du)dx$$

$$= \int_0^1 k_\lambda(u,1)u^{\lambda_1+\frac{\varepsilon}{q}-1}du$$

$$+\varepsilon\int_1^\infty (\int_u^\infty x^{-1-\varepsilon}dx)k_\lambda(u,1)u^{\lambda_1+\frac{\varepsilon}{q}-1}du$$

$$= \int_0^1 k_\lambda(u,1)u^{\lambda_1+\frac{\varepsilon}{q}-1}du + \int_1^\infty k_\lambda(u,1)u^{\lambda_1-\frac{\varepsilon}{p}-1}du.$$

By Fatou lemma, we have

$$k \geq \lim_{\varepsilon\to 0^+}\int_0^1 k_\lambda(u,1)u^{\lambda_1+\frac{\varepsilon}{q}-1}du$$

$$+\lim_{\varepsilon\to 0^+}\int_1^\infty k_\lambda(u,1)u^{\lambda_1-\frac{\varepsilon}{p}-1}du$$

$$\geq \int_0^1 \lim_{\varepsilon\to 0^+} k_\lambda(u,1)u^{\lambda_1+\frac{\varepsilon}{q}-1}du$$

$$+\int_1^\infty \lim_{\varepsilon\to 0^+} k_\lambda(u,1)u^{\lambda_1-\frac{\varepsilon}{p}-1}du = k(\lambda_1).$$

Hence $k = k(\lambda_1)$ is the best value of (4.2.10) and (4.2.12), and then $\| T \|_{p,\psi^{1-p}} = k(\lambda_1)$. □

Note 4.2.4 If $\lambda_i \leq 1(i=1,2)$, then the decreasing property condition of $k_\lambda(x,y)\frac{1}{x^{1-\lambda_1}}$ and $k_\lambda(x,y)\frac{1}{y^{1-\lambda_2}}$ in Theorem 4.1.4 may change for " $k_\lambda(x,y)$ is decreasing for $x(y)$ and strictly decreasing for a subinterval".

Theorem 4.2.5 Let the assumptions of Theorem 4.2.2 (or Theorem 4.2.3) be fulfilled and additionally, $a \in l_\phi^p, b \in l_\psi^q$,

$$0 < \| a \|_{p,\phi} = \{\sum_{n=1}^\infty n^{p(1-\lambda_1)-1}a_n^p\}^{\frac{1}{p}} < \infty$$

and

$$0 < \| b \|_{q,\psi} = \{\sum_{n=1}^\infty n^{q(1-\lambda_2)-1}b_n^q\}^{\frac{1}{q}} < \infty,$$

then we have the following equivalent inequalities:

$$(Ta,b) = \sum_{n=1}^\infty\sum_{m=1}^\infty k_\lambda(m,n)a_m b_n$$

$$< k(\lambda_1)\| a \|_{p,\phi}\| b \|_{q,\psi}, \quad (4.2.14)$$

$$\| Ta \|_{p,\psi^{1-p}} = \sum_{n=1}^\infty n^{p\lambda_2-1}[\sum_{m=1}^\infty k_\lambda(m,n)a_m]^p$$

$$< k^p(\lambda_1)\| a \|_{p,\phi}^p, \quad (4.2.15)$$

where the constant factors $k(\lambda_1)$ and $k^p(\lambda_1)$ are the best possible.

Proof Since $0 < \| a \|_{p,\phi} < \infty$, then the last inequality of (4.2.6) takes the form of strict-sign inequality, and we have (4.2.15). By (4.2.15), the middle inequality of (4.2.10) takes the form of strict-sign inequality, and we have (4.2.14).

On the other-hand, suppose that (4.2.14) is valid. By (4.2.6), for $\tilde{k}(\lambda_2) = k'(\lambda_1) = k(\lambda_1)$, we have $\| Ta \|_{p,\psi^{1-p}} < \infty$. If $\| Ta \|_{p,\psi^{1-p}} = 0$, then (4.2.15) is naturally valid; if $\| Ta \|_{p,\psi^{1-p}} > 0$, setting

$$b_n = n^{p\lambda_2-1}[\sum_{m=1}^\infty k_\lambda(m,n)a_m]^{p-1} > 0,$$

by (4.2.14), we have

$$0 < \| b \|_{q,\psi}^q = \| Ta \|_{p,\psi^{1-p}}^p$$

$$= (Ta,b) < k(\lambda_1)\| a \|_{p,\phi}\| b \|_{q,\psi} < \infty, \quad (4.2.16)$$

$$0 < \| b \|_{q,\psi}^{q-1} = \| Ta \|_{p,\psi^{1-p}}^{1/p} < k(\lambda_1)\| a \|_{p,\phi}. \quad (4.2.17)$$

Hence (4.2.15) is valid, which is equivalent to (4.2.14).

By the proof of Theorem 4.2.3 and Theorem 4.2.4, it follows that the constant factors in (4.2.14) and (4.2.15) are the best possible. □

Note 4.2.6 Theorem 4.2.3, Theorem 4.2.4 and Theorem 4.2.6 are respectively extensions of Theorem 4.1.2, Theorem 4.1.3 and Theorem 4.1.5. Some early results of Theorem 4.2.3, Theorem 4.2.4 and Theorem 4.2.6 are consulted in (Yang JIA 2009)[4], (Yang PMD 2010)[5].

4.2.2 SOME EXAMPLES FOR APPLYING THEOREM 4.2.3 AND THEOREM 4.2.5

Example 4.2.7 If $\alpha, \lambda_1, \lambda_2, \lambda \in \mathbf{R}, \lambda_1 + \lambda_2 = \lambda$, $0 < \alpha + \lambda_i \leq 1(i=1,2)$,

$$k_\lambda(x,y) = \frac{(\min\{x,y\})^\alpha}{(\max\{x,y\})^{\lambda+\alpha}}, x, y > 0,$$

then we find that $k_\lambda(x,y)\frac{1}{x^{1-\lambda_2}}$ is strict decreasing for $x > 0$, and $k_\lambda(x,y)\frac{1}{y^{1-\lambda_1}}$ is strict decreasing for $y > 0$, and

$$k(\lambda_1) = \int_0^\infty \frac{(\min\{u,1\})^\alpha}{(\max\{u,1\})^{\lambda+\alpha}}u^{\lambda_1-1}du$$

$$= \int_0^1 u^{\alpha+\lambda_1-1}du + \int_1^\infty u^{-\lambda_2-\alpha-1}du = \frac{2\alpha+\lambda}{(\alpha+\lambda_1)(\alpha+\lambda_2)}.$$

By Theorem 4.2.3 and Theorem 4.2.5, we have the following equivalent inequalities:

$$\sum_{n=1}^{\infty}\sum_{m=1}^{\infty}\frac{(\min\{m,n\})^{\alpha}}{(\max\{m,n\})^{\lambda+\alpha}}a_m b_n$$
$$<\frac{2\alpha+\lambda}{(\alpha+\lambda_1)(\alpha+\lambda_2)}\|a\|_{p,\phi}\|b\|_{q,\psi}, \quad (4.2.18)$$
$$\sum_{n=1}^{\infty}n^{p\lambda_2-1}[\sum_{m=1}^{\infty}\frac{(\min\{m,n\})^{\alpha}}{(\max\{m,n\})^{\lambda+\alpha}}a_m]^p$$
$$<[\frac{2\alpha+\lambda}{(\alpha+\lambda_1)(\alpha+\lambda_2)}]^p\|a\|_{p,\phi}^p, \quad (4.2.19)$$
where the constant factors are the best possible.

In the following examples, we suppose that $p,r>1, \frac{1}{p}+\frac{1}{q}=1, \frac{1}{r}+\frac{1}{s}=1, \quad \lambda>0, \quad \lambda_1=\frac{\lambda}{r}$, $\lambda_2=\frac{\lambda}{s}, \quad k(\lambda_1)=k(\frac{\lambda}{r}), \quad \phi(x)=x^{p(1-\frac{\lambda}{r})-1}$, $\psi(x)=x^{q(1-\frac{\lambda}{s})-1} \quad (x\in(0,\infty))$, $a\in l_{\phi}^p, b\in l_{\psi}^q$,
$$0<\|a\|_{p,\phi}=\{\sum_{n=1}^{\infty}n^{p(1-\frac{\lambda}{r})-1}a_n^p\}^{\frac{1}{p}}<\infty$$
and
$$0<\|b\|_{q,\psi}=\{\sum_{n=1}^{\infty}n^{q(1-\frac{\lambda}{s})-1}b_n^q\}^{\frac{1}{q}}<\infty.$$
The words that the constant factors are the best possible are omitted.

Example 4.2.8 If $\alpha>0, 0<\lambda\le\min\{r,s\}$, $k_{\lambda}(x,y)=\frac{1}{(x^{\alpha}+y^{\alpha})^{\lambda/\alpha}}$, then we find
$$k(\tfrac{\lambda}{r})=\tfrac{1}{\alpha}B(\tfrac{\lambda}{r\alpha},\tfrac{\lambda}{s\alpha}).$$
By Theorem 4.2.3 and Theorem 4.2.5, we have the following equivalent inequalities:
$$\sum_{n=1}^{\infty}\sum_{m=1}^{\infty}\frac{a_m b_n}{(m^{\alpha}+n^{\alpha})^{\lambda/\alpha}}$$
$$<\tfrac{1}{\alpha}B(\tfrac{\lambda}{r\alpha},\tfrac{\lambda}{s\alpha})\|a\|_{p,\phi}\|b\|_{q,\psi}, \quad (4.2.20)$$
$$\sum_{n=1}^{\infty}n^{\frac{p\lambda}{s}-1}[\sum_{m=1}^{\infty}\frac{1}{(m^{\alpha}+n^{\alpha})^{\lambda/\alpha}}a_m]^p$$
$$<[\tfrac{1}{\alpha}B(\tfrac{\lambda}{r\alpha},\tfrac{\lambda}{s\alpha})]^p\|a\|_{p,\phi}^p. \quad (4.2.21)$$

In particular, (1) for $0<\alpha=\lambda\le\min\{r,s\}$, we have the following equivalent inequalities:
$$\sum_{n=1}^{\infty}\sum_{m=1}^{\infty}\frac{1}{m^{\lambda}+n^{\lambda}}a_m b_n$$
$$<\frac{\pi}{\lambda\sin(\pi/r)}\|a\|_{p,\phi}\|b\|_{q,\psi}, \quad (4.2.22)$$
$$\sum_{n=1}^{\infty}n^{\frac{p\lambda}{s}-1}(\sum_{m=1}^{\infty}\frac{1}{m^{\lambda}+n^{\lambda}}a_m)^p$$
$$<[\frac{\pi}{\lambda\sin(\pi/r)}]^p\|a\|_{p,\phi}^p; \quad (4.2.23)$$
(2) for $\alpha=1, 0<\lambda\le\min\{r,s\}$, we have the following equivalent inequalities:
$$\sum_{n=1}^{\infty}\sum_{m=1}^{\infty}\frac{1}{(m+n)^{\lambda}}a_m b_n$$
$$<B(\tfrac{\lambda}{r},\tfrac{\lambda}{s})\|a\|_{p,\phi}\|b\|_{q,\psi}, \quad (4.2.24)$$
$$\sum_{n=1}^{\infty}n^{\frac{p\lambda}{s}-1}[\sum_{m=1}^{\infty}\frac{1}{(m+n)^{\lambda}}a_m]^p$$
$$<[B(\tfrac{\lambda}{r},\tfrac{\lambda}{s})]^p\|a\|_{p,\phi}^p. \quad (4.2.25)$$

Note 4.2.9 Some early results of (4.2.20)-(4.2.25) are consulted in (Yang CMA 2002)[6]-(Xu AM 2007)[22].

Example 4.2.10 If $0<\alpha\le\lambda\le\min\{r,s\}$,
$$k_{\lambda}(x,y)=\frac{1}{(x+y)^{\lambda-\alpha}(\max\{x,y\})^{\alpha}},$$
then we find
$$k(\tfrac{\lambda}{r})=k_{\lambda}(r,\alpha):=\sum_{k=0}^{\infty}\binom{\alpha-\lambda}{k}\frac{(\lambda+2k)rs}{(\lambda+rk)(\lambda+sk)}.$$
By Theorem 4.2.3 and Theorem 4.2.5, we have the following equivalent inequalities:
$$\sum_{n=1}^{\infty}\sum_{m=1}^{\infty}\frac{1}{(m+n)^{\lambda-\alpha}(\max\{m,n\})^{\alpha}}a_m b_n$$
$$<k_{\lambda}(r,\alpha)\|a\|_{p,\phi}\|b\|_{q,\psi}, \quad (4.2.26)$$
$$\sum_{n=1}^{\infty}n^{\frac{p\lambda}{s}-1}[\sum_{m=1}^{\infty}\frac{1}{(m+n)^{\lambda-\alpha}(\max\{m,n\})^{\alpha}}a_m]^p$$
$$<[k_{\lambda}(r,\alpha)]^p\|a\|_{p,\phi}^p. \quad (4.2.27)$$

In particular, (1) for $0<\alpha=\lambda\le\min\{r,s\}$,
$$k_{\lambda}(r,\lambda)=\sum_{k=0}^{\infty}\binom{0}{k}\frac{(\lambda+2k)rs}{(\lambda+rk)(\lambda+sk)}=\frac{rs}{\lambda},$$
we have the following equivalent inequalities:
$$\sum_{n=1}^{\infty}\sum_{m=1}^{\infty}\frac{a_m b_n}{(\max\{m,n\})^{\lambda}}<\frac{rs}{\lambda}\|a\|_{p,\phi}\|b\|_{q,\psi}, \quad (4.2.28)$$
$$\sum_{n=1}^{\infty}n^{\frac{p\lambda}{s}-1}[\sum_{m=1}^{\infty}\frac{a_m}{(\max\{m,n\})^{\lambda}}]^p<(\frac{rs}{\lambda})^p\|a\|_{p,\phi}^p; \quad (4.2.29)$$
(2) for $\alpha=\frac{1}{2}, \lambda=1, r=s=2$, we find
$$k_1(r,\tfrac{1}{2})=4\sum_{k=0}^{\infty}\binom{-\frac{1}{2}}{k}\frac{1}{1+2k}=4\int_0^1\frac{1}{(1+x^2)^{1/2}}dx$$
$$=4\ln(x+\sqrt{x^2+1})|_0^1=4\ln(1+\sqrt{2}),$$
and have the following equivalent inequalities:
$$\sum_{n=1}^{\infty}\sum_{m=1}^{\infty}\frac{a_m b_n}{\sqrt{(m+n)\max\{m,n\}}}$$
$$<4\ln(1+\sqrt{2})\{\sum_{n=1}^{\infty}n^{\frac{p}{2}-1}a_n^p\}^{\frac{1}{p}}\{\sum_{n=1}^{\infty}n^{\frac{q}{2}-1}b_n^q\}^{\frac{1}{q}},$$
(4.2.30)

$$\sum_{n=1}^{\infty} n^{\frac{p}{2}-1}[\sum_{m=1}^{\infty}\frac{a_m}{\sqrt{(m+n)\max\{m,n\}}}]^p$$
$$<[4\ln(1+\sqrt{2})]^p\sum_{n=1}^{\infty} n^{\frac{p}{2}-1}a_n^p. \quad (4.2.31)$$

Example 4.2.11 If $0\le\alpha<\lambda\le\min\{r,s\}$, $k_\lambda(x,y)=\frac{1}{(x^{\lambda-\alpha}+y^{\lambda-\alpha})(\max\{x,y\})^{\alpha}}$, then we find $k(\frac{\lambda}{r})=\tilde{k}_\lambda(r,\alpha):=\sum_{k=0}^{\infty}\frac{(-1)^k[(2k+1)\lambda-2k\alpha]rs}{[(rk+1)\lambda-rk\alpha][(sk+1)\lambda-sk\alpha]}$. By Theorem 4.2.3 and Theorem 4.2.5, we have the following equivalent inequalities:

$$\sum_{n=1}^{\infty}\sum_{m=1}^{\infty}\frac{1}{(m^{\lambda-\alpha}+n^{\lambda-\alpha})(\max\{m,n\})^{\alpha}}a_m b_n$$
$$<\tilde{k}_\lambda(r,\alpha)\|a\|_{p,\phi}\|b\|_{q,\psi}, \quad (4.2.32)$$
$$\sum_{n=1}^{\infty} n^{\frac{p\lambda}{s}-1}[\sum_{m=1}^{\infty}\frac{1}{(m^{\lambda-\alpha}+n^{\lambda-\alpha})(\max\{m,n\})^{\alpha}}a_m]^p$$
$$<[\tilde{k}_\lambda(r,\alpha)]^p\|a\|_{p,\phi}^p. \quad (4.2.33)$$

In particular, (1) for $0=\alpha<\lambda\le\min\{r,s\}$, we find

$$\tilde{k}_\lambda(r,0)=\frac{1}{\lambda}\sum_{k=0}^{\infty}\frac{(-1)^k(2k+1)rs}{(rk+1)(sk+1)}$$
$$=\frac{1}{\lambda}\int_0^1\frac{v^{1/r-1}+v^{1/s-1}}{v+1}dv=\frac{\pi}{\lambda\sin(\pi/r)}$$

and then we have (4.2.22) and (4.2.23); (2) for $\alpha=\frac{1}{2},\lambda=1$, $r=s=2$, we find

$$k_1(2,\tfrac{1}{2})=4\sum_{k=0}^{\infty}\frac{(-1)^k}{1+k}=4\ln 2$$

and then we have the following equivalent inequalities:

$$\sum_{n=1}^{\infty}\sum_{m=1}^{\infty}\frac{a_m b_n}{(\sqrt{m}+\sqrt{n})\sqrt{\max\{m,n\}}}$$
$$<4\ln 2\{\sum_{n=1}^{\infty} n^{\frac{p}{2}-1}a_n^p\}^{\frac{1}{p}}\{\sum_{n=1}^{\infty} n^{\frac{q}{2}-1}b_n^q\}^{\frac{1}{q}}, \quad (4.2.34)$$
$$\sum_{n=1}^{\infty} n^{\frac{p}{2}-1}(\sum_{m=1}^{\infty}\frac{a_m}{(\sqrt{m}+\sqrt{n})\sqrt{\max\{m,n\}}})^p$$
$$<(4\ln 2)^p\sum_{n=1}^{\infty} n^{\frac{p}{2}-1}a_n^p. \quad (4.2.35)$$

Example 4.2.12 If $0<\alpha\le\lambda\le\min\{r,s\}$, $k_\lambda(x,y)=\frac{\ln(x/y)}{(x^\alpha-y^\alpha)(\max\{x,y\})^{\lambda-\alpha}}$, then we find

$$k(\tfrac{\lambda}{r})=k_{\lambda,\alpha}(r):=\int_0^\infty\frac{u^{\lambda/r-1}\ln u}{(u^\alpha-1)(\max\{u,1\})^{\lambda-\alpha}}du$$
$$=\int_0^1\frac{-\ln u}{1-u^\alpha}u^{\frac{\lambda}{r}-1}du+\int_1^\infty\frac{\ln u}{1-u^{-\alpha}}u^{-\frac{\lambda}{s}-1}du$$
$$=\int_0^1(-\ln u)\sum_{k=0}^{\infty}u^{k\alpha+\frac{\lambda}{r}-1}du$$
$$+\int_1^\infty\ln u\sum_{k=0}^{\infty}u^{-k\alpha-\frac{\lambda}{s}-1}du$$
$$=\sum_{k=0}^{\infty}\frac{1}{k\alpha+\lambda/r}\int_0^1(-\ln u)du^{k\alpha+\frac{\lambda}{r}}$$
$$+\sum_{k=0}^{\infty}\frac{1}{-k\alpha-\lambda/s}\int_1^\infty(\ln u)du^{-k\alpha-\frac{\lambda}{s}}$$
$$=\sum_{k=0}^{\infty}[\frac{1}{(k\alpha+\lambda/r)^2}+\frac{1}{(k\alpha+\lambda/s)^2}].$$

By Theorem 4.2.3 and Theorem 4.2.5, we have the following equivalent inequalities:

$$\sum_{n=1}^{\infty}\sum_{m=1}^{\infty}\frac{\ln(m/n)}{(m^\alpha-n^\alpha)(\max\{m,n\})^{\lambda-\alpha}}a_m b_n$$
$$<k_{\lambda,\alpha}(r)\|a\|_{p,\phi}\|b\|_{q,\psi}, \quad (4.2.36)$$
$$\sum_{n=1}^{\infty} n^{\frac{p\lambda}{s}-1}[\sum_{m=1}^{\infty}\frac{\ln(m/n)}{(m^\alpha-n^\alpha)(\max\{m,n\})^{\lambda-\alpha}}a_m]^p$$
$$<[k_{\lambda,\alpha}(r)]^p\|a\|_{p,\phi}^p. \quad (4.2.37)$$

In particular, (1) for $0<\alpha=\lambda\le\min\{r,s\}$, since $k_{\lambda,\lambda}(r)=\frac{1}{\lambda^2}\sum_{k=0}^{\infty}[\frac{1}{(k+\frac{1}{r})^2}+\frac{1}{(k+\frac{1}{s})^2}]=[\frac{\pi}{\lambda\sin(\pi/r)}]^2$, then we have the following equivalent inequalities (Yang JMI 2008)[23]:

$$\sum_{n=1}^{\infty}\sum_{m=1}^{\infty}\frac{\ln(m/n)}{m^\lambda-n^\lambda}a_m b_n$$
$$<[\frac{\pi}{\lambda\sin(\pi/r)}]^2\|a\|_{p,\phi}\|b\|_{q,\psi}, \quad (4.2.38)$$
$$\sum_{n=1}^{\infty} n^{\frac{p\lambda}{s}-1}[\sum_{m=1}^{\infty}\frac{\ln(m/n)}{m^\lambda-n^\lambda}a_m]^p$$
$$<[\frac{\pi}{\lambda\sin(\pi/r)}]^{2p}\|a\|_{p,\phi}^p; \quad (4.2.39)$$

(2) for $\alpha=\frac{1}{2},\lambda=1,r=s=2$, we find

$$k_{1,\frac{1}{2}}(2,\tfrac{1}{2})=8\sum_{k=0}^{\infty}\frac{1}{(1+k)^2}=\frac{4\pi^2}{3}$$

and then we have the following equivalent inequalities:

$$\sum_{n=1}^{\infty}\sum_{m=1}^{\infty}\frac{\ln(m/n)a_m b_n}{(\sqrt{m}-\sqrt{n})\sqrt{\max\{m,n\}}}$$
$$<\frac{4\pi^2}{3}\{\sum_{n=1}^{\infty} n^{\frac{p}{2}-1}a_n^p\}^{\frac{1}{p}}\{\sum_{n=1}^{\infty} n^{\frac{q}{2}-1}b_n^q\}^{\frac{1}{q}}, \quad (4.2.40)$$
$$\sum_{n=1}^{\infty} n^{\frac{p}{2}-1}(\sum_{m=1}^{\infty}\frac{\ln(m/n)a_m}{(\sqrt{m}-\sqrt{n})\sqrt{\max\{m,n\}}})^p$$

$$<(\tfrac{4\pi^2}{3})^p \sum_{n=1}^{\infty} n^{\frac{p}{2}-1} a_n^p . \quad (4.2.41)$$

Example 4.2.13 If $0<\lambda\le\min\{r,s\}$, $\max\{\frac{\lambda}{r},\frac{\lambda}{s}\}<\alpha\le\lambda$, $k_\lambda(x,y)=\frac{\ln(x/y)}{(x^\alpha-y^\alpha)(\min\{x,y\})^{\lambda-\alpha}}$, then we find

$$k(\tfrac{\lambda}{r})=\tilde{k}_{\lambda,\alpha}(r):=\int_0^\infty \frac{u^{\lambda/r-1}\ln u}{(u^\alpha-1)(\min\{u,1\})^{\lambda-\alpha}}du$$

$$=\int_0^1 \frac{-\ln u}{1-u^\alpha} u^{\alpha-\frac{\lambda}{s}-1}du+\int_1^\infty \frac{\ln u}{1-u^{-\alpha}} u^{-\alpha+\frac{\lambda}{r}-1}du$$

$$=\int_0^1(-\ln u)\sum_{k=0}^{\infty} u^{(k+1)\alpha-\frac{\lambda}{s}-1}du+\int_1^\infty \ln u\sum_{k=0}^{\infty} u^{-(k+1)\alpha+\frac{\lambda}{r}-1}du$$

$$=\sum_{k=0}^{\infty}\frac{1}{(k+1)\alpha-\lambda/s}\int_0^1(-\ln u)du^{(k+1)\alpha-\frac{\lambda}{s}}+\sum_{k=0}^{\infty}\frac{1}{-(k+1)\alpha+\lambda/r}\int_1^\infty(\ln u)du^{-(k+1)\alpha+\frac{\lambda}{r}}$$

$$=\sum_{k=0}^{\infty}\{\frac{1}{[(k+1)\alpha-\lambda/s]^2}+\frac{1}{[(k+1)\alpha-\lambda/r]^2}\}.$$

By Theorem 4.2.3 and Theorem 4.2.5, we have the following equivalent inequalities:

$$\sum_{n=1}^{\infty}\sum_{m=1}^{\infty}\frac{\ln(m/n)}{(m^\alpha-n^\alpha)(\min\{m,n\})^{\lambda-\alpha}}a_m b_n<\tilde{k}_{\lambda,\alpha}(r)\|a\|_{p,\phi}\|b\|_{q,\psi}, \quad (4.2.42)$$

$$\sum_{n=1}^{\infty} n^{\frac{p\lambda}{s}-1}[\sum_{m=1}^{\infty}\frac{\ln(m/n)}{(m^\alpha-n^\alpha)(\min\{m,n\})^{\lambda-\alpha}}a_m]^p<[\tilde{k}_{\lambda,\alpha}(r)]^p\|a\|_{p,\phi}^p . \quad (4.2.43)$$

In particular, for $\alpha=\lambda$, we find $0<\lambda\le\min\{r,s\}$,

$$\tilde{k}_{\lambda,\lambda}(r)=[\frac{\pi}{\lambda\sin(\pi/r)}]^2$$

and then we have (4.2.38) and (4.2.39).

Example 4.2.14 If

$$0\le\alpha\max\{r,s\}<\lambda\le\min\{r,s\},$$

$k_\lambda(x,y)=\frac{1}{(x+y)^{\lambda-\alpha}(\min\{x,y\})^\alpha}$, then we find

$$k(\tfrac{\lambda}{r})=K_\lambda(r,\alpha):=\sum_{k=0}^{\infty}\binom{\alpha-\lambda}{k}\frac{\lambda-2\alpha+2k}{(\frac{\lambda}{r}-\alpha+k)(\frac{\lambda}{s}-\alpha+k)}.$$

By Theorem 4.2.3 and Theorem 4.2.5, we have the following equivalent inequalities:

$$\sum_{n=1}^{\infty}\sum_{m=1}^{\infty}\frac{1}{(m+n)^{\lambda-\alpha}(\min\{m,n\})^\alpha}a_m b_n<K_\lambda(r,\alpha)\|a\|_{p,\phi}\|b\|_{q,\psi}, \quad (4.2.44)$$

$$\sum_{n=1}^{\infty} n^{\frac{p\lambda}{s}-1}[\sum_{m=1}^{\infty}\frac{1}{(m+n)^{\lambda-\alpha}(\min\{m,n\})^\alpha}a_m]^p<[K_\lambda(r,\alpha)]^p\|a\|_{p,\phi}^p . \quad (4.2.45)$$

For $0<\lambda\le\min\{r,s\}$, we find

$$K_\lambda(r,0):=\sum_{k=0}^{\infty}\binom{-\lambda}{k}\frac{\lambda+2k}{(\lambda/r+k)(\lambda/s+k)}=B(\tfrac{\lambda}{r},\tfrac{\lambda}{s})$$

and then we have (4.2.24) and (4.2.25).

Example 4.2.15 If

$$0\le\alpha\max\{r,s\}<\lambda\le\min\{r,s\},$$

$k_\lambda(x,y)=\frac{1}{(x^{\lambda-\alpha}+y^{\lambda-\alpha})(\min\{x,y\})^\alpha}$, then we find

$$k(\tfrac{\lambda}{r})=\tilde{K}_\lambda(r,\alpha):=\sum_{k=0}^{\infty}\frac{(-1)^k[(2k+1)\lambda-2(k+1)\alpha]}{[(\frac{1}{r}+k)\lambda-(k+1)\alpha][(\frac{1}{s}+k)\lambda-(k+1)\alpha]}.$$

By Theorem 4.2.3 and Theorem 4.2.5, we have the following equivalent inequalities:

$$\sum_{n=1}^{\infty}\sum_{m=1}^{\infty}\frac{1}{(m^{\lambda-\alpha}+n^{\lambda-\alpha})(\min\{m,n\})^\alpha}a_m b_n<\tilde{K}_\lambda(r,\alpha)\|a\|_{p,\phi}\|b\|_{q,\psi}, \quad (4.2.46)$$

$$\sum_{n=1}^{\infty} n^{\frac{p\lambda}{s}-1}[\sum_{m=1}^{\infty}\frac{1}{(m^{\lambda-\alpha}+n^{\lambda-\alpha})(\min\{m,n\})^\alpha}a_m]^p<[\tilde{K}_\lambda(r,\alpha)]^p\|a\|_{p,\phi}^p . \quad (4.2.47)$$

In particular, for $\alpha=0, 0<\lambda\le\min\{r,s\}$, we find $\tilde{K}_\lambda(r,0)=\frac{\pi}{\lambda\sin(\pi/r)}$ and then we have (4.2.22) and (4.2.23).

Note 4.2.16 Some early results of (4.2.28) and (4.2.38) are consulted in (Yang JJU 2004)[24]-(Yang JMAA 2007)[29].

4.2.3 SOME EXAMPLES FOR APPLYING THEOREM 4.2.2 AND THEOREM 4.2.5

Example 4.2.17 If $0<\lambda\le 2\min\{r,s\}$, $k_\lambda(x,y)=\frac{1}{(x+y)^\lambda}$, then we define the following weight coefficients:

$$\varpi(\tfrac{\lambda}{s},m):=\sum_{n=1}^{\infty}\frac{1}{(m+n)^\lambda}\frac{m^{\lambda/r}}{n^{1-\lambda/s}},$$

$$\vartheta(\tfrac{\lambda}{r},n):=\sum_{m=1}^{\infty}\frac{1}{(m+n)^\lambda}\frac{n^{\lambda/s}}{m^{1-\lambda/r}},\quad m,n\in\mathbf{N}. \quad (4.2.48)$$

Setting $f_{\lambda,m}(y)=\frac{1}{(m+y)^\lambda y^{1-\lambda/s}}$, by (3.1.43), we have

$$\varpi(\tfrac{\lambda}{s},m)=m^{\frac{\lambda}{r}}\sum_{n=1}^{\infty}f_{\lambda,m}(n)$$
$$=m^{\frac{\lambda}{r}}[\int_1^{\infty}f_{\lambda,m}(y)dy+\tfrac{1}{2}f_{\lambda,m}(1)$$
$$+\int_1^{\infty}P_1(y)f'_{\lambda,m}(y)dy]$$
$$=m^{\frac{\lambda}{r}}\int_0^{\infty}f_{\lambda,m}(y)dy-m^{\frac{\lambda}{r}}\theta(\tfrac{\lambda}{s},m), \quad (4.2.49)$$
$$\theta(\tfrac{\lambda}{s},m)=\int_0^1 f_{\lambda,m}(y)dy-\tfrac{1}{2}f_{\lambda,m}(1)$$
$$-\int_1^{\infty}P_1(y)f'_{\lambda,m}(y)dy. \quad (4.2.50)$$

We find $-\frac{1}{2}f_{\lambda,m}(1)=-\frac{1}{2(m+1)^{\lambda}}$. Setting $u=\frac{y}{m}$, since $0<\lambda\le 2s$, we obtain

$$m^{\frac{\lambda}{r}}\int_0^{\infty}f_{\lambda,m}(y)dy=\int_0^{\infty}\frac{m^{\lambda/r}}{(m+y)^{\lambda}y^{1-\lambda/s}}dy$$
$$=\int_0^{\infty}\frac{1}{(u+1)^{\lambda}}u^{\frac{\lambda}{s}-1}du=B(\tfrac{\lambda}{r},\tfrac{\lambda}{s}), \quad (4.2.51)$$
$$\int_0^1 f_{\lambda,m}(y)dy=\tfrac{s}{\lambda}\int_0^1\frac{1}{(m+y)^{\lambda}}dy^{\frac{\lambda}{s}}$$
$$=\frac{s}{\lambda(m+1)^{\lambda}}+s\int_0^1\frac{1}{(m+y)^{\lambda+1}}y^{\frac{\lambda}{s}}dy$$
$$=\frac{s}{\lambda(m+1)^{\lambda}}+\frac{s^2}{s+\lambda}\int_0^1\frac{1}{(m+y)^{\lambda+1}}dy^{\frac{\lambda}{s}+1}$$
$$=\frac{s}{\lambda(m+1)^{\lambda}}+\frac{s^2}{(s+\lambda)(m+1)^{\lambda+1}}$$
$$+\frac{s^2(\lambda+1)}{(s+\lambda)}\int_0^1\frac{1}{(m+y)^{\lambda+2}}y^{\frac{\lambda}{s}+1}dy$$
$$=\frac{s}{\lambda(m+1)^{\lambda}}+\frac{s^2}{(s+\lambda)(m+1)^{\lambda+1}}+\frac{s^3(\lambda+1)}{(s+\lambda)(2s+\lambda)(m+1)^{\lambda+2}}$$
$$+\frac{s^3(\lambda+1)(\lambda+2)}{(s+\lambda)(2s+\lambda)}\int_0^1\frac{1}{(m+y)^{\lambda+3}}y^{\frac{\lambda}{s}+2}dy$$
$$>\frac{s}{\lambda(m+1)^{\lambda}}+\frac{s^2}{(s+\lambda)(m+1)^{\lambda+1}}+\frac{s^3(\lambda+1)}{(s+\lambda)(2s+\lambda)(m+1)^{\lambda+2}}$$
$$>\frac{s}{\lambda(m+1)^{\lambda}}+\frac{s}{3(m+1)^{\lambda+1}}+\frac{s(\lambda+1)}{12(m+1)^{\lambda+2}},$$
$$f'_{\lambda,m}(y)=-\frac{\lambda}{(m+y)^{\lambda+1}y^{1-\lambda/s}}-\frac{(1-\lambda/s)}{(m+y)^{\lambda}y^{2-\lambda/s}}$$
$$=-\frac{(1+\lambda/r)}{(m+y)^{\lambda}y^{2-\lambda/s}}+\frac{m\lambda}{(m+y)^{\lambda+1}y^{2-\lambda/s}}.$$

By (3.1.46), (3.1.44) and (3.1.48), since $0<\lambda\le 2\min\{r,s\}\le 4$, we find

$$-\int_1^{\infty}P_1(x)f'_{\lambda,n}(x)dx=\int_1^{\infty}P_1(y)\frac{(1+\lambda/r)dy}{(m+y)^{\lambda}y^{2-\lambda/s}}$$
$$-\int_1^{\infty}P_1(y)\frac{m\lambda dy}{(m+y)^{\lambda+1}y^{2-\lambda/s}} \quad (4.2.52)$$
$$>-\frac{(1+\lambda/r)}{12(m+1)^{\lambda}}+\frac{m\lambda}{12(m+1)^{\lambda+1}}$$
$$-\frac{m\lambda}{720}[\frac{(\lambda+1)(\lambda+2)}{(m+1)^{\lambda+3}}+\frac{2(\lambda+1)(2s-\lambda)}{s(m+1)^{\lambda+2}}+\frac{(2s-\lambda)(3s-\lambda)}{s^2(m+1)^{\lambda+1}}]$$
$$>-\frac{(1+\lambda/r)}{12(m+1)^{\lambda}}+\frac{(m+1)\lambda-\lambda}{12(m+1)^{\lambda+1}}$$
$$-\frac{4(m+1)}{720}[\frac{(\lambda+1)(\lambda+2)}{(m+1)^{\lambda+3}}+\frac{2(\lambda+1)(2s-\lambda)}{s(m+1)^{\lambda+2}}+\frac{(2s-\lambda)(3s-\lambda)}{s^2(m+1)^{\lambda+1}}]$$
$$>(-\tfrac{1}{12}+\tfrac{\lambda}{12s})\frac{1}{(m+1)^{\lambda}}-\frac{\lambda}{12(m+1)^{\lambda+1}}$$
$$-\tfrac{1}{180}[\frac{6(\lambda+1)}{(m+1)^{\lambda+2}}+\frac{12s}{(m+1)^{\lambda+1}}+\frac{6}{(m+1)^{\lambda}}].$$

In view of (4.2.50), we obtain

$$\theta(\tfrac{\lambda}{s},m)>[\tfrac{s}{\lambda}-\tfrac{7}{12}+\tfrac{\lambda}{12s}-\tfrac{1}{30}]\frac{1}{(m+1)^{\lambda}}$$
$$+(\tfrac{s}{3}-\tfrac{\lambda}{12}-\tfrac{s}{15})\frac{1}{(m+1)^{\lambda+1}}+(\tfrac{s}{12}-\tfrac{1}{30})\frac{(\lambda+1)}{(m+1)^{\lambda+2}}$$
$$\ge\tfrac{1}{60\lambda s}[60s^2-37s\lambda+5\lambda^2]\frac{1}{(m+1)^{\lambda}}$$
$$+(\tfrac{s}{3}-\tfrac{s}{6}-\tfrac{s}{15})\frac{1}{(m+1)^{\lambda+1}}+(\tfrac{s}{12}-\tfrac{1}{30})\frac{(\lambda+1)}{(m+1)^{\lambda+2}}$$
$$>\tfrac{1}{60\lambda s}[60s^2-37s\lambda+5\lambda^2]\frac{1}{(m+1)^{\lambda}}$$
$$=\frac{g(\lambda)}{60\lambda s}\cdot\frac{1}{(m+1)^{\lambda}}, \quad (4.2.53)$$

where, define $g(\lambda):=60s^2-37s\lambda+5\lambda^2$. Since

$$g'(\lambda)=[60s^2-37s\lambda+5\lambda^2]'_{\lambda}$$
$$=10\lambda-37s\le 20s-37s=-17s<0,$$
$$g(\lambda)\ge[60s^2-37s\lambda+5\lambda^2]_{\lambda=2s}$$
$$=60s^2-74s^2+20s^2=6s^2>0,$$

By (4.2.50), we have $\theta(\frac{\lambda}{s},m)>0$. In view of (4.2.51) and (4.2.49), we have $\varpi(\frac{\lambda}{s},m)<B(\frac{\lambda}{r},\frac{\lambda}{s})$. By the same way, it follows $\vartheta(\frac{\lambda}{r},n)<B(\frac{\lambda}{r},\frac{\lambda}{s})$.

Since we find

$$\int_0^1 f_{\lambda,m}(y)dy<\int_0^1\frac{1}{m^{\lambda}}y^{\frac{\lambda}{s}-1}dy=\frac{s}{\lambda m^{\lambda}},$$

then by (4.2.52) and (4.1.46), we have

$$-\int_1^{\infty}P_1(y)f'_{\lambda,m}(y)dy<\frac{m\lambda}{12(m+1)^{\lambda+1}}.$$

Hence by (4.1.30), we find

$$0<m^{\frac{\lambda}{r}}\theta(\tfrac{\lambda}{s},m)$$
$$<m^{\frac{\lambda}{r}}[\frac{s}{\lambda m^{\lambda}}-\frac{1}{2(m+1)^{\lambda}}+\frac{m\lambda}{12(m+1)^{\lambda+1}}]<(\tfrac{2}{\lambda}+\tfrac{\lambda}{12})\frac{1}{m^{\lambda/s}}.$$

It is obvious that

$$m^{\frac{\lambda}{r}}\theta(\tfrac{\lambda}{s},m)=O(\tfrac{1}{m^{\lambda/s}})\ (m\to\infty).$$

Then by (4.2.49) and (4.2.51), we have (4.2.9).

For $0<\varepsilon<\min\{\frac{\lambda}{r},1\}$, setting $\delta=\frac{\lambda}{s}+\varepsilon>\frac{\lambda}{s}$, we have $\delta<\min\{\frac{\lambda}{s}+1,\lambda\}$ and

$$k_{\lambda}(1,u)=\frac{1}{(1+u)^{\lambda}}=O(\tfrac{1}{u^{\delta}})\ (u\to\infty).$$

By Theorem 4.2.2 and Theorem 4.2.5, for $0<\lambda\le 2\min\{r,s\}$, we have the following equivalent inequalities:

$$\sum_{n=1}^{\infty}\sum_{m=1}^{\infty}\frac{a_m b_n}{(m+n)^{\lambda}}<B(\tfrac{\lambda}{r},\tfrac{\lambda}{s})\|a\|_{p,\phi}\|b\|_{q,\psi}, \quad (4.2.54)$$

$$\sum_{n=1}^{\infty} n^{\frac{p\lambda}{s}-1}[\sum_{m=1}^{\infty}\frac{a_m}{(m+n)^{\lambda}}]^p$$
$$<[B(\tfrac{\lambda}{r},\tfrac{\lambda}{s})]^p \| a \|_{p,\phi}^p . \quad (4.2.55)$$

Example 4.2.18 If $0<\lambda\le\min\{r,s\}$,

$$k_\lambda(x,y)=\frac{|\ln(x/y)|}{(\max\{x,y\})^{\lambda}},$$

then we define the following weight coefficients:

$$\varpi(\tfrac{\lambda}{s},m):=\sum_{n=1}^{\infty}\frac{|\ln m/n)|}{(\max\{m,n\})^{\lambda}}\frac{m^{\lambda/r}}{n^{1-\lambda/s}},$$

$$\vartheta(\tfrac{\lambda}{r},n):=\sum_{m=1}^{\infty}\frac{|\ln(m/n)|}{(\max\{m,n\})^{\lambda}}\frac{n^{\lambda/s}}{m^{1-\lambda/r}},\quad m,n\in\mathbf{N}. \quad (4.2.56)$$

Setting $f_{\lambda,m}(y)=\frac{|\ln(m/y)|}{(\max\{m,y\})^{\lambda}y^{1-\lambda/s}}$, by (3.1.43), it follows

$$\varpi(\tfrac{\lambda}{s},m)=m^{\frac{\lambda}{r}}\sum_{n=1}^{\infty}f_{\lambda,m}(n)$$
$$=m^{\frac{\lambda}{r}}[\int_1^{\infty}f_{\lambda,m}(y)dy+\tfrac{1}{2}f_{\lambda,m}(1)$$
$$+\int_1^{\infty}P_1(y)f'_{\lambda,m}(y)dy]$$
$$=m^{\frac{\lambda}{r}}\int_0^{\infty}f_{\lambda,m}(y)dy-m^{\frac{\lambda}{r}}\theta(\tfrac{\lambda}{s},m), \quad (4.2.57)$$

$$\theta(\tfrac{\lambda}{s},m)=\int_0^1 f_{\lambda,m}(y)dy$$
$$-\tfrac{1}{2}f_{\lambda,m}(1)-\int_1^{\infty}P_1(y)f'_{\lambda,m}(y)dy. \quad (4.2.58)$$

We have $-\frac{1}{2}f_{\lambda,m}(1)=-\frac{\ln m}{2m^{\lambda}}$. Setting $u=\frac{y}{m}$, we find

$$m^{\frac{\lambda}{r}}\int_0^{\infty}f_{\lambda,m}(y)dy=\int_0^{\infty}\frac{|\ln(m/y)|m^{\lambda/r}}{(\max\{m,y\})^{\lambda}y^{1-\lambda/s}}dy$$
$$=\int_0^{\infty}\frac{|\ln u|du}{(\max\{1,u\})^{\lambda}u^{1-\lambda/s}}=\int_0^1\frac{-\ln udu}{u^{1-\lambda/s}}+\int_1^{\infty}\frac{\ln udu}{u^{1+\lambda/r}}$$
$$=\tfrac{s}{\lambda}\int_0^1(-\ln u)du^{\frac{\lambda}{s}}-\tfrac{r}{\lambda}\int_1^{\infty}\ln udu^{\frac{-\lambda}{r}}$$
$$=\tfrac{s^2+r^2}{\lambda^2}, \quad (4.2.59)$$

$$\int_0^1 f_{\lambda,m}(y)dy=\frac{1}{m^{\lambda/r}}\int_0^{\frac{1}{m}}\frac{-\ln udu}{u^{1-\lambda/s}}$$
$$=\frac{s}{\lambda m^{\lambda/r}}\int_0^{\frac{1}{m}}(-\ln u)du^{\frac{\lambda}{s}}=\frac{s}{\lambda m^{\lambda}}(\ln m+\tfrac{s}{\lambda}).$$

For $1\le y\le m$, since $f_{\lambda,m}(y)=\frac{-\ln(y/m)}{m^{\lambda}y^{1-\lambda/s}}$, we have

$$f'_{\lambda,m}(y)=[\frac{-1}{y^{2-\lambda/s}}+(1-\tfrac{\lambda}{s})\frac{\ln(y/m)}{y^{2-\lambda/s}}]\frac{1}{m^{\lambda}}$$
$$=[\frac{-1}{y^{2-\lambda/s}}-(1-\tfrac{\lambda}{s})\frac{m\ln(y/m)}{(y-m)y^{2-\lambda/s}}$$
$$+(1-\tfrac{\lambda}{s})\frac{\ln(y/m)}{(y-m)y^{1-\lambda/s}}]\frac{1}{m^{\lambda}}; \quad (4.2.60)$$

for $y\ge m$, since $f_{\lambda,m}(y)=\frac{\ln(y/m)}{y^{1+\lambda/r}}$, we find

$$f'_{\lambda,m}(y)=\frac{1}{y^{2+\lambda/r}}-(1+\tfrac{\lambda}{r})\frac{\ln(y/m)}{y^{2+\lambda/r}}$$
$$=\frac{1}{y^{2+\lambda/r}}+(1+\tfrac{\lambda}{r})\frac{m\ln(y/m)}{(y-m)y^{2+\lambda/r}}$$
$$-(1+\tfrac{\lambda}{r})\frac{\ln(y/m)}{(y-m)y^{1+\lambda/r}}. \quad (4.2.61)$$

Then by (3.1.45), (3.1.46), since $0<\lambda\le s$, we have $\varepsilon_i\in(0,1)$ $(i=1,2,3,4)$, and

$$\int_1^{\infty}P_1(y)f'_{\lambda,m}(y)dy$$
$$=\int_1^{m}P_1(y)f'_{\lambda,m}(y)dy+\int_m^{\infty}P_1(y)f'_{\lambda,m}(y)dy$$
$$=\frac{\varepsilon_1}{12m^{\lambda}}[\frac{-1}{y^{2-\lambda/s}}-(1-\tfrac{\lambda}{s})\frac{m\ln(y/m)}{(y-m)y^{2-\lambda/s}}]_1^m$$
$$+\tfrac{\varepsilon_2}{12}(1+\tfrac{\lambda}{r})\frac{\ln(y/m)}{(y-m)y^{1+\lambda/r}}\Big|_{y=m}$$
$$+\tfrac{\varepsilon_3}{12}[(1-\tfrac{\lambda}{s})\frac{\ln(y/m)}{(y-m)y^{1-\lambda/s}}]_1^m\frac{1}{m^{\lambda}}$$
$$-\tfrac{\varepsilon_4}{12}[\frac{1}{y^{2+\lambda/r}}+(1+\tfrac{\lambda}{r})\frac{m\ln(y/m)}{(y-m)y^{2+\lambda/r}}]_{y=m} \quad (4.2.62)$$
$$<\frac{1}{12m^{\lambda}}[\frac{-1}{y^{2-\lambda/s}}-(1-\tfrac{\lambda}{s})\frac{m\ln(y/m)}{(y-m)y^{2-\lambda/s}}]_1^m$$
$$+\tfrac{1}{12}(1+\tfrac{\lambda}{r})\frac{\ln(y/m)}{(y-m)y^{1+\lambda/r}}\Big|_{y=m}$$
$$=\frac{1}{12m^{\lambda}}-\tfrac{1}{12}(2-\tfrac{\lambda}{s})\frac{1}{m^{2+\lambda/r}}+\tfrac{1}{12}(1-\tfrac{\lambda}{s})\frac{m\ln m}{(m-1)m^{\lambda}}$$
$$+\tfrac{1}{12}(1+\tfrac{\lambda}{r})\frac{1}{m^{2+\lambda/r}}.$$

Hence we find

$$m^{\frac{\lambda}{r}}\theta(\tfrac{\lambda}{s},m)>\rho_\lambda(s,m)$$
$$:=(\tfrac{s}{\lambda}-\tfrac{1}{2})\frac{\ln m}{m^{\lambda/s}}+(\tfrac{s^2}{\lambda^2}-\tfrac{1}{12})\frac{1}{m^{\lambda/s}}$$
$$+\tfrac{1}{12}(1-\lambda)\frac{1}{m^2}-\tfrac{1}{12}(1-\tfrac{\lambda}{s})\frac{m\ln m}{(m-1)m^{\lambda/s}}.$$

For $m=1$, we obtain

$$\rho_\lambda(s,1)=(\tfrac{s^2}{\lambda^2}-\tfrac{1}{12})+\tfrac{1}{12}(1-\lambda)-\tfrac{1}{12}(1-\tfrac{\lambda}{s})$$
$$=\tfrac{s^2}{\lambda^2}-\tfrac{1}{12}-\tfrac{\lambda}{12r}\ge 1-\tfrac{1}{12}-\tfrac{1}{12}>0,$$
$$0<\lambda\le\min\{r,s\};$$

for $m\ge 2$, since $\frac{\lambda}{s}\le 1$ and $-\frac{1}{m-1}\ge-\frac{2}{m}$, we have

$$\rho_\lambda(s,m)\ge\tfrac{1}{\lambda}(s-\tfrac{2}{3}\lambda+\tfrac{\lambda^2}{6s})\frac{\ln m}{m^{\lambda/s}}+(\tfrac{s^2}{\lambda^2}-\tfrac{\lambda}{12})\frac{1}{m^2}. \quad (4.2.63)$$

Since $h(\lambda)=s-\frac{2\lambda}{3}+\frac{\lambda^2}{6s}$ is decreasing in $0<\lambda\le s$, in fact, $h'(\lambda)=-\frac{2}{3}+\frac{\lambda}{3s}<0$, then we find

$$h(\lambda)\ge h(s)=s-\tfrac{2}{3}s+\tfrac{s^2}{6s}=\tfrac{s}{2}>0.$$

Since we obtain

$$\tfrac{s^2}{\lambda^2}-\tfrac{\lambda}{12}\ge 1-\tfrac{1}{6}=\tfrac{5}{6}>0(0<\lambda\le\min\{r,s\}\le 2),$$

then by (4.2.61), we have $\rho_\lambda(s,m)>0$. And by (4.2.59) and (4.2.57), we have $\varpi(\frac{\lambda}{s},m)<\frac{r^2+s^2}{\lambda^2}$. By the same way, we have $\vartheta(\frac{\lambda}{r},n)<\frac{r^2+s^2}{\lambda^2}$, and then (4.2.8) is valid.

Since by (4.2.62), we have

$$\int_1^\infty P_1(y)f'_{\lambda,m}(y)dy$$
$$>\frac{1}{12m^\lambda}[(1-\frac{\lambda}{s})\frac{\ln(y/m)}{(y-m)y^{1-\lambda/s}}]_1^m$$
$$-\frac{1}{12}[\frac{1}{y^{2+\lambda/r}}+(1+\frac{\lambda}{r})\frac{m\ln(y/m)}{(y-m)y^{2+\lambda/r}}]_{y=m}$$
$$=-\frac{1}{12}(1+\lambda)\frac{1}{m^{2+\lambda/r}}-\frac{1}{12}(1-\frac{\lambda}{s})\frac{\ln m}{(m-1)m^\lambda},$$

then by (4.2.58) and the above results, we find

$$0<m^{\frac{\lambda}{r}}\theta(\frac{\lambda}{s},m)<\frac{s}{\lambda m^{\lambda/s}}(\ln m+\frac{s}{\lambda})-\frac{\ln m}{2m^{\lambda/s}}$$
$$+\frac{1}{12}(1+\lambda)\frac{1}{m^2}+\frac{1}{12}(1-\frac{\lambda}{s})\frac{\ln m}{(m-1)m^{\lambda/s}}$$
$$<[\frac{s}{\lambda}\ln m+\frac{s^2}{\lambda^2}$$
$$+\frac{1}{12}(1+\lambda)+\frac{1}{12}(1-\frac{\lambda}{s})\frac{\ln m}{(m-1)}]\frac{1}{m^{\lambda/s}},$$

and $m^{\frac{\lambda}{r}}\theta(\frac{\lambda}{s},m)=O(\frac{1}{m^{\lambda/(2s)}})$ $(m\to\infty)$. Hence by (4.2.57) and (4.2.59), we have (4.2.9).

For $0<\varepsilon<\min\{\frac{\lambda}{r},1\}$, setting $\delta=\frac{\lambda}{s}+\varepsilon>\frac{\lambda}{s}$, then we have $\delta<\min\{\frac{\lambda}{s}+1,\lambda\}$ and

$$k_\lambda(1,u)=\frac{|\ln u|}{(\max\{1,u\})^\lambda}=O(\frac{1}{u^\delta})\ (u\to\infty).$$

By Theorem 4.2.3 and Theorem 4.2.6, for $0<\lambda\le\min\{r,s\}$, we have the following equivalent inequalities:

$$\sum_{n=1}^\infty\sum_{m=1}^\infty\frac{|\ln(m/n)|a_m b_n}{(\max\{m,n\})^\lambda}<\frac{r^2+s^2}{\lambda^2}\|a\|_{p,\phi}\|b\|_{q,\psi},\quad(4.2.64)$$

$$\sum_{n=1}^\infty n^{\frac{p\lambda}{s}-1}[\sum_{m=1}^\infty\frac{|\ln(m/n)|a_m}{(\max\{m,n\})^\lambda}]^p$$
$$<(\frac{r^2+s^2}{\lambda^2})^p\|a\|_{p,\phi}^p.\quad(4.2.65)$$

4. 3 SOME REVERSE HILBERT-TYPE INEQUALITIES

4.3.1 MAIN RESULTS

If $0<p<1,\frac{1}{p}+\frac{1}{q}=1,\lambda_1,\lambda_2,\lambda\in\mathbf{R},\lambda_1+\lambda_2=\lambda$, $k_\lambda(x,y)(\ge0)$ is a finite homogeneous function of degree $-\lambda$ in $\mathbf{R}_+^2$, setting $\phi(x)=x^{p(1-\lambda_1)-1}$, $\psi(x)=x^{q(1-\lambda_2)-1},[\psi(x)]^{1-p}=x^{p\lambda_2-1}$ $(x\in(0,\infty))$, then we define the following weight coefficients $\varpi(\lambda_2,m)$ and $\vartheta(\lambda_1,n)$:

$$\varpi(\lambda_2,m):=\sum_{n=1}^\infty k_\lambda(m,n)\frac{m^{\lambda_1}}{n^{1-\lambda_2}},$$
$$\vartheta(\lambda_1,n):=\sum_{m=1}^\infty k_\lambda(m,n)\frac{n^{\lambda_2}}{m^{1-\lambda_1}},m,n\in\mathbf{N}.\quad(4.3.1)$$

We have the following theorem:

Theorem 4.3.1 Assuming that $0<p<1,\frac{1}{p}+\frac{1}{q}=1$, $\lambda_1,\lambda_2,\lambda\in\mathbf{R},\lambda_1+\lambda_2=\lambda$, $k_\lambda(x,y)(\ge0)$ is a finite homogeneous function of degree $-\lambda$ in $\mathbf{R}_+^2$, satisfying

$$0<k(\lambda_1)=\int_0^\infty k_\lambda(u,1)u^{\lambda_1-1}du<\infty,$$
$$\varpi(\lambda_2,m)<k(\lambda_1),\vartheta(\lambda_1,n)<k(\lambda_1),$$
$$m,n\in\mathbf{N},\quad(4.3.2)$$

and there exist constants $\lambda'>0$ and $c_\lambda>0$, for any $m\in\mathbf{N}$,

$$c_\lambda\le k(\lambda_1)[1-\theta(\lambda_1,m)]\le\varpi(\lambda_2,m),\quad(4.3.3)$$

where, $\theta(\lambda_1,m)=O(\frac{1}{m^{\lambda'}})$ $(m\to\infty)$. If $a=\{a_m\}_{m=1}^\infty,\ b=\{b_n\}_{n=1}^\infty,a_m,b_n\ge0$, such that

$$0<\|a\|_{p,\phi}=\{\sum_{m=1}^\infty m^{p(1-\lambda_1)-1}a_m^p\}^{\frac{1}{p}}<\infty$$

and

$$0<\|b\|_{q,\psi}=\{\sum_{n=1}^\infty n^{q(1-\lambda_2)-1}b_n^q\}^{\frac{1}{q}}<\infty,$$

then we have the following equivalent inequalities:

$$I_\lambda:=\sum_{n=1}^\infty\sum_{m=1}^\infty k_\lambda(m,n)a_m b_n>k(\lambda_1)$$
$$\times\{\sum_{m=1}^\infty[1-\theta(\lambda_1,m)]m^{p(1-\lambda_1)-1}a_m^p\}^{\frac{1}{p}}\|b\|_{q,\psi},\quad(4.3.4)$$

$$J_\lambda:=\sum_{n=1}^\infty n^{p\lambda_2-1}(\sum_{m=1}^\infty k_\lambda(m,n)a_m)^p$$
$$>k^p(\lambda_1)\sum_{m=1}^\infty[1-\theta(\lambda_1,m)]m^{p(1-\lambda_1)-1}a_m^p,\quad(4.3.5)$$

$$L_\lambda:=\sum_{m=1}^\infty\frac{m^{q\lambda_1-1}}{[1-\theta(\lambda_1,m)]^{q-1}}(\sum_{n=1}^\infty k_\lambda(m,n)b_n)^q$$
$$<k^q(\lambda_1)\|b\|_{q,\psi}^q,\quad(4.3.6)$$

where the constant factors $k(\lambda_1)$, $k^p(\lambda_1)$ and $k^q(\lambda_1)$ are all the best possible.

Proof In view of (4.3.2) and (4.3.3), $0<\|a\|_{p,\phi}<\infty$ is equivalent to

$$0<\sum_{m=1}^{\infty}[1-\theta(\lambda_1,m)]m^{p(1-\lambda_1)-1}a_m^p<\infty.$$

By the reverse Hölder's inequality and (4.3.1), we have

$$I_\lambda=\sum_{n=1}^{\infty}\sum_{m=1}^{\infty}k_\lambda(m,n)[\tfrac{m^{(1-\lambda_1)/q}}{n^{(1-\lambda_2)/p}}a_m][\tfrac{n^{(1-\lambda_2)/p}}{m^{(1-\lambda_1)/q}}b_n]$$

$$\geq\{\sum_{m=1}^{\infty}\sum_{n=1}^{\infty}k_\lambda(m,n)\tfrac{m^{(1-\lambda_1)(p-1)}}{n^{1-\lambda_2}}a_m^p\}^{\frac{1}{p}}$$

$$\times\{\sum_{n=1}^{\infty}\sum_{m=1}^{\infty}k_\lambda(m,n)\tfrac{n^{(1-\lambda_2)(q-1)}}{m^{1-\lambda_1}}b_n^q\}^{\frac{1}{q}}$$

$$=\{\sum_{m=1}^{\infty}\varpi(\lambda_2,m)m^{p(1-\lambda_1)-1}a_m^p\}^{\frac{1}{p}}$$

$$\times\{\sum_{n=1}^{\infty}\vartheta(\lambda_1,n)n^{q(1-\lambda_2)-1}b_n^q\}^{\frac{1}{q}}.$$

Since $0<p<1,q<0$, by (4.3.2) and (4.3.3), we have (4.3.4).

Since $\|a\|_{p,\phi}>0$, it is obvious that $J_\lambda>0$. If $J_\lambda=\infty$, then (4.3.5) is naturally valid; if $0<J_\lambda<\infty$, setting

$$b_n=n^{p\lambda_2-1}(\sum_{m=1}^{\infty}k_\lambda(m,n)a_m)^{p-1}>0\ (n\in\mathbf{N}),$$

then by (4.3.4), we find

$$\infty>\|b\|_{q,\psi}^q=J_\lambda=I_\lambda>k(\lambda_1)$$

$$\times\{\sum_{n=1}^{\infty}[1-\theta(\lambda_1,m)]m^{p(1-\lambda_1)-1}a_m^p\}^{\frac{1}{p}}\|b\|_{q,\psi}>0,$$

$$J_\lambda=\|b\|_{q,\psi}^q$$

$$>k^p(\lambda_1)\sum_{m=1}^{\infty}[1-\theta(\lambda_1,m)]m^{p(1-\lambda_1)-1}a_m^p.$$

Hence we have (4.3.5).

On the other-hand, suppose that (4.3.5) is valid. By the reverse Hölder's inequality, we find

$$I_\lambda=\sum_{n=1}^{\infty}[n^{\frac{-1}{p}+\lambda_2}\sum_{m=1}^{\infty}k_\lambda(m,n)a_m][n^{\frac{1}{p}-\lambda_2}b_n]$$

$$\geq J_\lambda^{\frac{1}{p}}\|b\|_{q,\psi}. \tag{4.3.7}$$

Then by (4.3.5), in view of $0<\|b\|_{q,\psi}<\infty$, we have (4.3.4), which is equivalent to (4.3.5).

If there exists a constant $K\geq k(\lambda_1)$, such that (4.3.4) is valid as we replace $k(\lambda_1)$ by k, then for any $N\in\mathbf{N}$, setting $\tilde{a}=\{\tilde{a}_n\}_{n=1}^{\infty},\tilde{b}=\{\tilde{b}_n\}_{n=1}^{\infty}$ as

$$\tilde{a}_n=n^{\lambda_1-1},\tilde{b}_n=n^{\lambda_2-1}n\leq N;\ \tilde{a}_n=\tilde{b}_n=0,n>N,$$

it follows

$$\tilde{I}:=\sum_{n=1}^{\infty}\sum_{m=1}^{\infty}k_\lambda(m,n)\tilde{a}_m\tilde{b}_n$$

$$>K\{\sum_{m=1}^{\infty}[1-\theta(\lambda_1,m)]m^{p(1-\lambda_1)-1}\tilde{a}_m^p\}^{\frac{1}{p}}\|\tilde{b}\|_{q,\psi}$$

$$=K\{\sum_{m=1}^{N}[\tfrac{1}{m}-\tfrac{1}{m}\theta(\lambda_1,m)]\}^{\frac{1}{p}}\{\sum_{n=1}^{N}\tfrac{1}{n}\}^{\frac{1}{q}}$$

$$=K\{\sum_{m=1}^{N}\tfrac{1}{m}-\sum_{m=1}^{N}\tfrac{1}{m}\theta(\lambda_1,m)]\}^{\frac{1}{p}}\{\sum_{m=1}^{N}\tfrac{1}{m}\}^{\frac{1}{q}}$$

$$=K\{1-(\sum_{m=1}^{N}\tfrac{1}{m})^{-1}\sum_{m=1}^{N}\tfrac{1}{m}O(\tfrac{1}{m^{\lambda'}})]\}^{\frac{1}{p}}\sum_{m=1}^{N}\tfrac{1}{m}. \tag{4.3.8}$$

Since by (4.3.2), we have

$$\tilde{I}=\sum_{m=1}^{N}\tfrac{1}{m}[\sum_{n=1}^{N}k_\lambda(m,n)\tfrac{m^{\lambda_1}}{n^{1-\lambda_2}}]$$

$$<\sum_{m=1}^{N}\tfrac{1}{m}\varpi(\lambda_2,m)<k(\lambda_1)\sum_{m=1}^{N}\tfrac{1}{m},$$

then by (4.3.8), we find

$$k(\lambda_1)\sum_{m=1}^{N}\tfrac{1}{m}$$

$$>K\{1-(\sum_{m=1}^{N}\tfrac{1}{m})^{-1}\sum_{m=1}^{N}O(\tfrac{1}{m^{1+\lambda'}})\}^{\frac{1}{p}}\sum_{m=1}^{N}\tfrac{1}{m},$$

$$k(\lambda_1)>K[1-(\sum_{m=1}^{N}\tfrac{1}{m})^{-1}\sum_{m=1}^{N}O(\tfrac{1}{m^{1+\lambda'}})]^{\frac{1}{p}}. \tag{4.3.9}$$

For $N\to\infty$ in (4.3.9), since

$$0\leq\sum_{m=1}^{\infty}|O(\tfrac{1}{m^{1+\lambda'}})|<\infty(\lambda'>0),$$

and $\sum_{m=1}^{\infty}\frac{1}{m}=\infty$, we have $k(\lambda_1)\geq K$. Therefore $K=k(\lambda_1)$ is the best value of (4.3.4).

The constant factor in (4.3.5) is the best possible, otherwise, by (4.3.7), we can get a contradiction that the constant factor in (4.3.4) is not the best possible.

If $L_\lambda=0$, then (4.3.6) is naturally valid; if $L_\lambda>0$, then there exists $n_0\in\mathbf{N}$, such that for $N\geq n_0$, $\{\sum_{n=1}^{N}n^{q(1-\lambda_2)-1}b_n^q\}^{\frac{1}{q}}>0$ and

$$L_\lambda(N):=\sum_{m=1}^{N}\tfrac{m^{q\lambda_1-1}}{[1-\theta(\lambda_1,m)]^{q-1}}(\sum_{n=1}^{N}k_\lambda(m,n)b_n)^q>0,$$

$$a_m(N):=\tfrac{m^{q\lambda_1-1}}{[1-\theta(\lambda_1,m)]^{q-1}}(\sum_{n=1}^{N}k_\lambda(m,n)b_n)^{q-1}>0,$$

$$m \in \mathbf{N}, N \geq n_0.$$

Then by (4.3.4), in view of $q < 0$, we have

$$\infty > \sum_{m=1}^{N} [1-\theta(\lambda_1, m)] m^{p(1-\lambda_1)-1} a_m^p(N)$$

$$= L_\lambda(N) = \sum_{n=1}^{N} \sum_{m=1}^{N} k_\lambda(m,n) a_m(N) b_n$$

$$> k(\lambda_1) \{\sum_{m=1}^{N} [1-\theta(\lambda_1, m)] m^{p(1-\lambda_1)-1} a_m^p(N)\}^{\frac{1}{p}}$$

$$\times \{\sum_{n=1}^{N} n^{q(1-\lambda_2)-1} b_n^q\}^{\frac{1}{q}} > 0,$$

$$0 < \sum_{m=1}^{N} [1-\theta(\lambda_1, m)] m^{p(1-\lambda_1)-1} a_m^p(N)$$

$$= L_\lambda(N) < k^q(\lambda_1) \| b \|_{q,\psi}.$$

Hence it follows

$$0 < \sum_{m=1}^{\infty} [1-\theta(\lambda_1, m)] m^{p(1-\lambda_1)-1} a_m^p(\infty) < \infty.$$

For $N \to \infty$, still using (4.3.4), both of the above inequalities keep the forms of strict-sign inequalities, and we have (4.3.6).

On the other-hand, if (4.3.6) is valid, then by the reverse Hölder's inequality, we have

$$I_\lambda = \sum_{m=1}^{\infty} \{[1-\theta(\lambda_1, m)]^{\frac{1}{p}} m^{\frac{1}{q}-\lambda_1} a_m\}$$

$$\times \{\frac{m^{\lambda_1 - 1/q}}{[1-\theta(\lambda_1, m)]^{1/p}} \sum_{n=1}^{\infty} k_\lambda(m,n) b_n\}$$

$$\geq \{\sum_{m=1}^{\infty} [1-\theta(\lambda_1, m)] m^{p(1-\lambda_1)-1} a_m^p\}^{\frac{1}{p}} L_\lambda^{\frac{1}{q}}. \quad (4.3.10)$$

Then by (4.3.6), in view of $q < 0$, we have (4.3.4), which is equivalent to (4.3.6).

We affirm that the constant factor in (4.3.6) is the best possible, otherwise we can get a contradiction by (4.3.10) that the constant factor in (4.3.4) is not the best possible. It is obvious that (4.3.4), (4.3.5) and (4.3.6) are equivalent. □

Corollary 4.3.2 Assuming that $\lambda_1, \lambda_2, \lambda \in \mathbf{R}$, $\lambda_1 + \lambda_2 = \lambda$, $k_\lambda(x,y)(\geq 0)$ is a finite homogeneous function of degree $-\lambda$ in $\mathbf{R}_+^2$ and

$$0 < k(\lambda_1) = \int_0^\infty k_\lambda(u,1) u^{\lambda_1 - 1} du < \infty,$$

if $k_\lambda(x,y)\frac{1}{x^{1-\lambda_2}}$ is decreasing for $x > 0$, $k_\lambda(x,y)\frac{1}{y^{1-\lambda_1}}$ is decreasing for $y > 0$, and they are strict decreasing in a subinterval respectively, $k_\lambda(u,1)$ is positive in an interval containing 1 with

$$k_\lambda(u,1) = O(\tfrac{1}{u^\alpha}) \ (\alpha > \lambda_1; u \to \infty),$$

then for $0 < p < 1, \frac{1}{p} + \frac{1}{q} = 1$, we still have equivalent inequalities (4.3.4)-(4.3.6) with the best constant factors.

Proof By Theorem 4.2.3, we have (4.3.2). Setting $u = \frac{m}{y}$, we find

$$\varpi(\lambda_2, m) = \sum_{n=1}^{\infty} k_\lambda(m,n) \frac{m^{\lambda_1}}{n^{1-\lambda_2}}$$

$$\geq \int_1^\infty k_\lambda(m,y) \frac{m^{\lambda_1}}{y^{1-\lambda_2}} dy = \int_0^m k_\lambda(u,1) u^{\lambda_1 - 1} du$$

$$= k(\lambda_1)[1 - \frac{1}{k(\lambda_1)} \int_m^\infty k_\lambda(u,1) u^{\lambda_1 - 1} du]. \quad (4.3.11)$$

Putting $c_\lambda = \int_0^1 k_\lambda(u,1) u^{\lambda_1 - 1} du > 0$ and

$$\theta(\lambda_1, m) = \frac{1}{k(\lambda_1)} \int_m^\infty k_\lambda(u,1) u^{\lambda_1 - 1} du,$$

there exists $L > 0$, such that

$$0 \leq k_\lambda(u,1) = L(\tfrac{1}{u^\alpha}) \ (u \in [1, \infty)).$$

Hence we have

$$0 \leq \theta(\lambda_1, m) \leq \frac{L}{k(\lambda_1)} \int_m^\infty u^{-\alpha + \lambda_1 - 1} du$$

$$= \frac{L}{k(\lambda_1)(\alpha - \lambda_1)} \cdot \frac{1}{m^{\alpha - \lambda_1}}.$$

Setting $\lambda' = \alpha - \lambda_1 > 0$, then it follows

$$\theta(\lambda_1, m) = O(\tfrac{1}{m^{\lambda'}}) \ (m \to \infty),$$

and by (4.3.11), we have (4.3.3). Hence we can prove all the results of Theorem 4.3.1. □

Note 4.3.3 For $0 < \lambda \leq \min\{r, s\}$, the decreasing property condition of $k_\lambda(x,y)\frac{1}{x^{1-\lambda_1}}$ and $k_\lambda(x,y)\frac{1}{y^{1-\lambda_2}}$ in Corollary 4.3.2 may change for " $k_\lambda(x,y)$ is decreasing for $x(y)$ and strictly decreasing for a subinterval".

4.3.2 SOME EXAMPLES FOR APPLYING COROLLARY 4.3.2

Example 4.3.4 If $\alpha, \lambda_1, \lambda_2, \lambda \in \mathbf{R}, \lambda_1 + \lambda_2 = \lambda$, $0 < \alpha + \lambda_i \leq 1 (i = 1, 2)$,

$$k_\lambda(x,y) = \frac{(\min\{x,y\})^\alpha}{(\max\{x,y\})^{\lambda+\alpha}}, x, y > 0,$$

then we find

$$k(\lambda_1) = \int_0^\infty \frac{(\min\{u,1\})^\alpha}{(\max\{u,1\})^{\lambda+\alpha}} u^{\lambda_1 - 1} du$$

$$=\int_0^1 u^{\alpha+\lambda_1-1}du+\int_1^\infty u^{-\lambda_2-\alpha-1}du=\frac{2\alpha+\lambda}{(\alpha+\lambda_1)(\alpha+\lambda_2)}.$$

Since we have

$$c_\lambda=\int_0^1 k_\lambda(u,1)u^{\lambda_1-1}du$$

$$=\int_0^1 u^{\alpha+\lambda_1-1}du=\frac{1}{\alpha+\lambda_1}>0,$$

$$\theta(\lambda_1,m)=\frac{1}{k(\lambda_1)}\int_m^\infty \frac{(\min\{u,1\})^\alpha}{(\max\{u,1\})^{\lambda+\alpha}}u^{\lambda_1-1}du$$

$$=\frac{1}{k(\lambda_1)}\int_m^\infty u^{-\alpha-\lambda_2-1}du=\frac{\alpha+\lambda_1}{2\alpha+\lambda}(\frac{1}{m})^{\alpha+\lambda_2},$$

then by Example 4.2.7 and Corollary 4.3.2, we have the following equivalent inequalities:

$$\sum_{n=1}^\infty\sum_{m=1}^\infty \frac{(\min\{m,n\})^\alpha}{(\max\{m,n\})^{\lambda+\alpha}}a_m b_n>\frac{2\alpha+\lambda}{(\alpha+\lambda_1)(\alpha+\lambda_2)}$$

$$\times\{\sum_{m=1}^\infty[1-\frac{\alpha+\lambda_1}{(2\alpha+\lambda)m^{\alpha+\lambda_2}}]m^{p(1-\lambda_1)-1}a_m^p\}^{\frac{1}{p}}\|b\|_{q,\tilde{\psi}},\tag{4.3.12}$$

$$\sum_{n=1}^\infty n^{p\lambda_2-1}[\sum_{m=1}^\infty \frac{(\min\{m,n\})^\alpha a_m}{(\max\{m,n\})^{\lambda+\alpha}}]^p>[\frac{2\alpha+\lambda}{(\alpha+\lambda_1)(\alpha+\lambda_2)}]^p$$

$$\times\sum_{m=1}^\infty[1-\frac{\alpha+\lambda_1}{(2\alpha+\lambda)m^{\alpha+\lambda_2}}]m^{p(1-\lambda_1)-1}a_m^p.\tag{4.3.13}$$

$$\sum_{m=1}^\infty \frac{m^{q\lambda_1-1}}{[1-\frac{(\alpha+\lambda_1)}{(2\alpha+\lambda)m^{\alpha+\lambda_2}}]^{q-1}}[\sum_{n=1}^\infty \frac{(\min\{m,n\})^\alpha}{(\max\{m,n\})^{\lambda+\alpha}}b_n]^q$$

$$<[\frac{2\alpha+\lambda}{(\alpha+\lambda_1)(\alpha+\lambda_2)}]^q\|b\|_{q,\tilde{\psi}}^q,\tag{4.3.14}$$

where the constant factor are all the best possible.

In the following examples, we suppose that $0<p<1,r>1,\frac{1}{p}+\frac{1}{q}=1,\frac{1}{r}+\frac{1}{s}=1,\lambda>0,$ $\lambda_1=\frac{\lambda}{r},\ \lambda_2=\frac{\lambda}{s},\ k(\lambda_1)=k_\lambda(r),\ \phi(x)=x^{p(1-\frac{\lambda}{r})-1}$ $\psi(x)=x^{q(1-\frac{\lambda}{s})-1}\ (x\in(0,\infty)),a\in l_\phi^p,b\in l_\psi^q,$

$$0<\|a\|_{p,\phi}=\{\sum_{n=1}^\infty n^{p(1-\frac{\lambda}{r})-1}a_n^p\}^{\frac{1}{p}}<\infty$$

and $0<\|b\|_{q,\psi}=\{\sum_{n=1}^\infty n^{q(1-\frac{\lambda}{s})-1}b_n^q\}^{\frac{1}{q}}<\infty$. The words that the constant factors are the best possible are omitted.

Example 4.3.5 If $\alpha>0,0<\lambda\le\min\{r,s\},$ $k_\lambda(x,y)=\frac{1}{(x^\alpha+y^\alpha)^{\lambda/\alpha}}$, then by Corollary 4.3.2, setting

$$\theta(\frac{\lambda}{r},m)=\frac{\alpha}{B(\frac{\lambda}{r\alpha},\frac{\lambda}{s\alpha})}\int_m^\infty \frac{1}{(u^\alpha+1)^{\lambda/\alpha}}u^{\frac{\lambda}{r}-1}du$$

and $c_\lambda=\int_0^1 \frac{1}{(u^\alpha+1)^{\lambda/\alpha}}u^{\frac{\lambda}{r}-1}du>0$, we have the following equivalent inequalities:

$$\sum_{n=1}^\infty\sum_{m=1}^\infty \frac{a_m b_n}{(m^\alpha+n^\alpha)^{\lambda/\alpha}}>\frac{1}{\alpha}B(\frac{\lambda}{r\alpha},\frac{\lambda}{s\alpha})$$

$$\times\{\sum_{m=1}^\infty[1-\theta(\frac{\lambda}{r},m)]m^{p(1-\frac{\lambda}{r})-1}a_m^p\}^{\frac{1}{p}}\|b\|_{q,\psi},\tag{4.3.15}$$

$$\sum_{n=1}^\infty n^{\frac{p\lambda}{s}-1}[\sum_{m=1}^\infty \frac{a_m}{(m^\alpha+n^\alpha)^{\lambda/\alpha}}]^p>[\frac{1}{\alpha}B(\frac{\lambda}{r\alpha},\frac{\lambda}{s\alpha})]^p$$

$$\times\sum_{m=1}^\infty[1-\theta(\frac{\lambda}{r},m)]m^{p(1-\frac{\lambda}{r})-1}a_m^p,\tag{4.3.16}$$

$$\sum_{m=1}^\infty \frac{m^{q\lambda/r-1}}{[1-\theta(\frac{\lambda}{r},m)]^{q-1}}[\sum_{n=1}^\infty \frac{1}{(m^\alpha+n^\alpha)^{\lambda/\alpha}}b_n]^q$$

$$<[\frac{1}{\alpha}B(\frac{\lambda}{r\alpha},\frac{\lambda}{s\alpha})]^q\|b\|_{q,\psi}^q.\tag{4.3.17}$$

Example 4.3.6 If $0<\lambda\le\min\{r,s\}$, $k_\lambda(x,y)=\frac{1}{(\max\{x,y\})^\lambda}$, then by Corollary 4.3.2, setting

$$\theta(\frac{\lambda}{r},m)=\frac{\lambda}{rs}\int_m^\infty k_\lambda(u,1)u^{\frac{\lambda}{r}-1}du$$

$$=\frac{\lambda}{rs}\int_m^\infty u^{\frac{-\lambda}{s}-1}du=\frac{1}{rm^{\lambda/s}},$$

$c_\lambda=\int_0^1 u^{\frac{\lambda}{r}-1}du=\frac{r}{\lambda}>0$, we have the following equivalent inequalities:

$$\sum_{n=1}^\infty\sum_{m=1}^\infty \frac{a_m b_n}{(\max\{m,n\})^\lambda}$$

$$>\frac{rs}{\lambda}\{\sum_{m=1}^\infty[1-\frac{1}{rm^{\lambda/s}}]m^{p(1-\frac{\lambda}{r})-1}a_m^p\}^{\frac{1}{p}}\|b\|_{q,\psi},\tag{4.3.18}$$

$$\sum_{n=1}^\infty n^{\frac{p\lambda}{s}-1}[\sum_{m=1}^\infty \frac{1}{(\max\{m,n\})^\lambda}a_m]^p$$

$$>(\frac{rs}{\lambda})^p\sum_{m=1}^\infty[1-\frac{1}{sm^{\lambda/r}}]m^{p(1-\frac{\lambda}{r})-1}a_m^p,\tag{4.3.19}$$

$$\sum_{m=1}^\infty \frac{m^{q\lambda/r-1}}{[1-\frac{1}{rm^{\lambda/s}}]^{q-1}}[\sum_{n=1}^\infty \frac{1}{(\max\{m,n\})^\lambda}b_n]^q$$

$$<(\frac{rs}{\lambda})^q\|b\|_{q,\psi}^q.\tag{4.3.20}$$

Example 4.3.7 If $0<\lambda\le\min\{r,s\}$, $k_\lambda(x,y)=\frac{\ln(x/y)}{x^\lambda-y^\lambda}$, then we find $k_\lambda(r)=[\frac{\pi}{\lambda\sin(\pi/r)}]^2$. Setting $0<\delta<\lambda,\alpha=\lambda-\delta>0$, since

$$k_\lambda(u,1)u^\alpha=\frac{\ln u}{u^\delta}\cdot\frac{u^\lambda}{u^\lambda-1}\to 0(u\to\infty),$$

and $k_\lambda(u,1)=O(\frac{1}{u^\alpha})\ (u\to\infty)$, by Corollary 4.3.2, setting $\theta(\frac{\lambda}{r},m)=[\frac{\lambda\sin(\pi/r)}{\pi}]^2\int_m^\infty \frac{\ln u}{u^\lambda-1}u^{\frac{\lambda}{r}-1}du$, we have the following equivalent inequalities:

$$\sum_{n=1}^{\infty}\sum_{m=1}^{\infty}\frac{\ln(m/n)a_m b_n}{m^{\lambda}-n^{\lambda}}>[\frac{\pi}{\lambda\sin(\frac{\pi}{r})}]^2$$

$$\times\{\sum_{m=1}^{\infty}[1-\theta(\tfrac{\lambda}{r},m)]m^{p(1-\frac{\lambda}{r})-1}a_m^p\}^{\frac{1}{p}}\,||\,b\,||_{q,\psi}, \quad (4.3.21)$$

$$\sum_{n=1}^{\infty}n^{\frac{p\lambda}{s}-1}[\sum_{m=1}^{\infty}\frac{\ln(m/n)}{m^{\lambda}-n^{\lambda}}a_m]^p>[\frac{\pi}{\lambda\sin(\pi/r)}]^{2p}$$

$$\times\sum_{m=1}^{\infty}[1-\theta(\tfrac{\lambda}{r},m)]m^{p(1-\frac{\lambda}{r})-1}a_m^p, \quad (4.3.22)$$

$$\sum_{m=1}^{\infty}\frac{m^{q\lambda/r-1}}{[1-\theta(\frac{\lambda}{r},m)]^{q-1}}[\sum_{n=1}^{\infty}\frac{\ln(m/n)}{m^{\lambda}-n^{\lambda}}b_n]^q$$

$$<[\frac{\pi}{\lambda\sin(\pi/r)}]^{2q}\,||\,b\,||_{q,\psi}^q. \quad (4.3.23)$$

Note 4.3.8 Some early results of (4.3.21) are consulted in (Yang JIPAM 2005)[29]

4.3.2 SOME EXAMPLES FOR APPLYING THEOREM 4.3.1

Example 4.3.9 If $0<\lambda\le 2\min\{r,s\}$, $k_{\lambda}(x,y)=\frac{1}{(x+y)^{\lambda}}$, then we have $k(\frac{\lambda}{r})=B(\frac{\lambda}{r},\frac{\lambda}{s})$, and by Example 4.2.17, it follows

$$\theta(\tfrac{\lambda}{r},m)=1-\frac{1}{B(\frac{\lambda}{r},\frac{\lambda}{s})}\varpi(\tfrac{\lambda}{s},m)$$

$$=O(\frac{1}{m^{\lambda/s}})(m\to\infty).$$

Setting $f(y)=\frac{1}{(m+y)^{\lambda}y^{1-\lambda/s}}$ $(y>0)$, then for $0<\lambda\le s$, $f(y)$ is decreasing; for $s<\lambda\le 2s$, we can find that $f(y)$ is decreasing for $y\ge m$. In fact, we have

$$f'(m)=\frac{-1}{(m+y)^{\lambda+1}y^{2-\lambda/s}}$$

$$\times[(\tfrac{\lambda}{r}+1)y-(\tfrac{\lambda}{s}-1)m]|_{y=m}<0.$$

We obtain

$$\varpi(\tfrac{\lambda}{s},m)=\sum_{n=1}^{\infty}\frac{1}{(m+n)^{\lambda}}\cdot\frac{m^{\lambda/r}}{n^{1-\lambda/s}}$$

$$\ge\sum_{n=m}^{\infty}\frac{1}{(m+n)^{\lambda}}\cdot\frac{m^{\lambda/r}}{n^{1-\lambda/s}}$$

$$\ge\int_m^{\infty}\frac{1}{(m+y)^{\lambda}}\frac{m^{\lambda/r}dy}{y^{1-\lambda/s}}=c_{\lambda}:=\int_1^{\infty}\frac{u^{\frac{\lambda}{s}-1}}{(1+u)^{\lambda}}du>0.$$

Then by Theorem 4.3.1, we have the following equivalent inequalities:

$$\sum_{n=1}^{\infty}\sum_{m=1}^{\infty}\frac{a_m b_n}{(m+n)^{\lambda}}>B(\tfrac{\lambda}{r},\tfrac{\lambda}{s})$$

$$\times\{\sum_{m=1}^{\infty}[1-\theta(\tfrac{\lambda}{r},m)]m^{p(1-\frac{\lambda}{r})-1}a_m^p\}^{\frac{1}{p}}\,||\,b\,||_{q,\psi}, \quad (4.3.24)$$

$$\sum_{n=1}^{\infty}n^{\frac{p\lambda}{s}-1}[\sum_{m=1}^{\infty}\frac{1}{(m+n)^{\lambda}}a_m]^p>[B(\tfrac{\lambda}{r},\tfrac{\lambda}{s})]^p$$

$$\times\sum_{m=1}^{\infty}[1-\theta(\tfrac{\lambda}{r},m)]m^{p(1-\frac{\lambda}{r})-1}a_m^p, \quad (4.3.25)$$

$$\sum_{m=1}^{\infty}\frac{m^{q\lambda/r-1}}{[1-\theta(\frac{\lambda}{r},m)]^{q-1}}[\sum_{n=1}^{\infty}\frac{1}{(m+n)^{\lambda}}b_n]^q$$

$$<[B(\tfrac{\lambda}{r},\tfrac{\lambda}{s})]^q\,||\,b\,||_{q,\psi}^q. \quad (4.3.26)$$

Example 4.3.10 If $0<\lambda\le\min\{r,s\}$, $k_{\lambda}(x,y)=\frac{|\ln(x/y)|}{(\max\{x,y\})^{\lambda}}$, then we find $k_{\lambda}(r)=\frac{r^2+s^2}{\lambda^2}$. By Example 4.2.18, setting

$$\theta(\tfrac{\lambda}{r},m)=1-\frac{\lambda^2}{r^2+s^2}\varpi(\tfrac{\lambda}{s},m)$$

$$=O(\frac{1}{m^{\lambda/(2s)}})\ (m\to\infty),$$

since we find

$$\varpi(\tfrac{\lambda}{s},m)=\sum_{n=1}^{\infty}\frac{|\ln m/n)|}{(\max\{m,n\})^{\lambda}}\frac{m^{\lambda/r}}{n^{1-\lambda/s}}$$

$$\ge\sum_{n=2m}^{\infty}\frac{\ln(n/m)}{(\max\{m,n\})^{\lambda}}\frac{m^{\lambda/r}}{n^{1-\lambda/s}}\ge\sum_{n=2m}^{\infty}\frac{\ln 2}{n^{\lambda}}\frac{m^{\lambda/r}}{n^{1-\lambda/s}}$$

$$\ge m^{\frac{\lambda}{r}}\ln 2\int_{2m}^{\infty}\frac{1}{y^{1+\lambda/r}}dy=c_{\lambda}:=\frac{r\ln 2}{2^{\lambda/r}\lambda}>0,$$

in view of Theorem 4.3.1, we have the following equivalent inequalities:

$$\sum_{n=1}^{\infty}\sum_{m=1}^{\infty}\frac{|\ln(m/n)|a_m b_n}{(\max\{m,n\})^{\lambda}}>\frac{r^2+s^2}{\lambda^2}$$

$$\times\{\sum_{m=1}^{\infty}[1-\theta(\tfrac{\lambda}{r},m)]m^{p(1-\frac{\lambda}{r})-1}a_m^p\}^{\frac{1}{p}}\,||\,b\,||_{q,\psi}, \quad (4.3.27)$$

$$\sum_{n=1}^{\infty}n^{\frac{p\lambda}{s}-1}[\sum_{m=1}^{\infty}\frac{|\ln(m/n)|}{(\max\{m,n\})^{\lambda}}a_m]^p>(\frac{r^2+s^2}{\lambda^2})^p$$

$$\times\sum_{m=1}^{\infty}[1-\theta(\tfrac{\lambda}{r},m)]m^{p(1-\frac{\lambda}{r})-1}a_m^p, \quad (4.3.28)$$

$$\sum_{m=1}^{\infty}\frac{m^{q\lambda/r-1}}{[1-\theta(\frac{\lambda}{r},m)]^{q-1}}[\sum_{n=1}^{\infty}\frac{|\ln(m/n)|}{(\max\{m,n\})^{\lambda}}b_n]^q$$

$$<(\frac{r^2+s^2}{\lambda^2})^q\,||\,b\,||_{q,\psi}^q. \quad (4.3.29)$$

4.4 HILBERT-TYPE INEQUALITIES WITH COMPOSITE VARIABLES

4.4.1 MAIN RESULTS

Lemma 4.4.1 If $n_0\in\mathbf{N}_0$, $u(x)$ and $v(y)$ are strict increasing differentiable functions in $[n_0,\infty)$, such that $u(n_0),v(n_0)>0,u(\infty)=v(\infty)=\infty,u'(x)$

and $v'(x)$ are positive decreasing functions in $[n_0,\infty)$, and $\ln\frac{v(x)}{u(x)}=\tilde{O}(1)(x\to\infty)$, then for $N>n_0 (N\in\mathbf{N})$, $\lambda'>0$, we have $\sum_{m=n_0}^{\infty}\frac{u'(m)}{u(m)}=\infty$, and

$$\sum_{m=n_0}^{\infty}\frac{u'(m)}{[u(m)]^{1+\lambda'}}\le\frac{u'(n_0)}{[u(n_0)]^{1+\lambda'}}+\frac{1}{\lambda'[u(n_0)]^{\lambda'}},\tag{4.4.1}$$

$$\sum_{n=n_0}^{N}\frac{v'(n)}{v(n)}=\sum_{m=n_0}^{N}\frac{u'(m)}{u(m)}+O(1)(N\to\infty).\tag{4.4.2}$$

Proof Since $\frac{1}{[u(x)]^{\varepsilon}}$ and $\frac{u'(x)}{u(x)}$ are positive decreasing, we find

$$\sum_{m=n_0}^{\infty}\frac{u'(m)}{u(m)}\ge\int_{n_0}^{\infty}\frac{u'(x)}{u(x)}dx=[\ln u(x)]_{n_0}^{\infty}=\infty,$$

$$\sum_{m=n_0}^{\infty}\frac{u'(m)}{[u(m)]^{1+\lambda'}}=\frac{u'(n_0)}{[u(n_0)]^{1+\lambda'}}+\sum_{m=n_0+1}^{\infty}\frac{u'(m)}{[u(m)]^{1+\lambda'}}$$

$$\le\frac{u'(n_0)}{[u(n_0)]^{1+\lambda'}}+\int_{n_0}^{\infty}\frac{u'(x)dx}{[u(x)]^{1+\lambda'}}=\frac{u'(n_0)}{[u(n_0)]^{1+\lambda'}}+\frac{1}{\lambda'[u(n_0)]^{\lambda'}},$$

and (4.4.1) is valid. We find

$$\sum_{n=n_0}^{N}\frac{v'(n)}{v(n)}-\sum_{m=n_0}^{N}\frac{u'(m)}{u(m)}$$

$$=\frac{v'(n_0)}{v(n_0)}+\sum_{n=n_0+1}^{N}\frac{v'(n)}{v(n)}-\sum_{m=n_0}^{N}\frac{u'(m)}{u(m)}$$

$$\le\frac{v'(n_0)}{v(n_0)}+\int_{n_0}^{N}\frac{v'(y)}{v(y)}dy-\int_{n_0}^{N}\frac{u'(x)}{u(x)}dx$$

$$=\frac{v'(n_0)}{v(n_0)}+[\ln v(y)]_{n_0}^{N}-[\ln u(x)]_{n_0}^{N}$$

$$=\frac{v'(n_0)}{v(n_0)}+\ln\frac{u(n_0)}{v(n_0)}+\ln\frac{v(N)}{u(N)};$$

$$\sum_{n=n_0}^{N}\frac{v'(n)}{v(n)}-\sum_{m=n_0}^{N}\frac{u'(m)}{u(m)}$$

$$=\sum_{n=n_0}^{N}\frac{v'(n)}{v(n)}-\frac{u'(n_0)}{u(n_0)}-\sum_{m=n_0+1}^{N}\frac{u'(m)}{u(m)}$$

$$\ge\int_{n_0}^{N}\frac{v'(y)}{v(y)}dy-\frac{u'(n_0)}{u(n_0)}-\int_{n_0}^{N}\frac{u'(x)}{u(x)}dx$$

$$=-\frac{u'(n_0)}{u(n_0)}+\ln\frac{u(n_0)}{v(n_0)}+\ln\frac{v(N)}{u(N)}.$$

Since $\ln\frac{v(N)}{u(N)}=\tilde{O}(1)$, then we have (4.4.2). □

Let the assumptions of Lemma 4.4.1 be fulfilled and additionally, $p>0\ (p\ne1)$, $\frac{1}{p}+\frac{1}{q}=1$, $\lambda_1,\lambda_2,\lambda\in\mathbf{R}$, $\lambda_1+\lambda_2=\lambda$, $k_\lambda(x,y)(\ge0)$ is a finite homogeneous function of degree $-\lambda$ in $\mathbf{R}_+^2$ and

$$0<k(\lambda_1)=\int_0^{\infty}k_\lambda(u,1)u^{\lambda_1-1}du<\infty,$$

setting $\tilde{k}_\lambda(m,n):=k_\lambda(u(m),v(n))$, then we define the weight coefficients $\tilde{w}(\lambda_2,m)$ and $\tilde{\vartheta}(\lambda_1,n)$ as follows

$$\tilde{w}(\lambda_2,m):=\sum_{n=n_0}^{\infty}\tilde{k}_\lambda(m,n)\frac{[u(m)]^{\lambda_1}}{[v(n)]^{1-\lambda_2}}v'(n),$$

$$\tilde{\vartheta}(\lambda_1,n):=\sum_{m=n_0}^{\infty}\tilde{k}_\lambda(m,n)\frac{[v(n)]^{\lambda_2}}{[u(m)]^{1-\lambda_1}}u'(m),$$

$$m,n\ge n_0.\tag{4.4.3}$$

Putting $\tilde{\phi}(x)=[u(x)]^{p(1-\lambda_1)-1}[u'(x)]^{1-p}$, $\tilde{\psi}(x)=[v(x)]^{q(1-\lambda_2)-1}[v'(x)]^{1-q}$ $(x\in[n_0,\infty))$, if $a=\{a_m\}_{m=n_0}^{\infty}, b=\{b_n\}_{n=n_0}^{\infty}, a_m,b_n\ge0$, such that

$$0<\|a\|_{p,\tilde{\phi}}=\{\sum_{m=n_0}^{\infty}\frac{[u(m)]^{p(1-\lambda_1)-1}}{[u'(m)]^{p-1}}a_m^p\}^{\frac{1}{p}}<\infty,$$

$$0<\|b\|_{q,\tilde{\psi}}=\{\sum_{n=n_0}^{\infty}\frac{[v(n)]^{q(1-\lambda_2)-1}}{[v'(n)]^{q-1}}b_n^q\}^{\frac{1}{q}}<\infty,$$

then we have the following two theorems:

Theorem 4.4.2 For $p>1$, if there exists $\lambda'>0$, such that for any $m,n\ge n_0$, $\tilde{\vartheta}(\lambda_1,n)<k(\lambda_1)$, $\tilde{w}(\lambda_2,m)<k(\lambda_1)$ and

$$k(\lambda_1)[1-\tilde{\theta}(\lambda_1,m)]\le\tilde{w}(\lambda_2,m),\tag{4.4.4}$$

where $\tilde{\theta}(\lambda_1,m)=O(\frac{1}{(u(m))^{\lambda'}})(m\to\infty)$,

$$k_\lambda(1,t)=O(\tfrac{1}{t^{\delta}})$$
$$(\lambda_2<\delta<\lambda_2+1;t\to\infty),\tag{4.4.5}$$

and $\ln\frac{v(x)}{u(x)}=\tilde{O}(1)(x\to\infty)$, then we have the following equivalent inequalities:

$$\tilde{I}_\lambda:=\sum_{n=n_0}^{\infty}\sum_{m=n_0}^{\infty}\tilde{k}_\lambda(m,n)a_mb_n$$
$$<k(\lambda_1)\|a\|_{p,\tilde{\phi}}\|b\|_{q,\tilde{\psi}},\tag{4.4.6}$$

$$\tilde{J}_\lambda:=\sum_{n=n_0}^{\infty}[v(n)]^{p\lambda_2-1}v'(n)(\sum_{m=n_0}^{\infty}\tilde{k}_\lambda(m,n)a_m)^p$$
$$<k^p(\lambda_1)\|a\|_{p,\tilde{\phi}}^p,\tag{4.4.7}$$

where the constant factors $k(\lambda_1)$ and $k^p(\lambda_1)$ are the best possible.

Proof By Hölder's inequality and (4.4.3), we have

$$\tilde{I}_\lambda = \sum_{n=n_0}^{\infty}\sum_{m=n_0}^{\infty}\tilde{k}_\lambda(m,n)[\frac{(u(m))^{(1-\lambda_1)/q}(v'(n))^{1/p}}{(v(n))^{(1-\lambda_2)/p}(u'(m))^{1/q}}a_m]$$
$$\times[\frac{(v(n))^{(1-\lambda_2)/p}(u'(m))^{1/q}}{(u(m))^{(1-\lambda_1)/q}(v'(n))^{1/p}}b_n]$$
$$\le \{\sum_{m=n_0}^{\infty}\tilde{w}(\lambda_2,m)\tilde{\phi}(m)a_m^p\}^{\frac{1}{p}}$$
$$\times\{\sum_{n=n_0}^{\infty}\tilde{\vartheta}(\lambda_1,n)\tilde{\psi}(n)b_n^q\}^{\frac{1}{q}}.$$

By (4.4.4) , in view of $0<\|a\|_{p,\tilde{\phi}}<\infty$ and $0<\|b\|_{q,\tilde{\psi}}<\infty$, we have (4.4.6).

For large enough $N\in\mathbf{N}$, such that

$$k_\lambda(1,t)\le L(\frac{1}{t^\delta})(t\ge N;L>0),$$

by (4.4.5), since $u'(x)$ and $v'(x)$ are decreasing, we find

$$\sum_{m=n_0}^{N}\frac{u'(m)}{u(m)}\sum_{n=N+1}^{\infty}\tilde{k}_\lambda(m,n)\frac{[u(m)]^{\lambda_1}}{[v(n)]^{1-\lambda_2}}v'(n)$$
$$\le L\sum_{m=n_0}^{N}\frac{u'(m)}{[u(m)]^{1+\lambda_2-\delta}}\sum_{n=N+1}^{\infty}\frac{v'(n)}{[v(n)]^{1-\lambda_2+\delta}}$$
$$=L\{\frac{u'(n_0)}{[u(n_0)]^{1+\lambda_2-\delta}}+\sum_{m=n_0+1}^{N}\frac{u'(m)}{[u(m)]^{1+\lambda_2-\delta}}\}$$
$$\times\sum_{n=N+1}^{\infty}\frac{v'(n)}{[v(n)]^{1-\lambda_2+\delta}}$$
$$\le L\{\frac{u'(n_0)}{[u(n_0)]^{1+\lambda_2-\delta}}+\int_{n_0}^{N}\frac{u'(x)dx}{[u(x)]^{1+\lambda_2-\delta}}\}\int_N^\infty\frac{v'(y)dy}{[v(y)]^{1-\lambda_2+\delta}}$$
$$\le\frac{L}{\delta-\lambda_2}\{\frac{u'(n_0)[u(N)]^{\lambda_2-\delta}}{[u(n_0)]^{1+\lambda_2-\delta}}+\frac{1}{\delta-\lambda_2}\}$$
$$\times[\frac{v(N)}{u(N)}]^{\lambda_2-\delta}\le M\ \ (M>0). \qquad (4.4.8)$$

Setting $\tilde{a}=\{\tilde{a}_m\}_{m=n_0}^{\infty},\tilde{b}=\{\tilde{b}_n\}_{n=n_0}^{\infty}$ as follows:

$$\tilde{a}_m=[u(m)]^{\lambda_1-1}u'(m),b_n=[v(n)]^{\lambda_2-1}v'(n),$$
$$m,n\le N;\tilde{a}_m=\tilde{b}_n=0,m,n>N,$$

by (4.4.4) and (4.4.8), we find

$$\tilde{I}'_\lambda:=\sum_{n=n_0}^{\infty}\sum_{m=n_0}^{\infty}\tilde{k}_\lambda(m,n)\tilde{a}_m\tilde{b}_n$$
$$=\sum_{m=n_0}^{N}\frac{u'(m)}{u(m)}\sum_{n=n_0}^{N}\tilde{k}_\lambda(m,n)\frac{[u(m)]^{\lambda_1}}{[v(n)]^{1-\lambda_2}}v'(n)$$
$$=\sum_{m=n_0}^{N}\frac{u'(m)}{u(m)}\tilde{w}(\lambda_2,m)$$
$$-\sum_{m=n_0}^{N}\frac{u'(m)}{u(m)}\sum_{n=N+1}^{\infty}\tilde{k}_\lambda(m,n)\frac{[u(m)]^{\lambda_1}}{[v(n)]^{1-\lambda_2}}v'(n)$$
$$\ge k(\lambda_1)\sum_{m=n_0}^{N}\frac{u'(m)}{u(m)}[1-\tilde{\theta}(\lambda_1,m)]-M$$
$$=k(\lambda_1)[\sum_{m=n_0}^{N}\frac{u'(m)}{u(m)}-\sum_{m=n_0}^{N}\frac{u'(m)}{u(m)}\tilde{\theta}(\lambda_1,m)]-M$$
$$=k(\lambda_1)\sum_{m=n_0}^{N}\frac{u'(m)}{u(m)}[1-(\sum_{m=n_0}^{N}\frac{u'(m)}{u(m)})^{-1}$$
$$\times\sum_{m=n_0}^{N}\frac{u'(m)}{u(m)}O(\frac{1}{[u(m)]^{\lambda'}})]-M. \qquad (4.4.9)$$

If there exists a constant $0<k\le k(\lambda_1)$, such that (4.4.6) is still valid as we replace $k(\lambda_1)$ by k, then in particular, by (4.4.2), we have

$$\tilde{I}'_\lambda<k\|\tilde{a}\|_{p,\tilde{\phi}}\|\tilde{b}\|_{q,\tilde{\psi}}$$
$$=k\sum_{m=n_0}^{N}\frac{u'(m)}{u(m)}\{1+(\sum_{m=n_0}^{N}\frac{u'(m)}{u(m)})^{-1}O(1)\}^{\frac{1}{q}},$$

and by (4.4.9), we find

$$k(\lambda_1)[1-(\sum_{m=n_0}^{N}\frac{u'(m)}{u(m)})^{-1}$$
$$\times\sum_{m=n_0}^{N}\frac{u'(m)}{u(m)}O(\frac{1}{[u(m)]^{\lambda'}})]-(\sum_{m=n_0}^{N}\frac{u'(m)}{u(m)})^{-1}M$$
$$<k\{1+(\sum_{m=n_0}^{N}\frac{u'(m)}{u(m)})^{-1}O(1)\}^{\frac{1}{q}}.$$

For $N\to\infty$, by (4.4.1), we have $k(\lambda_1)\le k$. Hence $k=k(\lambda_1)$ is the best value of (4.4.6).

Since $\|a\|_{p,\tilde{\phi}}>0$, then there exists $N\ge m_0>n_0$, $\sum_{m=n_0}^{N}[u(m)]^{p(1-\lambda_1)-1}[u'(m)]^{1-p}a_m^p>0$. Setting

$$b_n(N):=[v(n)]^{p\lambda_2-1}v'(n)$$
$$\times[\sum_{m=n_0}^{N}\tilde{k}_\lambda(m,n)a_m]^{p-1}>0,N\ge m_0,n\ge n_0,$$

and

$$\tilde{J}_\lambda(N):=\sum_{n=n_0}^{N}[v(n)]^{p\lambda_2-1}v'(n)$$
$$\times[\sum_{m=n_0}^{N}\tilde{k}_\lambda(m,n)a_m]^p,$$

by (4.4.6), we have

$$0<\sum_{n=n_0}^{N}[v(n)]^{q(1-\lambda_2)-1}[v'(n)]^{1-q}b_n^q(N)$$
$$=\tilde{J}_\lambda(N)=\sum_{n=n_0}^{N}\sum_{m=n_0}^{N}\tilde{k}_\lambda(m,n)a_m b_n(N)$$
$$<k(\lambda_1)\{\sum_{n=n_0}^{N}[u(n)]^{p(1-\lambda_1)-1}[u'(n)]^{1-p}a_n^p\}^{\frac{1}{p}}$$
$$\times\{\sum_{n=n_0}^{N}[v(n)]^{q(1-\lambda_2)-1}[v'(n)]^{1-q}b_n^q(N)\}^{\frac{1}{q}},$$
$$\sum_{n=n_0}^{N}[v(n)]^{q(1-\lambda_2)-1}[v'(n)]^{1-q}b_n^q(N)$$
$$=\tilde{J}_\lambda(N)<k^p(\lambda_1)$$
$$\times\sum_{n=n_0}^{\infty}[u(n)]^{p(1-\lambda_1)-1}[u'(n)]^{1-p}a_n^p.$$

Hence it follows

$$0<\sum_{n=n_0}^{\infty}[v(n)]^{q(1-\lambda_2)-1}[v'(n)]^{1-q}b_n^q(\infty)<\infty.$$

For $N\to\infty$, by (4.4.6), both the above inequalities still keep the forms of strict-sign inequalities, and we have (4.4.7).

On the other-hand, if (4.4.7) is valid, by Hölder's inequality, we have

$$\tilde{I}_\lambda=\sum_{n=n_0}^{\infty}\{[v(n)]^{\lambda_2-\frac{1}{p}}[v'(n)]^{\frac{1}{p}}\sum_{m=n_0}^{\infty}\tilde{k}_\lambda(m,n)a_m\}$$
$$\times\{\frac{[v(n)]^{\frac{1}{p}-\lambda_2}}{[v'(n)]^{1/p}}b_n\}\le\tilde{J}_\lambda^{\frac{1}{p}}\|b\|_{q,\tilde{\psi}}. \tag{4.4.10}$$

Then by (4.4.7), we have (4.4.6), which is equivalent to (4.4.7).

We affirm that the constant factor in (4.4.7) is the best possible, otherwise we can get a contradiction by (4.4.10) that the constant factor in (4.4.5) is not the best possible. □

Theorem 4.4.3 For $0<p<1$, if there exist $\lambda'>0$ and $c_\lambda>0$, such that for any $m,n\ge n_0$ ($m,n\in\mathbf{N}_0$), $\tilde{\vartheta}(\lambda_1,n)<k(\lambda_1)$, $\tilde{w}(\lambda_2,m)<k(\lambda_1)$ and

$$c_\lambda\le k(\lambda_1)[1-\tilde{\theta}(\lambda_1,m)]\le\tilde{w}(\lambda_2,m), \tag{4.4.11}$$

where, $\tilde{\theta}(\lambda_1,m)=O(\frac{1}{(u(m))^{\lambda'}})\ (m\to\infty)$, and

$$\ln\frac{v(x)}{u(x)}=\tilde{O}(1)(x\to\infty),$$

then we have the following equivalent inequalities:

$$\sum_{n=n_0}^{\infty}\sum_{m=n_0}^{\infty}\tilde{k}_\lambda(m,n)a_m b_n>k(\lambda_1)$$
$$\times\{\sum_{m=n_0}^{\infty}[1-\tilde{\theta}(\lambda_1,m)]\tilde{\phi}(m)a_m^p\}^{\frac{1}{p}}\|b\|_{q,\tilde{\psi}}, \tag{4.4.12}$$
$$\sum_{n=n_0}^{\infty}[v(n)]^{p\lambda_2-1}v'(n)[\sum_{m=n_0}^{\infty}\tilde{k}_\lambda(m,n)a_m]^p$$
$$>k^p(\lambda_1)\sum_{m=n_0}^{\infty}[1-\tilde{\theta}(\lambda_1,m)]\tilde{\phi}(m)a_m^p, \tag{4.4.13}$$
$$\sum_{m=n_0}^{\infty}\frac{[u(m)]^{q\lambda_1-1}u'(m)}{[1-\tilde{\theta}(\lambda_1,m)]^{q-1}}[\sum_{n=n_0}^{\infty}\tilde{k}_\lambda(m,n)b_n]^q$$
$$<k^q(\lambda_1)\|b\|_{q,\tilde{w}}^q, \tag{4.4.14}$$

where the constant factors $k(\lambda_1)$, $k^p(\lambda_1)$ and $k^q(\lambda_1)$ are the best possible.

Proof We only prove that the constant factor in (4.4.12) is the best possible. By the same way of Theorem 4.3.1, we can prove the others.

For $N>n_0(N\in\mathbf{N})$, setting $\tilde{a}=\{\tilde{a}_m\}_{m=n_0}^{\infty}$, $\tilde{b}=\{\tilde{b}_n\}_{n=n_0}^{\infty}$ as follows:

$$\tilde{a}_m=[u(m)]^{\lambda_1-1}u'(m),b_n=[v(n)]^{\lambda_2-1}v'(n),$$
$$m,n\le N;\tilde{a}_m=\tilde{b}_n=0,m,n>N,$$

by (4.4.4) and (4.4.2), we find

$$\tilde{I}'_\lambda:=\sum_{n=n_0}^{\infty}\sum_{m=n_0}^{\infty}\tilde{k}_\lambda(m,n)\tilde{a}_m\tilde{b}_n$$
$$=\sum_{m=n_0}^{N}\frac{u'(m)}{u(m)}\sum_{n=n_0}^{N}\tilde{k}_\lambda(m,n)\frac{[u(m)]^{\lambda_1}}{[v(n)]^{1-\lambda_2}}v'(n)$$
$$\le\sum_{m=n_0}^{N}\frac{u'(m)}{u(m)}\tilde{w}(\lambda_2,m)<k(\lambda_1)\sum_{m=n_0}^{N}\frac{u'(m)}{u(m)}, \tag{4.4.15}$$
$$\{\sum_{m=n_0}^{\infty}[1-\tilde{\theta}(\lambda_1,m)]\tilde{\phi}(m)\tilde{a}_m^p\}^{\frac{1}{p}}\|\tilde{b}\|_{q,\tilde{\psi}}$$
$$=\{\sum_{m=n_0}^{N}[1-\tilde{\theta}(\lambda_1,m)]\frac{u'(m)}{u(m)}\}^{\frac{1}{p}}\{\sum_{n=n_0}^{N}\frac{v'(n)}{v(n)}\}^{\frac{1}{q}}$$
$$=\{\sum_{m=n_0}^{N}\frac{u'(m)}{u(m)}-\sum_{m=n_0}^{N}O(\frac{1}{(u(m))^{\lambda'}})\frac{u'(m)}{u(m)}\}^{\frac{1}{p}}$$
$$\times\{\sum_{m=n_0}^{N}\frac{u'(m)}{u(m)}+O(1)\}^{\frac{1}{q}}$$
$$=\{1-(\sum_{m=n_0}^{N}\frac{u'(m)}{u(m)})^{-1}\sum_{m=n_0}^{N}O(\frac{u'(m)}{(u(m))^{1+\lambda'}})\}^{\frac{1}{p}}$$

$$\times\{1+(\sum_{m=n_0}^{N}\tfrac{u'(m)}{u(m)})^{-1}O(1)\}^{\frac{1}{q}}\sum_{m=n_0}^{N}\tfrac{u'(m)}{u(m)}. \qquad (4.4.16)$$

If there exists a constant $K \ge k(\lambda_1)$, such that (4.4.12) is still valid as we replace $k(\lambda_1)$ by K, then in particular, by (4.4.15) and (4.4.16), we have

$$k(\lambda_1)\sum_{m=n_0}^{N}\tfrac{u'(m)}{u(m)} > \tilde{I}'_\lambda$$

$$> K\{\sum_{m=n_0}^{\infty}[1-\tilde{\theta}(\lambda_1,m)]\tilde{\phi}(m)\tilde{a}_m^p\}^{\frac{1}{p}}\|\tilde{b}\|_{q,\tilde{\psi}}$$

$$= K\{1-(\sum_{m=n_0}^{N}\tfrac{u'(m)}{u(m)})^{-1}\sum_{m=n_0}^{N}O(\tfrac{u'(m)}{(u(m))^{1+\lambda'}})\}^{\frac{1}{p}}$$

$$\times\{1+(\sum_{m=n_0}^{N}\tfrac{u'(m)}{u(m)})^{-1}O(1)\}^{\frac{1}{q}}\sum_{m=n_0}^{N}\tfrac{u'(m)}{u(m)},$$

$$k(\lambda_1) > K\{1-(\sum_{m=n_0}^{N}\tfrac{u'(m)}{u(m)})^{-1}\sum_{m=n_0}^{N}O(\tfrac{u'(m)}{(u(m))^{1+\lambda'}})\}^{\frac{1}{p}}$$

$$\times\{1+(\sum_{m=n_0}^{N}\tfrac{u'(m)}{u(m)})^{-1}O(1)\}^{\frac{1}{q}}.$$

For $N\to\infty$, we have $k(\lambda_1)\ge K$. Hence $K=k(\lambda_1)$ is the best value of (4.4.12). □

Note 4.4.4 For $u(x)=v(x)=1, n_0=1$, Theorem 4.4.2 and Theorem 4.4.3 reduce to Theorem 4.2.3 and Theorem 4.3.1.

Theorem 4.4.5 Exchanging the conditions (4.4.4), (4.4.5) and $\ln\frac{v(x)}{u(x)}=\tilde{O}(1)(x\to\infty)$ in Theorem 4.4.2 for the condition that there exists an interval **I**$=[\lambda_1,\lambda_1+\delta)(\delta>0)$, such that for any $\tilde{\lambda}_1\in$ **I**,

$$k(\tilde{\lambda}_1)[1-O(\tfrac{1}{(u(m))^{\tilde{\lambda}'}})]$$

$$\le \tilde{w}(\tilde{\lambda}_2,m)(\tilde{\lambda}'>0), \qquad (4.4.17)$$

then (4.4.6) - (4.4.7) are still valid with the best constant factors and keep the equivalent property.

Proof We only show that the constant factor in (4.4.6) is still the best possible. For $\varepsilon>0$, satisfying $\tilde{\lambda}_1=\lambda_1+\frac{\varepsilon}{q}\in$ **I**, setting $\tilde{a}_m,\tilde{b}_n$ as follows:

$$\tilde{a}_m=\frac{u'(m)}{[u(m)]^{-\lambda_1+\frac{\varepsilon}{p}+1}},\ \tilde{b}_n=\frac{v'(n)}{[v(n)]^{-\lambda_2+\frac{\varepsilon}{q}+1}}, m,n\ge n_0.$$

by (4.4.17), we find

$$\tilde{I}'_\lambda=\sum_{n=n_0}^{\infty}\sum_{m=n_0}^{\infty}\tilde{k}_\lambda(m,n)\tilde{a}_m\tilde{b}_n$$

$$=\sum_{m=n_0}^{\infty}\frac{u'(m)}{[u(m)]^{1+\varepsilon}}\sum_{n=n_0}^{\infty}\tilde{k}_\lambda(m,n)\frac{[u(m)]^{\lambda_1+\frac{\varepsilon}{q}}}{[v(n)]^{1-(\lambda_2-\frac{\varepsilon}{q})}}v'(n)$$

$$=\sum_{m=n_0}^{\infty}\frac{u'(m)}{[u(m)]^{1+\varepsilon}}\tilde{w}(\lambda_2-\tfrac{\varepsilon}{q},m)$$

$$\ge k(\lambda_1+\tfrac{\varepsilon}{q})\sum_{m=n_0}^{\infty}\frac{u'(m)}{[u(m)]^{1+\varepsilon}}[1-O(\tfrac{1}{(u(m))^{\tilde{\lambda}'}})]$$

$$=k(\lambda_1+\tfrac{\varepsilon}{q})[\sum_{m=n_0}^{\infty}\frac{u'(m)}{[u(m)]^{1+\varepsilon}}-\sum_{m=n_0}^{\infty}O(\frac{u'(m)}{[u(m)]^{1+\tilde{\lambda}'+\varepsilon}})]$$

$$=k(\lambda_1+\tfrac{\varepsilon}{q})\sum_{m=n_0}^{\infty}\frac{u'(m)}{[u(m)]^{1+\varepsilon}}$$

$$\times[1-(\sum_{m=n_0}^{\infty}\frac{u'(m)}{[u(m)]^{1+\varepsilon}})^{-1}\sum_{m=n_0}^{\infty}O(\frac{u'(m)}{[u(m)]^{1+\tilde{\lambda}'+\varepsilon}})]. \qquad (4.4.18)$$

If there exists a positive constant $k\le k(\lambda_1)$, such that (4.4.6) is still valid as we replace $k(\lambda_1)$ by k, then in particular, we obtain

$$\tilde{I}'_\lambda < k\|\tilde{a}\|_{p,\tilde{\phi}}\|\tilde{b}\|_{q,\tilde{\psi}}$$

$$=k\{\sum_{m=n_0}^{\infty}\frac{u'(m)}{[u(m)]^{1+\varepsilon}}\}^{\frac{1}{p}}\{\sum_{n=n_0}^{\infty}\frac{v'(n)}{[v(n)]^{1+\varepsilon}}\}^{\frac{1}{q}}.$$

In view of (4.4.18), we have

$$k(\lambda_1+\tfrac{\varepsilon}{q})[1-(\sum_{m=n_0}^{\infty}\frac{u'(m)}{[u(m)]^{1+\varepsilon}})^{-1}\sum_{m=n_0}^{\infty}O(\frac{u'(m)}{[u(m)]^{1+\tilde{\lambda}'+\varepsilon}})]$$

$$<k\{(\sum_{m=n_0}^{\infty}\frac{u'(m)}{[u(m)]^{1+\varepsilon}})^{-1}\sum_{n=n_0}^{\infty}\frac{v'(n)}{[v(n)]^{1+\varepsilon}}\}^{\frac{1}{q}}$$

$$\le k\{(\int_{n_0}^{\infty}\frac{u'(x)dx}{[u(x)]^{1+\varepsilon}})^{-1}[\frac{v'(n_0)}{[v(n_0)]^{1+\varepsilon}}+\int_{n_0}^{\infty}\frac{v'(y)dy}{[v(y)]^{1+\varepsilon}}]\}^{\frac{1}{q}}$$

$$=k\{\varepsilon\frac{[u(n_0)]^{\varepsilon}v'(n_0)}{[v(n_0)]^{1+\varepsilon}}+[\frac{u(n_0)}{v(n_0)}]^{\varepsilon}\}^{\frac{1}{q}}. \qquad (4.4.19)$$

For $\varepsilon\to0^+$ in (4.4.19), by Fatou lemma, we find

$$k\ge\lim_{\varepsilon\to0^+}k(\lambda_1+\tfrac{\varepsilon}{q})$$

$$=\lim_{\varepsilon\to0^+}\int_0^{\infty}k_\lambda(u,1)u^{\lambda_1+\frac{\varepsilon}{q}-1}du$$

$$\ge\int_0^{\infty}\lim_{\varepsilon\to0^+}k_\lambda(u,1)u^{\lambda_1+\frac{\varepsilon}{q}-1}du=k(\lambda_1).$$

Hence $k=k(\lambda_1)$ is the best value of (4.4.6). □

Theorem 4.4.6 Exchanging the conditions $\tilde{w}(\lambda_2,m)<k(\lambda_1)$ and $\ln\frac{v(x)}{u(x)}=\tilde{O}(1)(x\to\infty)$ in Theorem 4.4.3 for the condition that there exists an interval **I**$_1=(\lambda_1-\delta,\lambda_1](\delta>0)$, such that for any

$\tilde{\lambda}_1 \in \mathbf{I}_1$, $\tilde{w}(\tilde{\lambda}_2, m) \le k(\tilde{\lambda}_1) < \infty$, then (4.4.12) - (4.4.14) are still valid with the best constant factors and keep the equivalent property.

Proof We only show that the constant factor in (4.4.12) is still the best possible. For $0 < \varepsilon < -\frac{q\delta}{2}$, $\tilde{\lambda}_1 = \lambda_1 + \frac{\varepsilon}{q} \in \mathbf{I}_1$, setting $\tilde{a}_m, \tilde{b}_n$ as follows:

$$\tilde{a}_m = \frac{u'(m)}{[u(m)]^{-\lambda_1 + \frac{\varepsilon}{p} + 1}}, \tilde{b}_n = \frac{v'(n)}{[v(n)]^{-\lambda_2 + \frac{\varepsilon}{q} + 1}}, m, n \ge n_0,$$

we find

$$\tilde{I}'_\lambda = \sum_{n=n_0}^{\infty} \sum_{m=n_0}^{\infty} \tilde{k}_\lambda(m,n)\tilde{a}_m \tilde{b}_n$$

$$= \sum_{m=n_0}^{\infty} \frac{u'(m)}{[u(m)]^{1+\varepsilon}} \sum_{n=n_0}^{\infty} \tilde{k}_\lambda(m,n) \frac{[u(m)]^{\lambda_1 + \frac{\varepsilon}{q}}}{[v(n)]^{1-(\lambda_2 - \frac{\varepsilon}{q})}} v'(n)$$

$$= \sum_{m=n_0}^{\infty} \frac{u'(m)}{[u(m)]^{1+\varepsilon}} \tilde{w}(\lambda_2 - \tfrac{\varepsilon}{q}, m)$$

$$\le k(\lambda_1 + \tfrac{\varepsilon}{q}) \sum_{m=n_0}^{\infty} \frac{u'(m)}{[u(m)]^{1+\varepsilon}}. \quad (4.4.20)$$

If there exists a positive constant $K \ge k(\lambda_1)$, such that (4.4.12) is still valid as we replace $k(\lambda_1)$ by K, then in particular, we obtain

$$\tilde{I}'_\lambda > K\{\sum_{m=n_0}^{\infty} [1 - \tilde{\theta}(\lambda_1, m)]\tilde{\phi}(m)\tilde{a}_m^p\}^{\frac{1}{p}} \| \tilde{b} \|_{q,\tilde{\psi}}$$

$$= K\{\sum_{m=n_0}^{\infty} [1 - \tilde{\theta}(\lambda_1, m)]\frac{u'(m)}{[u(m)]^{1+\varepsilon}}\}^{\frac{1}{p}} \{\sum_{n=n_0}^{\infty} \frac{v'(n)}{[v(n)]^{1+\varepsilon}}\}^{\frac{1}{q}}$$

$$= K\{\sum_{m=n_0}^{\infty} \frac{u'(m)}{[u(m)]^{1+\varepsilon}} - \sum_{m=n_0}^{\infty} O(\frac{u'(m)}{[u(m)]^{1+\lambda'+\varepsilon}})\}^{\frac{1}{p}}$$

$$\times \{\sum_{n=n_0}^{\infty} \frac{v'(n)}{[v(n)]^{1+\varepsilon}}\}^{\frac{1}{q}}$$

$$= K\{1 - (\sum_{m=n_0}^{\infty} \frac{u'(m)}{[u(m)]^{1+\varepsilon}})^{-1} \sum_{m=n_0}^{\infty} O(\frac{u'(m)}{[u(m)]^{1+\lambda'+\varepsilon}})\}^{\frac{1}{p}}$$

$$\times \{\sum_{m=n_0}^{\infty} \frac{u'(m)}{[u(m)]^{1+\varepsilon}}\}^{\frac{1}{p}} \{\sum_{n=n_0}^{\infty} \frac{v'(n)}{[v(n)]^{1+\varepsilon}}\}^{\frac{1}{q}}. \quad (4.4.21)$$

In view of (4.4.19) , (4.4.20) and (4.4.21), we have

$$k(\lambda_1 + \tfrac{\varepsilon}{q}) \ge K\{(\sum_{m=n_0}^{\infty} \frac{u'(m)}{[u(m)]^{1+\varepsilon}})^{-1} \sum_{n=n_0}^{\infty} \frac{v'(n)}{[v(n)]^{1+\varepsilon}}\}^{\frac{1}{q}}$$

$$\times \{1 - (\sum_{m=n_0}^{\infty} \frac{u'(m)}{[u(m)]^{1+\varepsilon}})^{-1} \sum_{m=n_0}^{\infty} O(\frac{u'(m)}{[u(m)]^{1+\lambda'+\varepsilon}})\}^{\frac{1}{p}}$$

$$\ge K\{\varepsilon \frac{[u(n_0)]^{\varepsilon} v'(n_0)}{[v(n_0)]^{1+\varepsilon}} + [\frac{u(n_0)}{v(n_0)}]^{\varepsilon}\}^{\frac{1}{q}}$$

$$\times \{1 - (\sum_{m=n_0}^{\infty} \frac{u'(m)}{[u(m)]^{1+\varepsilon}})^{-1} \sum_{m=n_0}^{\infty} O(\frac{u'(m)}{[u(m)]^{1+\lambda'+\varepsilon}})\}^{\frac{1}{p}}. \quad (4.4.22)$$

Since for $q < 0, 0 < t \le 1$, we find

$$k_\lambda(t,1)t^{\lambda_1 + \frac{\varepsilon}{q} - 1} \le k_\lambda(t,1)t^{\lambda_1 - \frac{\delta}{2} - 1},$$

$$\int_0^1 k_\lambda(u,1)u^{\lambda_1 - \frac{\delta}{2} - 1} du \le k(\lambda_1 - \tfrac{\delta}{2}) < \infty,$$

Then by Lebesgue control convergent theorem, we obtain

$$k(\lambda_1 + \tfrac{\varepsilon}{q}) = \int_0^1 k_\lambda(u,1)u^{\lambda_1 + \frac{\varepsilon}{q} - 1} du$$

$$+ \int_1^{\infty} k_\lambda(u,1)u^{\lambda_1 + \frac{\varepsilon}{q} - 1} du$$

$$\le \int_0^1 k_\lambda(u,1)u^{\lambda_1 + \frac{\varepsilon}{q} - 1} du + \int_1^{\infty} k_\lambda(u,1)u^{\lambda_1 - 1} du$$

$$\to \int_0^1 k_\lambda(u,1)u^{\lambda_1 - 1} du + \int_1^{\infty} k_\lambda(u,1)u^{\lambda_1 - 1} du$$

$$= k(\lambda_1)(\varepsilon \to 0^+). \quad (4.4.23)$$

For $\varepsilon \to 0^+$ in (4.4.22), by (4.4.23), it follows $k(\lambda_1) \ge K$. Hence $K = k(\lambda_1)$ is the best value of (4.4.12). □

Example 4.4.7 If $0 < \alpha_1, \alpha_2 \le 1$, $u(x) = x^{\alpha_1}$, $v(x) = x^{\alpha_2}, x > 0$, $n_0 = 1, r > 1, \frac{1}{r} + \frac{1}{s} = 1$, $0 < \lambda \le \min\{r, s\}$, $\lambda_1 = \frac{\lambda}{r}, \lambda_2 = \frac{\lambda}{s}$, $k_\lambda(x,y) = \frac{1}{(x+y)^\lambda}$, we find $k(\frac{\lambda}{r}) = B(\frac{\lambda}{r}, \frac{\lambda}{s})$,

$$\tilde{w}(\tfrac{\lambda}{s}, m) = \sum_{n=1}^{\infty} \frac{\alpha_2}{(m^{\alpha_1} + n^{\alpha_2})^\lambda} \frac{m^{\lambda\alpha_1/r}}{n^{1-\lambda\alpha_2/s}}$$

$$< \int_0^{\infty} \frac{\alpha_2}{(m^{\alpha_1} + y^{\alpha_2})^\lambda} \frac{m^{\lambda\alpha_1/r}}{y^{1-\lambda\alpha_2/s}} dy = B(\tfrac{\lambda}{r}, \tfrac{\lambda}{s}),$$

$$\tilde{\vartheta}(\tfrac{\lambda}{r}, n) < B(\tfrac{\lambda}{r}, \tfrac{\lambda}{s}).$$

Setting $\tilde{\phi}(x) = x^{p(1-\frac{\lambda\alpha_1}{r})-1}, \tilde{\psi}(x) = x^{q(1-\frac{\lambda\alpha_2}{s})-1}$,

(a) for $\tilde{\lambda}_1 = \frac{\lambda}{\tilde{r}} \in \mathbf{I} = [\frac{\lambda}{r}, \lambda)$,

$$\tilde{w}(\tilde{\lambda}_2, m) = \sum_{n=1}^{\infty} \frac{\alpha_2}{(m^{\alpha_1} + n^{\alpha_2})^\lambda} \frac{m^{\lambda\alpha_1/\tilde{r}}}{n^{1-\lambda\alpha_2/\tilde{s}}}$$

$$\ge \int_1^{\infty} \frac{\alpha_2}{(m^{\alpha_1} + y^{\alpha_2})^\lambda} \frac{m^{\lambda\alpha_1/\tilde{r}}}{y^{1-\lambda\alpha_2/\tilde{s}}} dy = \int_{\frac{1}{m^{\alpha_1}}}^{\infty} \frac{t^{\frac{\lambda}{\tilde{s}}-1}}{(1+t)^\lambda} dt$$

$$= B(\tfrac{\lambda}{\tilde{r}}, \tfrac{\lambda}{\tilde{s}})\{1 - [B(\tfrac{\lambda}{\tilde{r}}, \tfrac{\lambda}{\tilde{s}})]^{-1} \int_0^{\frac{1}{m^{\alpha_1}}} \frac{t^{\frac{\lambda}{\tilde{s}}-1}}{(1+t)^\lambda} dt\},$$

$$0 < [B(\tfrac{\lambda}{\tilde{r}}, \tfrac{\lambda}{\tilde{s}})]^{-1} \int_0^{\frac{1}{m^{\alpha_1}}} \frac{t^{\frac{\lambda}{\tilde{s}}-1}}{(1+t)^\lambda} dt$$

$$\leq [B(\tfrac{\lambda}{\tilde{r}},\tfrac{\lambda}{\tilde{s}})]^{-1}\int_0^{\frac{1}{m^{\alpha_1}}} t^{\frac{\lambda}{\tilde{s}}-1}dt$$
$$= [B(\tfrac{\lambda}{\tilde{r}},\tfrac{\lambda}{\tilde{s}})]^{-1}\tfrac{\tilde{s}}{\lambda}(\tfrac{1}{m^{\lambda\alpha_1/\tilde{s}}}).$$

Then by Theorem 4.4.5, for $p>1$, we have the following equivalent inequalities with the best constant factors:

$$\sum_{n=1}^{\infty}\sum_{m=1}^{\infty}\frac{a_m b_n}{(m^{\alpha_1}+n^{\alpha_2})^{\lambda}} < \frac{B(\frac{\lambda}{r},\frac{\lambda}{s})}{\alpha_1^{1/q}\alpha_2^{1/p}}\|a\|_{p,\tilde{\phi}}\|b\|_{q,\tilde{\psi}}, \tag{4.4.24}$$

$$\sum_{n=1}^{\infty} n^{p\frac{\lambda\alpha_2}{s}-1}[\sum_{m=1}^{\infty}\frac{a_m}{(m^{\alpha_1}+n^{\alpha_2})^{\lambda}}]^p < [\frac{B(\frac{\lambda}{r},\frac{\lambda}{s})}{\alpha_1^{1/q}\alpha_2^{1/p}}]^p \|a\|_{p,\tilde{\phi}}^p; \tag{4.4.25}$$

(b) for $\tilde{\lambda}_1=\frac{\lambda}{\tilde{r}}\in \mathbf{I}_1=(\frac{\lambda}{r}-\delta,\frac{\lambda}{r}](0<\delta<\frac{\lambda}{r})$, we find

$$\tilde{w}(\tfrac{\lambda}{\tilde{s}},m)=\sum_{n=1}^{\infty}\frac{\alpha_2}{(m^{\alpha_1}+n^{\alpha_2})^{\lambda}}\frac{m^{\lambda\alpha_1/\tilde{r}}}{n^{1-\lambda\alpha_2/\tilde{s}}}<B(\tfrac{\lambda}{\tilde{r}},\tfrac{\lambda}{\tilde{s}}).$$

Setting $\tilde{\theta}(\frac{\lambda}{r},m)$ as follows:

$$\tilde{\theta}(\tfrac{\lambda}{r},m)=[B(\tfrac{\lambda}{r},\tfrac{\lambda}{s})]^{-1}\int_0^{\frac{1}{m^{\alpha_1}}}\frac{t^{\frac{\lambda}{s}-1}}{(1+t)^{\lambda}}dt,$$

we find $\tilde{\theta}(\frac{\lambda}{r},m)=(\frac{1}{m^{\lambda\alpha_1/s}})$. Since for $c_\lambda=\int_0^1\frac{t^{\frac{\lambda}{r}-1}}{(1+t)^{\lambda}}dt$,

$$c_\lambda \leq B(\tfrac{\lambda}{r},\tfrac{\lambda}{s})[1-\tilde{\theta}(\tfrac{\lambda}{r},m)]\leq \tilde{w}(\tfrac{\lambda}{s},m),$$

then by Theorem 4.4.6, for $0<p<1$, we have the following equivalent inequalities with the best constant factors:

$$\sum_{n=1}^{\infty}\sum_{m=1}^{\infty}\frac{a_m b_n}{(m^{\alpha_1}+n^{\alpha_2})^{\lambda}} > \frac{B(\frac{\lambda}{r},\frac{\lambda}{s})}{\alpha_1^{1/q}\alpha_2^{1/p}}$$
$$\times\{\sum_{m=1}^{\infty}[1-\tilde{\theta}(\tfrac{\lambda}{r},m)]m^{p(1-\frac{\lambda\alpha_1}{r})-1}a_m^p\}^{\frac{1}{p}}\|b\|_{q,\tilde{\psi}}, \tag{4.4.26}$$

$$\sum_{n=1}^{\infty} n^{p\frac{\lambda\alpha_2}{s}-1}[\sum_{m=1}^{\infty}\frac{a_m}{(m^{\alpha_1}+n^{\alpha_2})^{\lambda}}]^p > [\frac{B(\frac{\lambda}{r},\frac{\lambda}{s})}{\alpha_1^{1/q}\alpha_2^{1/p}}]^p$$
$$\times\sum_{m=1}^{\infty}[1-\tilde{\theta}(\tfrac{\lambda}{r},m)]m^{p(1-\frac{\lambda\alpha_1}{r})-1}a_m^p, \tag{4.4.27}$$

$$\sum_{m=1}^{\infty}\frac{m^{q\lambda\alpha_1/r-1}}{[1-\tilde{\theta}(\frac{\lambda}{r},m)]^{q-1}}[\sum_{n=1}^{\infty}\frac{b_n}{(m^{\alpha_1}+n^{\alpha_2})^{\lambda}}]^q$$
$$<[\frac{B(\frac{\lambda}{r},\frac{\lambda}{s})}{\alpha_1^{1/q}\alpha_2^{1/p}}]^q\|b\|_{q,\tilde{w}}^q. \tag{4.4.28}$$

Note 4.4.8 For $0<\alpha_1\neq\alpha_2\leq 1$, $u(x)=x^{\alpha_1}$, $v(x)=x^{\alpha_2}, x>0$, it is obvious that the expression $\ln\frac{v(x)}{u(x)}=\tilde{O}(1)(x\to\infty)$ is not value. Hence we can not use Theorem 4.4.2 and Theorem 4.4.3 to Example 4.4.7.

4.4.2 SOME EXAMPLES FOR APPLYING THEOREM 4.4.2 AND THEOREM 4.4.3

In the following examples, we continue to using the signs and assumptions of Theorem 4.4.2 and Theorem 4.4.3. If $\lambda>0$, then we set $\lambda_1=\frac{\lambda}{r},\lambda_2=\frac{\lambda}{s}$ ($r>1$, $\frac{1}{r}+\frac{1}{s}=1$). The words that the constant factors are the best possible are omitted.

First we review the following inequalities for estimating the weight coefficients:

If $n_1<n_0(\in\mathbf{N}_0)$, $f(x)\in C^4[n_1,\infty)$,

$$(-1)^i f^{(i)}(x)>0 \ (i=1,2,3,4)$$

and both $\sum_{k=n_0}^{\infty}f(k)$ and $\int_{n_0}^{\infty}f(x)dx$ are convergence, by (3.1.50), we have

$$\sum_{k=n_0}^{\infty}f(k)<\int_{n_0}^{\infty}f(x)dx+\tfrac{1}{2}f(n_0)-\tfrac{1}{12}f'(n_0). \tag{4.4.29}$$

In particular, if $\int_{n_1}^{\infty}f(x)dx$ is convergence, then we have the following inequality:

$$\sum_{k=n_0}^{\infty}f(k)<\int_{n_1}^{\infty}f(x)dx$$
$$-[\int_{n_1}^{n_0}f(x)dx-\tfrac{1}{2}f(n_0)+\tfrac{1}{12}f'(n_0)]. \tag{4.4.30}$$

Example 4.4.9 If $0<\lambda\leq\min\{r,s\}$, $k(x,y)=\frac{1}{(x+y)^{\lambda}}, u(x)=v(x)=x+\frac{1}{2}$ $(x\in[0,\infty))$, then we define

$$\tilde{w}(\tfrac{\lambda}{s},m):=\sum_{n=0}^{\infty}\frac{1}{(m+n+1)^{\lambda}}\cdot\frac{(m+\frac{1}{2})^{\lambda/r}}{(n+\frac{1}{2})^{1-\lambda/s}},\quad m\in\mathbf{N}_0,$$

$$\tilde{\vartheta}(\tfrac{\lambda}{r},n):=\sum_{m=0}^{\infty}\frac{1}{(m+n+1)^{\lambda}}\cdot\frac{(n+\frac{1}{2})^{\lambda/s}}{(m+\frac{1}{2})^{1-\lambda/r}},\quad n\in\mathbf{N}_0.$$

Setting $f_m(y):=\frac{1}{(m+y+1)^{\lambda}}\frac{(m+\frac{1}{2})^{\lambda/r}}{(y+\frac{1}{2})^{1-\lambda/s}}$, we find

$$k(\tfrac{\lambda}{r})=B(\tfrac{\lambda}{r},\tfrac{\lambda}{s})=\int_{-\frac{1}{2}}^{\infty}f_m(y)dy,$$

$\frac{-1}{2}f_m(0)=\frac{-(2m+1)^{\lambda/r}}{(2m+2)^{\lambda}}$, and

$$\int_{-\frac{1}{2}}^{0}f_m(y)dy=\int_{-\frac{1}{2}}^{0}\frac{1}{(m+y+1)^{\lambda}}\frac{(m+\frac{1}{2})^{\lambda/r}}{(y+\frac{1}{2})^{1-\lambda/s}}dy$$
$$=\int_0^{\frac{1}{2m+1}}\frac{u^{\frac{\lambda}{s}-1}}{(1+u)^{\lambda}}du=\frac{s}{\lambda}\int_0^{\frac{1}{2m+1}}\frac{du^{\frac{\lambda}{s}}}{(1+u)^{\lambda}}$$

$$=\frac{s(2m+1)^{\lambda/r}}{\lambda(2m+2)^{\lambda}}+\frac{s^2}{\lambda+s}\int_0^{\frac{1}{2m+1}}\frac{1}{(1+u)^{\lambda+1}}du^{\frac{\lambda}{s}+1}$$
$$=\frac{s(2m+1)^{\lambda/r}}{\lambda(2m+2)^{\lambda}}+\frac{s^2(2m+1)^{\lambda/r}}{(\lambda+s)(2m+2)^{\lambda+1}}$$
$$+\frac{s^2(\lambda+1)}{(\lambda+s)}\int_0^{\frac{1}{2m+1}}\frac{1}{(1+u)^{\lambda+2}}u^{\frac{\lambda}{s}+1}du$$
$$>[\frac{s}{\lambda(2m+2)^{\lambda}}+\frac{s^2}{(\lambda+s)(2m+2)^{\lambda+1}}](2m+1)^{\lambda/r},$$
$$\frac{1}{12}f_m'(0)=[\frac{-\lambda}{3(2m+2)^{\lambda+1}}-\frac{(1-\lambda/s)}{3(2m+2)^{\lambda}}](2m+1)^{\frac{\lambda}{r}}.$$

Setting $R_\lambda(s,m)$ as follows:

$$R_\lambda(s,m):=\int_{-\frac{1}{2}}^{0}f_m(y)dy-\frac{1}{2}f_m(0)+\frac{1}{12}f_m'(0),$$

we have

$$\frac{R_\lambda(s,m)}{(2m+1)^{\lambda/r}}>\frac{s}{\lambda(2m+2)^{\lambda}}+\frac{s^2}{(\lambda+s)(2m+2)^{\lambda+1}}-\frac{1}{(2m+2)^{\lambda}}$$
$$-\frac{\lambda}{3(2m+2)^{\lambda+1}}-\frac{(1-\lambda/s)}{3(2m+2)^{\lambda}}$$
$$=\frac{1}{\lambda}[s-\frac{4}{3}\lambda+\frac{\lambda^2}{3s}]\frac{1}{(2m+2)^{\lambda}}$$
$$+(\frac{s^2}{\lambda+s}-\frac{\lambda}{3})\frac{1}{(2m+2)^{\lambda+1}}. \quad (4.4.31)$$

It is obvious that $g(\lambda):=s-\frac{4}{3}\lambda+\frac{\lambda^2}{3s}$,

$$g'(\lambda)=-\frac{4}{3}+\frac{2\lambda}{3s}\le-\frac{4}{3}+\frac{2s}{3s}<0$$

$(0<\lambda\le s)$, $g(\lambda)\ge g(s)=s-\frac{4}{3}s+\frac{s^2}{3s}=0$ and $\frac{s^2}{\lambda+s}-\frac{\lambda}{3}\ge\frac{\lambda^2}{\lambda}-\frac{\lambda}{3}>0$.

By (4.4.31), we find $R_\lambda(s,m)>0$. By (4.4.29) and (4.4.30), we have

$$\tilde{w}(\tfrac{\lambda}{s},m)<B(\tfrac{\lambda}{r},\tfrac{\lambda}{s}),\ \ m\in\mathbf{N}_0.$$

By the same way, we have

$$\vartheta(\tfrac{\lambda}{r},n)<B(\tfrac{\lambda}{r},\tfrac{\lambda}{s}),\ \ n\in\mathbf{N}_0.$$

Since it follows

$$\tilde{w}(\tfrac{\lambda}{s},m)>\int_0^{\infty}f_m(y)dy$$
$$=B(\tfrac{\lambda}{r},\tfrac{\lambda}{s})[1-\frac{1}{B(\frac{\lambda}{r},\frac{\lambda}{s})}\int_0^{\frac{1}{2m+1}}\frac{u^{\frac{\lambda}{s}-1}}{(1+u)^{\lambda}}du],$$

setting the constant c_λ as follows:

$$0<c_\lambda=\int_0^{\infty}f_0(y)dy$$
$$=\int_1^{\infty}\frac{u^{\frac{\lambda}{s}-1}}{(1+u)^{\lambda}}du<\tilde{w}(\tfrac{\lambda}{s},m)$$

and

$$\tilde{\theta}(\tfrac{\lambda}{r},m):=\frac{1}{B(\frac{\lambda}{r},\frac{\lambda}{s})}\int_0^{\frac{1}{2m+1}}\frac{u^{\frac{\lambda}{s}-1}}{(1+u)^{\lambda}}du,$$

we have $\lambda'=\frac{\lambda}{s}>0$, such that

$$\tilde{\theta}(\tfrac{\lambda}{r},m)=O(\frac{1}{(m+1/2)^{\lambda'}})\ (m\to\infty).$$

In fact, we find

$$0<\tilde{\theta}(\tfrac{\lambda}{r},m)(m+\tfrac{1}{2})^{\lambda'}$$
$$<\frac{(m+\frac{1}{2})^{\lambda'}}{B(\frac{\lambda}{r},\frac{\lambda}{s})}\int_0^{\frac{1}{2m+1}}u^{\frac{\lambda}{s}-1}du=\frac{s}{2^{\lambda'}\lambda B(\frac{\lambda}{r},\frac{\lambda}{s})}.$$

Since for $0<\varepsilon<\frac{\lambda}{r}\le1,\delta=\frac{\lambda}{s}+\varepsilon<\lambda$,

$$k_\lambda(1,u)=\frac{1}{(1+u)^{\lambda}}=O(\frac{1}{u^{\delta}})\ (u\to\infty),$$

then by Theorem 4.4.2 and Theorem 4.4.3. setting $0<\lambda\le\min\{r,s\}$, and

$$\tilde{\phi}(x)=(x+\tfrac{1}{2})^{p(1-\frac{\lambda}{r})-1},\ \ \tilde{\psi}(x)=(x+\tfrac{1}{2})^{q(1-\frac{\lambda}{s})-1},$$

(1) for $p>1$, we have the following equivalent inequalities:

$$\sum_{n=0}^{\infty}\sum_{m=0}^{\infty}\frac{a_mb_n}{(m+n+1)^{\lambda}}<B(\tfrac{\lambda}{r},\tfrac{\lambda}{s})\|a\|_{p,\tilde{\phi}}\|b\|_{q,\tilde{\psi}}, \quad (4.4.32)$$

$$\sum_{n=0}^{\infty}(n+\tfrac{1}{2})^{\frac{p\lambda}{s}-1}[\sum_{m=0}^{\infty}\frac{a_m}{(m+n+1)^{\lambda}}]^p$$
$$<[B(\tfrac{\lambda}{r},\tfrac{\lambda}{s})]^p\|a\|_{p,\tilde{\phi}}^p; \quad (4.4.33)$$

(2) for $0<p<1$, we have the following equivalent inequalities:

$$\sum_{n=0}^{\infty}\sum_{m=0}^{\infty}\frac{a_mb_n}{(m+n+1)^{\lambda}}>B(\tfrac{\lambda}{r},\tfrac{\lambda}{s})$$
$$\times\{\sum_{m=0}^{\infty}[1-\tilde{\theta}(\tfrac{\lambda}{r},m)]\tilde{\phi}(m)a_m^p\}^{\frac{1}{p}}\|b\|_{q,\tilde{\psi}}, \quad (4.4.34)$$

$$\sum_{n=0}^{\infty}(n+\tfrac{1}{2})^{\frac{p\lambda}{s}-1}[\sum_{m=0}^{\infty}\frac{a_m}{(m+n+1)^{\lambda}}]^p>[B(\tfrac{\lambda}{r},\tfrac{\lambda}{s})]^p$$
$$\times\{\sum_{m=0}^{\infty}[1-\tilde{\theta}(\tfrac{\lambda}{r},m)]\tilde{\phi}(m)a_m^p\}^{\frac{1}{p}}, \quad (4.4.35)$$

$$\sum_{m=0}^{\infty}\frac{(m+\frac{1}{2})^{q\lambda/r-1}}{[1-\tilde{\theta}(\frac{\lambda}{r},m)]^{q-1}}[\sum_{n=0}^{\infty}\frac{b_n}{(m+n+1)^{\lambda}}]^q$$
$$<[B(\tfrac{\lambda}{r},\tfrac{\lambda}{s})]^q\|b\|_{q,\tilde{\psi}}^q. \quad (4.4.36)$$

Note 4.4.10 Some early results of (4.4.32) and (4.4.33) are consulted in (Yang JMAA 1999)[31]-(Xi JIA 2006)[38].

Example 4.4.11 If $0<\lambda\le1,k(x,y)=\frac{1}{x^{\lambda}+y^{\lambda}}$, $u(x)=v(x)=x+\frac{1}{2}\ (x\in[0,\infty))$, then we define

$$\tilde{w}(\tfrac{\lambda}{s},m):=\sum_{n=0}^{\infty}\frac{1}{(m+\frac{1}{2})^{\lambda}+(n+\frac{1}{2})^{\lambda}}\cdot\frac{(m+\frac{1}{2})^{\lambda/r}}{(n+\frac{1}{2})^{1-\lambda/s}},\ \ m\in\mathbf{N}_0,$$

$$\vartheta(\tfrac{\lambda}{r},n):=\sum_{m=0}^{\infty}\frac{1}{(m+\frac{1}{2})^{\lambda}+(n+\frac{1}{2})^{\lambda}}\cdot\frac{(n+\frac{1}{2})^{\lambda/s}}{(m+\frac{1}{2})^{1-\lambda/r}},\ \ n\in\mathbf{N}_0.$$

Setting $f_m(y):=\frac{1}{(m+\frac{1}{2})^\lambda+(y+\frac{1}{2})^\lambda}\frac{(m+\frac{1}{2})^{\lambda/r}}{(y+\frac{1}{2})^{1-\lambda/s}}$, then for $0<\lambda\le 1$, $f_m(y)$ satisfies the condition of (4.4.29). We find

$$k(\tfrac{\lambda}{r})=\frac{\pi}{\lambda\sin(\pi/r)}=\int_{-\frac{1}{2}}^{\infty}f_m(y)dy,$$

$$\frac{-1}{2}f_m(0)=\frac{-(2m+1)^{\lambda/r}}{(2m+1)^\lambda+1},$$

$$\int_{-\frac{1}{2}}^{0}f_m(y)dy=\int_{-\frac{1}{2}}^{0}\frac{1}{(m+\frac{1}{2})^\lambda+(y+\frac{1}{2})^\lambda}\frac{(m+\frac{1}{2})^{\lambda/r}dy}{(y+\frac{1}{2})^{1-\lambda/s}}$$

$$=\frac{1}{\lambda}\int_0^{\frac{1}{(2m+1)^\lambda}}\frac{u^{\frac{1}{s}-1}}{1+u}du=\frac{s}{\lambda}\int_0^{\frac{1}{(2m+1)^\lambda}}\frac{du^{\frac{1}{s}}}{1+u}$$

$$=\frac{s(2m+1)^{\lambda/r}}{\lambda[(2m+1)^\lambda+1]}+\frac{s^2}{\lambda(1+s)}\int_0^{\frac{1}{(2m+1)^\lambda}}\frac{1}{(1+u)^2}du^{\frac{1}{s}+1}$$

$$>\frac{s(2m+1)^{\lambda/r}}{\lambda[(2m+1)^\lambda+1]}+\frac{s^2(2m+1)^{\lambda/r}}{\lambda(1+s)[(2m+1)^\lambda+1]^2};$$

$$\frac{f_m'(0)}{12}=\{\frac{-\lambda}{3[(2m+1)^\lambda+1]^2}-\frac{(1-\lambda/s)}{3[(2m+1)^\lambda+1]}\}(2m+1)^{\frac{\lambda}{r}}.$$

Setting $R_\lambda(s,m)$ as follows:

$$R_\lambda(s,m):=\int_{-\frac{1}{2}}^{0}f_m(y)dy-\frac{1}{2}f_m(0)+\frac{1}{12}f_m'(0),$$

we obtain the following inequality:

$$\frac{R_\lambda(s,m)}{(2m+1)^{\lambda/r}}>\frac{s}{\lambda[(2m+1)^\lambda+1]}$$
$$+\frac{s^2}{\lambda(1+s)[(2m+1)^\lambda+1]^2}-\frac{1}{(2m+1)^\lambda+1}$$
$$-\frac{\lambda}{3[(2m+1)^\lambda+1]^2}-\frac{(1-\lambda/s)}{3[(2m+1)^\lambda+1]}$$
$$=\frac{1}{\lambda}(s+\frac{\lambda^2}{3s}-\frac{4}{3}\lambda)\frac{1}{(2m+1)^\lambda+1}$$
$$+\frac{1}{\lambda}(\frac{s^2}{1+s}-\frac{\lambda^2}{3})\frac{1}{[(2m+1)^\lambda+1]^2}.$$

Setting $F(\lambda):=s+\frac{\lambda^2}{3s}-\frac{4\lambda}{3}(0<\lambda\le 1)$, since

$$F'(\lambda)=\frac{2\lambda}{3s}-\frac{4}{3}<\frac{2}{3}-\frac{4}{3}<0,$$

Then we find

$$F(\lambda)\ge F(1)=g(s):=s+\frac{1}{3s}-\frac{4}{3}.$$

Since $g'(s)=1-\frac{1}{3s^2}<0$, we have

$$F(\lambda)\ge g(s)>g(1)=0.$$

Then it follows

$$\frac{R_\lambda(s,m)}{(2m+1)^{\lambda/r}}>\frac{1}{\lambda}F(\lambda)\frac{1}{(2m+1)^\lambda+1}$$
$$+\frac{1}{\lambda}(\frac{s}{2}-\frac{1}{3})\frac{1}{[(2m+1)^\lambda+1]^2}>0.$$

By (4.4.29), we have

$$\tilde{w}(\tfrac{\lambda}{s},m)<\frac{\pi}{\lambda\sin(\pi/r)},\quad m\in\mathbf{N}_0.$$

By the same way, we have

$$\vartheta(\tfrac{\lambda}{r},n)<\frac{\pi}{\lambda\sin(\pi/r)},n\in\mathbf{N}_0.$$

Since we find

$$\tilde{w}(\tfrac{\lambda}{s},m)>\int_0^\infty f_m(y)dy$$

$$=\frac{\pi}{\lambda\sin(\pi/r)}[1-\frac{\sin(\pi/r)}{\pi}\int_0^{\frac{1}{(2m+1)^\lambda}}\frac{u^{\frac{1}{s}-1}}{1+u}du],$$

setting the constant c_λ as follows

$$0<c_\lambda=\int_0^\infty f_0(y)dy$$
$$=\frac{1}{\lambda}\int_1^\infty\frac{u^{\frac{1}{s}-1}}{1+u}du<\tilde{w}(\tfrac{\lambda}{s},m)$$

and

$$\tilde{\theta}(\tfrac{\lambda}{r},m):=\frac{\pi}{\sin(\pi/r)}\int_0^{\frac{1}{(2m+1)^\lambda}}\frac{u^{\frac{1}{s}-1}}{1+u}du,$$

then we have $\lambda'=\frac{\lambda}{s}>0$, such that

$$\tilde{\theta}(\tfrac{\lambda}{r},m)=O(\frac{1}{(m+1/2)^{\lambda'}})\ (m\to\infty).$$

In fact, we find

$$0<\tilde{\theta}(\tfrac{\lambda}{r},m)(m+\tfrac{1}{2})^{\lambda'}$$
$$<\frac{(m+\frac{1}{2})^{\lambda'}\sin(\pi/r)}{\pi}\int_0^{\frac{1}{(2m+1)^\lambda}}u^{\frac{1}{s}-1}du=\frac{s\sin(\pi/r)}{2^{\lambda'}\pi}.$$

Since for $0<\varepsilon<\frac{\lambda}{r}\le 1,\delta=\frac{\lambda}{s}+\varepsilon<\lambda$,

$$k_\lambda(1,u)=\frac{1}{1+u^\lambda}=O(\frac{1}{u^\delta})\ (u\to\infty),$$

then by Theorem 4.4.2 and 4.4.3, setting $0<\lambda\le 1$,

$$\tilde{\phi}(x)=(x+\tfrac{1}{2})^{p(1-\frac{\lambda}{r})-1},\quad \tilde{\psi}(x)=(x+\tfrac{1}{2})^{q(1-\frac{\lambda}{s})-1},$$

(1) for $p>1$, we have the following equivalent inequalities:

$$\sum_{n=0}^{\infty}\sum_{m=0}^{\infty}\frac{a_mb_n}{(m+\frac{1}{2})^\lambda+(n+\frac{1}{2})^\lambda}<\frac{\pi}{\lambda\sin(\pi/r)}||a||_{p,\tilde{\phi}}||b||_{q,\tilde{\psi}},\tag{4.4.37}$$

$$\sum_{n=0}^{\infty}(n+\tfrac{1}{2})^{\frac{p\lambda}{s}-1}[\sum_{m=0}^{\infty}\frac{a_m}{(m+\frac{1}{2})^\lambda+(n+\frac{1}{2})^\lambda}]^p<[\frac{\pi}{\lambda\sin(\pi/r)}]^p||a||_{p,\tilde{\phi}}^p;\tag{4.4.38}$$

(2) for $0<p<1$, we have the following equivalent inequalities:

$$\sum_{n=0}^{\infty}\sum_{m=0}^{\infty}\frac{a_mb_n}{(m+\frac{1}{2})^\lambda+(n+\frac{1}{2})^\lambda}>\frac{\pi}{\lambda\sin(\pi/r)}\times\{\sum_{m=0}^{\infty}[1-\tilde{\theta}(\tfrac{\lambda}{r},m)]\tilde{\phi}(m)a_m^p\}^{\frac{1}{p}}||b||_{q,\tilde{\psi}},\tag{4.4.39}$$

$$\sum_{n=0}^{\infty}(n+\tfrac{1}{2})^{\frac{p\lambda}{s}-1}[\sum_{m=0}^{\infty}\frac{a_m}{(m+\frac{1}{2})^\lambda+(n+\frac{1}{2})^\lambda}]^p>[\frac{\pi}{\lambda\sin(\pi/r)}]^p\sum_{m=0}^{\infty}[1-\tilde{\theta}(\tfrac{\lambda}{r},m)]\tilde{\phi}(m)a_m^p,\tag{4.4.40}$$

$$\sum_{m=0}^{\infty}\frac{(m+\frac{1}{2})^{q\lambda/r-1}}{[1-\tilde{\theta}(\frac{\lambda}{r},m)]^{q-1}}[\sum_{n=0}^{\infty}\frac{b_n}{(m+\frac{1}{2})^\lambda+(n+\frac{1}{2})^\lambda}]^q<[\frac{\pi}{\lambda\sin(\pi/r)}]^q||b||_{q,\tilde{\psi}}^q.\tag{4.4.41}$$

Note 4.4.12 Some early results of (4.4.37) and (4.4.39) are consulted in (Yang AM 2006)[39].

Example 4.4.13 If $0<\lambda\le\min\{r,s\}$, $\alpha\ge e^{7/6}$, $k(x,y)=\frac{1}{(x+y)^{\lambda}}$, $u(x)=v(x)=\ln\sqrt{\alpha}x, x\ge 1$, then we define

$$\tilde{w}(\tfrac{\lambda}{s},m):=\sum_{n=1}^{\infty}\frac{1}{(\ln\alpha mn)^{\lambda}}\cdot\frac{(\ln\sqrt{\alpha}m)^{\lambda/r}}{(\ln\sqrt{\alpha}n)^{1-\lambda/s}n},\quad m\in\mathbf{N},$$

$$\vartheta(\tfrac{\lambda}{r},n):=\sum_{m=1}^{\infty}\frac{1}{(\ln\alpha mn)^{\lambda}}\cdot\frac{(\ln\sqrt{\alpha}n)^{\lambda/s}}{(\ln\sqrt{\alpha}m)^{1-\lambda/r}m},\quad n\in\mathbf{N}.$$

Setting $f_m(y)=\frac{(\ln\sqrt{\alpha}m)^{\lambda/r}}{(\ln\alpha my)^{\lambda}(\ln\sqrt{\alpha}y)^{1-\frac{\lambda}{s}}y}$, we find

$$k(\tfrac{\lambda}{r})=B(\tfrac{\lambda}{r},\tfrac{\lambda}{s})=\int_{\sqrt{\alpha}^{-1}}^{\infty}f_m(y)dy,$$

$$-\frac{1}{2}f_m(1)=-\frac{(\ln\sqrt{\alpha})^{\lambda-1}}{2(\ln\alpha m)^{\lambda}}(\frac{\ln\sqrt{\alpha}m}{\ln\sqrt{\alpha}})^{\lambda/r}.$$

Setting $u=(\ln\sqrt{\alpha}y)/(\ln\sqrt{\alpha}m)$, we have

$$\int_{\sqrt{\alpha}^{-1}}^{1}f_m(y)dy=\int_{\sqrt{\alpha}^{-1}}^{1}\frac{(\ln\sqrt{\alpha}m)^{\lambda/r}dy}{(\ln\alpha my)^{\lambda}(\ln\sqrt{\alpha}y)^{1-\lambda/s}y}$$

$$=\int_0^{\frac{\ln\sqrt{\alpha}}{\ln\sqrt{\alpha}m}}\frac{u^{\frac{\lambda}{s}-1}}{(1+u)^{\lambda}}du=\frac{s}{\lambda}\int_0^{\frac{\ln\sqrt{\alpha}}{\ln\sqrt{\alpha}m}}\frac{du^{\frac{\lambda}{s}}}{(1+u)^{\lambda}}$$

$$=\frac{s(\ln\sqrt{\alpha})^{\lambda}}{\lambda(\ln\alpha m)^{\lambda}}(\frac{\ln\sqrt{\alpha}m}{\ln\sqrt{\alpha}})^{\lambda/r}+\frac{s^2}{\lambda+s}\int_0^{\frac{\ln\sqrt{\alpha}}{\ln\sqrt{\alpha}m}}\frac{1}{(1+u)^{\lambda+1}}du^{\frac{\lambda}{s}+1}$$

$$>\frac{s(\ln\sqrt{\alpha})^{\lambda}}{\lambda(\ln\alpha m)^{\lambda}}(\frac{\ln\sqrt{\alpha}m}{\ln\sqrt{\alpha}})^{\lambda/r}+\frac{s^2}{\lambda+s}\frac{(\ln\sqrt{\alpha})^{\lambda+1}}{(\ln\alpha m)^{\lambda+1}}(\frac{\ln\sqrt{\alpha}m}{\ln\sqrt{\alpha}})^{\lambda/r},$$

$$\frac{1}{12}f'_m(1)=(\frac{\ln\sqrt{\alpha}m}{\ln\sqrt{\alpha}})^{\lambda/r}$$

$$\times[-\frac{\lambda(\ln\sqrt{\alpha})^{\lambda-1}}{12(\ln\alpha m)^{\lambda+1}}-\frac{1}{12}(1-\frac{\lambda}{s})\frac{(\ln\sqrt{\alpha})^{\lambda-2}}{(\ln\alpha m)^{\lambda}}-\frac{(\ln\sqrt{\alpha})^{\lambda-1}}{12(\ln\alpha m)^{\lambda}}].$$

Setting $R_{\lambda}(s,m)$ as follows:

$$R_{\lambda}(s,m):=\int_{\sqrt{\alpha}^{-1}}^{1}f_m(y)dy-\frac{f_m(1)}{2}+\frac{f'_m(1)}{12},$$

by the above results and $\alpha\ge e^{7/6}$, we obtain

$$(\frac{\ln\sqrt{\alpha}}{\ln\sqrt{\alpha}m})^{\lambda/r}R_{\lambda}(s,m)$$

$$>\frac{s(\ln\sqrt{\alpha})^{\lambda}}{\lambda(\ln\alpha m)^{\lambda}}+\frac{s^2}{\lambda+s}\frac{(\ln\sqrt{\alpha})^{\lambda+1}}{(\ln\alpha m)^{\lambda+1}}-\frac{(\ln\sqrt{\alpha})^{\lambda-1}}{2(\ln\alpha m)^{\lambda}}$$

$$-\frac{\lambda(\ln\sqrt{\alpha})^{\lambda-1}}{12(\ln\alpha m)^{\lambda+1}}-\frac{1}{12}(1-\frac{\lambda}{s})\frac{(\ln\sqrt{\alpha})^{\lambda-2}}{(\ln\alpha m)^{\lambda}}-\frac{(\ln\sqrt{\alpha})^{\lambda-1}}{12(\ln\alpha m)^{\lambda}}$$

$$=\frac{1}{\lambda}[s\ln\sqrt{\alpha}-\frac{\lambda}{12}(7+\frac{1}{\ln\sqrt{\alpha}})+\frac{\lambda^2}{12s\ln\sqrt{\alpha}}]\frac{(\ln\sqrt{\alpha})^{\lambda-1}}{(\ln\alpha m)^{\lambda}}$$

$$+[\frac{s^2(\ln\sqrt{\alpha})^2}{\lambda+s}-\frac{\lambda}{12}]\frac{(\ln\sqrt{\alpha})^{\lambda-1}}{(\ln\alpha m)^{\lambda+1}}$$

$$\ge\frac{1}{\lambda}[\frac{7}{12}s-\frac{1}{12}(7+\frac{1}{\ln\sqrt{\alpha}})s+\frac{s^2}{12s\ln\sqrt{\alpha}}]\frac{(\ln\sqrt{\alpha})^{\lambda-1}}{(\ln\alpha m)^{\lambda}}$$

$$+[\frac{49s^2}{144(s+s)}-\frac{s}{12}]\frac{(\ln\sqrt{\alpha})^{\lambda-1}}{(\ln\alpha m)^{\lambda+1}}>0.$$

By (4.4.29) and (4.4.30), we have

$$\tilde{w}(\tfrac{\lambda}{s},m)<B(\tfrac{\lambda}{r},\tfrac{\lambda}{s}),\quad m\in\mathbf{N}.$$

By the same way, it follows

$$\vartheta(\tfrac{\lambda}{r},n)<B(\tfrac{\lambda}{r},\tfrac{\lambda}{s}),\quad n\in\mathbf{N}.$$

Since we find

$$\tilde{w}(\tfrac{\lambda}{s},m)>\int_0^{\infty}f_m(y)dy$$

$$=B(\tfrac{\lambda}{r},\tfrac{\lambda}{s})[1-\frac{1}{B(\frac{\lambda}{r},\frac{\lambda}{s})}\int_0^{\frac{\ln\sqrt{\alpha}}{\ln\sqrt{\alpha}m}}\frac{u^{\frac{\lambda}{s}-1}}{(1+u)^{\lambda}}du],$$

setting the constant c_{λ} as follows

$$0<c_{\lambda}=\int_0^{\infty}f_1(y)dy=\int_1^{\infty}\frac{u^{\frac{\lambda}{s}-1}du}{(1+u)^{\lambda}}<\tilde{w}(\tfrac{\lambda}{s},m)$$

and

$$\tilde{\theta}(\tfrac{\lambda}{r},m):=\frac{1}{B(\frac{\lambda}{r},\frac{\lambda}{s})}\int_0^{\frac{\ln\sqrt{\alpha}}{\ln\sqrt{\alpha}m}}\frac{u^{\frac{\lambda}{s}-1}}{(1+u)^{\lambda}}du,\qquad(4.4.42)$$

then we have $\lambda'=\frac{\lambda}{s}>0$, such that

$$\tilde{\theta}(\tfrac{\lambda}{r},m)=O(\frac{1}{(\ln\sqrt{\alpha}m)^{\lambda'}})\ (m\to\infty).$$

In fact, we find

$$0<\tilde{\theta}(\tfrac{\lambda}{r},m)(\ln\sqrt{\alpha}m)^{\lambda'}$$

$$<\frac{(\ln\sqrt{\alpha}m)^{\lambda'}}{B(\frac{\lambda}{r},\frac{\lambda}{s})}\int_0^{\frac{\ln\sqrt{\alpha}}{\ln\sqrt{\alpha}m}}u^{\frac{\lambda}{s}-1}du=\frac{s(\ln\sqrt{\alpha})^{\lambda/s}}{\lambda B(\frac{\lambda}{r},\frac{\lambda}{s})}.$$

Since for $0<\varepsilon<\frac{\lambda}{r}\le 1,\delta=\frac{\lambda}{s}+\varepsilon<\lambda$,

$$k_{\lambda}(1,u)=\frac{1}{(1+u)^{\lambda}}=O(\frac{1}{u^{\delta}})\ (u\to\infty),$$

then by Theorem 4.4.2 and Theorem 4.4.3, setting $0<\lambda\le\min\{r,s\}$, $\alpha\ge e^{7/6}$,

$$\tilde{\phi}(x)=x^{p-1}(\ln\sqrt{\alpha}x)^{p(1-\frac{\lambda}{r})-1},$$

$$\tilde{\psi}(x)=x^{q-1}(\ln\sqrt{\alpha}x)^{q(1-\frac{\lambda}{s})-1},$$

(1) for $p>1$, we have the following equivalent inequalities:

$$\sum_{n=1}^{\infty}\sum_{m=1}^{\infty}\frac{a_m b_n}{(\ln\alpha mn)^{\lambda}}<B(\tfrac{\lambda}{r},\tfrac{\lambda}{s})\|a\|_{p,\tilde{\phi}}\|b\|_{q,\tilde{\psi}},\qquad(4.4.43)$$

$$\sum_{n=1}^{\infty}\frac{1}{n}(\ln\sqrt{\alpha}n)^{\frac{p\lambda}{s}-1}[\sum_{m=1}^{\infty}\frac{a_m}{(\ln\alpha mn)^{\lambda}}]^p$$

$$<[B(\tfrac{\lambda}{r},\tfrac{\lambda}{s})]^p\|a\|_{p,\tilde{\phi}}^p;\qquad(4.4.44)$$

(2) for $0<p<1$, we have the following equivalent inequalities:

$$\sum_{n=1}^{\infty}\sum_{m=1}^{\infty}\frac{a_m b_n}{(\ln\alpha mn)^{\lambda}}>B(\tfrac{\lambda}{r},\tfrac{\lambda}{s})$$

$$\times\{\sum_{m=1}^{\infty}[1-\tilde{\theta}(\tfrac{\lambda}{r},m)]\tilde{\phi}(m)a_m^p\}^{\frac{1}{p}}\|b\|_{q,\tilde{\psi}},\qquad(4.4.45)$$

$$\sum_{n=1}^{\infty}\frac{1}{n}(\ln\sqrt{\alpha}n)^{\frac{p\lambda}{s}-1}[\sum_{m=1}^{\infty}\frac{a_m}{(\ln\alpha mn)^{\lambda}}]^p$$

$$>[B(\tfrac{\lambda}{r},\tfrac{\lambda}{s})]^p\sum_{m=1}^{\infty}[1-\tilde{\theta}(\tfrac{\lambda}{r},m)]\tilde{\phi}(m)a_m^p, \quad (4.4.46)$$

$$\sum_{m=1}^{\infty}\frac{(\ln\sqrt{\alpha}m)^{\frac{q\lambda}{r}-1}}{m[1-\tilde{\theta}(\frac{\lambda}{r},m)]^{q-1}}[\sum_{n=1}^{\infty}\frac{b_n}{(\ln\alpha mn)^{\lambda}}]^q$$
$$<[B(\tfrac{\lambda}{r},\tfrac{\lambda}{s})]^q\|b\|_{q,\tilde{\psi}}^q. \quad (4.4.47)$$

Note 4.4.14 Some early results of (4.4.43) and (4.4.45) are consulted in (Yang AMS 2007)[40].

Example 4.4.15 If $0<\lambda\le\min\{r,s\}$, $\alpha\ge\frac{1+\sqrt{5}}{4}$, $k(x,y)=\frac{1}{(\max\{x,y\})^{\lambda}}$, $u(x)=v(x)=x+\alpha, x\ge 0$, then we define

$$\tilde{w}(\tfrac{\lambda}{s},m):=\sum_{n=0}^{\infty}\frac{1}{(\max\{m,n\}+\alpha)^{\lambda}}\frac{(m+\alpha)^{\lambda/r}}{(n+\alpha)^{1-\lambda/s}}$$
$$=\frac{1}{(m+\alpha)^{\lambda/s}}\sum_{n=0}^{m-1}\frac{1}{(n+\alpha)^{1-\lambda/s}}+(m+\alpha)^{\frac{\lambda}{r}}\sum_{n=m}^{\infty}\frac{1}{(n+\alpha)^{1+\lambda/r}},$$
$$\vartheta(\tfrac{\lambda}{r},n):=\sum_{m=0}^{\infty}\frac{1}{(\max\{m,n\}+\alpha)^{\lambda}}\frac{(n+\alpha)^{\lambda/s}}{(m+\alpha)^{1-\lambda/r}}, m,n\in\mathbf{N}_0. \quad (4.4.48)$$

By (3.1.49) and (3.1.50), we find

$$\frac{1}{(m+\alpha)^{\lambda/s}}\sum_{n=0}^{m-1}\frac{1}{(n+\alpha)^{1-\lambda/s}}$$
$$=\frac{1}{(m+\alpha)^{\lambda/s}}\sum_{n=0}^{m}\frac{1}{(n+\alpha)^{1-\lambda/s}}-\frac{1}{m+\alpha}$$
$$<\frac{1}{(m+\alpha)^{\lambda/s}}\{\int_0^m\frac{dy}{(y+\alpha)^{1-\lambda/s}}+\frac{1}{2}[\frac{1}{(m+\alpha)^{1-\lambda/s}}+\frac{1}{\alpha^{1-\lambda/s}}]$$
$$-\frac{(1-\lambda/s)}{12}\frac{1}{(y+\alpha)^{2-\lambda/s}}\Big|_0^m\}-\frac{1}{m+\alpha}$$
$$=\frac{s}{\lambda}-\frac{1}{2(m+\alpha)}-(\frac{1}{12}-\frac{\lambda}{12s})\frac{1}{(m+\alpha)^2}$$
$$+\frac{\alpha^{\lambda/s}}{(m+\alpha)^{\lambda/s}}[\frac{-s}{\lambda}+\frac{1}{2\alpha}+\frac{1}{12\alpha^2}-\frac{\lambda}{12\alpha^2 s}],$$

$$(m+\alpha)^{\lambda/r}\sum_{n=m}^{\infty}\frac{1}{(n+\alpha)^{1+\lambda/r}}$$
$$<(m+\alpha)^{\frac{\lambda}{r}}[\int_m^{\infty}\frac{1}{(y+\alpha)^{1+\lambda/r}}dy$$
$$+\frac{1}{2(m+\alpha)^{1+\lambda/r}}+\frac{1+\lambda/r}{12(m+\alpha)^{2+\lambda/r}}]$$
$$=\frac{r}{\lambda}+\frac{1}{2(m+\alpha)}+(\frac{1}{12}+\frac{\lambda}{12r})\frac{1}{(m+\alpha)^2}.$$

Then by (4.4.45), we have

$$\tilde{w}(\tfrac{\lambda}{s},m)<\tfrac{rs}{\lambda}-R_{\lambda}(s,m), m\in\mathbf{N}_0, \quad (4.4.49)$$
$$R_{\lambda}(s,m):=-\frac{\lambda}{12}\frac{1}{(m+\alpha)^2}$$
$$+\frac{\alpha^{\lambda/s}}{(m+\alpha)^{\lambda/s}}[\frac{s}{\lambda}-\frac{1}{2\alpha}-\frac{1}{12\alpha^2}+\frac{\lambda}{12\alpha^2 s}].$$

Since we find

$$\frac{1}{(m+\alpha)^2}=\frac{1}{(m+\alpha)^{\lambda/s}(m+\alpha)^{2-\lambda/s}}\le\frac{\alpha^{\lambda/s}}{(m+\alpha)^{\lambda/s}\alpha^2}$$

and $0<\lambda\le\min\{r,s\}\le 2, \alpha\ge\frac{1+\sqrt{5}}{4}$, then we have

$$R_{\lambda}(s,m)\ge\frac{\alpha^{\lambda/s}}{\lambda(m+\alpha)^{\lambda/s}}$$
$$\times[s+(-\frac{1}{2\alpha}-\frac{1}{12\alpha^2})\lambda+(\frac{1}{12\alpha^2 s}-\frac{1}{12\alpha^2})\lambda^2]$$
$$\ge\frac{\alpha^{\lambda/s}}{\lambda(m+\alpha)^{\lambda/s}}[s+(-\frac{1}{2\alpha}-\frac{1}{12\alpha^2})s$$
$$+(\frac{1}{12\alpha^2 s}-\frac{1}{12\alpha^2})2s]$$
$$=\frac{s\alpha^{\lambda/s}}{\lambda(m+\alpha)^{\lambda/s}}(1-\frac{1}{2\alpha}-\frac{1}{12\alpha^2}-\frac{1}{6\alpha^2 r})$$
$$>\frac{s\alpha^{\lambda/s}}{4\alpha^2\lambda(m+\alpha)^{\lambda/s}}(4\alpha^2-2\alpha-1)\ge 0.$$

By (4.4.29) and (4.4.30), we find

$$\tilde{w}(\tfrac{\lambda}{s},m)<\tfrac{rs}{\lambda},\quad m\in\mathbf{N}_0.$$

By the same way, it follows

$$\vartheta(\tfrac{\lambda}{r},n)<\tfrac{rs}{\lambda}, n\in\mathbf{N}_0.$$

Since we obtain

$$\tilde{w}(\tfrac{\lambda}{s},m)>\int_0^{\infty}\frac{(m+\alpha)^{\lambda/r}}{(\max\{m,y\}+\alpha)^{\lambda}(y+\alpha)^{1-\lambda/s}}dy$$
$$=\int_{\frac{\alpha}{m+\alpha}}^{\infty}\frac{u^{\frac{\lambda}{s}-1}du}{(\max\{1,u\})^{\lambda}}=\frac{rs}{\lambda}[1-\frac{\lambda}{rs}\int_0^{\frac{\alpha}{m+\alpha}}\frac{u^{\frac{\lambda}{s}-1}du}{(\max\{1,u\})^{\lambda}}]$$
$$=\frac{rs}{\lambda}[1-\frac{1}{r}(\frac{\alpha}{m+\alpha})^{\lambda/s}],$$

setting

$$0<c_{\lambda}=\int_1^{\infty}\frac{u^{\frac{\lambda}{s}-1}du}{(\max\{1,u\})^{\lambda}}=\frac{r}{\lambda}<\tilde{w}(\tfrac{\lambda}{s},m)$$

and $\tilde{\theta}(\frac{\lambda}{r},m):=\frac{1}{r}(\frac{\alpha}{m+\alpha})^{\lambda/s}$, then we have $\lambda'=\frac{\lambda}{s}>0$, satisfying

$$\tilde{\theta}(\tfrac{\lambda}{r},m)=O(\frac{1}{(m+\alpha)^{\lambda'}})\ (m\to\infty).$$

Since for $0<\varepsilon<\frac{\lambda}{r}\le 1, \delta=\frac{\lambda}{s}+\varepsilon<\lambda$,

$$k_{\lambda}(1,u)=\frac{1}{(\max\{1,u\})^{\lambda}}=O(\frac{1}{u^{\delta}})\ (u\to\infty),$$

then by Theorem 4.4.2 and Theorem 4.4.3, setting $0<\lambda\le\min\{r,s\}, \alpha\ge\frac{1+\sqrt{5}}{4}=0.8090^{+}$,

$$\tilde{\phi}(x)=(x+\alpha)^{p(1-\frac{\lambda}{r})-1},$$
$$\tilde{\psi}(x)=(x+\alpha)^{q(1-\frac{\lambda}{s})-1},$$

(1) for $p>1$, we have the following equivalent inequalities:

$$\sum_{n=0}^{\infty}\sum_{m=0}^{\infty}\frac{a_m b_n}{(\max\{m,n\}+\alpha)^{\lambda}}<\frac{rs}{\lambda}\|a\|_{p,\tilde{\phi}}\|b\|_{q,\tilde{\psi}}, \quad (4.4.50)$$

$$\sum_{n=0}^{\infty}(n+\alpha)^{\frac{p\lambda}{s}-1}[\sum_{m=0}^{\infty}\frac{a_m}{(\max\{m,n\}+\alpha)^{\lambda}}]^p$$
$$<(\tfrac{rs}{\lambda})^p\|a\|_{p,\tilde{\phi}}^p; \quad (4.4.51)$$

(2) for $0<p<1$, we have the following equivalent inequalities:

$$\sum_{n=0}^{\infty}\sum_{m=0}^{\infty}\frac{a_m b_n}{(\max\{m,n\}+\alpha)^{\lambda}} > \frac{rs}{\lambda}$$

$$\times\{\sum_{m=0}^{\infty}[1-\tfrac{1}{r}(\tfrac{\lambda}{m+\alpha})^{\frac{\lambda}{s}}]\tilde{\phi}(m)a_m^p\}^{\frac{1}{p}}\,\|b\|_{q,\tilde{\psi}}, \quad (4.4.52)$$

$$\sum_{n=0}^{\infty}(n+\alpha)^{\frac{p\lambda}{s}-1}[\sum_{m=0}^{\infty}\frac{a_m}{(\max\{m,n\}+\alpha)^{\lambda}}]^p > (\tfrac{rs}{\lambda})^p$$

$$\times\sum_{m=0}^{\infty}[1-\tfrac{1}{r}(\tfrac{\alpha}{m+\alpha})^{\frac{\lambda}{s}}]\tilde{\phi}(m)a_m^p, \quad (4.4.53)$$

$$\sum_{m=0}^{\infty}\frac{(m+\alpha)^{\frac{q\lambda}{r}-1}}{[1-\frac{1}{r}(\frac{\alpha}{m+\alpha})^{\lambda/s}]^{q-1}}[\sum_{n=0}^{\infty}\frac{b_n}{(\max\{m,n\}+\alpha)^{\lambda}}]^q$$

$$< (\tfrac{rs}{\lambda})^q\,\|b\|_{q,\tilde{\psi}}^q. \quad (4.4.54)$$

Example 4.4.16 If $0<\lambda\le\min\{r,s\}$,

$$k(x,y)=\frac{\ln(x/y)}{x^{\lambda}-y^{\lambda}},\ \alpha\ge\tfrac{\sqrt{3}}{3}=0.57735^{+},$$

$u(x)=v(x)=x+\alpha$, then we define

$$\tilde{w}(\tfrac{\lambda}{s},m):=\sum_{n=0}^{\infty}\frac{\ln(\frac{m+\alpha}{n+\alpha})}{(m+\alpha)^{\lambda}-(n+\alpha)^{\lambda}}\cdot\frac{(m+\alpha)^{\lambda/r}}{(n+\alpha)^{1-\lambda/s}},\ m\in\mathbf{N}_0,$$

$$\vartheta(\tfrac{\lambda}{r},n):=\sum_{m=0}^{\infty}\frac{\ln(\frac{m+\alpha}{n+\alpha})}{(m+\alpha)^{\lambda}-(n+\alpha)^{\lambda}}\cdot\frac{(n+\alpha)^{\lambda/s}}{(m+\alpha)^{1-\lambda/r}},\ n\in\mathbf{N}_0. \quad (4.4.55)$$

Setting $g(u)=\frac{\ln u}{u-1}$, and

$$f_m(y):=\frac{\ln(\frac{m+\alpha}{y+\alpha})}{(m+\alpha)^{\lambda}-(y+\alpha)^{\lambda}}\frac{(m+\alpha)^{\lambda/r}}{(y+\alpha)^{1-\lambda/s}}$$

$$=\frac{1}{\lambda(m+\alpha)}g((\tfrac{y+\alpha}{m+\alpha})^{\lambda})(\tfrac{y+\alpha}{m+\alpha})^{\frac{\lambda}{s}-1},$$

then by Example 2.2.2, $g'(u)<0, g''(u)>0$, and for $0<\lambda\le\min\{r,s\}$, $f_m(y)$ satisfies the condition of (4.4.29). We find

$$k_{\lambda}(r)=[\frac{\pi}{\lambda\sin(\pi/r)}]^2=\int_{-\alpha}^{\infty}f_m(y)dy,$$

$$\tfrac{-1}{2}f_m(0)=\tfrac{-1}{2\lambda\alpha}g((\tfrac{\alpha}{m+\alpha})^{\lambda})(\tfrac{\alpha}{m+\alpha})^{\frac{\lambda}{s}},$$

$$\int_{-\alpha}^{0}f_m(y)dy$$

$$=\int_{-\alpha}^{0}\frac{1}{\lambda(m+\alpha)}g((\tfrac{y+\alpha}{m+\alpha})^{\lambda})(\tfrac{y+\alpha}{m+\alpha})^{\frac{\lambda}{s}-1}dy$$

$$=\frac{1}{\lambda^2}\int_0^{(\frac{\alpha}{m+\alpha})^{\lambda}}g(u)u^{\frac{1}{s}-1}du$$

$$=\frac{s}{\lambda^2}\int_0^{(\frac{\alpha}{m+\alpha})^{\lambda}}g(u)du^{\frac{1}{s}}$$

$$=\frac{s}{\lambda^2}[g((\tfrac{\alpha}{m+\alpha})^{\lambda})(\tfrac{\alpha}{m+\alpha})^{\frac{\lambda}{s}}$$

$$-\frac{s}{s+1}\int_0^{(\frac{\alpha}{m+\alpha})^{\lambda}}g'(u)du^{\frac{1}{s}+1}]$$

$$>\frac{s}{\lambda^2}g((\tfrac{\alpha}{m+\alpha})^{\lambda})(\tfrac{\alpha}{m+\alpha})^{\frac{\lambda}{s}}$$

$$-\frac{s^2}{\lambda^2(s+1)}g'((\tfrac{\alpha}{m+\alpha})^{\lambda})(\tfrac{\alpha}{m+\alpha})^{\lambda(\frac{1}{s}+1)},$$

$$\tfrac{1}{12}f_m'(0)=\frac{1}{12\alpha^2}g'((\tfrac{\alpha}{m+\alpha})^{\lambda})(\tfrac{\alpha}{m+\alpha})^{\lambda(1+\frac{1}{s})}$$

$$+(\tfrac{1}{12s}-\tfrac{1}{12\lambda})\tfrac{1}{\alpha^2}g((\tfrac{\alpha}{m+\alpha})^{\lambda})(\tfrac{\alpha}{m+\alpha})^{\frac{\lambda}{s}}.$$

Setting $R_{\lambda}(s,m)$ as follows:

$$R_{\lambda}(s,m):=\int_{-\alpha}^{0}f_m(y)dy$$

$$-\tfrac{1}{2}f_m(0)+\tfrac{1}{12}f_m'(0),$$

by the above results and $\alpha\ge\frac{\sqrt{3}}{3}$, we have

$$R_{\lambda}(s,m)>g((\tfrac{\alpha}{m+\alpha})^{\lambda})(\tfrac{\alpha}{m+\alpha})^{\frac{\lambda}{s}}$$

$$\times\tfrac{1}{\lambda^2}[s-(\tfrac{1}{2\alpha}+\tfrac{1}{12\alpha^2})\lambda+\tfrac{\lambda^2}{12s\alpha^2}]$$

$$+[\tfrac{s^2}{\lambda^2(s+1)}-\tfrac{1}{12\alpha^2}][-g'((\tfrac{\alpha}{m+\alpha})^{\lambda})](\tfrac{\alpha}{m+\alpha})^{\lambda(\frac{1}{s}+1)}$$

$$\ge\tfrac{1}{\lambda^2}[s-(\tfrac{1}{2\alpha}+\tfrac{1}{12\alpha^2})s+\tfrac{s^2}{12s\alpha^2}]$$

$$\times g((\tfrac{\alpha}{m+\alpha})^{\lambda})(\tfrac{\alpha}{m+\alpha})^{\frac{\lambda}{s}}$$

$$+[\tfrac{1}{4}-\tfrac{1}{12\alpha^2}][-g'((\tfrac{\alpha}{m+\alpha})^{\lambda})](\tfrac{\alpha}{m+\alpha})^{\lambda(\frac{1}{s}+1)}>0.$$

By (4.4.29) and (4.4.30), we find

$$\tilde{w}(\tfrac{\lambda}{s},m)<[\frac{\pi}{\lambda\sin(\pi/r)}]^2,\quad m\in\mathbf{N}_0.$$

By the same way, it follows

$$\vartheta(\tfrac{\lambda}{r},n)<[\frac{\pi}{\lambda\sin(\pi/r)}]^2.$$

Since we find

$$\tilde{w}(\tfrac{\lambda}{s},m)>\int_0^{\infty}f_m(y)dy$$

$$=[\frac{\pi}{\lambda\sin(\pi/r)}]^2\{1-[\frac{\sin(\pi/r)}{\pi}]^2\int_0^{(\frac{\alpha}{m+\alpha})^{\lambda}}\frac{(\ln u)u^{\frac{1}{s}-1}}{u-1}du\},$$

setting the constant c_{λ} as follows:

$$0<c_{\lambda}=\int_0^{\infty}f_0(y)dy$$

$$=\frac{1}{\lambda^2}\int_1^{\infty}\frac{(\ln u)u^{\frac{1}{s}-1}}{u-1}du<\tilde{w}(\tfrac{\lambda}{s},m)$$

and

$$\tilde{\theta}(\tfrac{\lambda}{r},m):=[\frac{\pi}{\sin(\pi/r)}]^2\int_0^{(\frac{\alpha}{m+\alpha})^{\lambda}}\frac{(\ln u)u^{\frac{1}{s}-1}}{u-1}du, \quad (4.4.56)$$

we have $\lambda'=\frac{\lambda}{2s}>0$, such that

$$\tilde{\theta}(\tfrac{\lambda}{r},m)=O(\frac{1}{(m+\alpha)^{\lambda'}})\ (m\to\infty).$$

In fact, we have

$$\lim_{y\to\infty}\tilde{\theta}(\tfrac{\lambda}{r},y)(y+\alpha)^{\lambda'}$$

$$=[\frac{\pi}{\sin(\pi/r)}]^2\lim_{y\to\infty}\frac{\int_0^{(\frac{\alpha}{y+\alpha})^{\lambda}}\frac{(\ln u)u^{\frac{1}{s}-1}}{u-1}du}{(y+\alpha)^{-\lambda'}}$$

$$=[\frac{\pi}{\sin(\pi/r)}]^2\lim_{y\to\infty}\frac{\lambda g((\frac{\alpha}{y+\alpha})^{\lambda})\alpha^{\lambda/s}}{\lambda'(y+\alpha)^{-\lambda'+\frac{\lambda}{s}}}=0.$$

Since for $0<\varepsilon<\frac{\lambda}{r}\le 1, \delta=\frac{\lambda}{s}+\varepsilon<\lambda$,

$$k_\lambda(1,u)=\frac{\ln u}{u^\lambda-1}=O(\frac{1}{u^\delta})\ (u\to\infty),$$

by Theorem 4.4.2 and Theorem 4.4.3, setting

$$0<\lambda\le\min\{r,s\}, \alpha\ge\frac{\sqrt{3}}{3}=0.57735^+,$$

$$\tilde{\phi}(x)=(x+\alpha)^{p(1-\frac{\lambda}{r})-1},$$

$$\tilde{\psi}(x)=(x+\alpha)^{q(1-\frac{\lambda}{s})-1},$$

(1) for $p>1$, we have the following equivalent inequalities:

$$\sum_{n=0}^\infty\sum_{m=0}^\infty\frac{\ln(\frac{m+\alpha}{n+\alpha})a_m b_n}{(m+\alpha)^\lambda-(n+\alpha)^\lambda}<[\frac{\pi}{\lambda\sin(\pi/r)}]^2\,||a||_{p,\tilde{\phi}}||b||_{q,\tilde{\psi}}, \tag{4.4.57}$$

$$\sum_{n=0}^\infty(n+\alpha)^{\frac{p\lambda}{s}-1}[\sum_{m=0}^\infty\frac{\ln(\frac{m+\alpha}{n+\alpha})a_m}{(m+\alpha)^\lambda-(n+\alpha)^\lambda}]^p<[\frac{\pi}{\lambda\sin(\pi/r)}]^{2p}\,||a||_{p,\tilde{\phi}}^p; \tag{4.4.58}$$

(2) for $0<p<1$, we have the following equivalent inequalities:

$$\sum_{n=0}^\infty\sum_{m=0}^\infty\frac{\ln(\frac{m+\alpha}{n+\alpha})a_m b_n}{(m+\alpha)^\lambda-(n+\alpha)^\lambda}>[\frac{\pi}{\lambda\sin(\pi/r)}]^2\times\{\sum_{m=0}^\infty[1-\tilde{\theta}(\frac{\lambda}{r},m)]\tilde{\phi}(m)a_m^p\}^{\frac{1}{p}}\,||b||_{q,\tilde{\psi}}, \tag{4.4.59}$$

$$\sum_{n=0}^\infty(n+\alpha)^{\frac{p\lambda}{s}-1}[\sum_{m=0}^\infty\frac{\ln(\frac{m+\alpha}{n+\alpha})a_m}{(m+\alpha)^\lambda-(n+\alpha)^\lambda}]^p>[\frac{\pi}{\lambda\sin(\pi/r)}]^{2p}\sum_{m=0}^\infty[1-\tilde{\theta}(\frac{\lambda}{r},m)]\tilde{\phi}(m)a_m^p, \tag{4.4.60}$$

$$\sum_{m=0}^\infty\frac{(m+\alpha)^{\frac{q\lambda}{r}-1}}{[1-\tilde{\theta}(\frac{\lambda}{r},m)]^{q-1}}[\sum_{n=0}^\infty\frac{\ln(\frac{m+\alpha}{n+\alpha})b_n}{(m+\alpha)^\lambda-(n+\alpha)^\lambda}]^q<[\frac{\pi}{\lambda\sin(\pi/r)}]^{2q}\,||b||_{q,\tilde{\psi}}^q. \tag{4.4.61}$$

Note 4.4.17 Some early results are consulted in (Yang AMS 2006)[41]-(Yang MIA 2003)[42].

Example 4.4.18 If $0<\lambda\le\min\{r,s\}$, $\alpha\ge\frac{1}{2}$,

$$k(x,y)=(\min\{x,y\})^\lambda,$$

$$u(x)=v(x)=x+\alpha,\ x\ge 0,$$

then we define

$$\tilde{w}(\frac{-\lambda}{s},m):=\sum_{n=0}^\infty(\min\{m,n\}+\alpha)^\lambda\frac{(m+\alpha)^{-\lambda/r}}{(n+\alpha)^{1+\lambda/s}},$$

$$\vartheta(\frac{-\lambda}{r},n):=\sum_{m=0}^\infty(\min\{m,n\}+\alpha)^\lambda\frac{(n+\alpha)^{-\lambda/s}}{(m+\alpha)^{1+\lambda/r}},$$

$$m,n\in\mathbf{N}_0. \tag{4.4.62}$$

For $y>-\alpha$, setting

$$f_m(y):=(\min\{m,y\}+\alpha)^\lambda\frac{(m+\alpha)^{-\lambda/r}}{(y+\alpha)^{1+\lambda/s}},$$

it follows

$$f_m(y)=\begin{cases}\frac{(m+\alpha)^{-\lambda/r}}{(y+\alpha)^{1-\lambda/r}}, & -\alpha<y\le m;\\ \frac{(m+\alpha)^{\lambda/s}}{(y+\alpha)^{1+\lambda/s}}, & y>m.\end{cases}$$

It is obvious that

$$(-1)^i f_m^{(i)}(y)\ge 0, 0<y\le m;$$

$$(-1)^i f_m^{(i)}(y)>0, y>m(i=1,2).$$

By Hadamard's inequality, we find

$$\tilde{w}(\frac{-\lambda}{s},m)=\sum_{n=0}^m f_m(n)+\sum_{n=m+1}^\infty f_m(n)$$

$$<\int_{-\frac{1}{2}}^{m+\frac{1}{2}}f_m(y)dy+\int_{m+\frac{1}{2}}^\infty f_m(y)dy$$

$$=\int_{-\frac{1}{2}}^\infty f_m(y)dy\le\int_{-\alpha}^\infty f_m(y)dy$$

$$=\int_{-\alpha}^m\frac{(m+\alpha)^{-\lambda/r}}{(y+\alpha)^{1-\lambda/r}}dy+\int_m^\infty\frac{(m+\alpha)^{\lambda/s}}{(y+\alpha)^{1+\lambda/s}}dy=\frac{rs}{\lambda}.$$

By the same way, we have $\vartheta(\frac{-\lambda}{r},n)<\frac{rs}{\lambda}$. Since

$$\tilde{w}(\frac{-\lambda}{s},m)\ge\int_0^\infty f_m(y)dy$$

$$=\int_{-\alpha}^\infty f_m(y)dy-\int_{-\alpha}^0 f_m(y)dy$$

$$=\frac{rs}{\lambda}-\int_{-\alpha}^0\frac{(m+\alpha)^{-\lambda/r}}{(y+\alpha)^{1-\lambda/r}}dy$$

$$=\frac{rs}{\lambda}[1-\frac{1}{s}(\frac{\alpha}{m+\alpha})^{\frac{\lambda}{r}}]\ge c_\lambda=\frac{r}{\lambda}>0,$$

Since for $0<\varepsilon<\min\{\frac{\lambda}{s},1\}, \delta=\frac{-\lambda}{s}+\varepsilon<0$,

$$k_\lambda(1,u)=(\min\{1,u\})^\lambda=O(\frac{1}{u^\delta})\ (u\to\infty),$$

by Theorem 4.4.2 and Theorem 4.4.3, setting

$$\phi(x)=(x+\alpha)^{p(1+\frac{\lambda}{r})-1},$$

$$\psi(x)=(x+\alpha)^{q(1+\frac{\lambda}{s})-1},$$

(1) for $p>1$, we have the following equivalent inequalities:

$$\sum_{n=0}^\infty\sum_{m=0}^\infty(\min\{m,n\}+\alpha)^\lambda a_m b_n<\frac{rs}{\lambda}||a||_{p,\phi}||b||_{q,\psi}, \tag{4.4.63}$$

$$\sum_{n=0}^\infty(n+\alpha)^{\frac{-p\lambda}{s}-1}[\sum_{m=0}^\infty(\min\{m,n\}+\alpha)^\lambda a_m]^p<(\frac{rs}{\lambda})^p\,||a||_{p,\phi}^p; \tag{4.4.64}$$

(2) for $0<p<1$, we have the following equivalent inequalities:

$$\sum_{n=0}^\infty\sum_{m=0}^\infty(\min\{m,n\}+\alpha)^\lambda a_m b_n>\frac{rs}{\lambda}$$

$$\times\{\sum_{m=0}^{\infty}[1-\tfrac{1}{s}(\tfrac{\lambda}{m+\alpha})^{\frac{\lambda}{r}}]\phi(m)a_m^p\}^{\frac{1}{p}}\|b\|_{q,\psi}, \quad (4.4.65)$$

$$\sum_{n=0}^{\infty}(n+\alpha)^{\frac{-p\lambda}{s}-1}[\sum_{m=0}^{\infty}(\min\{m,n\}+\alpha)^{\lambda}a_m]^p$$

$$>(\tfrac{rs}{\lambda})^p\sum_{m=0}^{\infty}[1-\tfrac{1}{s}(\tfrac{\alpha}{m+\alpha})^{\frac{\lambda}{r}}]\phi(m)a_m^p, \quad (4.4.66)$$

$$\sum_{m=0}^{\infty}\frac{(m+\alpha)^{-\frac{q\lambda}{r}-1}}{[1-\frac{1}{s}(\frac{\alpha}{m+\alpha})^{\lambda/r}]^{q-1}}[\sum_{n=0}^{\infty}(\min\{m,n\}+\alpha)^{\lambda}b_n]^q$$

$$<(\tfrac{rs}{\lambda})^q\|b\|_{q,\psi}^q. \quad (4.4.67)$$

Example 4.4.19 If $0<\lambda\le\min\{r,s\}$, $\alpha\ge\beta\ge\frac{1}{2}$, $k(x,y)=\frac{1}{(x+y)^{\lambda}}$, $u(x)=x+\alpha$, $v(x)=x+\beta$, $x\ge1, n_0=1$, then we define

$$\tilde{w}(\tfrac{\lambda}{s},m):=\sum_{n=1}^{\infty}\frac{1}{\ln^{\lambda}(m+\alpha)(n+\beta)}\frac{[\ln(m+\alpha)]^{\lambda/r}}{[\ln(n+\beta)]^{1-\lambda/s}(n+\beta)},$$

$$\vartheta(\tfrac{\lambda}{r},n):=\sum_{m=1}^{\infty}\frac{1}{\ln^{\lambda}(m+\alpha)(n+\beta)}\frac{[\ln(n+\beta)]^{\lambda/s}}{[\ln(m+\alpha)]^{1-\lambda/r}(m+\alpha)},$$

$$m,n\in\mathbf{N}. \quad (4.4.68)$$

For $y>-\beta$, setting

$$f_m(y):=\frac{1}{\ln^{\lambda}(m+\alpha)(y+\beta)}\frac{[\ln(m+\alpha)]^{\lambda/r}}{[\ln(y+\beta)]^{1-\lambda/s}(y+\beta)},$$

It is obvious that

$$(-1)^i f_m^{(i)}(y)>0 \quad (i=1,2).$$

By Hadamard's inequality, setting $u=\frac{\ln(y+\beta)}{\ln(m+\alpha)}$, we find

$$\tilde{w}(\tfrac{\lambda}{s},m)=\sum_{n=1}^{\infty}f_m(n)<\int_{\frac{1}{2}}^{\infty}f_m(y)dy$$

$$\le\int_{1-\beta}^{\infty}\frac{1}{\ln^{\lambda}(m+\alpha)(y+\beta)}\frac{[\ln(m+\alpha)]^{\lambda/r}}{[\ln(y+\beta)]^{1-\lambda/s}(y+\beta)}dy$$

$$=B(\tfrac{\lambda}{r},\tfrac{\lambda}{s}).$$

By the same way, it follows $\vartheta(\frac{\lambda}{r},n)<B(\frac{\lambda}{r},\frac{\lambda}{s})$.

In view of the decreasing property of $f_m(y)$, we find

$$\tilde{w}(\tfrac{\lambda}{s},m)\ge\int_1^{\infty}f_m(y)dy$$

$$=\int_{1-\beta}^{\infty}f_m(y)dy-\int_{1-\beta}^{1}f_m(y)dy$$

$$=B(\tfrac{\lambda}{r},\tfrac{\lambda}{s})-\int_{1-\beta}^{1}\frac{1}{\ln^{\lambda}(m+\alpha)(y+\beta)}\frac{[\ln(m+\alpha)]^{\lambda/r}dy}{[\ln(y+\beta)]^{1-\lambda/s}(y+\beta)}$$

$$=B(\tfrac{\lambda}{r},\tfrac{\lambda}{s})-\int_0^{\frac{\ln(1+\beta)}{\ln(m+\alpha)}}\frac{1}{(1+u)^{\lambda}}u^{\frac{\lambda}{s}-1}du.$$

For $m=1$, we have

$$\int_0^{\frac{\ln(1+\beta)}{\ln(1+\alpha)}}\frac{1}{(1+u)^{\lambda}}u^{\frac{\lambda}{s}-1}du=\theta_{\lambda}\int_0^{\frac{\ln(1+\beta)}{\ln(1+\alpha)}}u^{\frac{\lambda}{s}-1}du$$

$$=\tfrac{s}{\lambda}\theta_{\lambda}[\tfrac{\ln(1+\beta)}{\ln(1+\alpha)}]^{\frac{\lambda}{s}}(0<\theta_{\lambda}<1);$$

for $m\ge2$, we have

$$\int_0^{\frac{\ln(1+\beta)}{\ln(m+\alpha)}}\frac{1}{(1+u)^{\lambda}}u^{\frac{\lambda}{s}-1}du<\int_0^{\frac{\ln(1+\beta)}{\ln(m+\alpha)}}u^{\frac{\lambda}{s}-1}du$$

$$=\tfrac{s}{\lambda}[\tfrac{\ln(1+\beta)}{\ln(m+\alpha)}]^{\frac{\lambda}{s}}.$$

Setting

$$\tilde{\theta}_{\lambda}=\begin{cases}\theta_{\lambda}, m=1,\\ 1, \quad m\ge2,\end{cases}$$

$$c_{\lambda}=\begin{cases}B(\frac{\lambda}{r},\frac{\lambda}{s})-\frac{s}{\lambda}\theta_{\lambda}, m=1,\\ B(\frac{\lambda}{r},\frac{\lambda}{s})-\frac{s}{\lambda}[\frac{\ln(1+\beta)}{\ln(2+\alpha)}]^{\frac{\lambda}{s}}, m\ge2,\end{cases}$$

since for $\lambda\le r$, we find

$$B(\tfrac{\lambda}{r},\tfrac{\lambda}{s})=\int_0^1(1-t)^{\frac{\lambda}{r}-1}t^{\frac{\lambda}{s}-1}dt$$

$$\ge\int_0^1 t^{\frac{\lambda}{s}-1}dt=\tfrac{s}{\lambda},$$

then $c_{\lambda}>0$. By the above results and $\beta\le\alpha$,

$$\tilde{w}(\tfrac{\lambda}{s},m)\ge B(\tfrac{\lambda}{r},\tfrac{\lambda}{s})$$

$$\times\{1-\tfrac{s}{\lambda}(B(\tfrac{\lambda}{r},\tfrac{\lambda}{s}))^{-1}\tilde{\theta}_{\lambda}[\tfrac{\ln(1+\beta)}{\ln(m+\alpha)}]^{\frac{\lambda}{s}}\}\ge c_{\lambda}.$$

Since for $0<\varepsilon<\frac{\lambda}{r}\le1, \delta=\frac{\lambda}{s}+\varepsilon<\lambda$,

$$k_{\lambda}(1,u)=\tfrac{1}{(1+u)^{\lambda}}=O(\tfrac{1}{u^{\delta}})\ (u\to\infty),$$

by Theorem 4.4.2 and Theorem 4.4.3, setting

$$\tilde{\phi}(x)=[\ln(x+\alpha)]^{p(1-\frac{\lambda}{r})-1}(x+\alpha)^{p-1},$$

$$\tilde{\psi}(x)=[\ln(x+\beta)]^{q(1-\frac{\lambda}{s})-1}(x+\beta)^{p-1},$$

(1) for $p>1$, we have the following equivalent inequalities:

$$\sum_{n=1}^{\infty}\sum_{m=1}^{\infty}\frac{a_m b_n}{\ln^{\lambda}(m+\alpha)(n+\beta)}$$

$$<B(\tfrac{\lambda}{r},\tfrac{\lambda}{s})\|a\|_{p,\tilde{\phi}}\|b\|_{q,\tilde{\psi}}, \quad (4.4.69)$$

$$\sum_{n=1}^{\infty}\frac{[\ln(n+\beta)]^{\frac{p\lambda}{s}-1}}{n+\beta}[\sum_{m=1}^{\infty}\frac{a_m}{\ln^{\lambda}(m+\alpha)(n+\beta)}]^p$$

$$<[B(\tfrac{\lambda}{r},\tfrac{\lambda}{s})]^p\|a\|_{p,\tilde{\phi}}^p; \quad (4.4.70)$$

(2) for $0<p<1$, we have the following equivalent inequalities:

$$\sum_{n=1}^{\infty}\sum_{m=1}^{\infty}\frac{a_m b_n}{\ln^{\lambda}(m+\alpha)(n+\beta)}>B(\tfrac{\lambda}{r},\tfrac{\lambda}{s})$$

$$\times\{\sum_{m=1}^{\infty}[1-\tfrac{s}{\lambda}(B(\tfrac{\lambda}{r},\tfrac{\lambda}{s}))^{-1}\tilde{\theta}_{\lambda}[\tfrac{\ln(1+\beta)}{\ln(m+\alpha)}]^{\frac{\lambda}{s}}]$$

$$\times\tilde{\phi}(m)a_m^p\}^{\frac{1}{p}}\|b\|_{q,\tilde{\psi}}, \quad (4.4.71)$$

$$\sum_{n=1}^{\infty}\frac{[\ln(n+\beta)]^{\frac{p\lambda}{s}-1}}{n+\beta}[\sum_{m=1}^{\infty}\frac{a_m}{\ln^{\lambda}(m+\alpha)(n+\beta)}]^p > [B(\frac{\lambda}{r},\frac{\lambda}{s})]^p$$

$$\times\sum_{m=1}^{\infty}\{1-\frac{s}{\lambda}(B(\frac{\lambda}{r},\frac{\lambda}{s}))^{-1}\tilde{\theta}_{\lambda}[\frac{\ln(1+\beta)}{\ln(m+\alpha)}]^{\frac{\lambda}{s}}\}\tilde{\phi}(m)a_m^p, \quad (4.4.72)$$

$$\sum_{m=1}^{\infty}\frac{[\ln(m+\alpha)]^{\frac{q\lambda}{r}-1}}{\{1-\frac{s}{\lambda}(B(\frac{\lambda}{r},\frac{\lambda}{s}))^{-1}\tilde{\theta}_{\lambda}[\frac{\ln(1+\beta)}{\ln(m+\alpha)}]^{\frac{\lambda}{s}}\}^{q-1}(m+\alpha)}$$

$$\times[\sum_{n=1}^{\infty}\frac{b_n}{\ln^{\lambda}(m+\alpha)(n+\beta)}]^q < [B(\frac{\lambda}{r},\frac{\lambda}{s})]^q \| b \|_{q,\tilde{\psi}}^q . \quad (4.4.73)$$

Note 4.4.20 for $\lambda = \alpha = \beta = 1, r = q, s = p$, (4.4.69) reduces to Mulholland's inequality (3.4.38).

4.5 REFERENCES

1. Yang BC. On a generalization of Hilbert double series theorem. Journal of Nanjing University Mathematical Biquarterly, 2001,18(1):145-151.
2. Zhang KW. A bilinear inequality. J. Math. Anal. Appl.,2002,271:288-296.
3. Hardy GH, Littlewood JE, Polya G. Inequalities. Cambridge : Cambridge University Press,1934.
4. Yang BC. On a Hilbert-type operator with a class of homogeneous kernels. Journal of Inequalities and Applications, Volume 2009, Article ID 572176, 9 pages, doi 10.1155/2009/572176.
5. Yang BC. A new Hilbert-type operator and applications. Publ. Math. Debreen, 2010,70(1-2): 147-156.
6. Wilhelm M. On the spectrum of Hilbert's matrix .Amer. J. Math., 1950,72: 699~704.
7. Yang BC. On an extension of Hardy-Hilbert's inequality. Chinese Mathematical Annal, 2002,23A(2):247-254.
8. Yang BC. On the extended Hardy-Hilbert's inequality. J. Math. Anal. Appl., 2002, 272: 187-199.
9. Yang BC. On new generalizations of Hilbert's inequality. J. Math. Anal. Appl., 2002, 248: 29-40.
10. Yang BC. On a generalization of Hilbert's inequality . J. Pure Math., 2002,19:1-11.
11. Yang BC. On new extensions of Hilbert's inequality. Acta Math. Hungar, 2005,104 (4): 291-299.
12. Yang BC. On generalization of Hardy-Hilbert's inequality and their equivalent forms. Journal of Mathematics, 2004,24(1):24-30.
13. Yang BC. On Hilbert's inequality with some parameters. Acta Mathematica Sinica, Chinese Series, 2006,49(5):1121-1126.
14. Yang BC. On an extended Hardy-Hilbert's inequality and some reversed form. International Mathematical Forum,2006,1(39):1905-1912.
15. Hong Y. A extension and improvement of Hardy-Hilbert's double series inequality. Mathematics in Practice and Theory, 2002,32(5):850-854.
16. Yang BC, Debnath L. On a new extension of Hilbert's double series theorem and applications. Journal of Interdisciplinary Mathematics, 2005, 8(2):265-275.
17. Salem SR. Some new Hilbert type inequalities. Kyungpook Math. J., 2006,46:19-29.
18. Brnetic I, Pecaric J. Generalization of inequalities of Hardy-Hilbert type. Mathematical Inequalities and Appl., 2004,7(2):217-225.
19. Jia WJ, Gao MZ, Gao XM. On an extension of the hardy-Hilbert theorem. Studies Scientiarum Mathematicarum Hungrica, 2005,42(1):21-35.
20. Yang BC. On an extension of Hardy-Hilbert's inequality. Kyungpook Math. J., 2006, 46:425-431.
21. Yang BC. On an extension of Hilbert's inequality and applications. Chin. Quart. Of Math., 2006,21(1):96-102.
22. Xu JS. Hardy-Hilbert's inequalities with two parameters. Advances in Mathematics, 2007,36(2):189-202.
23. Yang BC. An extension of the Hilbert-type inequality and its reverse. Journal of Mathematical Inequalities, 2008,2(1):139-149.
24. Yang BC. Best generalization of Hilbert's type of inequality. Journal of Jilin University (Science Edition), 2004,42(1):30-34.
25. Yang BC. On a generalization of the Hilbert's type inequality and its applications. Chinese Journal of Engineering Mathematics, 2004, 21(5):821-824.
26. Yang BC. Generalization of Hilbert's type inequality with best constant factor and its applications, Journal of Mathematical Research and Exposition, 2005,25(2):341-346.
27. Yang BC. A reverse of the Hardy-Hilbert's type inequality. Journal of Southwest China Normal University (Science), 2005,30(6): 1012-1015.
28. Sun BJ. Best generalization of a Hilbert type inequality. Journal of Inequalities in Pure and Applied Math., 2006,7(3): Article 113,1-7.
29. Yang BC. On the norm of a Hilbert's type linear operator and applications . J. Math. Anal. Appl., 2007,325:529-541.
30. Yang BC. On best extensions of Hardy-Hilbert's inequality with two parameters. Journal of Inequalities in Pure and Applied Mathematics, 2005,6(3): Article 81,1-15.
31. Yang BC. On a new generalization of Hardy-Hilbert's inequality and its applications. J. Math. Anal. Appl., 1999,233,484-497.
32. Yang BC. 2002. On a generalization of Hilbert's double series theorem . Mathematical Inequalities and Applications,2002,5(2):197-204.
33. Yang BC. On a new extension of Hardy-Hilbert's inequality and its applications. International Journal of Pure and Applied Mathematics, 2003,5(1):57-66.

34. Yang BC, Rassias TM. On a new extension of Hilbert's inequality. Mathematical Inequalities and Applications,2005,8(4):575-582.
35. Yang B C. On a new extension of Hilbert's inequality with some parameters . Acta Math. Hungar., 2005,108(4):337-350.
36 Yang BC. On a dual hardy-Hilbert's inequality and its generalization. Analysis Mathematica, 2005,31:151-161.
37. Yang BC. An extension of Hardy-Hilbert's inequality. Applied Mathematics Journal Chinese University, Series A, 2005,20(3):351-357.
38. Xi GW. A reverse Hardy-Hilbert-type inequality. Journal of Inequalities and Applications, Vol.2006, Article ID 79758, 7 pages, doi:1155/2007/79758.
39. Yang BC. A dual Hardy-Hilbert inequality and generalizations. Advances in Mathematics, 2006, 35(1):102-108.
40. Yang BC. On an extension of Hardy-Hilbert's type inequality and a reversion. Acta Mathematica Sinica, Chinene Series, 2007,50(4): 529-541.
41. Yang BC. On a relation between Hardy-Hilbert's inequality and Mulholland's inequality. Acta Mathematica Sinica, Chinese Series, 2006,49(3): 559-566.
42. Krnic M, Gao MZ, Pecaric J, Gao XMi. On the best constant in Hilbert's inequality. Mathematical Inequalities and Applications, 2005,8(2):317-329.
43. Lu ZX. On new generalizations of Hilbert's inequalities. Tamkang Journal of Mathematics, 2004,35(1):77-86.
44. He LP, Jia WJ, Gao MZ. A Hardy-Hilbert's type inequality with gamma function and its applications. Integral Transforms and Special Functions, 2006,17(5):355-363.
45. Gao MZ. A new Hardy-Hilbert's type inequality for double series and its applications . The Australian Journal of Mathematical Analysis and Applications, 2006,3(1): Article 13, 1-10.
46. Yang BC, Rassias TM. On the way of weight coefficient and research for the Hilbert-type inequalities. Mathematical Inequalities and Applications, 2003,6(4): 625-658.

CHAPTER 5

Some Innovative Hilbert-Type Inequalities

Abstract: In this chapter, based on some theorems of Chapter 4, by using the technique of real analysis and applying the improved Euler-Macraurin's summation formula mentioned in Chapter 2, we discuss how to use some particular parameters to denote some new Hilbert-type inequalities and the reverses with the best constant factors. A class of Hilbert-type inequalities with the general measurable kernels is considered.

5.1. SOME PARTICULAR HILBERT-TYPE INEQUALITIES WITH THE HOMOGENEOUS KERNEL OF DEGREE 0 AND EXTENSIONS

5.1.1 SOME COROLLARIES

Setting $\lambda_1 = \lambda_2 = \lambda = 0$, $\tilde{\phi}_\rho(x) = x^{\rho-1}$ $(x > 0;$ $\rho = p,q)$ in Theorem 4.2.3, Theorem 4.2.5 and Corollary 4.3.2, we have

Corollary 5.1.1 Assuming that $p > 0(p \neq 1)$, $\frac{1}{p}+\frac{1}{q}=1$, $k_0(x,y)(\geq 0)$ is a finite homogeneous function of degree 0 in $(0,\infty)\times(0,\infty)$, and there exists an interval **I** of 0, satisfying for any $\delta \in$ **I**,

$$k_\delta = k(\delta) = \int_0^\infty k_0(u,1)u^{\delta-1}du$$

is a positive number. If $k_0(x,y)\frac{1}{x}$ is decreasing for x and $k_0(x,y)\frac{1}{y}$ is decreasing for y , and respectively strictly decreasing in a subinterval, $a \in l^p_{\tilde{\phi}_p}, b \in l^q_{\tilde{\phi}_q}$ are non-negative sequences, such that

$$0 < \|a\|_{p,\tilde{\phi}_p} = \{\sum_{n=1}^\infty n^{p-1}a_n^p\}^{\frac{1}{p}} < \infty$$

and $0 < \|b\|_{q,\tilde{\phi}_q} < \infty$, then (1) for $p > 1$, we have the following equivalent inequalities:

$$\sum_{n=1}^\infty\sum_{m=1}^\infty k_0(m,n)a_m b_n < k_0 \|a\|_{p,\tilde{\phi}_p}\|b\|_{q,\tilde{\phi}_q}, \quad (5.1.1)$$

$$\sum_{n=1}^\infty \frac{1}{n}[\sum_{m=1}^\infty k_0(m,n)a_m]^p < k_0^p \|a\|_{p,\tilde{\phi}_p}^p; \quad (5.1.2)$$

(2) for $0 < p < 1$, if $k_0(u,1)$ is positive in an interval containing 1with

$$k_0(u,1) = O(\frac{1}{u^\alpha}) \ (\alpha > 0; u \to \infty),$$

then we have the following reverse equivalent inequalities:

$$\sum_{n=1}^\infty\sum_{m=1}^\infty k_0(m,n)a_m b_n > k_0\{\sum_{m=1}^\infty[1-\tilde{\theta}_0(m)]m^{p-1}a_m^p\}^{\frac{1}{p}}\|b\|_{q,\tilde{\phi}_q}, \quad (5.1.3)$$

$$\sum_{n=1}^\infty \frac{1}{n}(\sum_{m=1}^\infty k_0(m,n)a_m)^p > k_0^p\sum_{m=1}^\infty[1-\tilde{\theta}_0(m)]m^{p-1}a_m^p, \quad (5.1.4)$$

$$\sum_{m=1}^\infty \frac{[1-\tilde{\theta}_0(m)]^{1-q}}{m}(\sum_{n=1}^\infty k_0(m,n)b_n)^q < k_0^q\|b\|_{q,\tilde{\phi}_q}^q, \quad (5.1.5)$$

where the constant factors k_0, k_0^p and k_0^q are all the best possible.

In Theorem 4.2.3, Theorem 4.2.5 and Corollary 4.3.2, for $\lambda_1 = \lambda_2 = \frac{\lambda}{2}$, $\phi_\rho(x) = x^{\rho(1-\frac{\lambda}{2})-1}$ $(x > 0;$ $\rho = p,q)$, we have the following corollary:

Corollary 5.1.2 Assuming that $p > 0(\neq 1)$, $\frac{1}{p}+\frac{1}{q}=1$, $\lambda \in \mathbf{R}$, $k_\lambda(x,y)(\geq 0)$ is a homogeneous function of degree $-\lambda$ in $(0,\infty)\times(0,\infty)$, there exists an interval **I** of $\frac{\lambda}{2}$, satisfying for any $\alpha \in$ **I**,

$$k_\alpha = k(\tfrac{\alpha}{2}) = \int_0^\infty k_\lambda(u,1)u^{\frac{\alpha}{2}-1}du$$

is a positive number. If for $0 < \lambda \leq 2$, $k_\lambda(x,y)$ is respectively decreasing for x and y , and strictly decreasing for a subinterval, $a = \{a_n\}_{n=1}^\infty \in l^p_{\phi_p}$, $b = \{b_n\}_{n=1}^\infty \in l^q_{\phi_q}$ are non-negative sequences, such that

$$0 < \|a\|_{p,\phi_p} = \{\sum_{n=1}^\infty n^{p(1-\frac{\lambda}{2})-1}a_n^p\}^{\frac{1}{p}} < \infty,$$

$$0<\|b\|_{q,\phi_q}=\{\sum_{n=1}^{\infty}n^{q(1-\frac{\lambda}{2})-1}b_n^q\}^{\frac{1}{q}}<\infty,$$

then (1) for $p>1$, we have the following equivalent inequalities:

$$\sum_{n=1}^{\infty}\sum_{m=1}^{\infty}k_\lambda(m,n)a_m b_n$$
$$<k_\lambda\|a\|_{p,\phi_p}\|b\|_{q,\phi_q}, \tag{5.1.6}$$

$$\sum_{n=1}^{\infty}n^{\frac{p\lambda}{2}-1}[\sum_{m=1}^{\infty}k_\lambda(m,n)a_m]^p$$
$$<k_\lambda^p\|a\|_{p,\phi_p}^p; \tag{5.1.7}$$

(2) for $0<p<1$, if $k_\lambda(u,1)$ is positive in a subinterval containing 1 with

$$k_\lambda(u,1)=O(\tfrac{1}{u^\alpha})(\alpha>\tfrac{\lambda}{2};u\to\infty),$$

then we have the following equivalent inequalities:

$$\sum_{n=1}^{\infty}\sum_{m=1}^{\infty}k_\lambda(m,n)a_m b_n$$
$$>k_\lambda\{\sum_{n=1}^{\infty}[1-\theta_\lambda(n)]n^{p(1-\frac{\lambda}{2})-1}a_n^p\}^{\frac{1}{p}}\|b\|_{q,\phi_q}, \tag{5.1.8}$$

$$\sum_{n=1}^{\infty}n^{\frac{p\lambda}{2}-1}[\sum_{m=1}^{\infty}k_\lambda(m,n)a_m]^p$$
$$>k_\lambda^p\sum_{n=1}^{\infty}[1-\theta_\lambda(n)]n^{p(1-\frac{\lambda}{2})-1}a_n^p, \tag{5.1.9}$$

$$\sum_{m=1}^{\infty}\frac{m^{q\lambda/2-1}}{[1-\theta_\lambda(m)]^{q-1}}(\sum_{n=1}^{\infty}k_\lambda(m,n)b_n)^q$$
$$<k_\lambda^q\|b\|_{q,\phi_q}^q, \tag{5.1.10}$$

where,

$$0<\theta_\lambda(n)=\frac{1}{k_\lambda}\int_n^\infty k(u,1)u^{\frac{\lambda}{2}-1}du$$
$$=O(\frac{1}{n^{\alpha-\lambda/2}})(n\to\infty),$$

and the constant factors k_λ and $k_\lambda^\rho(\rho=p,q)$ are all the best possible.

5.1.2 SOME HILBERT-TYPE INEQUALITIES WITH THE PARTICULAR HOMOGENEOUS KERNEL OF DEGREE 0

Example 5.1.3 If $0<\alpha\le 1, A\ge -1$,

$$k_0(x,y)=\frac{(\min\{x,y\})^\alpha}{x^\alpha+y^\alpha+A(\min\{x,y\})^\alpha},$$

then it follows that

$$k_0(x,y)\frac{1}{y}=\begin{cases}\frac{y^{\alpha-1}}{x^\alpha+(A+1)y^\alpha}, 0<y<x\\ \frac{x^\alpha}{(x^\alpha+y^\alpha+Ax^\alpha)y}, y\ge x\end{cases}$$

is decreasing for $y>0$; so is $k_0(x,y)\frac{1}{x}$ for $x>0$. Since for $|\delta|<\alpha$, we find

$$0<k_\delta(A):=\int_0^\infty k_0(u,1)u^{\delta-1}du$$
$$=\int_0^1\frac{u^{\alpha-1}}{(A+1)u^\alpha+1}(u^\delta+u^{-\delta})du$$
$$\le\int_0^1(u^{\delta+\alpha-1}+u^{-\delta+\alpha-1})du$$
$$=\frac{1}{\alpha-\delta}+\frac{1}{\alpha+\delta}<\infty,$$
$$k_0(-1)=2\int_0^1u^{\alpha-1}du=\frac{2}{\alpha};$$
$$k_0(A)=2\int_0^1\frac{u^{\alpha-1}}{(A+1)u^\alpha+1}du$$
$$=\frac{2}{\alpha(A+1)}\int_0^1\frac{d((A+1)u^\alpha+1)}{(A+1)u^\alpha+1}$$
$$=\frac{2\ln(A+2)}{\alpha(A+1)}(A>-1), \tag{5.1.11}$$

and

$$k_0(u,1)=\frac{1}{u^\alpha+1+A}=O(\frac{1}{u^\alpha})(u>1;u\to\infty),$$

then by Corollary 5.1.1, (1) for $p>1$, we have the following equivalent inequalities:

$$\sum_{n=1}^{\infty}\sum_{m=1}^{\infty}\frac{(\min\{m,n\})^\alpha}{m^\alpha+n^\alpha+A(\min\{m,n\})^\alpha}a_m b_n$$
$$<k_0(A)\|a\|_{p,\phi_p}\|b\|_{q,\phi_q}, \tag{5.1.12}$$

$$\sum_{n=1}^{\infty}\frac{1}{n}[\sum_{m=1}^{\infty}\frac{(\min\{m,n\})^\alpha}{m^\alpha+n^\alpha+A(\min\{m,n\})^\alpha}a_m]^p$$
$$<k_0^p(A)\|a\|_{p,\phi_p}^p; \tag{5.1.13}$$

(2) for $0<p<1$, we have the following reverse equivalent inequalities:

$$\sum_{n=1}^{\infty}\sum_{m=1}^{\infty}\frac{(\min\{m,n\})^\alpha}{m^\alpha+n^\alpha+A(\min\{m,n\})^\alpha}a_m b_n$$
$$>k_0(A)\{\sum_{n=1}^{\infty}[1-\tilde{\theta}_0(n)]n^{p-1}a_n^p\}^{\frac{1}{p}}\|b\|_{q,\phi_q}, \tag{5.1.14}$$

$$\sum_{n=1}^{\infty}\frac{1}{n}[\sum_{m=1}^{\infty}\frac{(\min\{m,n\})^\alpha}{m^\alpha+n^\alpha+A(\min\{m,n\})^\alpha}a_m]^p$$
$$>k_0^p(A)\sum_{n=1}^{\infty}[1-\tilde{\theta}_0(n)]n^{p-1}a_n^p, \tag{5.1.15}$$

$$\sum_{m=1}^{\infty}\frac{[1-\tilde{\theta}_0(m)]^{1-q}}{m}(\sum_{n=1}^{\infty}\frac{(\min\{m,n\})^\alpha}{m^\alpha+n^\alpha+A(\min\{m,n\})^\alpha}b_n)^q$$
$$<k_0^q(A)\|b\|_{q,\phi_q}^q, \tag{5.1.16}$$

where, $0<\tilde{\theta}_0(n)=O(\frac{1}{n^{\alpha}})(n\to\infty)$.

Example 5.1.4 If $0<\alpha\le 1, A>-1$,

$$k_0(x,y)=\frac{(\min\{x,y\})^{\alpha}}{x^{\alpha}+y^{\alpha}+A(\max\{x,y\})^{\alpha}},$$

then it follows that

$$k_0(x,y)\frac{1}{y}=\begin{cases}\frac{y^{\alpha-1}}{y^{\alpha}+(A+1)x^{\alpha}}, & 0<y<x\\ \frac{x^{\alpha}}{[x^{\alpha}+(A+1)y^{\alpha}]y}, & y\ge x\end{cases}$$

is decreasing for $y>0$; so is $k_0(x,y)\frac{1}{x}$ for $x>0$. Since for $|\delta|<\alpha$, we find

$$0<\tilde{k}_{\delta}(A):=\int_0^{\infty}k_0(u,1)u^{\delta-1}du$$

$$=\int_0^1\frac{u^{\alpha-1}}{u^{\alpha}+A+1}(u^{\delta}+u^{-\delta})du$$

$$\le\frac{1}{A+1}\int_0^1(u^{\delta+\alpha-1}+u^{-\delta+\alpha-1})du$$

$$=\frac{1}{A+1}(\frac{1}{\alpha-\delta}+\frac{1}{\alpha+\delta})<\infty,$$

$$\tilde{k}_0(A)=2\int_0^1 k_0(u,1)u^{-1}du$$

$$=2\int_0^1\frac{u^{\alpha-1}}{u^{\alpha}+A+1}du$$

$$=\frac{2}{\alpha}\int_0^1\frac{d(u^{\alpha}+A+1)}{u^{\alpha}+A+1}=\frac{2}{\alpha}\ln(\frac{A+2}{A+1}), \quad (5.1.17)$$

and $k_0(u,1)=\frac{1}{(A+1)u^{\alpha}+1}=O(\frac{1}{u^{\alpha}})(u>1;u\to\infty)$,

then by Corollary 5.1.1, (1) for $p>1$, we have the following equivalent inequalities (Yang JSCNU 2010) [1]:

$$\sum_{n=1}^{\infty}\sum_{m=1}^{\infty}\frac{(\min\{m,n\})^{\alpha}}{m^{\alpha}+n^{\alpha}+A(\max\{m,n\})^{\alpha}}a_m b_n$$

$$<\frac{2}{\alpha}\ln(\frac{A+2}{A+1})\|a\|_{p,\phi_p}\|b\|_{q,\phi_q}, \quad (5.1.18)$$

$$\sum_{n=1}^{\infty}\frac{1}{n}[\sum_{m=1}^{\infty}\frac{(\min\{m,n\})^{\alpha}}{m^{\alpha}+n^{\alpha}+A(\max\{m,n\})^{\alpha}}a_m]^p$$

$$<[\frac{2}{\alpha}\ln(\frac{A+2}{A+1})]^p\|a\|_{p,\phi_p}^p; \quad (5.1.19)$$

(2) for $0<p<1$, we have the following reverse equivalent inequalities:

$$\sum_{n=1}^{\infty}\sum_{m=1}^{\infty}\frac{(\min\{m,n\})^{\alpha}}{m^{\alpha}+n^{\alpha}+A(\max\{m,n\})^{\alpha}}a_m b_n>\frac{2}{\alpha}\ln(\frac{A+2}{A+1})$$

$$\times\{\sum_{n=1}^{\infty}[1-\tilde{\theta}_0(n)]n^{p-1}a_n^p\}^{\frac{1}{p}}\|b\|_{q,\phi_q}, \quad (5.1.20)$$

$$\sum_{n=1}^{\infty}\frac{1}{n}[\sum_{m=1}^{\infty}\frac{(\min\{m,n\})^{\alpha}}{m^{\alpha}+n^{\alpha}+A(\max\{m,n\})^{\alpha}}a_m]^p$$

$$>[\frac{2}{\alpha}\ln(\frac{A+2}{A+1})]^p$$

$$\times\sum_{n=1}^{\infty}[1-\tilde{\theta}_0(n)]n^{p-1}a_n^p, \quad (5.1.21)$$

$$\sum_{m=1}^{\infty}\frac{[1-\tilde{\theta}_0(m)]^{1-q}}{m}(\sum_{n=1}^{\infty}\frac{(\min\{m,n\})^{\alpha}}{m^{\alpha}+n^{\alpha}+A(\max\{m,n\})^{\alpha}}b_n)^q$$

$$<[\frac{2}{\alpha}\ln(\frac{A+2}{A+1})]^q\|b\|_{q,\phi_q}^q, \quad (5.1.22)$$

where, $0<\tilde{\theta}_0(n)=O(\frac{1}{n^{\alpha}})(n\to\infty)$.

Example 5.1.5 If $0<\alpha\le 1$,

$$k_0(x,y)=\frac{(\min\{x,y\})^{\alpha}}{\sqrt{(x^{\alpha}+y^{\alpha})(\max\{x,y\})^{\alpha}}},$$

then it follows that

$$k_0(x,y)\frac{1}{y}=\begin{cases}\frac{y^{\alpha-1}}{\sqrt{(x^{\alpha}+y^{\alpha})x^{\alpha}}}, & 0<y<x\\ \frac{x^{\alpha}}{y\sqrt{(x^{\alpha}+y^{\alpha})y^{\alpha}}}, & y\ge x\end{cases}$$

is decreasing for $y>0$; so is $k_0(x,y)\frac{1}{x}$ for $x>0$. Since for $|\delta|<\alpha/2$,

$$0<K_{\delta}:=\int_0^{\infty}k_0(u,1)u^{\delta-1}du$$

$$=\int_0^1\frac{u^{\alpha}}{\sqrt{u^{\alpha}+1}}(u^{\delta}+u^{-\delta})du$$

$$\le\int_0^1(u^{\delta+\frac{\alpha}{2}-1}+u^{-\delta+\frac{\alpha}{2}-1})du$$

$$=\frac{2}{\alpha-2\delta}+\frac{2}{\alpha+2\delta}<\infty,$$

$$K_0=2\int_0^1 k_0(u,1)u^{-1}du$$

$$=2\int_0^1\frac{u^{\alpha-1}}{\sqrt{u^{\alpha}+1}}du=\frac{2}{\alpha}\int_0^1\frac{1}{\sqrt{u^{\alpha}+1}}d(u^{\alpha}+1)$$

$$=\frac{4}{\alpha}(\sqrt{2}-1), \quad (5.1.23)$$

and

$$k_0(u,1)=\frac{1}{\sqrt{(u^{\alpha}+1)u^{\alpha}}}=O(\frac{1}{u^{\alpha}})(u>1;u\to\infty),$$

then by Corollary 5.1.1, (1) for $p>1$, we have the following equivalent inequalities:

$$\sum_{n=1}^{\infty}\sum_{m=1}^{\infty}\frac{(\min\{m,n\})^{\alpha}}{\sqrt{(m^{\alpha}+n^{\alpha})(\max\{m,n\})^{\alpha}}}a_m b_n$$

$$<\frac{4}{\alpha}(\sqrt{2}-1)\|a\|_{p,\phi_p}\|b\|_{q,\phi_q}, \quad (5.1.24)$$

$$\sum_{n=1}^{\infty}\frac{1}{n}[\sum_{m=1}^{\infty}\frac{(\min\{m,n\})^{\alpha}}{\sqrt{(m^{\alpha}+n^{\alpha})(\max\{m,n\})^{\alpha}}}a_m]^p$$

$$<[\frac{4}{\alpha}(\sqrt{2}-1)]^p\|a\|_{p,\phi_p}^p; \quad (5.1.25)$$

(2) for $0<p<1$, we have the following reverse equivalent inequalities:

$$\sum_{n=1}^{\infty}\sum_{m=1}^{\infty}\frac{(\min\{m,n\})^{\alpha}}{\sqrt{(m^{\alpha}+n^{\alpha})(\max\{m,n\})^{\alpha}}}a_m b_n>\frac{4}{\alpha}(\sqrt{2}-1)$$

$$\times\{\sum_{n=1}^{\infty}[1-\theta_0(n)]n^{p-1}a_n^p\}^{\frac{1}{p}}\|b\|_{q,\phi_q}, \quad (5.1.26)$$

$$\sum_{n=1}^{\infty}\frac{1}{n}[\sum_{m=1}^{\infty}\frac{(\min\{m,n\})^{\alpha}}{\sqrt{(m^{\alpha}+n^{\alpha})(\max\{m,n\})^{\alpha}}}a_m]^p$$

$$>[\tfrac{4}{\alpha}(\sqrt{2}-1)]^p\sum_{n=1}^{\infty}[1-\theta_0(n)]n^{p-1}a_n^p, \quad (5.1.27)$$

$$\sum_{m=1}^{\infty}\frac{[1-\theta_0(m)]^{1-q}}{m}(\sum_{n=1}^{\infty}\frac{(\min\{m,n\})^{\alpha}}{\sqrt{(m^{\alpha}+n^{\alpha})(\max\{m,n\})^{\alpha}}}b_n)^q$$

$$<[\tfrac{4}{\alpha}(\sqrt{2}-1)]^q\|b\|_{q,\phi_q}^q, \quad (5.1.28)$$

where, $0<\tilde{\theta}_0(n)=O(\frac{1}{n^{\alpha}})(n\to\infty)$.

Example 5.1.6 If $0<\alpha\le 1$,

$$k_0(x,y)=\frac{(\min\{x,y\})^{\alpha}}{(x^{\alpha/2}+y^{\alpha/2})(\max\{x,y\})^{\alpha/2}},$$

then it follows that

$$k_0(x,y)\frac{1}{y}=\begin{cases}\frac{y^{\alpha-1}}{(x^{\alpha/2}+y^{\alpha/2})x^{\alpha/2}}, 0<y<x\\ \frac{(\min\{x,y\})^{\alpha}}{(x^{\alpha/2}+y^{\alpha/2})y^{1+\alpha/2}}, y\ge x\end{cases}$$

is decreasing for $y>0$; so is $k_0(x,y)\frac{1}{x}$ for $x>0$. Since for $|\delta|<\alpha/2$,

$$0<K_{\delta}:=\int_0^{\infty}k_0(u,1)u^{\delta-1}du$$
$$=\int_0^1\frac{u^{\alpha}}{u^{\alpha/2+1}}(u^{\delta}+u^{-\delta})du$$
$$\le\int_0^1(u^{\delta+\frac{\alpha}{2}-1}+u^{-\delta+\frac{\alpha}{2}-1})du$$
$$=\frac{2}{\alpha-2\delta}+\frac{2}{\alpha+2\delta}<\infty,$$
$$K_0=2\int_0^1 k_0(u,1)u^{-1}du$$
$$=2\int_0^1\frac{u^{\alpha-1}}{u^{\alpha/2}+1}du=\frac{4}{\alpha}\int_0^1(1-\frac{1}{v+1})dv$$
$$=\frac{4}{\alpha}(1-\ln 2), \quad (5.1.29)$$

and

$$k_0(u,1)=\frac{1}{(u^{\alpha/2}+1)u^{\alpha/2}}=O(\frac{1}{u^{\alpha}})(u>1;u\to\infty),$$

then by Corollary 5.1.1, (1) for $p>1$, we have the following equivalent inequalities:

$$\sum_{n=1}^{\infty}\sum_{m=1}^{\infty}\frac{(\min\{m,n\})^{\alpha}}{(\sqrt{m^{\alpha}}+\sqrt{n^{\alpha}})\sqrt{(\max\{m,n\})^{\alpha}}}a_m b_n$$

$$<\tfrac{4}{\alpha}(1-\ln 2)\|a\|_{p,\phi_p}\|b\|_{q,\phi_q}, \quad (5.1.30)$$

$$\sum_{n=1}^{\infty}\frac{1}{n}[\sum_{m=1}^{\infty}\frac{(\min\{m,n\})^{\alpha}}{(\sqrt{m^{\alpha}}+\sqrt{n^{\alpha}})\sqrt{(\max\{m,n\})^{\alpha}}}a_m]^p$$

$$<[\tfrac{4}{\alpha}(1-\ln 2)]^p\|a\|_{p,\phi_p}^p; \quad (5.1.31)$$

(2) for $0<p<1$, we have the following reverse equivalent inequalities:

$$\sum_{n=1}^{\infty}\sum_{m=1}^{\infty}\frac{(\min\{m,n\})^{\alpha}}{(\sqrt{m^{\alpha}}+\sqrt{n^{\alpha}})\sqrt{(\max\{m,n\})^{\alpha}}}a_m b_n>\tfrac{4}{\alpha}(1-\ln 2)$$

$$\times\{\sum_{n=1}^{\infty}[1-\tilde{\theta}_0(n)]n^{p-1}a_n^p\}^{\frac{1}{p}}\|b\|_{q,\phi_q}, \quad (5.1.32)$$

$$\sum_{n=1}^{\infty}\frac{1}{n}[\sum_{m=1}^{\infty}\frac{(\min\{m,n\})^{\alpha}}{(\sqrt{m^{\alpha}}+\sqrt{n^{\alpha}})\sqrt{(\max\{m,n\})^{\alpha}}}a_m]^p$$

$$>[\tfrac{4}{\alpha}(1-\ln 2)]^p\sum_{n=1}^{\infty}[1-\tilde{\theta}_0(n)]n^{p-1}a_n^p, \quad (5.1.33)$$

$$\sum_{m=1}^{\infty}\frac{[1-\tilde{\theta}_0(m)]^{1-q}}{m}(\sum_{n=1}^{\infty}\frac{(\min\{m,n\})^{\alpha}}{(\sqrt{m^{\alpha}}+\sqrt{n^{\alpha}})\sqrt{(\max\{m,n\})^{\alpha}}}b_n)^q$$

$$<[\tfrac{4}{\alpha}(1-\ln 2)]^q\|b\|_{q,\phi_q}^q, \quad (5.1.34)$$

where, $0<\tilde{\theta}_0(n)=O(\frac{1}{n^{\alpha}})(n\to\infty)$.

Example 5.1.7 If $0<\alpha\le 1$,

$$k_0(x,y)=\frac{(\min\{x,y\})^{\alpha}}{x^{\alpha}-y^{\alpha}}\ln(\tfrac{x}{y}),$$

then it follows that

$$k_0(x,y)\frac{1}{y}=\begin{cases}\frac{y^{\alpha-1}}{x^{\alpha}-y^{\alpha}}\ln(\frac{x}{y}), 0<y<x\\ \frac{x^{\alpha}}{(x^{\alpha}-y^{\alpha})y}\ln(\frac{x}{y}), y\ge x\end{cases}$$

is decreasing for $y>0$; so is $k_0(x,y)\frac{1}{x}$ for $x>0$. Since for $|\delta|<\alpha/2$,

$$\frac{u^{\alpha/2}}{u^{\alpha}-1}(-\ln u)\to 0(u\to 0^+);$$
$$\frac{u^{\alpha/2}}{u^{\alpha}-1}(-\ln u)\to\frac{1}{\alpha}(u\to 1^-),$$
$$0<\frac{u^{\alpha/2}}{u^{\alpha}-1}(-\ln u)\le M(u\in(0,1]),$$
$$0<K_{\delta}:=\int_0^{\infty}k_0(u,1)u^{\delta-1}du$$
$$=\int_0^1\frac{u^{\alpha-1}}{u^{\alpha}-1}(-\ln u)(u^{\delta}+u^{-\delta})du$$
$$\le M\int_0^1(u^{\delta+\frac{\alpha}{2}-1}+u^{-\delta+\frac{\alpha}{2}-1})du$$
$$=M(\frac{2}{\alpha-2\delta}+\frac{2}{\alpha+2\delta})<\infty,$$
$$K_0=2\int_0^1 k_0(u,1)u^{-1}du$$
$$=2\int_0^1\frac{u^{\alpha-1}}{u^{\alpha}-1}(-\ln u)du$$
$$=2\int_0^1\sum_{k=0}^{\infty}u^{\alpha(k+1)-1}(-\ln u)du$$
$$=2\sum_{k=0}^{\infty}\int_0^1 u^{\alpha(k+1)-1}(-\ln u)du$$

$$=\frac{2}{\alpha^2}\sum_{k=0}^{\infty}\frac{1}{(k+1)^2}=\frac{\pi^2}{3\alpha^2}, \quad (5.1.35)$$

and

$$k_0(u,1)=\frac{1}{u^{\alpha}-1}\ln u=O(\frac{1}{u^{\alpha'}})$$

$(u>1;0<\alpha'<\alpha,u\to\infty)$, then by Corollary 5.1.1, (1) for $p>1$, we have the following equivalent inequalities:

$$\sum_{n=1}^{\infty}\sum_{m=1}^{\infty}\frac{(\min\{m,n\})^{\alpha}}{m^{\alpha}-n^{\alpha}}\ln(\frac{m}{n})a_m b_n$$

$$<\frac{\pi^2}{3\alpha^2}\|a\|_{p,\phi_p}\|b\|_{q,\phi_q}, \quad (5.1.36)$$

$$\sum_{n=1}^{\infty}\frac{1}{n}[\sum_{m=1}^{\infty}\frac{(\min\{m,n\})^{\alpha}}{m^{\alpha}-n^{\alpha}}\ln(\frac{m}{n})a_m]^p$$

$$<(\frac{\pi^2}{3\alpha^2})^p\|a\|_{p,\phi_p}^p; \quad (5.1.37)$$

(2) for $0<p<1$, we have the following reverse equivalent inequalities:

$$\sum_{n=1}^{\infty}\sum_{m=1}^{\infty}\frac{(\min\{m,n\})^{\alpha}}{m^{\alpha}-n^{\alpha}}\ln(\frac{m}{n})a_m b_n$$

$$>\frac{\pi^2}{3\alpha^2}\{\sum_{n=1}^{\infty}[1-\theta_0(n)]n^{p-1}a_n^p\}^{\frac{1}{p}}\|b\|_{q,\phi_q}, \quad (5.1.38)$$

$$\sum_{n=1}^{\infty}\frac{1}{n}[\sum_{m=1}^{\infty}\frac{(\min\{m,n\})^{\alpha}}{m^{\alpha}-n^{\alpha}}\ln(\frac{m}{n})a_m]^p$$

$$>(\frac{\pi^2}{3\alpha^2})^p\sum_{n=1}^{\infty}[1-\theta_0(n)]n^{p-1}a_n^p, \quad (5.1.39)$$

$$\sum_{m=1}^{\infty}\frac{[1-\theta_0(m)]^{1-q}}{m}(\sum_{n=1}^{\infty}\frac{(\min\{m,n\})^{\alpha}}{m^{\alpha}-n^{\alpha}}\ln(\frac{m}{n})b_n)^q$$

$$<(\frac{\pi^2}{3\alpha^2})^q\|b\|_{q,\phi_q}^q, \quad (5.1.40)$$

where, $0<\theta_0(n)=O(\frac{1}{n^{\alpha/2}})(n\to\infty)$.

Note 5.1.8 The constant factors in the above inequalities are all the best possible.

5.2. SOME HILBERT-TYPE INEQUALITIES WITH THE HOMOGENEOUS KERNELS OF DEGREE -1 AND EXTENSIONS

5.2.1. A RELATION TO HILBERT'S INEQUALITY AND A H-L-P INEQUALITY

If $x,y>0$, then we have the following inequalities about the mean values:

$$\min\{x,y\}\le\frac{2}{x^{-1}+y^{-1}}\le(xy)^{1/2}$$

$$\le\frac{x+y}{2}\le\max\{x,y\}.$$

Example 5.2.1 (Yang SP 2009) [2] If $0<\lambda\le 2$, $B,C\ge 0$, $A>-\min\{B,C\}$, and

$$k_{\lambda}(x,y)=\frac{1}{A\max\{x^{\lambda},y^{\lambda}\}+Bx^{\lambda}+Cy^{\lambda}}.$$

It is obvious that $k_{\lambda}(x,y)$ is strictly decreasing respectively for x and y,

$$k_{\lambda}=\int_0^{\infty}k_{\lambda}(u,1)u^{\frac{\lambda}{2}-1}du=\int_0^{\infty}\frac{u^{(\lambda/2)-1}}{A\max\{u^{\lambda},1\}+Bu^{\lambda}+C}du$$

$$=\int_0^1\frac{1}{Bu^{\lambda}+(A+C)}u^{\frac{\lambda}{2}-1}du$$

$$+\int_1^{\infty}\frac{1}{(A+B)u^{\lambda}+C}u^{\frac{\lambda}{2}-1}du. \quad (5.2.1)$$

(1) If $B,C>0,A>-\min\{B,C\}$, then setting $v=\sqrt{\frac{B}{A+C}}u^{\lambda/2}$ and $v=\sqrt{\frac{C}{A+B}}u^{-1/2}$ in (5.2.1), we find

$$k_{\lambda}=\frac{1}{\lambda}[\frac{2}{\sqrt{B(A+C)}}\int_0^{\sqrt{\frac{B}{A+C}}}\frac{1}{1+v^2}dv$$

$$+\frac{2}{\sqrt{C(A+B)}}\int_0^{\sqrt{\frac{C}{A+B}}}\frac{1}{1+v^2}dv]$$

$$=\frac{2}{\lambda}[\frac{1}{\sqrt{B(A+C)}}\arctan\sqrt{\frac{B}{A+C}}$$

$$+\frac{1}{\sqrt{C(A+B)}}\arctan\sqrt{\frac{C}{A+B}}];$$

(2) if $B=0,C>0,A>0$, then setting $v=\sqrt{\frac{C}{A}}u^{-\lambda/2}$ in the second integral of (5.2.1), we find

$$k_{\lambda}=\frac{1}{A+C}\int_0^1u^{\frac{\lambda}{2}-1}du+\frac{2}{\lambda\sqrt{CA}}\int_0^{\sqrt{\frac{C}{A}}}\frac{1}{1+v^2}dv$$

$$=\frac{2}{\lambda}(\frac{1}{A+C}+\frac{1}{\sqrt{AC}}\arctan\sqrt{\frac{C}{A}});$$

(3) if $C=0,B>0,A>0$, then by the symmetric property, it follows $k_{\lambda}=\frac{2}{\lambda}(\frac{1}{A+B}+\frac{1}{\sqrt{AB}}\arctan\sqrt{\frac{B}{A}})$;

(4) if $B=C=0,A>0$ in (5.1.6), then we find $k_{\lambda}=\frac{4}{\lambda A}$. Hence we have

$$k_{\lambda}:=\begin{cases}\frac{2}{\lambda}[\frac{\arctan\sqrt{\frac{B}{A+C}}}{\sqrt{B(A+C)}}+\frac{\arctan\sqrt{\frac{C}{A+B}}}{\sqrt{C(A+B)}}], & B,C>0, \\ & A>-\min\{B,C\}, \\ \frac{2}{\lambda}(\frac{1}{A+C}+\frac{\arctan\sqrt{C/A}}{\sqrt{AC}}), & B=0,C,A>0, \\ \frac{2}{\lambda}(\frac{1}{A+B}+\frac{\arctan\sqrt{B/A}}{\sqrt{AB}}), & C=0,B,A>0, \\ \frac{4}{\lambda A}, & B=C=0,A>0.\end{cases} \quad (5.2.2)$$

By Corollary 5.1.2, (1) for $p>1$, we have the following equivalent inequalities:

$$\sum_{n=1}^{\infty}\sum_{m=1}^{\infty}\frac{a_m b_n}{A\max\{m^{\lambda},n^{\lambda}\}+Bm^{\lambda}+Cn^{\lambda}}$$

$$<k_{\lambda}\|a\|_{p,\phi_p}\|b\|_{q,\phi_q}, \quad (5.2.3)$$

$$\sum_{n=1}^{\infty} n^{\frac{p\lambda}{2}-1}[\sum_{m=1}^{\infty}\frac{a_m}{A\max\{m^\lambda,n^\lambda\}+Bm^\lambda+Cn^\lambda}]^p$$
$$<k_\lambda^p \|a\|_{p,\phi_p}^p; \quad (5.2.4)$$

(2) for $0<p<1$, we have the following reverse equivalent inequalities:

$$\sum_{n=1}^{\infty}\sum_{m=1}^{\infty}\frac{a_m b_n}{A\max\{m^\lambda,n^\lambda\}+Bm^\lambda+Cn^\lambda}>k_\lambda$$
$$\times\{\sum_{n=1}^{\infty}[1-\tilde{\theta}_\lambda(n)]n^{p(1-\frac{\lambda}{2})-1}a_n^p\}^{\frac{1}{p}}\|b\|_{q,\phi_q}, \quad (5.2.5)$$

$$\sum_{n=1}^{\infty} n^{\frac{p\lambda}{2}-1}[\sum_{m=1}^{\infty}\frac{a_m}{A\max\{m^\lambda,n^\lambda\}+Bm^\lambda+Cn^\lambda}]^p>k_\lambda^p$$
$$\times\sum_{n=1}^{\infty}[1-\tilde{\theta}_\lambda(n)]n^{p(1-\frac{\lambda}{2})-1}a_n^p, \quad (5.2.6)$$

$$\sum_{m=1}^{\infty}\frac{m^{q\lambda/2-1}}{[1-\tilde{\theta}_\lambda(m)]^{q-1}}(\sum_{n=1}^{\infty}\frac{b_n}{A\max\{m^\lambda,n^\lambda\}+Bm^\lambda+Cn^\lambda})^q$$
$$<k_\lambda^q\|b\|_{q,\phi_q}^q, \quad (5.2.7)$$

where ,

$$0<\tilde{\theta}_\lambda(n)=\frac{1}{k_\lambda}\int_n^\infty\frac{1}{(A+B)u^\lambda+C}u^{\frac{\lambda}{2}-1}du$$
$$\le\frac{2}{\lambda k_\lambda(A+B)n^{\lambda/2}}.$$

Note 5.2.2

For $\lambda=1, A=0, B=C=1, p=q=2$, (5.2.3) reduces the following Hilbert's inequality:

$$\sum_{n=1}^{\infty}\sum_{m=1}^{\infty}\frac{a_m b_n}{m+n}<\pi\{\sum_{n=1}^{\infty}a_n^2\sum_{n=1}^{\infty}b_n^2\}^{\frac{1}{2}}; \quad (5.2.8)$$

for $\lambda=1$, $A=1$, $B=C=0, p=q=2$, (5.2.3) reduces to the following H-L-P inequality:

$$\sum_{n=1}^{\infty}\sum_{m=1}^{\infty}\frac{a_m b_n}{\max\{m,n\}}<4\{\sum_{n=1}^{\infty}a_n^2\sum_{n=1}^{\infty}b_n^2\}^{\frac{1}{2}}. \quad (5.2.9)$$

Hence (5.2.3) is a well relation to (5.2.8) and (5.2.9).

Particular case 5.2.3 Since we have

$$\max\{x,y\}-\tilde{A}|x-y|$$
$$=(1-2\tilde{A})\max\{x,y\}+\tilde{A}(x+y),$$

setting $B=C=\tilde{A}, A=1-2\tilde{A}$ in (5.1.7), for $0<\lambda\le 2,$ we have

$$\tilde{k}_\lambda(\tilde{A}):=\begin{cases}\frac{4}{\lambda\sqrt{\tilde{A}(1-\tilde{A})}}\arctan\sqrt{\frac{\tilde{A}}{1-\tilde{A}}}, & 0<\tilde{A}<1,\\ \frac{4}{\lambda}, & \tilde{A}=0.\end{cases} \quad (5.2.10)$$

Then by (5.2.3)-(5.2.7), (1) for $p>1$, we have the following equivalent inequalities:

$$\sum_{n=1}^{\infty}\sum_{m=1}^{\infty}\frac{a_m b_n}{\max\{m^\lambda,n^\lambda\}-\tilde{A}|m^\lambda-n^\lambda|}$$
$$<\tilde{k}_\lambda(\tilde{A})\|a\|_{p,\phi_p}\|b\|_{q,\phi_q}, \quad (5.2.11)$$

$$\sum_{n=1}^{\infty} n^{\frac{p\lambda}{2}-1}[\sum_{m=1}^{\infty}\frac{a_m}{\max\{m^\lambda,n^\lambda\}-\tilde{A}|m^\lambda-n^\lambda|}]^p$$
$$<\tilde{k}_\lambda^p(\tilde{A})\|a\|_{p,\phi_p}^p; \quad (5.2.12)$$

(2) for $0<p<1$, we have the following reverse equivalent inequalities:

$$\sum_{n=1}^{\infty}\sum_{m=1}^{\infty}\frac{a_m b_n}{\max\{m^\lambda,n^\lambda\}-\tilde{A}|m^\lambda-n^\lambda|}>\tilde{k}_\lambda(\tilde{A})$$
$$\times\{\sum_{n=1}^{\infty}[1-\tilde{\theta}_\lambda(n)]n^{p(1-\frac{\lambda}{2})-1}a_n^p\}^{\frac{1}{p}}\|b\|_{q,\phi_q}, \quad (5.2.13)$$

$$\sum_{n=1}^{\infty} n^{\frac{p\lambda}{2}-1}[\sum_{m=1}^{\infty}\frac{a_m}{\max\{m^\lambda,n^\lambda\}-\tilde{A}|m^\lambda-n^\lambda|}]^p>\tilde{k}_\lambda^p(\tilde{A})$$
$$\times\sum_{n=1}^{\infty}[1-\tilde{\theta}_\lambda(n)]n^{p(1-\frac{\lambda}{2})-1}a_n^p, \quad (5.2.14)$$

$$\sum_{m=1}^{\infty}\frac{m^{q\lambda/2-1}}{[1-\tilde{\theta}_\lambda(m)]^{q-1}}(\sum_{n=1}^{\infty}\frac{b_n}{\max\{m^\lambda,n^\lambda\}-\tilde{A}|m^\lambda-n^\lambda|})^q$$
$$<\tilde{k}_\lambda^q(\tilde{A})\|b\|_{q,\phi_q}^q, \quad (5.2.15)$$

where,

$$0<\tilde{\theta}_\lambda(n)=\frac{1}{\tilde{k}_\lambda(\tilde{A})}\int_n^\infty\frac{1}{(1-\tilde{A})u^\lambda+\tilde{A}}u^{\frac{\lambda}{2}-1}du$$
$$\le\frac{2}{\lambda\tilde{k}_\lambda(\tilde{A})(1-\tilde{A})n^{\lambda/2}}.$$

Particular case 5.2.4 Since

$$\max\{x,y\}+\tilde{A}\min\{x,y\}$$
$$=(1-\tilde{A})\max\{x,y\}+\tilde{A}(x+y),$$

setting $B=C=\tilde{A}, A=1-\tilde{A}$ in (5.1.7), for $0<\lambda\le 2,$ we find

$$\tilde{K}_\lambda(\tilde{A}):=\begin{cases}\frac{4}{\lambda\sqrt{\tilde{A}}}\arctan\sqrt{\tilde{A}}, & \tilde{A}>0,\\ \frac{4}{\lambda}, & \tilde{A}=0.\end{cases} \quad (5.2.16)$$

Then by (5.2.3)-(5.2.7), (1) for $p>1$, we have the following equivalent inequalities:

$$\sum_{n=1}^{\infty}\sum_{m=1}^{\infty}\frac{a_m b_n}{\max\{m^\lambda,n^\lambda\}+\tilde{A}\min\{m^\lambda,n^\lambda\}}$$
$$<\tilde{K}_\lambda(\tilde{A})\|a\|_{p,\phi_p}\|b\|_{q,\phi_q}, \quad (5.2.17)$$

$$\sum_{n=1}^{\infty} n^{\frac{p\lambda}{2}-1}[\sum_{m=1}^{\infty}\frac{a_m}{\max\{m^\lambda,n^\lambda\}+\tilde{A}\min\{m^\lambda,n^\lambda\}}]^p$$
$$<\tilde{K}_\lambda^p(\tilde{A})\|a\|_{p,\phi_p}^p; \quad (5.2.18)$$

(2) for $0<p<1$, we have the following reverse equivalent inequalities:

$$\sum_{n=1}^{\infty}\sum_{m=1}^{\infty}\frac{a_m b_n}{\max\{m^\lambda,n^\lambda\}+\tilde{A}\min\{m^\lambda,n^\lambda\}}>\tilde{K}_\lambda(\tilde{A})$$

$$\times\{\sum_{n=1}^{\infty}[1-\tilde{\theta}_\lambda(n)]n^{p(1-\frac{\lambda}{2})-1}a_n^p\}^{\frac{1}{p}}\|b\|_{q,\phi_q}, \quad (5.2.19)$$

$$\sum_{n=1}^{\infty}n^{\frac{p\lambda}{2}-1}[\sum_{m=1}^{\infty}\frac{a_m}{\max\{m^\lambda,n^\lambda\}+\tilde{A}\min\{m^\lambda,n^\lambda\}}]^p$$

$$>\tilde{K}_\lambda^p(\tilde{A})\sum_{n=1}^{\infty}[1-\tilde{\theta}_\lambda(n)]n^{p(1-\frac{\lambda}{2})-1}a_n^p, \quad (5.2.20)$$

$$\sum_{m=1}^{\infty}\frac{m^{q\lambda/2-1}}{[1-\tilde{\theta}_\lambda(m)]^{q-1}}(\sum_{n=1}^{\infty}\frac{b_n}{\max\{m^\lambda,n^\lambda\}+\tilde{A}\min\{m^\lambda,n^\lambda\}})^q$$

$$<\tilde{K}_\lambda^q(\tilde{A})\|b\|_{q,\phi_q}^q, \quad (5.2.21)$$

where,

$$0<\tilde{\theta}_\lambda(n)=\frac{1}{\tilde{K}_\lambda(\tilde{A})}\int_n^\infty\frac{1}{u^\lambda+\tilde{A}}u^{\frac{\lambda}{2}-1}du$$

$$\le\frac{2}{\lambda\tilde{K}_\lambda(\tilde{A})n^{\lambda/2}}.$$

Particular case 5.2.5 Since

$$x+y+\tilde{A}|x-y|$$

$$=2\tilde{A}\max\{x,y\}+(1-\tilde{A})(x+y),$$

setting $B=C=1-\tilde{A}, A=2\tilde{A}$, in (5.1.7), for $0<\lambda\le 2,$ we find

$$k_\lambda'(\tilde{A}):=\begin{cases}\frac{4}{\lambda\sqrt{1-\tilde{A}^2}}\arctan\sqrt{\frac{1-\tilde{A}}{1+\tilde{A}}}, & -1<\tilde{A}<1,\\ \frac{2}{\lambda}, & \tilde{A}=1.\end{cases} \quad (5.2.22)$$

Then by (5.2.3)-(5.2.7), (1) for $p>1$, we have the following equivalent inequalities:

$$\sum_{n=1}^{\infty}\sum_{m=1}^{\infty}\frac{a_m b_n}{m^\lambda+n^\lambda+\tilde{A}|m^\lambda-n^\lambda|}$$

$$<k_\lambda'(\tilde{A})\|a\|_{p,\phi_p}\|b\|_{q,\phi_q}, \quad (5.2.23)$$

$$\sum_{n=1}^{\infty}n^{\frac{p\lambda}{2}-1}(\sum_{m=1}^{\infty}\frac{a_m}{m^\lambda+n^\lambda+\tilde{A}|m^\lambda-n^\lambda|})^p$$

$$<k_\lambda'^p(\tilde{A})\|a\|_{p,\phi_p}^p; \quad (5.2.24)$$

(2) for $0<p<1$, we have the following reverse equivalent inequalities:

$$\sum_{n=1}^{\infty}\sum_{m=1}^{\infty}\frac{a_m b_n}{m^\lambda+n^\lambda+\tilde{A}|m^\lambda-n^\lambda|}>k_\lambda'(\tilde{A})$$

$$\times\{\sum_{n=1}^{\infty}[1-\tilde{\theta}_\lambda(n)]n^{p(1-\frac{\lambda}{2})-1}a_n^p\}^{\frac{1}{p}}\|b\|_{q,\phi_q}, \quad (5.2.25)$$

$$\sum_{n=1}^{\infty}n^{\frac{p\lambda}{2}-1}(\sum_{m=1}^{\infty}\frac{a_m}{m^\lambda+n^\lambda+\tilde{A}|m^\lambda-n^\lambda|})^p>k_\lambda'^p(\tilde{A})$$

$$\times\sum_{n=1}^{\infty}[1-\tilde{\theta}_\lambda(n)]n^{p(1-\frac{\lambda}{2})-1}a_n^p, \quad (5.2.26)$$

$$\sum_{m=1}^{\infty}\frac{m^{\frac{q\lambda}{2}-1}}{[1-\tilde{\theta}_\lambda(m)]^{q-1}}(\sum_{n=1}^{\infty}\frac{b_n}{m^\lambda+n^\lambda+\tilde{A}|m^\lambda-n^\lambda|})^q$$

$$<k_\lambda'^q(\tilde{A})\|b\|_{q,\phi_q}^q, \quad (5.2.27)$$

where,

$$0<\tilde{\theta}_\lambda(n)=\frac{1}{k_\lambda'(\tilde{A})}\int_n^\infty\frac{1}{(1+\tilde{A})u^\lambda+(1-\tilde{A})}u^{\frac{\lambda}{2}-1}du$$

$$\le\frac{2}{\lambda k_\lambda'(\tilde{A})(1+\tilde{A})n^{\lambda/2}}.$$

Particular case 5.2.6 Since

$$x+y+\tilde{A}\min\{x,y\}$$

$$=(1+\tilde{A})(x+y)-\tilde{A}\max\{x,y\},$$

setting $B=C=1+\tilde{A}, A=-\tilde{A}$ in (5.1.7), for $0<\lambda\le 2,$ we find

$$K_\lambda'(\tilde{A}):=\begin{cases}\frac{4}{\lambda\sqrt{1+\tilde{A}}}\arctan\sqrt{1+\tilde{A}}, & \tilde{A}>-1,\\ \frac{4}{\lambda}, & \tilde{A}=-1.\end{cases} \quad (5.2.28)$$

Then by (5.2.3)-(5.2.7), (1) for $p>1$, we have the following equivalent inequalities:

$$\sum_{n=1}^{\infty}\sum_{m=1}^{\infty}\frac{a_m b_n}{m^\lambda+n^\lambda+\tilde{A}\min\{m^\lambda,n^\lambda\}}$$

$$<K_\lambda'(\tilde{A})\|a\|_{p,\phi_p}\|b\|_{q,\phi_q}, \quad (5.2.29)$$

$$\sum_{n=1}^{\infty}n^{\frac{p\lambda}{2}-1}[\sum_{m=1}^{\infty}\frac{a_m}{m^\lambda+n^\lambda+\tilde{A}\min\{m^\lambda,n^\lambda\}}]^p$$

$$<K_\lambda'^p(\tilde{A})\|a\|_{p,\phi_p}^p; \quad (5.2.30)$$

(2) for $0<p<1$, we have the following reverse equivalent inequalities:

$$\sum_{n=1}^{\infty}\sum_{m=1}^{\infty}\frac{a_m b_n}{m^\lambda+n^\lambda+\tilde{A}\min\{m^\lambda,n^\lambda\}}>K_\lambda'(\tilde{A})$$

$$\times\{\sum_{n=1}^{\infty}[1-\tilde{\theta}_\lambda(n)]n^{p(1-\frac{\lambda}{2})-1}a_n^p\}^{\frac{1}{p}}\|b\|_{q,\phi_q}, \quad (5.2.31)$$

$$\sum_{n=1}^{\infty}n^{\frac{p\lambda}{2}-1}[\sum_{m=1}^{\infty}\frac{a_m}{m^\lambda+n^\lambda+\tilde{A}\min\{m^\lambda,n^\lambda\}}]^p>K_\lambda'^p(\tilde{A})$$

$$\times\sum_{n=1}^{\infty}[1-\tilde{\theta}_\lambda(n)]n^{p(1-\frac{\lambda}{2})-1}a_n^p, \quad (5.2.32)$$

$$\sum_{m=1}^{\infty}\frac{m^{q\lambda/2-1}}{[1-\tilde{\theta}_\lambda(m)]^{q-1}}(\sum_{n=1}^{\infty}\frac{b_n}{m^\lambda+n^\lambda+\tilde{A}\min\{m^\lambda,n^\lambda\}})^q$$

$$<K_\lambda'^q(\tilde{A})\|b\|_{q,\phi_q}^q, \quad (5.2.33)$$

where,

$$0<\tilde{\theta}_\lambda(n)=\frac{1}{K'_\lambda(\tilde{A})}\int_n^\infty \frac{1}{u^\lambda+1+\tilde{A}}u^{\frac{\lambda}{2}-1}du$$
$$\leq \frac{2}{\lambda K'_\lambda(\tilde{A})n^{\lambda/2}}.$$

5.2.2 SOME EXAMPLES WITH PARAMETERS

Example 5.2.7 If $0<\lambda\leq 2, A\geq 0$,

$$k_\lambda(x,y)=\frac{1}{\max\{x^\lambda,y^\lambda\}+A(xy)^{\lambda/2}},$$

then $k_\lambda(x,y)$ is strictly decreasing respectively for x and y, and

$$k_\lambda(A):=\int_0^\infty k_\lambda(u,1)u^{\frac{\lambda}{2}-1}du$$
$$=\int_0^\infty \frac{1}{\max\{u^\lambda,1\}+Au^{\lambda/2}}u^{\frac{\lambda}{2}-1}du$$
$$=2\int_0^1 \frac{u^{\frac{\lambda}{2}-1}}{1+Au^{\lambda/2}}du$$
$$=\begin{cases}\frac{4}{\lambda A}\ln(1+A), A>0,\\ \frac{4}{\lambda},\quad A=0.\end{cases} \tag{5.2.34}$$

Hence by Corollary 5.1.2, (1) for $p>1$, we have the following equivalent inequalities:

$$\sum_{n=1}^\infty\sum_{m=1}^\infty \frac{a_m b_n}{\max\{m^\lambda,n^\lambda\}+A(mn)^{\lambda/2}}$$
$$<k_\lambda(A)\|a\|_{p,\phi_p}\|b\|_{q,\phi_q}, \tag{5.2.35}$$

$$\sum_{n=1}^\infty n^{\frac{p\lambda}{2}-1}[\sum_{m=1}^\infty \frac{a_m}{\max\{m^\lambda,n^\lambda\}+A(mn)^{\lambda/2}}]^p$$
$$<k_\lambda^p(A)\|a\|_{p,\phi_p}^p; \tag{5.2.36}$$

(2) for $0<p<1$, we have the following reverse equivalent inequalities:

$$\sum_{n=1}^\infty\sum_{m=1}^\infty \frac{a_m b_n}{\max\{m^\lambda,n^\lambda\}+A(mn)^{\lambda/2}}>k_\lambda(A)$$
$$\times\{\sum_{n=1}^\infty[1-\tilde{\theta}_\lambda(n)]n^{p(1-\frac{\lambda}{2})-1}a_n^p\}^{\frac{1}{p}}\|b\|_{q,\phi_q}, \tag{5.2.37}$$

$$\sum_{n=1}^\infty n^{\frac{p\lambda}{2}-1}[\sum_{m=1}^\infty \frac{a_m}{\max\{m^\lambda,n^\lambda\}+A(mn)^{\lambda/2}}]^p>k_\lambda^p(A)$$
$$\times\sum_{n=1}^\infty[1-\tilde{\theta}_\lambda(n)]n^{p(1-\frac{\lambda}{2})-1}a_n^p, \tag{5.2.38}$$

$$\sum_{m=1}^\infty \frac{m^{q\lambda/2-1}}{[1-\tilde{\theta}_\lambda(m)]^{q-1}}[\sum_{n=1}^\infty \frac{b_n}{\max\{m^\lambda,n^\lambda\}+A(mn)^{\lambda/2}}]^q$$
$$<k_\lambda^q(A)\|b\|_{q,\phi_q}^q, \tag{5.2.39}$$

where,

$$\tilde{\theta}_\lambda(n)=\frac{1}{k_\lambda(A)}\int_n^\infty \frac{1}{u^\lambda+Au^{\lambda/2}}u^{\frac{\lambda}{2}-1}du$$
$$=\begin{cases}\frac{1}{2\ln(1+A)}\ln(1+\frac{A}{n^{\lambda/2}}),\ A>0,\\ \frac{1}{2n^{\lambda/2}},\quad A=0.\end{cases}$$

Example 5.2.8 If $0<\lambda\leq 2, A>-2$,

$$k_\lambda(x,y)=\frac{1}{x^\lambda+y^\lambda+A(xy)^{\lambda/2}},$$

then for $A\geq 0$, $k_\lambda(x,y)$ is strictly decreasing respectively for x and y, and

$$\tilde{k}_\lambda(A):=\int_0^\infty k_\lambda(u,1)u^{\frac{\lambda}{2}-1}du$$
$$=2\int_0^1 \frac{1}{u^\lambda+1+Au^{\lambda/2}}u^{\frac{\lambda}{2}-1}du=\frac{4}{\lambda}\int_0^1\frac{1}{v^2+Av+1}dv$$
$$=\begin{cases}\frac{8}{\lambda\sqrt{4-A^2}}(\arctan\sqrt{\frac{2+A}{2-A}}-\arctan\frac{A}{\sqrt{4-A^2}}),\\ \qquad -2<A<2,\\ \frac{2}{\lambda},\qquad A=2,\\ \frac{8}{\lambda\sqrt{A^2-4}}\ln(\frac{A+\sqrt{A^2-4}}{\sqrt{A+2}+\sqrt{A-2}}),\quad A>2.\end{cases} \tag{5.2.40}$$

Hence by Corollary 5.1.2, for $A\geq 0$, (1) if $p>1$, we then have the following equivalent inequalities:

$$\sum_{n=1}^\infty\sum_{m=1}^\infty \frac{a_m b_n}{m^\lambda+n^\lambda+A(mn)^{\lambda/2}}$$
$$<\tilde{k}_\lambda(A)\|a\|_{p,\phi_p}\|b\|_{q,\phi_q}, \tag{5.2.41}$$

$$\sum_{n=1}^\infty n^{\frac{p\lambda}{2}-1}[\sum_{m=1}^\infty \frac{a_m}{m^\lambda+n^\lambda+A(mn)^{\lambda/2}}]^p$$
$$<\tilde{k}_\lambda^p(A)\|a\|_{p,\phi_p}^p; \tag{5.2.42}$$

(2) if $0<p<1$, then we have the following reverse equivalent inequalities:

$$\sum_{n=1}^\infty\sum_{m=1}^\infty \frac{a_m b_n}{m^\lambda+n^\lambda+A(mn)^{\lambda/2}}>\tilde{k}_\lambda(A)$$
$$\times\{\sum_{n=1}^\infty[1-\tilde{\theta}_\lambda(n)]n^{p(1-\frac{\lambda}{2})-1}a_n^p\}^{\frac{1}{p}}\|b\|_{q,\phi_q}, \tag{5.2.43}$$

$$\sum_{n=1}^\infty n^{\frac{p\lambda}{2}-1}[\sum_{m=1}^\infty \frac{a_m}{m^\lambda+n^\lambda+A(mn)^{\lambda/2}}]^p>\tilde{k}_\lambda^p(A)$$
$$\times\sum_{n=1}^\infty[1-\tilde{\theta}_\lambda(n)]n^{p(1-\frac{\lambda}{2})-1}a_n^p, \tag{5.2.44}$$

$$\sum_{m=1}^\infty \frac{m^{q\lambda/2-1}}{[1-\tilde{\theta}_\lambda(m)]^{q-1}}[\sum_{n=1}^\infty \frac{b_n}{m^\lambda+n^\lambda+A(mn)^{\lambda/2}}]^q$$
$$<\tilde{k}_\lambda^q(A)\|b\|_{q,\phi_q}^q, \tag{5.2.45}$$

where,

$$0<\tilde{\theta}_\lambda(n)=\frac{1}{\tilde{k}_\lambda(A)}\int_n^\infty \frac{1}{u^\lambda+1+Au^{\lambda/2}}u^{\frac{\lambda}{2}-1}du$$
$$\le \frac{2}{\lambda\tilde{k}_\lambda(A)n^{\lambda/2}}.$$

Example 5.2.9 If $0<\lambda\le 2, A\ge -1$,
$$k_\lambda(x,y)=\frac{1}{\max\{x^\lambda,y^\lambda\}+A(x^{-\lambda}+y^{-\lambda})^{-1}},$$
then for $A\ge 0$, $k_\lambda(x,y)$ is strictly decreasing respectively for x and y, and
$$k'_\lambda(A):=\int_0^\infty k_\lambda(u,1)u^{\frac{\lambda}{2}-1}du$$
$$=\int_0^\infty \frac{1}{\max\{u^\lambda,1\}+A(u^{-\lambda}+1)^{-1}}u^{\frac{\lambda}{2}-1}du$$
$$=2\int_0^1 \frac{(u^\lambda+1)}{(1+A)u^\lambda+1}u^{\frac{\lambda}{2}-1}du$$
$$=\frac{4}{\lambda}\int_0^1 \frac{v^2+1}{(1+A)v^2+1}dv$$
$$=\begin{cases}\frac{4}{\lambda(1+A)}(1+\frac{A}{\sqrt{1+A}}\arctan\sqrt{1+A}), A>-1,\\ \frac{16}{3\lambda},\ A=-1.\end{cases} \quad (5.2.46)$$
Hence by Corollary 5.1.2, for $A\ge 0$, (1) if $p>1$, then we have the following equivalent inequalities:
$$\sum_{n=1}^\infty\sum_{m=1}^\infty \frac{a_m b_n}{\max\{m^\lambda,n^\lambda\}+A(m^{-\lambda}+n^{-\lambda})^{-1}}$$
$$<k'_\lambda(A)\|a\|_{p,\phi_p}\|b\|_{q,\phi_q}, \quad (5.2.47)$$
$$\sum_{n=1}^\infty n^{\frac{p\lambda}{2}-1}[\sum_{m=1}^\infty \frac{a_m}{\max\{m^\lambda,n^\lambda\}+A(m^{-\lambda}+n^{-\lambda})^{-1}}]^p$$
$$<k'^p_\lambda(A)\|a\|^p_{p,\phi_p}.; \quad (5.2.48)$$
(2) if $0<p<1$, then we have the following reverse equivalent inequalities:
$$\sum_{n=1}^\infty\sum_{m=1}^\infty \frac{a_m b_n}{\max\{m^\lambda,n^\lambda\}+A(m^{-\lambda}+n^{-\lambda})^{-1}}>k'_\lambda(A)$$
$$\times\{\sum_{n=1}^\infty[1-\tilde{\theta}_\lambda(n)]n^{p(1-\frac{\lambda}{2})-1}a_n^p\}^{\frac{1}{p}}\|b\|_{q,\phi_q}, \quad (5.2.49)$$
$$\sum_{n=1}^\infty n^{\frac{p\lambda}{2}-1}[\sum_{m=1}^\infty \frac{a_m}{\max\{m^\lambda,n^\lambda\}+A(m^{-\lambda}+n^{-\lambda})^{-1}}]^p$$
$$>k'^p_\lambda(A)\sum_{n=1}^\infty[1-\tilde{\theta}_\lambda(n)]n^{p(1-\frac{\lambda}{2})-1}a_n^p, \quad (5.2.50)$$
$$\sum_{m=1}^\infty \frac{m^{q\lambda/2-1}}{[1-\tilde{\theta}_\lambda(m)]^{q-1}}[\sum_{n=1}^\infty \frac{b_n}{\max\{m^\lambda,n^\lambda\}+A(m^{-\lambda}+n^{-\lambda})^{-1}}]^q$$
$$<k'^q_\lambda(A)\|b\|^q_{q,\phi_q}, \quad (5.2.51)$$
where,
$$0<\tilde{\theta}_\lambda(n)=\frac{1}{k'_\lambda(A)}\int_n^\infty \frac{1}{u^\lambda+A(u^{-\lambda}+1)^{-1}}u^{\frac{\lambda}{2}-1}du$$
$$\le \frac{2}{\lambda k'_\lambda(A)n^{\lambda/2}}.$$

Example 5.2.10 If $0<\lambda\le 2, A>-4$,
$$k_\lambda(x,y)=\frac{1}{x^\lambda+y^\lambda+A(x^{-\lambda}+y^{-\lambda})^{-1}},$$
then for $A\ge 0$, $k_\lambda(x,y)$ is strictly decreasing respectively for x and y, and
$$K_\lambda(A):=\int_0^\infty k_\lambda(u,1)u^{\frac{\lambda}{2}-1}du$$
$$=\int_0^\infty \frac{1}{u^\lambda+1+A(u^{-\lambda}+1)^{-1}}u^{\frac{\lambda}{2}-1}du$$
$$=\frac{1}{\lambda}\int_{-\infty}^\infty \frac{v^2+1}{v^4+(2+A)v^2+1}dv.$$
For $A>-4$, we obtain
$$v^4+(2+A)v^2+1=v^4+2v^2+1+Av^2$$
$$=(v^2+1)^2-(\sqrt{-A}v)^2$$
$$=(v^2+1+\sqrt{-A}v)(v^2+1-\sqrt{-A}v).$$
By the obtaining roots formula, we have the following 4 roots of the above polynomial:
$$v_1=\tfrac{1}{2}(\sqrt{-A}+\sqrt{4+A}i),$$
$$v_2=\tfrac{1}{2}(\sqrt{-A}-\sqrt{4+A}i),$$
$$v_3=\tfrac{1}{2}(-\sqrt{-A}+\sqrt{4+A}i),$$
$$v_4=\tfrac{1}{2}(-\sqrt{-A}-\sqrt{4+A}i).$$
It is obvious that for $A\ne 0$, the above four roots are different imaginary numbers and only v_1 and v_3 are in the upper half plane. Since
$$f(z)=\frac{z^2+1}{z^4+(2+A)z^2+1}=\frac{z^2+1}{(z-v_1)(z-v_2)(z-v_3)(z-v_4)},$$
then we find (Zhong HDP 2003) [3]
$$\operatorname{Res}_{z=v_1} f(z)=\frac{v_1^2+1}{(v_1-v_2)(v_1-v_3)(v_1-v_4)}$$
$$=\frac{-A+\sqrt{-A(A+4)}i}{2\sqrt{-A(A+4)}(\sqrt{-A}+\sqrt{A+4}i)i}=\frac{1}{2\sqrt{A+4}i},$$
$$\operatorname{Res}_{z=v_3} f(z)=\frac{v_3^2+1}{(v_3-v_1)(v_3-v_2)(v_3-v_4)}$$
$$=\frac{A+\sqrt{-A(A+4)}i}{2\sqrt{-A(A+4)}(-\sqrt{-A}+\sqrt{A+4}i)i}=\frac{1}{2\sqrt{A+4}i}.$$

Applying the theorem of obtaining real integral by the residue, we have
$$k_\lambda(A)=\frac{1}{\lambda}\int_{-\infty}^\infty f(v)dv$$
$$=\frac{1}{\lambda}2\pi i[\operatorname{Res}_{z=v_1} f(z)+\operatorname{Res}_{z=v_3} f(z)]$$
$$=\frac{2\pi i}{\lambda}(\frac{1}{2\sqrt{A+4}i}+\frac{1}{2\sqrt{A+4}i})=\frac{2\pi}{\lambda\sqrt{A+4}}. \quad (5.2.52)$$
We conclude that (5.2.52) is valid for $A=0$ by obtaining the integral straightway.

Hence by Corollary 5.1.2, for $A \geq 0$, (1) if $p > 1$, then we have the following equivalent inequalities:

$$\sum_{n=1}^{\infty}\sum_{m=1}^{\infty}\frac{a_m b_n}{m^\lambda+n^\lambda+A(m^{-\lambda}+n^{-\lambda})^{-1}}$$

$$< \frac{2\pi}{\lambda\sqrt{A+4}} \| a \|_{p,\phi_p} \| b \|_{q,\phi_q}, \quad (5.2.53)$$

$$\sum_{n=1}^{\infty} n^{\frac{p\lambda}{2}-1}[\sum_{m=1}^{\infty}\frac{a_m}{m^\lambda+n^\lambda+A(m^{-\lambda}+n^{-\lambda})^{-1}}]^p$$

$$< (\frac{2\pi}{\lambda\sqrt{A+4}})^p \| a \|_{p,\phi_p}^p; \quad (5.2.54)$$

(2) if $0 < p < 1$, then we have the following reverse equivalent inequalities:

$$\sum_{n=1}^{\infty}\sum_{m=1}^{\infty}\frac{a_m b_n}{m^\lambda+n^\lambda+A(m^{-\lambda}+n^{-\lambda})^{-1}} > \frac{2\pi}{\lambda\sqrt{A+4}}$$

$$\times\{\sum_{n=1}^{\infty}[1-\tilde{\theta}_\lambda(n)]n^{p(1-\frac{\lambda}{2})-1}a_n^p\}^{\frac{1}{p}} \| b \|_{q,\phi_q}, \quad (5.2.55)$$

$$\sum_{n=1}^{\infty} n^{\frac{p\lambda}{2}-1}[\sum_{m=1}^{\infty}\frac{a_m}{m^\lambda+n^\lambda+A(m^{-\lambda}+n^{-\lambda})^{-1}}]^p > (\frac{2\pi}{\lambda\sqrt{A+4}})^p$$

$$\times\sum_{n=1}^{\infty}[1-\tilde{\theta}_\lambda(n)]n^{p(1-\frac{\lambda}{2})-1}a_n^p, \quad (5.2.56)$$

$$\sum_{m=1}^{\infty}\frac{m^{q\lambda/2-1}}{[1-\tilde{\theta}_\lambda(m)]^{q-1}}[\sum_{n=1}^{\infty}\frac{b_n}{m^\lambda+n^\lambda+A(m^{-\lambda}+n^{-\lambda})^{-1}}]^q$$

$$< (\frac{2\pi}{\lambda\sqrt{A+4}})^q \| b \|_{q,\phi_q}^q, \quad (5.2.57)$$

where,

$$0 < \tilde{\theta}_\lambda(n) = \frac{1}{k_\lambda(A)}\int_n^\infty \frac{u^{\frac{\lambda}{2}-1}}{u^\lambda+1+A(u^{-\lambda}+1)^{-1}}du$$

$$\leq \frac{1}{\pi n^{\lambda/2}}\sqrt{A+4}.$$

5.2.3 ON A DECOMPOSITION OF HILBERT'S INEQUALITY

Example 5.2.11 If $0 < \lambda \leq 2$,

$$k_\lambda(x,y) = \frac{(\max\{x,y\})^\lambda}{(x^\lambda+y^\lambda)^2},$$

then setting $v = u^{\lambda/2}$, we find

$$k(\tfrac{\lambda}{2}) = \frac{1}{\lambda}\int_0^\infty \frac{\max\{u,1\}}{(u+1)^2}u^{\frac{1}{2}-1}du$$

$$= \frac{2}{\lambda}\int_0^\infty \frac{\max\{v^2,1\}}{(v^2+1)^2}dv = \frac{4}{\lambda}\int_0^1 \frac{dv}{(v^2+1)^2}$$

$$= \frac{4}{\lambda}\int_0^{\pi/4}\cos^2\theta d\theta$$

$$= \frac{2}{\lambda}[\tfrac{1}{2}\sin 2\theta+\theta]_0^{\pi/4} = \frac{1}{\lambda}(\tfrac{\pi}{2}+1).$$

Since we find

$$k_\lambda(x,y)\frac{1}{y^{1-\lambda/2}} = \begin{cases} \frac{x^\lambda}{(x^\lambda+y^\lambda)^2}\frac{1}{y^{1-(\lambda/2)}}, 0 < y \leq x, \\ \frac{y^{(3\lambda/2)-1}}{(x^\lambda+y^\lambda)^2}, y > x, \end{cases}$$

and for $y > x$,

$$(\frac{y^{(3\lambda/2)-1}}{(x^\lambda+y^\lambda)^2})'_y = \frac{[(3\lambda/2-1)x^\lambda-(1+\lambda/2)y^\lambda]y^{(3\lambda/2)-2}}{(x^\lambda+y^\lambda)^3}$$

$$< \frac{[(3\lambda/2-1)x^\lambda-(1+\lambda/2)x^\lambda]}{(x^\lambda+y^\lambda)^3 y^{2-(3\lambda/2)}} = \frac{(\lambda-2)x^\lambda}{(x^\lambda+y^\lambda)^3 y^{2-(3\lambda/2)}} \leq 0,$$

then it follows that $k_\lambda(x,y)\frac{1}{y^{1-\lambda/2}}$ is strictly decreasing for $y > 0$ and by the same way, $k_\lambda(x,y)\frac{1}{x^{1-\lambda/2}}$ is strictly decreasing for $x > 0$. Since

$$k_\lambda(u,1) = \frac{(\max\{u,1\})^\lambda}{(u^\lambda+1)^2} = O(\frac{1}{u^\lambda})(u \to \infty),$$

then by Theorem 4.2.5 and Corollary 4.3.2, (1) for $p > 1$, we have the following equivalent inequalities:

$$\sum_{n=1}^{\infty}\sum_{m=1}^{\infty}\frac{(\max\{m,n\})^\lambda}{(m^\lambda+n^\lambda)^2}a_m b_n$$

$$< \frac{1}{\lambda}(\frac{\pi}{2}+1) \| a \|_{p,\phi_p} \| b \|_{q,\phi_q}, \quad (5.2.58)$$

$$\sum_{n=1}^{\infty} n^{\frac{p\lambda}{2}-1}[\sum_{m=1}^{\infty}\frac{(\max\{m,n\})^\lambda a_m}{(m^\lambda+n^\lambda)^2}]^p$$

$$< [\frac{1}{\lambda}(\frac{\pi}{2}+1)]^p \| a \|_{p,\phi_p}^p; \quad (5.2.59)$$

(2) for $0 < p < 1$, we have the following reverse equivalent inequalities:

$$\sum_{n=1}^{\infty}\sum_{m=1}^{\infty}\frac{(\max\{m,n\})^\lambda a_m b_n}{(m^\lambda+n^\lambda)^2} > \frac{1}{\lambda}(\frac{\pi}{2}+1)$$

$$\times\{\sum_{n=1}^{\infty}[1-\tilde{\theta}_\lambda(n)]n^{p(1-\frac{\lambda}{2})-1}a_n^p\}^{\frac{1}{p}} \| b \|_{q,\phi_q}, \quad (5.2.60)$$

$$\sum_{n=1}^{\infty} n^{\frac{p\lambda}{2}-1}[\sum_{m=1}^{\infty}\frac{(\max\{m,n\})^\lambda a_m}{(m^\lambda+n^\lambda)^2}]^p > [\frac{1}{\lambda}(\frac{\pi}{2}+1)]^p$$

$$\times\sum_{n=1}^{\infty}[1-\tilde{\theta}_\lambda(n)]n^{p(1-\frac{\lambda}{2})-1}a_n^p, \quad (5.2.61)$$

$$\sum_{m=1}^{\infty}\frac{m^{q\lambda/2-1}}{[1-\tilde{\theta}_\lambda(m)]^{q-1}}[\sum_{n=1}^{\infty}\frac{(\max\{m,n\})^\lambda b_n}{(m^\lambda+n^\lambda)^2}]^q$$

$$< [\frac{1}{\lambda}(\frac{\pi}{2}+1)]^q \| b \|_{q,\phi_q}^q, \quad (5.2.62)$$

where,

$$0 < \tilde{\theta}_\lambda(n) = \frac{1}{k(\frac{\lambda}{2})}\int_n^\infty \frac{1}{(u^\lambda+1)^2}u^{\frac{3\lambda}{2}-1}du$$

$$\leq \frac{2}{(1+\pi/2)n^{\lambda/2}}.$$

Example 5.2.12 If $0 < \lambda \leq 1$,

$$k_\lambda(x,y) = \frac{(\min\{x,y\})^\lambda}{(x^\lambda+y^\lambda)^2},$$

then setting $v = u^{\lambda/2}$, we find

$$\tilde{k}(\tfrac{\lambda}{2}) = \int_0^\infty \frac{\min\{u^\lambda,1\}}{(u^\lambda+1)^2}u^{\frac{\lambda}{2}-1}du$$

$$=\frac{2}{\lambda}\int_0^{\infty}\frac{\min\{v^2,1\}}{(v^2+1)^2}dv=\frac{4}{\lambda}\int_1^{\infty}\frac{dv}{(v^2+1)^2}$$
$$=\frac{4}{\lambda}\int_{\pi/4}^{\pi/2}\cos^2\theta d\theta.$$
$$=\frac{2}{\lambda}[\frac{1}{2}\sin 2\theta+\theta]_{\pi/4}^{\pi/2}=\frac{1}{\lambda}(\frac{\pi}{2}-1).$$

We define the following weight coefficients:

$$\varpi(\frac{\lambda}{2},m)=\sum_{n=1}^{\infty}\frac{(\min\{m,n\})^{\lambda}m^{\lambda/2}}{(m^{\lambda}+n^{\lambda})^2 n^{1-\lambda/2}},$$
$$\vartheta(\frac{\lambda}{2},n)=\sum_{m=1}^{\infty}\frac{(\min\{m,n\})^{\lambda}n^{\lambda/2}}{(m^{\lambda}+n^{\lambda})^2 m^{1-\lambda/2}}, m,n\in\mathbf{N}. \quad (5.2.63)$$

Setting $f_m(y):=\frac{(\min\{m,y\})^{\lambda}m^{\lambda/2}}{(m^{\lambda}+y^{\lambda})^2 y^{1-\lambda/2}}$, it follows that

$$f_m(y)=\begin{cases}\frac{y^{(3\lambda/2)-1}m^{\lambda/2}}{(m^{\lambda}+y^{\lambda})^2}, 0<y<m,\\ \frac{m^{3\lambda/2}}{(m^{\lambda}+y^{\lambda})^2 y^{1-\lambda/2}}, y\ge m,\end{cases}$$

$$f_m'(y)=\begin{cases}\frac{(\frac{3\lambda}{2}-3)m^{\lambda/2}}{(m^{\lambda}+y^{\lambda})^2 y^{2-(3\lambda/2)}}+\frac{2m^{3\lambda/2}}{(m^{\lambda}+y^{\lambda})^3 y^{2-(3\lambda/2)}},\\ \qquad 0<y<m,\\ -\frac{(1-\frac{\lambda}{2})m^{3\lambda/2}}{(m^{\lambda}+y^{\lambda})^2 y^{2-(\lambda/2)}}-\frac{2m^{3\lambda/2}}{(m^{\lambda}+y^{\lambda})^3 y^{2-(3\lambda/2)}},\\ \qquad y\ge m.\end{cases}$$

By Euler-Maclaurin summation formula (2.2.13) (for $m=1,n\to\infty$), we have

$$\varpi(\frac{\lambda}{2},m)=\sum_{n=1}^{\infty}f_m(n)$$
$$=\int_1^{\infty}f_m(y)dy+\frac{1}{2}f(1)+\int_1^{\infty}P_1(y)f_m'(y)dy$$
$$=\int_0^{\infty}f_m(y)dy-\rho_{\lambda,m},$$
$$\rho_{\lambda,m}:=\int_0^1 f_m(y)dy-\frac{1}{2}f_m(1)$$
$$-\int_1^{\infty}P_1(y)f_m'(y)dy. \quad (5.2.64)$$

Setting $v=(y/m)^{\lambda/2}$, we obtain

$$\int_0^{\infty}f_m(y)dy=\int_0^{\infty}\frac{(\min\{m,y\})^{\lambda}m^{\lambda/2}}{(m^{\lambda}+y^{\lambda})^2 y^{1-\lambda/2}}dy$$
$$=\frac{2}{\lambda}\int_0^{\infty}\frac{\min\{v^2,1\}}{(v^2+1)^2}dv=\tilde{k}(\frac{\lambda}{2}),$$
$$\int_0^1 f_m(y)dy=\frac{2}{\lambda}\int_0^{(\frac{1}{m})^{\frac{\lambda}{2}}}\frac{v^2}{(v^2+1)^2}dv$$
$$>\frac{2}{\lambda}\int_0^{(\frac{1}{m})^{\frac{\lambda}{2}}}\frac{v^2}{(\frac{1}{m^{\lambda}}+1)^2}dv=\frac{2m^{\lambda/2}}{3\lambda(m^{\lambda}+1)^2},$$
$$\int_0^1 f_m(y)dy=\frac{2}{\lambda}\int_0^{(\frac{1}{m})^{\frac{\lambda}{2}}}\frac{v^2}{(v^2+1)^2}dv$$
$$<\frac{2}{\lambda}\int_0^{(\frac{1}{m})^{\frac{\lambda}{2}}}v^2dv=\frac{2}{3\lambda m^{3\lambda/2}},$$
$$\frac{-1}{2}f_m(1)=\frac{-m^{\lambda/2}}{2(m^{\lambda}+1)^2}.$$

By (2.2.2) for $q=0$, we have

$$-\int_1^{\infty}P_1(y)f_m'(y)dy$$
$$=-\int_1^{m}P_1(y)f_m'(y)dy-\int_m^{\infty}P_1(y)f_m'(y)dy$$
$$=(3-\frac{3\lambda}{2})\int_1^m P_1(y)\frac{m^{\lambda/2}}{(m^{\lambda}+y^{\lambda})^2 y^{2-(3\lambda/2)}}dy$$
$$-\int_1^m P_1(y)\frac{2m^{3\lambda/2}}{(m^{\lambda}+y^{\lambda})^3 y^{2-(3\lambda/2)}}dy$$
$$+\int_m^{\infty}P_1(y)[\frac{(1-\frac{\lambda}{2})m^{3\lambda/2}}{(m^{\lambda}+y^{\lambda})^2 y^{2-(\lambda/2)}}+\frac{2m^{3\lambda/2}}{(m^{\lambda}+y^{\lambda})^3 y^{2-(3\lambda/2)}}]dy$$
$$=\frac{\varepsilon_1}{12}(3-\frac{3\lambda}{2})[\frac{m^{\lambda/2}}{(m^{\lambda}+y^{\lambda})^2 y^{2-(3\lambda/2)}}]_1^m$$
$$-\frac{\varepsilon_2}{12}[\frac{2m^{3\lambda/2}}{(m^{\lambda}+y^{\lambda})^3 y^{2-(3\lambda/2)}}]_1^m$$
$$+\frac{\varepsilon_3}{12}[(1-\frac{\lambda}{2})\frac{m^{3\lambda/2}}{(m^{\lambda}+y^{\lambda})^2 y^{2-(\lambda/2)}}+\frac{2m^{3\lambda/2}}{(m^{\lambda}+y^{\lambda})^3 y^{2-(3\lambda/2)}}]_m^{\infty}$$
$$(\varepsilon_i\in(0,1),i=1,2,3).$$

Hence we find

$$-\int_1^{\infty}P_1(y)f_m'(y)dy<-\frac{1}{12}[\frac{2m^{3\lambda/2}}{(m^{\lambda}+y^{\lambda})^3 y^{2-(3\lambda/2)}}]_1^m$$
$$=\frac{m^{3\lambda/2}}{6(m^{\lambda}+1)^3}-\frac{1}{48m^2};$$
$$-\int_1^{\infty}P_1(y)f_m'(y)dy$$
$$>\frac{1}{12}(3-\frac{3\lambda}{2})[\frac{m^{\lambda/2}}{(m^{\lambda}+y^{\lambda})^2 y^{2-(3\lambda/2)}}]_1^m$$
$$+\frac{1}{12}[\frac{(1-\frac{\lambda}{2})m^{3\lambda/2}}{(m^{\lambda}+y^{\lambda})^2 y^{2-(\lambda/2)}}+\frac{2m^{3\lambda/2}}{(m^{\lambda}+y^{\lambda})^3 y^{2-(3\lambda/2)}}]_m^{\infty}$$
$$=-\frac{1}{4}(1-\frac{\lambda}{2})\frac{m^{\lambda/2}}{(m^{\lambda}+1)^2}-\frac{1}{48m^2}.$$

For $0<\lambda\le 1$, we find

$$\rho_{\lambda,m}>\frac{2m^{\lambda/2}}{3\lambda(m^{\lambda}+1)^2}-\frac{m^{\lambda/2}}{2(m^{\lambda}+1)^2}$$
$$-\frac{1}{4}(1-\frac{\lambda}{2})\frac{m^{\lambda/2}}{(m^{\lambda}+1)^2}+\frac{1-\lambda}{48m^2}$$
$$=(\frac{2}{3\lambda}-\frac{3}{4}+\frac{\lambda}{8})\frac{m^{\lambda/2}}{(m^{\lambda}+1)^2}+\frac{1-\lambda}{48m^2}>\frac{1}{24}\frac{m^{\lambda/2}}{(m^{\lambda}+1)^2}>0.$$
$$0<m^{3\lambda/2}\rho_{\lambda,m}<\frac{2}{3\lambda}-\frac{m^{2\lambda}}{2(m^{\lambda}+1)^2}+\frac{m^{3\lambda}}{6(m^{\lambda}+1)^3}-\frac{m^{3\lambda/2}}{48m^2}$$
$$\to\frac{2}{3\lambda}-\frac{1}{3}(m\to\infty).$$

Hence $\rho_{\lambda,m}=O(\frac{1}{m^{3\lambda/2}})$. Setting

$$\tilde{\theta}_{\lambda}(m)=\frac{1}{\tilde{k}(\lambda/2)}\rho_{\lambda,m}=\frac{\lambda}{(\frac{\pi}{2}-1)}\rho_{\lambda,m},$$

it follows that

$$\frac{1}{\lambda}(\frac{\pi}{2}-1)[1-\tilde{\theta}_{\lambda}(m)]$$
$$=\varpi(\frac{\lambda}{2},m)<\frac{1}{\lambda}(\frac{\pi}{2}-1).$$

By the same way, we have $\vartheta(\frac{\lambda}{2},n)<\frac{1}{\lambda}(\frac{\pi}{2}-1)$.

By Theorem 4.2.3 and Theorem 4.3.1, (1) for $p>1$, we have the following equivalent inequalities:

$$\sum_{n=1}^{\infty}\sum_{m=1}^{\infty}\frac{(\min\{m,n\})^{\lambda}}{(m^{\lambda}+n^{\lambda})^2}a_m b_n$$

$$<\frac{1}{\lambda}(\frac{\pi}{2}-1)\|a\|_{p,\phi_p}\|b\|_{q,\phi_q}, \quad (5.2.65)$$

$$\sum_{n=1}^{\infty}n^{\frac{p\lambda}{2}-1}[\sum_{m=1}^{\infty}\frac{(\min\{m,n\})^{\lambda}a_m}{(m^{\lambda}+n^{\lambda})^2}]^p$$

$$<[\frac{1}{\lambda}(\frac{\pi}{2}-1)]^p\|a\|_{p,\phi_p}^p; \quad (5.2.66)$$

(2) for $0<p<1$, we have the following reverse equivalent inequalities:

$$\sum_{n=1}^{\infty}\sum_{m=1}^{\infty}\frac{(\min\{m,n\})^{\lambda}a_m b_n}{(m^{\lambda}+n^{\lambda})^2}>\frac{1}{\lambda}(\frac{\pi}{2}-1)$$

$$\times\{\sum_{n=1}^{\infty}[1-\tilde{\theta}_{\lambda}(n)]n^{p(1-\frac{\lambda}{2})-1}a_n^p\}^{\frac{1}{p}}\|b\|_{q,\phi_q}, \quad (5.2.67)$$

$$\sum_{n=1}^{\infty}n^{\frac{p\lambda}{2}-1}[\sum_{m=1}^{\infty}\frac{(\min\{m,n\})^{\lambda}a_m}{(m^{\lambda}+n^{\lambda})^2}]^p>[\frac{1}{\lambda}(\frac{\pi}{2}-1)]^p$$

$$\times\sum_{n=1}^{\infty}[1-\tilde{\theta}_{\lambda}(n)]n^{p(1-\frac{\lambda}{2})-1}a_n^p, \quad (5.2.68)$$

$$\sum_{m=1}^{\infty}\frac{m^{q\lambda/2-1}}{[1-\tilde{\theta}_{\lambda}(m)]^{q-1}}[\sum_{n=1}^{\infty}\frac{(\min\{m,n\})^{\lambda}b_n}{(m^{\lambda}+n^{\lambda})^2}]^q$$

$$<[\frac{1}{\lambda}(\frac{\pi}{2}-1)]^q\|b\|_{q,\phi_q}^q, \quad (5.2.69)$$

where, $\tilde{\theta}_{\lambda}(n)=O(\frac{1}{n^{3\lambda/2}})$, the constant factors are the best possible.

Note 5.2.13 Since $\pi=(\frac{\pi}{2}+1)+(\frac{\pi}{2}-1)$ and

$$\frac{1}{m+n}=\frac{\max\{m,n\}}{(m+n)^2}+\frac{\min\{m,n\}}{(m+n)^2},\quad m,n\in\mathbf{N},$$

For $\lambda=1, p=q=2$ in (5.2.58) and (5.2.65), we have

$$\sum_{n=1}^{\infty}\sum_{m=1}^{\infty}\frac{\max\{m,n\}}{(m+n)^2}a_m b_n$$

$$<(\frac{\pi}{2}+1)(\sum_{n=1}^{\infty}a_n^2\sum_{n=1}^{\infty}b_n^2)^{\frac{1}{2}}, \quad (5.2.70)$$

$$\sum_{n=1}^{\infty}\sum_{m=1}^{\infty}\frac{\min\{m,n\}}{(m+n)^2}a_m b_n$$

$$<(\frac{\pi}{2}-1)(\sum_{n=1}^{\infty}a_n^2\sum_{n=1}^{\infty}b_n^2)^{\frac{1}{2}}, \quad (5.2.71)$$

which is a well decomposition of the following Hilbert's inequality

$$\sum_{n=1}^{\infty}\sum_{m=1}^{\infty}\frac{a_m b_n}{m+n}<\pi(\sum_{n=1}^{\infty}a_n^2\sum_{n=1}^{\infty}b_n^2)^{\frac{1}{2}} \quad (5.2.72)$$

(cf. (Yang JIPAM 2009) [4]).

5.3 SOME HILBERT-TYPE INEQUALITIES WITH THE HOMOGENEOUS KERNELS OF DEGREE -2 AND -3 AND EXTENSIONS

5.3.1 A HILBERT-TYPE INEQUALITY WITH THE HOMOGENEOUS KERNEL OF DEGREE -2 AND EXTENSIONS

Example 5.3.1 If $\lambda>0, C>0, B>-C$,

$$k_{2\lambda}(x,y)=\frac{1}{x^{2\lambda}+2Bx^{\lambda}y^{\lambda}+C^2y^{2\lambda}},$$

then setting $v=u^{\lambda}$, we have

$$k_{2\lambda}:=\int_0^{\infty}k_{2\lambda}(u,1)u^{\frac{2\lambda}{2}-1}du$$

$$=\int_0^{\infty}\frac{1}{u^{2\lambda}+2Bu^{\lambda}+C^2}u^{\lambda-1}du$$

$$=\frac{1}{\lambda}\int_0^{\infty}\frac{dv}{v^2+2Bv+C^2}$$

$$=\begin{cases}\frac{1}{\lambda\sqrt{C^2-B^2}}(\frac{\pi}{2}-\arctan\frac{B}{\sqrt{C^2-B^2}}), & |B|<C,\\ \frac{1}{\lambda B}, & B=C,\\ \frac{1}{\lambda\sqrt{B^2-C^2}}\ln\frac{B+\sqrt{B^2-C^2}}{C}, & B>C.\end{cases}$$

By Corollary 5.1.2, replacing λ by 2λ and $\phi_r(x)=x^{r(1-\frac{\lambda}{2})-1}$ by $\tilde{\phi}_r(x)=x^{r(1-\lambda)-1}$ $(r=p,q)$, for $0<\lambda\le 1, C>0, B\ge 0$, (1) if $p>1$, then we have the following equivalent inequalities:

$$\sum_{n=1}^{\infty}\sum_{m=1}^{\infty}\frac{a_m b_n}{m^{2\lambda}+2Bm^{\lambda}n^{\lambda}+C^2n^{2\lambda}}$$

$$<k_{2\lambda}\|a\|_{p,\tilde{\phi}_p}\|b\|_{q,\tilde{\phi}_q}, \quad (5.3.1)$$

$$\sum_{n=1}^{\infty}n^{p\lambda-1}(\sum_{m=1}^{\infty}\frac{a_m}{m^{2\lambda}+2Bm^{\lambda}n^{\lambda}+C^2n^{2\lambda}})^p$$

$$<k_{2\lambda}^p\|a\|_{p,\tilde{\phi}_p}^p; \quad (5.3.2)$$

(2) if $0<p<1$, then we have the following reverse equivalent inequalities:

$$\sum_{n=1}^{\infty}\sum_{m=1}^{\infty}\frac{a_m b_n}{m^{2\lambda}+2Bm^{\lambda}n^{\lambda}+C^2n^{2\lambda}}>k_{2\lambda}$$

$$\times\{\sum_{n=1}^{\infty}[1-\tilde{\theta}_{2\lambda}(n)]n^{p(1-\lambda)-1}a_n^p\}^{\frac{1}{p}}\|b\|_{q,\tilde{\phi}_q}, \quad (5.3.3)$$

$$\sum_{n=1}^{\infty}n^{p\lambda-1}(\sum_{m=1}^{\infty}\frac{a_m}{m^{2\lambda}+2Bm^{\lambda}n^{\lambda}+C^2n^{2\lambda}})^p>k_{2\lambda}^p$$

$$\times\sum_{n=1}^{\infty}[1-\tilde{\theta}_{2\lambda}(n)]n^{p(1-\lambda)-1}a_n^p, \quad (5.3.4)$$

$$\sum_{m=1}^{\infty}\frac{m^{q\lambda-1}}{[1-\tilde{\theta}_{2\lambda}(m)]^{q-1}}(\sum_{n=1}^{\infty}\frac{b_n}{m^{2\lambda}+2Bm^{\lambda}n^{\lambda}+C^2n^{2\lambda}})^q$$
$$< k_{2\lambda}^q\|b\|_{q,\tilde{\phi}_q}^q, \tag{5.3.5}$$

where,

$$0<\tilde{\theta}_{2\lambda}(n)=\frac{1}{k_{2\lambda}}\int_n^{\infty}\frac{1}{u^{2\lambda}+Bu^{\lambda}+C}u^{\lambda-1}du\le\frac{1}{\lambda k_{2\lambda}n^{\lambda}}.$$

5.3.2 A HILBERT-TYPE INEQUALITY WITH THE HOMOGENEOUS KERNEL OF DEGREE -3 AND EXTENSIONS

Example 5.3.2 If $\lambda>0, A>0, C>0, \ B>-C$,

$$k_{3\lambda}(x,y)=\frac{1}{(x^{\lambda}+Ay^{\lambda})(x^{2\lambda}+2Bx^{\lambda}y^{\lambda}+C^2y^{2\lambda})},$$

then setting $v=u^{\lambda/2}$, we have

$$k_{3\lambda}:=\int_0^{\infty}k_{3\lambda}(u,1)u^{\frac{3\lambda}{2}-1}du$$
$$=\int_0^{\infty}\frac{1}{(u^{\lambda}+A)(u^{2\lambda}+2Bu^{\lambda}+C^2)}u^{\frac{3\lambda}{2}-1}du$$
$$=\frac{1}{\lambda}\int_{-\infty}^{\infty}\frac{v^2}{(v^2+A)(v^4+2Bv^2+C^2)}dv. \tag{5.3.6}$$

In the following, we obtain the integral (5.3.6) by using the theory of residue:

(1) For $-C<B<C$,

$$v^4+2Bv^2+C^2=(v^2+\alpha)(v^2+\beta)$$

(α and β are imaginary numbers), since $\alpha+\beta=2B$ is a real number with $\operatorname{Im}\alpha+\operatorname{Im}\beta=0$, setting $\alpha=re^{i\theta}$, $\beta=re^{-i\theta}$ $(r>0, 0<\theta<\pi)$, and

$$\sqrt{\alpha}=\sqrt{r}e^{\frac{i\theta}{2}},\ \sqrt{\beta}=\sqrt{r}e^{\frac{-i\theta}{2}},$$

then we find

$$(v^2+\alpha)(v^2+\beta)$$
$$=(v+i\sqrt{\alpha})(v-i\sqrt{\alpha})(v+i\sqrt{\beta})(v-i\sqrt{\beta}),$$

where, $v_1=i\sqrt{\alpha}=\sqrt{r}e^{i\frac{\theta+\pi}{2}}$ and

$$v_2=i\sqrt{\beta}=\sqrt{r}e^{i\frac{-\theta+\pi}{2}}$$

are in the upper half plane. setting $v_0=i\sqrt{A}$, then applying the theorem of obtaining real integral by the residue, we find

$$k_{3\lambda}=\frac{1}{\lambda}\int_{-\infty}^{\infty}\frac{v^2}{(v^2+A)(v^4+2Bv^2+C^2)}dv$$
$$=\frac{1}{\lambda}2\pi i\sum_{i=0}^{2}\operatorname{Res}_{z=v_i}f(z)$$
$$=\frac{1}{\lambda}2\pi i[\frac{z^2}{2z(z^2+\alpha)(z^2+\beta)}|_{z=i\sqrt{A}}$$
$$+\frac{z^2}{2z(z^2+A)(z^2+\beta)}|_{z=i\sqrt{\alpha}}+\frac{z^2}{2z(z^2+A)(z^2+\alpha)}|_{z=i\sqrt{\beta}}]$$
$$=\frac{1}{\lambda}\pi[\frac{-\sqrt{A}}{(-A+\alpha)(-A+\beta)}$$
$$+\frac{-\sqrt{\alpha}}{(-\alpha+A)(-\alpha+\beta)}+\frac{-\sqrt{\beta}}{(-\beta+A)(-\beta+\alpha)}]$$
$$=\frac{\pi}{\lambda(\sqrt{\alpha}+\sqrt{A})(\sqrt{\beta}+\sqrt{A})(\sqrt{\beta}+\sqrt{\alpha})}. \tag{5.3.7}$$

(2) For $B\ge C$,

$$v^4+2Bv^2+C^2=(v^2+\alpha)(v^2+\beta)$$

(α and β are positive numbers), without loses of generality, suppose that $0<A\le\alpha\le\beta$. For $n\in\mathbf{N}$, setting $\tilde{\alpha}=\alpha+\frac{1}{2n}, \tilde{\beta}=\beta+\frac{1}{n}$, then it follows $0<A<\tilde{\alpha}<\tilde{\beta}$. Putting

$$\tilde{f}(z)=\frac{z^2}{(z^2+A)(z^2+\tilde{\alpha})(z^2+\tilde{\beta})},$$

by using the same way of (1), we find

$$\tilde{k}_{3\lambda}=\frac{1}{\lambda}\int_{-\infty}^{\infty}\tilde{f}(v)dv$$
$$=\frac{\pi}{\lambda(\sqrt{\tilde{\alpha}}+\sqrt{A})(\sqrt{\tilde{\beta}}+\sqrt{A})(\sqrt{\tilde{\beta}}+\sqrt{\tilde{\alpha}})}.$$

For $n\to\infty$, by Levi theorem, we have (5.3.6).

Since $u^2+2Bu+C^2=(u+\alpha)(u+\beta)$, we find $\alpha+\beta=2B, \alpha\beta=C^2$. In view of the cases of (1) and (2), we have $\sqrt{\alpha\beta}=C$,

$$(\sqrt{\beta}+\sqrt{\alpha})^2=\beta+\alpha+2\sqrt{\alpha\beta}=2(B+C)$$

and $\sqrt{\beta}+\sqrt{\alpha}=\sqrt{2(B+C)}$. Then by (5.3.7), we have

$$k_{3\lambda}=\frac{\pi}{\lambda[\sqrt{\alpha\beta}+(\sqrt{\beta}+\sqrt{\alpha})\sqrt{A}+A](\sqrt{\beta}+\sqrt{\alpha})}$$
$$=\frac{\pi}{\lambda(C+\sqrt{2(B+C)A}+A)\sqrt{2(B+C)}}. \tag{5.3.8}$$

By Corollary 5.1.2, replacing λ by 3λ, and $\phi_r(x)=x^{r(1-\frac{\lambda}{2})-1}$ by

$$\tilde{\varphi}_r(x)=x^{r(1-\frac{3\lambda}{2})-1}\ (r=p,q),$$

for $0<\lambda\le\frac{2}{3}, A>0, C>0, B\ge0$, (1) if $p>1$, then we have the following equivalent inequalities:

$$\sum_{n=1}^{\infty}\sum_{m=1}^{\infty}\frac{a_m b_n}{(m^{\lambda}+An^{\lambda})(m^{2\lambda}+2Bm^{\lambda}n^{\lambda}+C^2n^{2\lambda})}$$
$$<k_{3\lambda}\|a\|_{p,\tilde{\varphi}_p}\|b\|_{q,\tilde{\varphi}_q}, \tag{5.3.9}$$

$$\sum_{n=1}^{\infty}n^{\frac{3p\lambda}{2}-1}[\sum_{m=1}^{\infty}\frac{a_m}{(m^{\lambda}+An^{\lambda})(m^{2\lambda}+2Bm^{\lambda}n^{\lambda}+C^2n^{2\lambda})}]^p$$
$$<k_{3\lambda}^p\|a\|_{p,\tilde{\varphi}_p}^p; \tag{5.3.10}$$

(2) if $0<p<1$, then we have the following reverse equivalent inequalities:

$$\sum_{n=1}^{\infty}\sum_{m=1}^{\infty}\frac{a_m b_n}{(m^\lambda+An^\lambda)(m^{2\lambda}+2Bm^\lambda n^\lambda+C^2n^{2\lambda})}>k_{3\lambda}$$

$$\times\{\sum_{n=1}^{\infty}[1-\tilde{\theta}_{3\lambda}(n)]n^{p(1-\frac{3\lambda}{2})-1}a_n^p\}^{\frac{1}{p}}\|b\|_{q,\tilde{\varphi}_q}, \quad (5.3.11)$$

$$\sum_{n=1}^{\infty}n^{\frac{3p\lambda}{2}-1}[\sum_{m=1}^{\infty}\frac{a_m}{(m^\lambda+An^\lambda)(m^{2\lambda}+2Bm^\lambda n^\lambda+C^2n^{2\lambda})}]^p$$

$$>k_{3\lambda}^p\sum_{n=1}^{\infty}[1-\tilde{\theta}_{3\lambda}(n)]n^{p(1-\frac{3\lambda}{2})-1}a_n^p, \quad (5.3.12)$$

$$\sum_{m=1}^{\infty}\frac{m^{3q\lambda/2-1}}{[1-\tilde{\theta}_{3\lambda}(m)]^{q-1}}[\sum_{n=1}^{\infty}\frac{b_n}{(m^\lambda+An^\lambda)(m^{2\lambda}+2Bm^\lambda n^\lambda+C^2n^{2\lambda})}]^q$$

$$<k_{3\lambda}^q\|b\|_{q,\tilde{\varphi}_q}^q, \quad (5.3.13)$$

where,

$$0<\tilde{\theta}_{3\lambda}(n)=\frac{1}{k_{3\lambda}}\int_n^{\infty}\frac{u^{\frac{3\lambda}{2}-1}}{(u^\lambda+A)(u^{2\lambda}+2Bu^\lambda+C^2)}du$$

$$\leq\frac{2}{3\lambda k_{3\lambda}n^{(3\lambda)/2}}.$$

5.4 SOME INEQUALITIES WITH THE HOMOGENEOUS KERNEL OF DEGREE -4

5.4.1 HILBERT-TYPE INEQUALITY WITH THE KERNEL $\frac{1}{(m+An)(m+Bn)(m+Cn)(m+Dn)}$ AND EXTENSIONS

Example 5.4.1 If $\lambda>0, 0<A<B<C<D$,

$$k_{4\lambda}(x,y)=\frac{1}{(x^\lambda+Ay^\lambda)(x^\lambda+By^\lambda)(x^\lambda+Cy^\lambda)(x^\lambda+Dy^\lambda)},$$

then setting $v=u^\lambda$, we find

$$k_{4\lambda}:=\int_0^{\infty}k_{4\lambda}(u,1)u^{2\lambda-1}du$$

$$=\int_0^{\infty}\frac{u^{2\lambda-1}}{(u^\lambda+A)(u^\lambda+B)(u^\lambda+C)(u^\lambda+D)}du$$

$$=\frac{1}{\lambda}\int_0^{\infty}\frac{v}{(v+A)(v+B)(v+C)(v+D)}dv. \quad (5.4.1)$$

Setting

$$\frac{v}{(v+A)(v+B)(v+C)(v+D)}=\frac{\alpha}{v+A}+\frac{\beta}{v+B}+\frac{\chi}{v+C}+\frac{\delta}{v+D},$$

we obtain

$$v=\alpha(v+B)(v+C)(v+D)$$
$$+\beta(v+A)(v+C)(v+D)$$
$$+\chi(v+A)(v+B)(v+D)$$
$$+\delta(v+A)(v+B)(v+C).$$

Substitution of $v=-A,-B,-C,-D$ in the above equality, we find

$$\begin{cases}\alpha=\frac{-A}{(B-A)(C-A)(D-A)},\beta=\frac{B}{(B-A)(C-B)(D-B)},\\ \chi=\frac{-C}{(C-A)(C-B)(D-C)},\delta=\frac{D}{(D-A)(D-B)(D-C)}.\end{cases} \quad (5.4.2)$$

In view of $\alpha+\beta+\chi+\delta=0$, it follows

$$k_{4\lambda}=\frac{1}{\lambda}\int_0^{\infty}(\frac{\alpha}{v+A}+\frac{\beta}{v+B}+\frac{\chi}{v+C}+\frac{\delta}{v+D})dv$$

$$=\frac{1}{\lambda}\ln(v+A)^\alpha(v+B)^\beta(v+C)^\chi(v+D)^\delta|_0^\infty$$

$$=\frac{1}{\lambda}\ln(A^\alpha B^\beta C^\chi D^\delta)^{-1}$$

$$=\frac{1}{\lambda}[\frac{A\ln A}{(B-A)(C-A)(D-A)}-\frac{B\ln B}{(B-A)(C-B)(D-B)}$$

$$+\frac{C\ln C}{(C-A)(C-B)(D-C)}-\frac{D\ln D}{(D-A)(D-B)(D-C)}]. \quad (5.4.3)$$

By Corollary 5.1.2, replacing λ by 4λ, and $\phi_r(x)=x^{r(1-\frac{\lambda}{2})-1}$ by

$$\tilde{\psi}_r(x)=x^{r(1-2\lambda)-1}\ (r=p,q),$$

for $0<\lambda\leq\frac{1}{2}$, (1) if $p>1$, then we have the following equivalent inequalities:

$$\sum_{n=1}^{\infty}\sum_{m=1}^{\infty}\frac{a_m b_n}{(m^\lambda+An^\lambda)(m^\lambda+Bn^\lambda)(m^\lambda+Cn^\lambda)(m^\lambda+Dn^\lambda)}$$

$$<k_{4\lambda}\|a\|_{p,\tilde{\psi}_p}\|b\|_{q,\tilde{\psi}_q}, \quad (5.4.4)$$

$$\sum_{n=1}^{\infty}n^{2p\lambda-1}[\sum_{m=1}^{\infty}\frac{a_m}{(m^\lambda+An^\lambda)(m^\lambda+Bn^\lambda)(m^\lambda+Cn^\lambda)(m^\lambda+Dn^\lambda)}]^p$$

$$<k_{4\lambda}^p\|a\|_{p,\tilde{\psi}_p}^p; \quad (5.4.5)$$

(2) if $0<p<1$, then we have the following reverse equivalent inequalities:

$$\sum_{n=1}^{\infty}\sum_{m=1}^{\infty}\frac{a_m b_n}{(m^\lambda+An^\lambda)(m^\lambda+Bn^\lambda)(m^\lambda+Cn^\lambda)(m^\lambda+Dn^\lambda)}>k_{4\lambda}$$

$$\times\{\sum_{n=1}^{\infty}[1-\tilde{\theta}_{4\lambda}(n)]n^{p(1-2\lambda)-1}a_n^p\}^{\frac{1}{p}}\|b\|_{q,\tilde{\psi}_q}, \quad (5.4.6)$$

$$\sum_{n=1}^{\infty}n^{2p\lambda-1}[\sum_{m=1}^{\infty}\frac{a_m}{(m^\lambda+An^\lambda)(m^\lambda+Bn^\lambda)(m^\lambda+Cn^\lambda)(m^\lambda+Dn^\lambda)}]^p$$

$$>k_{4\lambda}^p\sum_{n=1}^{\infty}[1-\tilde{\theta}_{4\lambda}(n)]n^{p(1-2\lambda)-1}a_n^p, \quad (5.4.7)$$

$$\sum_{m=1}^{\infty}\frac{m^{2q\lambda-1}}{[1-\tilde{\theta}_{4\lambda}(m)]^{q-1}}[\sum_{n=1}^{\infty}\frac{b_n}{(m^\lambda+An^\lambda)(m^\lambda+Bn^\lambda)(m^\lambda+Cn^\lambda)(m^\lambda+Cn^\lambda)}]^q$$

$$<k_{4\lambda}^q\|b\|_{q,\tilde{\psi}_q}^q, \quad (5.4.8)$$

where,

$$0<\tilde{\theta}_{4\lambda}(n)=\frac{1}{k_{4\lambda}}\int_n^{\infty}\frac{u^{2\lambda-1}du}{(u^\lambda+A)(u^\lambda+B)(u^\lambda+C)(u^\lambda+D)}$$

$$\leq\frac{1}{2\lambda k_{4\lambda}n^{2\lambda}}.$$

Particular case 5.4.2 For $D=C$, in (5.4.3), setting $D=C+t\ (t>0)$, we find

$$\lim_{t\to 0^+}(\chi\ln C+\delta\ln D)$$

$$=\lim_{t\to 0^+}[\frac{-C\ln C}{(C-A)(C-B)t}+\frac{(t+C)\ln(t+C)}{(t+C-A)(t+C-B)t}]$$

$$=\lim_{t\to 0^+}\frac{-C\ln C}{(C-A)(C-B)}$$

$$\times[\frac{(t+C-A)(x+C-B)C\ln C-(C-A)(C-B)(t+C)\ln(t+C)}{t(t+C-A)(t+C-B)C\ln C}]$$

$$=\frac{-(C^2-AB)\ln C}{(C-A)^2(C-B)^2}+\frac{1}{(C-A)(C-B)},$$

$$\tilde{k}_{4\lambda}:=\lim_{t\to 0^+}k_{4\lambda}$$

$$=\frac{1}{\lambda}\lim_{t\to 0^+}[\ln(A^{\alpha}B^{\beta})^{-1}-\chi\ln C-\delta\ln D]$$

$$=\frac{1}{\lambda}[\frac{A\ln A}{(B-A)(C-A)^2}-\frac{B\ln B}{(B-A)(C-B)^2}$$

$$+\frac{(C^2-AB)\ln C}{(C-A)^2(C-B)^2}-\frac{1}{(C-A)(C-B)}]. \qquad (5.4.9)$$

In (5.4.1), setting $D=C+\frac{1}{n}$, by Levi theorem, we find

$$\int_0^{\infty}\frac{u^{2\lambda-1}}{(u^{\lambda}+A)(u^{\lambda}+B)(u^{\lambda}+C)^2}du$$

$$=\lim_{n\to\infty}\int_0^{\infty}\frac{u^{2\lambda-1}du}{(u^{\lambda}+A)(u^{\lambda}+B)(u^{\lambda}+C)(u^{\lambda}+C+\frac{1}{n})}=\tilde{k}_{4\lambda}. \qquad (5.4.10)$$

Similar to (5.4.4)-(5.4.8), for $0<\lambda\le\frac{1}{2}$, $0<A<B<C$, (1) if $p>1$, then we have the following equivalent inequalities:

$$\sum_{n=1}^{\infty}\sum_{m=1}^{\infty}\frac{a_m b_n}{(m^{\lambda}+An^{\lambda})(m^{\lambda}+Bn^{\lambda})(m^{\lambda}+Cn^{\lambda})^2}$$

$$<\tilde{k}_{4\lambda}\,||a||_{p,\tilde{\psi}_p}||b||_{q,\tilde{\varphi}_q}, \qquad (5.4.11)$$

$$\sum_{n=1}^{\infty}n^{2p\lambda-1}[\sum_{m=1}^{\infty}\frac{a_m}{(m^{\lambda}+An^{\lambda})(m^{\lambda}+Bn^{\lambda})(m^{\lambda}+Cn^{\lambda})^2}]^p$$

$$<\tilde{k}_{4\lambda}^{p}\,||a||_{p,\tilde{\psi}_p}^{p}; \qquad (5.4.12)$$

(2) if $0<p<1$, then we have the following reverse equivalent inequalities:

$$\sum_{n=1}^{\infty}\sum_{m=1}^{\infty}\frac{a_m b_n}{(m^{\lambda}+An^{\lambda})(m^{\lambda}+Bn^{\lambda})(m^{\lambda}+Cn^{\lambda})^2}>\tilde{k}_{4\lambda}$$

$$\times\{\sum_{n=1}^{\infty}[1-\tilde{\theta}_{4\lambda}(n)]n^{p(1-2\lambda)-1}a_n^p\}^{\frac{1}{p}}\,||b||_{q,\tilde{\varphi}_q} \qquad (5.4.13)$$

$$\sum_{n=1}^{\infty}n^{2p\lambda-1}[\sum_{m=1}^{\infty}\frac{a_m}{(m^{\lambda}+An^{\lambda})(m^{\lambda}+Bn^{\lambda})(m^{\lambda}+Cn^{\lambda})^2}]^p$$

$$>\tilde{k}_{4\lambda}^{p}\sum_{n=1}^{\infty}[1-\tilde{\theta}_{4\lambda}(n)]n^{p(1-2\lambda)-1}a_n^p \qquad (5.4.14)$$

$$\sum_{m=1}^{\infty}\frac{m^{2q\lambda-1}}{[1-\tilde{\theta}_{4\lambda}(m)]^{q-1}}[\sum_{n=1}^{\infty}\frac{b_n}{(m^{\lambda}+An^{\lambda})(m^{\lambda}+Bn^{\lambda})(m^{\lambda}+Cn^{\lambda})^2}]^q$$

$$<\tilde{k}_{4\lambda}^{q}\,||b||_{q,\tilde{\varphi}_q}^{q}, \qquad (5.4.15)$$

where,

$$0<\tilde{\theta}_{4\lambda}(n)=\frac{1}{\tilde{k}_{4\lambda}}\int_n^{\infty}\frac{u^{2\lambda-1}du}{(u^{\lambda}+A)(u^{\lambda}+B)(u^{\lambda}+C)^2}$$

$$\le\frac{1}{2\lambda\tilde{k}_{4\lambda}n^{2\lambda}}.$$

Particular case 5.4.3 For $C=B$, in (5.4.9), setting $C=B+t\ (t>0)$, we find

$$k'_{4\lambda}:=\lim_{t\to 0^+}\tilde{k}_{4\lambda}=\frac{1}{\lambda}\lim_{t\to 0^+}[\frac{A\ln A}{(B-A)(B+t-A)^2}$$

$$-\frac{B\ln B}{(B-A)t^2}+\frac{[(B+t)^2-AB]\ln(B+t)}{(B+t-A)^2t^2}-\frac{1}{(B+t-A)t}]$$

$$=\frac{1}{\lambda}[\frac{A\ln(A/B)}{(B-A)^3}+\frac{B+A}{2(B-A)^2B}]. \qquad (5.4.16)$$

In (5.3.10), setting $C=B+\frac{1}{n}$, by Levi theorem, we have

$$\int_0^{\infty}\frac{u^{2\lambda-1}}{(u^{\lambda}+A)(u^{\lambda}+B)^3}du$$

$$=\lim_{n\to\infty}\int_0^{\infty}\frac{u^{2\lambda-1}du}{(u^{\lambda}+A)(u^{\lambda}+B)(u^{\lambda}+B+\frac{1}{n})^2}=k'_{4\lambda}. \qquad (5.4.17)$$

Similar to (5.4.11)-(5.4.15), for $0<\lambda\le\frac{1}{2}$, $0<A<B$, (1) if $p>1$, then we have the following equivalent inequalities:

$$\sum_{n=1}^{\infty}\sum_{m=1}^{\infty}\frac{a_m b_n}{(m^{\lambda}+An^{\lambda})(m^{\lambda}+Bn^{\lambda})^3}$$

$$<k'_{4\lambda}\,||a||_{p,\tilde{\psi}_p}||b||_{q,\tilde{\varphi}_q}, \qquad (5.4.18)$$

$$\sum_{n=1}^{\infty}n^{2p\lambda-1}[\sum_{m=1}^{\infty}\frac{a_m}{(m^{\lambda}+An^{\lambda})(m^{\lambda}+Bn^{\lambda})^3}]^p$$

$$<k'^{p}_{4\lambda}\,||a||_{p,\tilde{\psi}_p}^{p}; \qquad (5.4.19)$$

(2) if $0<p<1$, then we have the following reverse equivalent inequalities:

$$\sum_{n=1}^{\infty}\sum_{m=1}^{\infty}\frac{a_m b_n}{(m^{\lambda}+An^{\lambda})(m^{\lambda}+Bn^{\lambda})^3}>k'_{4\lambda}$$

$$\times\{\sum_{n=1}^{\infty}[1-\tilde{\theta}_{4\lambda}(n)]n^{p(1-2\lambda)-1}a_n^p\}^{\frac{1}{p}}\,||b||_{q,\tilde{\varphi}_q}, \qquad (5.4.20)$$

$$\sum_{n=1}^{\infty}n^{2p\lambda-1}[\sum_{m=1}^{\infty}\frac{a_m}{(m^{\lambda}+An^{\lambda})(m^{\lambda}+Bn^{\lambda})^3}]^p>k'^{p}_{4\lambda}$$

$$\times\sum_{n=1}^{\infty}[1-\tilde{\theta}_{4\lambda}(n)]n^{p(1-2\lambda)-1}a_n^p, \qquad (5.4.21)$$

$$\sum_{m=1}^{\infty}\frac{m^{2q\lambda-1}}{[1-\tilde{\theta}_{4\lambda}(m)]^{q-1}}[\sum_{n=1}^{\infty}\frac{b_n}{(m^{\lambda}+An^{\lambda})(m^{\lambda}+Bn^{\lambda})^3}]^q$$

$$<k'^{q}_{4\lambda}\,||b||_{q,\tilde{\varphi}_q}^{q}, \qquad (5.4.22)$$

where,

$$0<\tilde{\theta}_{4\lambda}(n)=\frac{1}{k'_{4\lambda}}\int_n^{\infty}\frac{u^{2\lambda-1}}{(u^{\lambda}+A)(u^{\lambda}+B)^3}du$$

$$\le \frac{1}{2\lambda k'_{4\lambda} n^{2\lambda}}.$$

Particular case 5.4.4 For $B = A$, in (5.4.9), setting $B = A + t \ (t > 0)$, we find

$$K_{4\lambda} := \lim_{t \to 0^+} \tilde{k}_{4\lambda}$$

$$= \frac{1}{\lambda} \lim_{t \to 0^+} [\frac{A \ln A}{t(C-A)^2} - \frac{(A+t)\ln(A+t)}{t(C-A-t)^2}$$

$$+ \frac{[C^2 - A(A+t)]\ln C}{(C-A)^2(C-A-t)^2} - \frac{1}{(C-A)(C-A-t)}]$$

$$= \frac{1}{\lambda}[\frac{-2A\ln A + (C+A)\ln C}{(C-A)^3} - \frac{2 + \ln A}{(C-A)^2}]. \quad (5.4.23)$$

In (5.4.10), setting $B = A + \frac{1}{n}$, by Levi theorem, we have

$$\int_0^\infty \frac{u^{2\lambda-1}}{(u^\lambda + A)^2 (u^\lambda + C)^2} du$$

$$= \lim_{n \to \infty} \int_0^\infty \frac{u^{2\lambda-1} du}{(u^\lambda + A)(u^\lambda + A + \frac{1}{n})(u^\lambda + C)^2} = K_{4\lambda}. \quad (5.4.24)$$

Similar to (5.4.18)-(5.4.22), for $0 < \lambda \le \frac{1}{2}$, $0 < A < C$, (1) if $p > 1$, then we have the following equivalent inequalities:

$$\sum_{n=1}^\infty \sum_{m=1}^\infty \frac{a_m b_n}{(m^\lambda + An^\lambda)^2 (m^\lambda + Cn^\lambda)^2}$$

$$< K_{4\lambda} \| a \|_{p,\tilde{\psi}_p} \| b \|_{q,\tilde{\psi}_q}, \quad (5.4.25)$$

$$\sum_{n=1}^\infty n^{2p\lambda-1} [\sum_{m=1}^\infty \frac{a_m}{(m^\lambda + An^\lambda)^2 (m^\lambda + Cn^\lambda)^2}]^p$$

$$< K_{4\lambda}^p \| a \|_{p,\tilde{\psi}_p}^p; \quad (5.4.26)$$

(2) if $0 < p < 1$, then we have the following reverse equivalent inequalities:

$$\sum_{n=1}^\infty \sum_{m=1}^\infty \frac{a_m b_n}{(m^\lambda + An^\lambda)^2 (m^\lambda + Cn^\lambda)^2} > K_{4\lambda}$$

$$\times \{\sum_{n=1}^\infty [1 - \tilde{\theta}_{4\lambda}(n)] n^{p(1-2\lambda)-1} a_n^p\}^{\frac{1}{p}} \| b \|_{q,\tilde{\psi}_q}, \quad (5.4.27)$$

$$\sum_{n=1}^\infty n^{2p\lambda-1} [\sum_{m=1}^\infty \frac{a_m}{(m^\lambda + An^\lambda)^2 (m^\lambda + Cn^\lambda)^2}]^p > K_{4\lambda}^p$$

$$\times \sum_{n=1}^\infty [1 - \tilde{\theta}_{4\lambda}(n)] n^{p(1-2\lambda)-1} a_n^p, \quad (5.4.28)$$

$$\sum_{m=1}^\infty \frac{m^{2q\lambda-1}}{[1 - \tilde{\theta}_{4\lambda}(m)]^{q-1}} [\sum_{n=1}^\infty \frac{b_n}{(m^\lambda + An^\lambda)^2 (m^\lambda + Bn^\lambda)^2}]^q$$

$$< K_{4\lambda}^q \| b \|_{q,\tilde{\psi}_q}^q, \quad (5.4.29)$$

where,

$$0 < \tilde{\theta}_{4\lambda}(n) = \frac{1}{K_{4\lambda}} \int_n^\infty \frac{u^{2\lambda-1}}{(u^\lambda + A)^2 (u^\lambda + C)^2} du$$

$$\le \frac{1}{2\lambda K_{4\lambda} n^{2\lambda}}.$$

Particular case 5.4.5 For $B = A$, in (5.4.16), setting $B = A + t \ (t > 0)$, we find

$$\tilde{K}_{4\lambda} := \lim_{t \to 0^+} k'_{4\lambda}$$

$$= \frac{1}{\lambda} \lim_{t \to 0^+} [\frac{A \ln[A/(A+t)]}{t^3} + \frac{2A+t}{2t^2(A+t)}]$$

$$= \frac{1}{\lambda} \lim_{t \to 0^+} [\frac{2A(A+t)\ln[A/(A+t)] + (2A+t)t}{2t^3(A+t)}] = \frac{1}{6\lambda A^2}. \quad (5.4.30)$$

In (5.4.17), setting $B = A + \frac{1}{n}$, by Levi theorem (Kuang HEP 1996)[5], we find

$$\int_0^\infty \frac{u^{2\lambda-1} du}{(u^\lambda + A)^4} = \lim_{n \to \infty} \int_0^\infty \frac{u^{2\lambda-1} du}{(u^\lambda + A)(u^\lambda + A + \frac{1}{n})^3} = \tilde{K}_{4\lambda}. \quad (5.4.31)$$

Similar to (5.4.25)-(5.4.29), for $0 < \lambda \le \frac{1}{2}, A > 0$,

(1) if $p > 1$, then we have the following equivalent inequalities:

$$\sum_{n=1}^\infty \sum_{m=1}^\infty \frac{a_m b_n}{(m^\lambda + An^\lambda)^4} < \frac{1}{6\lambda A^2} \| a \|_{p,\tilde{\psi}_p} \| b \|_{q,\tilde{\psi}_q}, \quad (5.4.32)$$

$$\sum_{n=1}^\infty n^{2p\lambda-1} [\sum_{m=1}^\infty \frac{a_m}{(m^\lambda + An^\lambda)^4}]^p$$

$$< (\frac{1}{6\lambda A^2})^p \| a \|_{p,\tilde{\psi}_p}^p; \quad (5.4.33)$$

(2) if $0 < p < 1$, then we have the following reverse equivalent inequalities:

$$\sum_{n=1}^\infty \sum_{m=1}^\infty \frac{a_m b_n}{(m^\lambda + An^\lambda)^4} > \frac{1}{6\lambda A^2}$$

$$\times \{\sum_{n=1}^\infty [1 - \tilde{\theta}_{4\lambda}(n)] n^{p(1-2\lambda)-1} a_n^p\}^{\frac{1}{p}} \| b \|_{q,\tilde{\psi}_q}, \quad (5.4.34)$$

$$\sum_{n=1}^\infty n^{2p\lambda-1} [\sum_{m=1}^\infty \frac{a_m}{(m^\lambda + An^\lambda)^4}]^p > (\frac{1}{6\lambda A^2})^p$$

$$\times \sum_{n=1}^\infty [1 - \tilde{\theta}_{4\lambda}(n)] n^{p(1-2\lambda)-1} a_n^p, \quad (5.4.35)$$

$$\sum_{m=1}^\infty \frac{m^{2q\lambda-1}}{[1 - \tilde{\theta}_{4\lambda}(m)]^{q-1}} [\sum_{n=1}^\infty \frac{b_n}{(m^\lambda + An^\lambda)^4}]^q$$

$$< (\frac{1}{6\lambda A^2})^q \| b \|_{q,\tilde{\psi}_q}^q, \quad (5.4.36)$$

where,

$$0 < \tilde{\theta}_{4\lambda}(n) = 6\lambda A^2 \int_n^\infty \frac{u^{2\lambda-1} du}{(u^\lambda + A)^4} \le \frac{3A^2}{n^{2\lambda}}.$$

5.4.2 HILBERT-TYPE INEQUALITY WITH THE KERNEL $\frac{1}{(m+An)(m+Bn)(m^2+Cn^2)}$ AND EXTENSIONS

Example 5.4.6 If $\lambda > 0, 0 < A < B$, $C > 0$,

$$k_{4\lambda}(x, y) = \frac{1}{(x^\lambda + Ay^\lambda)(x^\lambda + By^\lambda)(x^{2\lambda} + Cy^{2\lambda})},$$

then setting $v = u^\lambda$, we find

$$k_{4\lambda} := \int_0^\infty k_{4\lambda}(u,1)u^{2\lambda-1}du$$
$$= \int_0^\infty \frac{u^{2\lambda-1}}{(u^\lambda + A)(u^\lambda + B)(u^{2\lambda}+C)}du$$
$$= \frac{1}{\lambda}\int_0^\infty \frac{v}{(v+A)(v+B)(v^2+C)}dv. \quad (5.4.37)$$

Putting

$$\frac{v}{(v+A)(v+B)(v^2+C)} = \frac{\alpha}{v+A} + \frac{\beta}{v+B} + \frac{\chi v+\delta}{v^2+C},$$

we have

$$v = \alpha(v+B)(v^2+C) + \beta(v+A)(v^2+C)$$
$$+(\chi v+\delta)(v+A)(v+B).$$

Substitution of $v = -A, -B$, we obtain

$$\begin{cases} \alpha = \frac{-A}{(B-A)(A^2+C)}, \beta = \frac{B}{(B-A)(B^2+C)}, \\ \chi = \frac{AB-C}{(A^2+C)(B^2+C)}, \delta = \frac{(B+A)C}{(A^2+C)(B^2+C)}. \end{cases} \quad (5.4.38)$$

Since $\alpha+\beta+\chi = 0$, then we find

$$k_{4\lambda} := \frac{1}{\lambda}\int_0^\infty \left[\frac{\alpha}{v+A} + \frac{\beta}{v+B} + \frac{\chi v+\delta}{v^2+C}\right]dv$$
$$= \frac{1}{\lambda}\left[\int_0^\infty \left(\frac{\alpha}{v+A} + \frac{\beta}{v+B} + \frac{\chi v}{v^2+C}\right)dv + \int_0^\infty \frac{\delta}{v^2+C}dv\right]$$
$$= \frac{1}{\lambda}[\ln(v+A)^\alpha (v+B)^\beta (v^2+C)^{\chi/2} |_0^\infty$$
$$+\frac{\delta}{\sqrt{C}}\arctan v|_0^\infty]$$
$$= \frac{1}{\lambda}[-\ln(A^\alpha B^\beta C^{\chi/2}) + \frac{\delta\pi}{2\sqrt{C}}]$$
$$= \frac{1}{\lambda}[-\alpha\ln A - \beta\ln B - \frac{\chi}{2}\ln C + \frac{\delta\pi}{2\sqrt{C}}]$$
$$= \frac{1}{\lambda}\left[\frac{A\ln A}{(B-A)(A^2+C)} - \frac{B\ln B}{(B-A)(B^2+C)}\right.$$
$$\left.-\frac{(AB-C)\ln C}{2(A^2+C)(B^2+C)} + \frac{(A+B)\sqrt{C}\pi}{2(A^2+C)(B^2+C)}\right]. \quad (5.4.39)$$

By Corollary 5.1.2, replacing λ by 4λ, and $\phi_r(x) = x^{r(1-\frac{\lambda}{2})-1}$ by

$$\tilde{\psi}_r(x) = x^{r(1-2\lambda)-1} \ (r = p,q),$$

for $0<\lambda\le\frac{1}{2}, 0<A<B$, $C>0$, (1) if $p>1$, then we have the following equivalent inequalities:

$$\sum_{n=1}^\infty\sum_{m=1}^\infty \frac{a_m b_n}{(m^\lambda + An^\lambda)(m^\lambda + Bn^\lambda)(m^{2\lambda}+Cn^{2\lambda})}$$
$$< k_{4\lambda}\|a\|_{p,\tilde{\psi}_p}\|b\|_{q,\tilde{\psi}_q}, \quad (5.4.40)$$

$$\sum_{n=1}^\infty n^{2p\lambda-1}\left[\sum_{m=1}^\infty \frac{a_m}{(m^\lambda + An^\lambda)(m^\lambda + Bn^\lambda)(m^{2\lambda}+Cn^{2\lambda})}\right]^p$$
$$< k_{4\lambda}^p\|a\|_{p,\tilde{\psi}_p}^p; \quad (5.4.41)$$

(2) if $0<p<1$, then we have the following reverse equivalent inequalities:

$$\sum_{n=1}^\infty\sum_{m=1}^\infty \frac{a_m b_n}{(m^\lambda + An^\lambda)(m^\lambda + Bn^\lambda)(m^{2\lambda}+Cn^{2\lambda})} > k_{4\lambda}$$
$$\times\{\sum_{n=1}^\infty[1-\tilde{\theta}_{4\lambda}(n)]n^{p(1-2\lambda)-1}a_n^p\}^{\frac{1}{p}}\|b\|_{q,\tilde{\psi}_q}, \quad (5.4.42)$$

$$\sum_{n=1}^\infty n^{2p\lambda-1}\left[\sum_{m=1}^\infty \frac{a_m}{(m^\lambda + An^\lambda)(m^\lambda + Bn^\lambda)(m^{2\lambda}+Cn^{2\lambda})}\right]^p$$
$$> k_{4\lambda}^p\sum_{n=1}^\infty[1-\tilde{\theta}_{4\lambda}(n)]n^{p(1-2\lambda)-1}a_n^p, \quad (5.4.43)$$

$$\sum_{m=1}^\infty \frac{m^{2q\lambda-1}}{[1-\tilde{\theta}_{4\lambda}(m)]^{q-1}}\left[\sum_{n=1}^\infty \frac{b_n}{(m^\lambda + An^\lambda)(m^\lambda + Bn^\lambda)(m^{2\lambda}+Cn^{2\lambda})}\right]^q$$
$$< k_{4\lambda}^q\|b\|_{q,\tilde{\psi}_q}^q, \quad (5.4.44)$$

where,

$$0<\tilde{\theta}_{4\lambda}(n) = \frac{1}{k_{4\lambda}}\int_n^\infty \frac{u^{2\lambda-1}du}{(u^\lambda+A)(u^\lambda+B)(u^{2\lambda}+C)}$$
$$\le \frac{1}{2\lambda k_{4\lambda}n^{2\lambda}}.$$

Particular case 5.4.7 For $B = A$ in (5.4.39), setting $B = A+t \ (t>0)$, we obtain

$$K'_{4\lambda} := \lim_{t\to 0^+} k_{4\lambda}$$
$$= \frac{1}{\lambda}\lim_{t\to 0^+}\left\{\frac{A\ln A}{t(A^2+C)} - \frac{(A+t)\ln(A+t)}{t[(A+t)^2+C]}\right.$$
$$\left.-\frac{[A(A+t)-C]\ln C}{2(A^2+C)[(A+t)^2+C]} + \frac{(2A+t)\sqrt{C}\pi}{2(A^2+C)[(A+t)^2+C]}\right\}$$
$$= \frac{1}{\lambda}\lim_{t\to 0^+}\left\{\frac{A\ln A[(A+t)^2+C]-(A^2+C)(A+t)\ln(A+t)}{t(A^2+C)[(A+t)^2+C]}\right.$$
$$\left.-\frac{[A(A+t)-C]\ln C}{2(A^2+C)[(A+t)^2+C]} + \frac{(2A+t)\sqrt{C}\pi}{2(A^2+C)[(A+t)^2+C]}\right\}$$
$$= \frac{1}{\lambda(A^2+C)^2}[(A^2-C)(\ln A - \frac{1}{2}\ln C)$$
$$-A^2 - C + A\sqrt{C}\pi]. \quad (5.4.45)$$

In (5.4.37), setting $B = A+\frac{1}{n}$, by Levi theorem, we have

$$\int_0^\infty \frac{u^{2\lambda-1}}{(u^\lambda+A)^2(u^{2\lambda}+C)}du$$
$$= \lim_{n\to\infty}\int_0^\infty \frac{u^{2\lambda-1}}{(u^\lambda+A)(u^\lambda+A+\frac{1}{n})(u^{2\lambda}+C)}du = K'_{4\lambda}.$$

Similar to (5.4.40)-(5.4.44), for $0<\lambda\le\frac{1}{2}, A, C>0$, (1) if $p>1$, then we have the following equivalent inequalities:

$$\sum_{n=1}^\infty\sum_{m=1}^\infty \frac{a_m b_n}{(m^\lambda + An^\lambda)^2(m^{2\lambda}+Cn^{2\lambda})}$$
$$< K'_{4\lambda}\|a\|_{p,\tilde{\psi}_p}\|b\|_{q,\tilde{\psi}_q}, \quad (5.4.46)$$

$$\sum_{n=1}^\infty n^{2p\lambda-1}\left[\sum_{m=1}^\infty \frac{a_m}{(m^\lambda + An^\lambda)^2(m^{2\lambda}+Cn^{2\lambda})}\right]^p$$
$$< K'^p_{4\lambda}\|a\|_{p,\tilde{\psi}_p}^p; \quad (5.4.47)$$

(2) if $0<p<1$, then we have the following reverse equivalent inequalities:

$$\sum_{n=1}^{\infty}\sum_{m=1}^{\infty}\frac{a_m b_n}{(m^{\lambda}+An^{\lambda})^2(m^{2\lambda}+Cn^{2\lambda})} > K'_{4\lambda}$$

$$\times\{\sum_{n=1}^{\infty}[1-\tilde{\theta}_{4\lambda}(n)]n^{p(1-2\lambda)-1}a_n^p\}^{\frac{1}{p}}\|b\|_{q,\tilde{\psi}_q}, \quad (5.4.48)$$

$$\sum_{n=1}^{\infty}n^{2p\lambda-1}[\sum_{m=1}^{\infty}\frac{a_m}{(m^{\lambda}+An^{\lambda})^2(m^{2\lambda}+Cn^{2\lambda})}]^p > K'^p_{4\lambda}$$

$$\times\sum_{n=1}^{\infty}[1-\tilde{\theta}_{4\lambda}(n)]n^{p(1-2\lambda)-1}a_n^p, \quad (5.4.49)$$

$$\sum_{m=1}^{\infty}\frac{m^{2q\lambda-1}}{[1-\tilde{\theta}_{4\lambda}(m)]^{q-1}}[\sum_{n=1}^{\infty}\frac{b_n}{(m^{\lambda}+An^{\lambda})^2(m^{2\lambda}+Cn^{2\lambda})}]^q$$

$$< K'^q_{4\lambda}\|b\|^q_{q,\tilde{\psi}_q}, \quad (5.4.50)$$

where,

$$0<\tilde{\theta}_{4\lambda}(n)=\frac{1}{K'_{4\lambda}}\int_n^{\infty}\frac{u^{2\lambda-1}du}{(u^{\lambda}+A)^2(u^{2\lambda}+C)}$$

$$\le\frac{1}{2\lambda K'_{4\lambda}n^{2\lambda}}.$$

5.4.3 HILBERT-TYPE INEQUALITY WITH THE KERNEL $\frac{1}{(m^2+An^2)(m^2+Bn^2)}$ AND EXTENSIONS

Example 5.4.8 If $\lambda>0, 0<A<B$,

$$k_{4\lambda}(x,y)=\frac{1}{(x^{2\lambda}+Ay^{2\lambda})(x^{2\lambda}+By^{2\lambda})},$$

then setting $v=u^{\lambda}$, we find

$$k_{4\lambda}=\int_0^{\infty}k_{4\lambda}(u,1)u^{2\lambda-1}du$$

$$=\int_0^{\infty}\frac{u^{2\lambda-1}}{(u^{2\lambda}+A)(u^{2\lambda}+B)}du$$

$$=\frac{1}{\lambda}\int_0^{\infty}\frac{v}{(v^2+A)(v^2+B)}dv$$

$$=\frac{1}{\lambda(B-A)}\int_0^{\infty}(\frac{v}{v^2+A}-\frac{v}{v^2+B})dv$$

$$=\frac{\ln(B/A)}{2\lambda(B-A)}. \quad (5.4.51)$$

By Corollary 5.1.2, replacing λ by 4λ, and $\phi_r(x)=x^{r(1-\frac{\lambda}{2})-1}$ by

$$\tilde{\psi}_r(x)=x^{r(1-2\lambda)-1}\ (r=p,q),$$

for $0<\lambda\le\frac{1}{2}, 0<A<B$, (1) if $p>1$, then we have the following equivalent inequalities:

$$\sum_{n=1}^{\infty}\sum_{m=1}^{\infty}\frac{a_m b_n}{(m^{2\lambda}+An^{2\lambda})(m^{2\lambda}+Bn^{2\lambda})}$$

$$<\frac{\ln(B/A)}{2\lambda(B-A)}\|a\|_{p,\tilde{\psi}_p}\|b\|_{q,\tilde{\psi}_q}, \quad (5.4.52)$$

$$\sum_{n=1}^{\infty}n^{2p\lambda-1}[\sum_{m=1}^{\infty}\frac{a_m}{(m^{2\lambda}+An^{2\lambda})(m^{2\lambda}+Bn^{2\lambda})}]^p$$

$$<[\frac{\ln(B/A)}{2\lambda(B-A)}]^p\|a\|^p_{p,\tilde{\psi}_p}; \quad (5.4.53)$$

(2) if $0<p<1$, then we have the following reverse equivalent inequalities:

$$\sum_{n=1}^{\infty}\sum_{m=1}^{\infty}\frac{a_m b_n}{(m^{2\lambda}+An^{2\lambda})(m^{2\lambda}+Bn^{2\lambda})} > \frac{\ln(B/A)}{2\lambda(B-A)}$$

$$\times\{\sum_{n=1}^{\infty}[1-\tilde{\theta}_{4\lambda}(n)]n^{p(1-2\lambda)-1}a_n^p\}^{\frac{1}{p}}\|b\|_{q,\tilde{\psi}_q}, \quad (5.4.54)$$

$$\sum_{n=1}^{\infty}n^{2p\lambda-1}[\sum_{m=1}^{\infty}\frac{a_m}{(m^{2\lambda}+An^{2\lambda})(m^{2\lambda}+Bn^{2\lambda})}]^p$$

$$>[\frac{\ln(B/A)}{2\lambda(B-A)}]^p\sum_{n=1}^{\infty}[1-\tilde{\theta}_{4\lambda}(n)]n^{p(1-2\lambda)-1}a_n^p, \quad (5.4.55)$$

$$\sum_{m=1}^{\infty}\frac{m^{2q\lambda-1}}{[1-\tilde{\theta}_{4\lambda}(m)]^{q-1}}[\sum_{n=1}^{\infty}\frac{b_n}{(m^{2\lambda}+An^{2\lambda})(m^{2\lambda}+Bn^{2\lambda})}]^q$$

$$<[\frac{\ln(B/A)}{2\lambda(B-A)}]^q\|b\|^q_{q,\tilde{\psi}_q}, \quad (5.4.56)$$

where,

$$0<\tilde{\theta}_{4\lambda}(n)=\frac{2\lambda(B-A)}{\ln(B/A)}\int_n^{\infty}\frac{u^{2\lambda-1}du}{(u^{2\lambda}+A)(u^{2\lambda}+B)}$$

$$\le\frac{B-A}{\ln(B/A)n^{2\lambda}}.$$

Particular case 5.4.9 For $B=A$, in (5.4.51), setting $B=A+t\ (t>0)$, we find

$$K'_{4\lambda}:=\lim_{t\to0^+}k_{4\lambda}=\lim_{t\to0^+}\frac{\ln[(A+t)/A]}{2\lambda t}=\frac{1}{2A\lambda}.$$

By Levi theorem, we have

$$\int_0^{\infty}\frac{u^{2\lambda-1}du}{(u^{2\lambda}+A)^2}=\lim_{n\to\infty}\int_0^{\infty}\frac{u^{2\lambda-1}du}{(u^{2\lambda}+A)(u^{2\lambda}+A+\frac{1}{n})}=K'_{4\lambda}.$$

Similar to (5.4.52)-(5.4.56), for $0<\lambda\le\frac{1}{2}, A>0$,

(1) if $p>1$, then we have the following equivalent inequalities:

$$\sum_{n=1}^{\infty}\sum_{m=1}^{\infty}\frac{a_m b_n}{(m^{2\lambda}+An^{2\lambda})^2}<\frac{1}{2\lambda A}\|a\|_{p,\tilde{\psi}_p}\|b\|_{q,\tilde{\psi}_q}, \quad (5.4.57)$$

$$\sum_{n=1}^{\infty}n^{2p\lambda-1}[\sum_{m=1}^{\infty}\frac{a_m}{(m^{2\lambda}+An^{2\lambda})^2}]^p$$

$$<\frac{1}{(2\lambda A)^p}\|a\|^p_{p,\tilde{\psi}_p}; \quad (5.4.58)$$

(2) if $0<p<1$, then we have the following reverse equivalent inequalities:

$$\sum_{n=1}^{\infty}\sum_{m=1}^{\infty}\frac{a_m b_n}{(m^{2\lambda}+An^{2\lambda})^2}>\frac{1}{2\lambda A}$$

$$\times\{\sum_{n=1}^{\infty}[1-\tilde{\theta}_{4\lambda}(n)]n^{p(1-2\lambda)-1}a_n^p\}^{\frac{1}{p}}\|b\|_{q,\tilde{\psi}_q}, \quad (5.4.59)$$

$$\sum_{n=1}^{\infty} n^{2p\lambda-1}[\sum_{m=1}^{\infty} \frac{a_m}{(m^{2\lambda}+An^{2\lambda})^2}]^p > \frac{1}{(2\lambda A)^p}$$

$$\times\sum_{n=1}^{\infty}[1-\tilde{\theta}_{4\lambda}(n)]n^{p(1-2\lambda)-1}a_n^p, \quad (5.4.60)$$

$$\sum_{m=1}^{\infty} \frac{m^{2q\lambda-1}}{[1-\tilde{\theta}_{4\lambda}(m)]^{q-1}}[\sum_{n=1}^{\infty} \frac{b_n}{(m^{2\lambda}+An^{2\lambda})^2}]^q$$

$$< \frac{1}{(2\lambda A)^q} \| b \|_{q,\tilde{\psi}_q}^q, \quad (5.4.61)$$

where,

$$0 < \tilde{\theta}_{4\lambda}(n) = 2\lambda A \int_n^{\infty} \frac{u^{2\lambda-1}du}{(u^{2\lambda}+A)^2} \le \frac{A}{n^{2\lambda}}.$$

5.5 SOME HILBERT-TYPE INEQUALITIES WITH THE GENERAL MEASURABLE KERNEL

5.5.1 SOME BASIC RESULTS

Theorem 5.5.1 Assuming that $p>0(p\ne 1)$, $\frac{1}{p}+\frac{1}{q}=1$, $h(t)(\ge 0)$ is a finite measurable function in $\mathbf{R}_+$, $\alpha\in\mathbf{R}$, satisfying

$$0<k_\alpha := \int_0^{\infty} h(t)t^{\alpha-1}dt < \infty,$$

we define the following weight coefficient:

$$\omega_\alpha(m) := \sum_{n=1}^{\infty} h(mn)\frac{m^\alpha}{n^{1-\alpha}}, m\in\mathbf{N}. \quad (5.5.1)$$

If $a_n, b_n \ge 0 (n\in \mathbf{N})$, and there exists $0<\theta_\alpha(m)<1$, such that

$$k_\alpha[1-\theta_\alpha(m)] \le \omega_\alpha(m) < k_\alpha,\ m\in\mathbf{N}, \quad (5.5.2)$$

then (1) for $p>1$, $0<\sum_{n=1}^{\infty} n^{p(1-\alpha)-1}a_n^p<\infty$ and $0<\sum_{n=1}^{\infty} n^{q(1-\alpha)-1}b_n^q<\infty$, we have the following equivalent inequalities:

$$I := \sum_{n=1}^{\infty}\sum_{m=1}^{\infty} h(mn)a_m b_n$$

$$< k_\alpha \{\sum_{n=1}^{\infty} n^{p(1-\alpha)-1}a_n^p\}^{\frac{1}{p}} \{\sum_{n=1}^{\infty} n^{q(1-\alpha)-1}b_n^q\}^{\frac{1}{q}}, \quad (5.5.3)$$

$$J := \sum_{n=1}^{\infty} n^{p\alpha-1}(\sum_{m=1}^{\infty} h(mn)a_m)^p$$

$$< k_\alpha^p \sum_{n=1}^{\infty} n^{p(1-\alpha)-1}a_n^p; \quad (5.5.4)$$

(2) for $0<p<1$,

$0<\sum_{n=1}^{\infty}[1-\theta_\alpha(n)]n^{p(1-\alpha)-1}a_n^p<\infty$ and $0<\sum_{n=1}^{\infty} n^{q(1-\alpha)-1}b_n^q<\infty$, we have the following reverse equivalent inequalities:

$$I > k_\alpha \{\sum_{n=1}^{\infty}[1-\theta_\alpha(n)]n^{p(1-\alpha)-1}a_n^p\}^{\frac{1}{p}}$$

$$\times\{\sum_{n=1}^{\infty} n^{q(1-\alpha)-1}b_n^q\}^{\frac{1}{q}}, \quad (5.5.5)$$

$$J > k_\alpha^p \sum_{n=1}^{\infty}[1-\theta_\alpha(n)]n^{p(1-\alpha)-1}a_n^p, \quad (5.5.6)$$

$$L := \sum_{m=1}^{\infty} \frac{m^{q\alpha-1}}{[1-\theta_\alpha(m)]^{q-1}}(\sum_{n=1}^{\infty} h(mn)b_n)^q$$

$$< k_\alpha^q \sum_{n=1}^{\infty} n^{q(1-\alpha)-1}b_n^q. \quad (5.5.7)$$

Proof (1) For $p>1$, by Hölder's inequality (Kuang SCP 2004)[6] and (5.5.1), we have

$$I = \sum_{n=1}^{\infty}\sum_{m=1}^{\infty} h(mn)[\frac{m^{(1-\alpha)/q}}{n^{(1-\alpha)/p}}a_m][\frac{n^{(1-\alpha)/p}}{m^{(1-\alpha)/q}}b_n]$$

$$\le \{\sum_{m=1}^{\infty}\sum_{n=1}^{\infty} h(mn)\frac{m^{(1-\alpha)(p-1)}}{n^{1-\alpha}}a_m^p\}^{\frac{1}{p}}$$

$$\times\{\sum_{n=1}^{\infty}\sum_{m=1}^{\infty} h(mn)\frac{n^{(1-\alpha)(q-1)}}{m^{1-\alpha}}b_n^q\}^{\frac{1}{q}}$$

$$= \{\sum_{m=1}^{\infty} \omega_\alpha(m)m^{p(1-\alpha)-1}a_m^p\}^{\frac{1}{p}}$$

$$\times\{\sum_{n=1}^{\infty} \omega_\alpha(n)n^{q(1-\alpha)-1}b_n^q\}^{\frac{1}{q}}. \quad (5.5.8)$$

By (5.5.2), we have (5.5.3).

If $J=0$, then (5.5.4) is naturally valid; if $J>0$, then there exists a $N\in\mathbf{N}$, satisfying

$$J(N) := \sum_{n=1}^{N} n^{p\alpha-1}(\sum_{m=1}^{N} h(mn)a_m)^p > 0,$$

$$\sum_{n=1}^{N} n^{p(1-\alpha)-1}a_n^p > 0,$$

and $b_n(N) = n^{p\alpha-1}(\sum_{m=1}^{N} h(mn)a_m)^{p-1} > 0$. By (5.5.3), we have the following inequalities:

$$0<\sum_{n=1}^{N} n^{q(1-\alpha)-1}b_n^q(N) = J(N)$$

$$= \sum_{n=1}^{N}\sum_{m=1}^{N} h(mn)a_m b_n(N)$$

$$< k_\alpha \{\sum_{n=1}^{N} n^{p(1-\alpha)-1}a_n^p\}^{\frac{1}{p}}$$

$$\times\{\sum_{n=1}^{N} n^{q(1-\alpha)-1} b_n^q(N)\}^{\frac{1}{q}} < \infty, \quad (5.5.9)$$

$$0 < \sum_{n=1}^{N} n^{q(1-\alpha)-1} b_n^q(N) = J(N)$$

$$< k_\alpha^p \sum_{n=1}^{\infty} n^{p(1-\alpha)-1} a_n^p. \quad (5.5.10)$$

It follows that $0 < \sum_{n=1}^{\infty} n^{q(1-\alpha)-1} b_n^q(\infty) < \infty$ and for $N \to \infty$, by using (5.5.3), both (5.5.9) and (5.5.10) still keep the strict-sign inequalities. Hence (5.5.4) is valid. On the other-hand, suppose (5.5.4) is valid. By Hölder's inequality, we have

$$I = \sum_{n=1}^{\infty} [n^{\frac{-1}{p}+\alpha} \sum_{m=1}^{\infty} h(mn) a_m][n^{\frac{1}{p}-\alpha} b_n]$$

$$\leq J^{\frac{1}{p}} \{\sum_{n=1}^{\infty} n^{q(1-\alpha)-1} b_n{}^q\}^{\frac{1}{q}}. \quad (5.5.11)$$

By (5.5.4), we have (5.5.3), which is equivalent to (5.5.4).

(2) For $0 < p < 1$, by the reverse Hölder's inequality, we find the reverse of (5.5.8), and then by (5.5.2), in view of $q < 0$, we have (5.5.5).

It is obvious that $J > 0$. If $J = \infty$, then (5.5.6) is naturally valid; if $0 < J < \infty$, then setting $b_n = n^{p\alpha-1}(\sum_{m=1}^{\infty} h(mn) a_m)^{p-1}$ ($n \in$ **N**), by (5.5.5), we have

$$\sum_{n=1}^{\infty} n^{q(1-\alpha)-1} b_n^q = J = I$$

$$> k_\alpha \{\sum_{n=1}^{\infty} [1-\theta_\alpha(n)] n^{p(1-\alpha)-1} a_n^p\}^{\frac{1}{p}}$$

$$\times\{\sum_{n=1}^{\infty} n^{q(1-\alpha)-1} b_n^q\}^{\frac{1}{q}}, \quad (5.5.12)$$

$$J^{\frac{1}{p}} = \{\sum_{n=1}^{\infty} n^{q(1-\alpha)-1} b_n^q\}^{\frac{1}{p}}$$

$$> k_\alpha \{\sum_{n=1}^{\infty} [1-\theta_\alpha(n)] n^{p(1-\alpha)-1} a_n^p\}^{\frac{1}{p}}. \quad (5.5.13)$$

Hence we have (5.5.6). On the other hand, suppose (5.5.6) is valid. By the reverse Hölder's inequality, we have the reverse of (5.5.11), then by (5.5.6), we have (5.5.5), which is equivalent to (5.5.6).

If $L = 0$, then (5.5.7) is naturally valid; if $L > 0$, then there exists a $N \in$ **N**, such that

$$L(N) := \sum_{m=1}^{N} \frac{m^{q\alpha-1}}{[1-\theta_\alpha(m)]^{q-1}} (\sum_{n=1}^{N} h(mn) b_n)^q > 0,$$

$$\sum_{n=1}^{N} n^{q(1-\alpha)-1} b_n^q > 0,$$

and

$$a_m(N) := \frac{m^{q\alpha-1}}{[1-\theta_\alpha(m)]^{q-1}} (\sum_{n=1}^{N} h(mn) b_n)^{q-1} > 0.$$

By (5.5.5), we have

$$\sum_{m=1}^{N} [1-\theta_\alpha(m)] m^{p(1-\alpha)-1} a_m^p(N) = L(N)$$

$$= \sum_{n=1}^{N} \sum_{m=1}^{N} h(mn) a_m(N) b_n$$

$$> k_\alpha \{\sum_{n=1}^{N} [1-\theta_\alpha(m)] m^{p(1-\alpha)-1} a_m^p(N)\}^{\frac{1}{p}}$$

$$\times\{\sum_{n=1}^{N} n^{q(1-\alpha)-1} b_n^q\}^{\frac{1}{q}} > 0,$$

$$\sum_{m=1}^{N} [1-\theta_\alpha(m)] m^{p(1-\alpha)-1} a_m^p(N) = L(N)$$

$$< k_\alpha \sum_{n=1}^{\infty} n^{q(1-\alpha)-1} b_n^q < \infty.$$

Hence it follows

$$0 < \sum_{m=1}^{\infty} [1-\theta_\alpha(m)] m^{p(1-\alpha)-1} a_m^p(\infty) < \infty.$$

By using (5.5.5), for $N \to \infty$, the above two inequalities still keep the forms of strict-sign inequalities. Therefore (5.5.7) is valid.

On the other-hand, suppose that (5.5.7) is valid. By the reverse Hölder's inequality, we have

$$I = \sum_{m=1}^{\infty} \{\frac{m^{\frac{-1}{q}+\alpha}}{[1-\theta_\alpha(m)]^{\frac{1}{p}}} \sum_{n=1}^{\infty} h(mn) b_n\}$$

$$\times\{[1-\theta_\alpha(m)]^{\frac{1}{p}} m^{\frac{1}{q}-\alpha} a_m\}$$

$$\geq L^{\frac{1}{q}} \{\sum_{m=1}^{\infty} [1-\theta_\alpha(m)] m^{p(1-\alpha)-1} a_m{}^p\}^{\frac{1}{p}}. \quad (5.5.14)$$

By (5.5.7), we have (5.5.5), which is equivalent to (5.5.7). Hence (5.5.5), (5.5.6) and (5.5.7) are equivalent. □

If $\lambda \in \mathbf{R}$, $k_\lambda(x,y)(\geq 0)$ is a finite homogeneous function of degree $-\lambda$ in $\mathbf{R}_+^2$, then setting $t = \frac{1}{v}$, we find

$$k_{\lambda/2} = \int_0^\infty k_\lambda(1,t) t^{\frac{\lambda}{2}-1} dt$$

$$= \int_0^\infty k_\lambda(v,1) v^{\frac{\lambda}{2}-1} dv.$$

For $h(t) = k_\lambda(1,t), \alpha = \frac{\lambda}{2}$ in Theorem 5.5.1, we have the following corollary:

Corollary 5.5.2 Assuming that $p > 0 (p \neq 1)$, $\frac{1}{p} + \frac{1}{q} = 1, 0 < k_{\lambda/2} = \int_0^\infty k_\lambda(1,t) t^{\frac{\lambda}{2}-1} dt < \infty$, we define the following weight coefficient:

$$\omega_{\lambda/2}(m) := \sum_{n=1}^\infty k_\lambda(1,mn) \frac{m^{\lambda/2}}{n^{1-\lambda/2}}, \quad m \in \mathbf{N}. \tag{5.5.15}$$

If $a_n, b_n \geq 0$, there exists $0 < \theta_{\lambda/2}(m) < 1$, such that

$$k_{\lambda/2}[1 - \theta_{\lambda/2}(m)] \leq \omega_{\lambda/2}(m) < k_{\lambda/2}, \ m \in \mathbf{N}, \tag{5.5.16}$$

then (1) for $p > 1$, $0 < \sum_{n=1}^\infty n^{p(1-\frac{\lambda}{2})-1} a_n^p < \infty$ and $0 < \sum_{n=1}^\infty n^{q(1-\frac{\lambda}{2})-1} b_n^q < \infty$, we have the following equivalent inequalities:

$$\tilde{I} := \sum_{n=1}^\infty \sum_{m=1}^\infty k_\lambda(1,mn) a_m b_n < k_{\frac{\lambda}{2}} \times \{\sum_{n=1}^\infty n^{p(1-\frac{\lambda}{2})-1} a_n^p\}^{\frac{1}{p}} \{\sum_{n=1}^\infty n^{q(1-\frac{\lambda}{2})-1} b_n^q\}^{\frac{1}{q}}, \tag{5.5.17}$$

$$\tilde{J} := \sum_{n=1}^\infty n^{\frac{p\lambda}{2}-1} (\sum_{m=1}^\infty k_\lambda(1,mn) a_m)^p < k_{\frac{\lambda}{2}}^p \sum_{n=1}^\infty n^{p(1-\frac{\lambda}{2})-1} a_n^p; \tag{5.5.18}$$

(2) for $0 < p < 1$, $0 < \sum_{n=1}^\infty [1 - \theta_{\lambda/2}(n)] n^{p(1-\frac{\lambda}{2})-1} a_n^p < \infty$ and $0 < \sum_{n=1}^\infty n^{q(1-\frac{\lambda}{2})-1} b_n^q < \infty$, we have the following reverse equivalent inequalities:

$$\tilde{I} > k_{\frac{\lambda}{2}} \{\sum_{n=1}^\infty [1 - \theta_{\frac{\lambda}{2}}(n)] n^{p(1-\frac{\lambda}{2})-1} a_n^p\}^{\frac{1}{p}} \times \{\sum_{n=1}^\infty n^{q(1-\frac{\lambda}{2})-1} b_n^q\}^{\frac{1}{q}}, \tag{5.5.19}$$

$$\tilde{J} > k_{\frac{\lambda}{2}}^p \sum_{n=1}^\infty [1 - \theta_{\frac{\lambda}{2}}(n)] n^{p(1-\frac{\lambda}{2})-1} a_n^p, \tag{5.5.20}$$

$$\tilde{L} := \sum_{m=1}^\infty \frac{m^{(q\lambda/2)-1}}{[1-\theta_{\lambda/2}(m)]^{q-1}} (\sum_{n=1}^\infty k_\lambda(1,mn) b_n)^q < k_{\frac{\lambda}{2}}^q \sum_{n=1}^\infty n^{q(1-\frac{\lambda}{2})-1} b_n^q; \tag{5.5.21}$$

(3) replacing $k_\lambda(1,mn)$ by $k_\lambda(mn,1)$ in (5.5.15) – (5.5.21), if (5.5.16) keeps valid, then (5.5.17)-(5.5.21) also keep valid and keep the same equivalent property.

Corollary 5.5.3 If $h(t)$ is a finite measurable function in R_+ and $\alpha \in \mathbf{R}$, $0 < k_\alpha = \int_0^\infty h(t) t^{\alpha-1} dt < \infty$, $h(t)t^{\alpha-1} (> 0)$ is decreasing for $t > 0$ and strictly decreasing in a subinterval, then we have the results of Theorem 5.5.1; if $h(t)(> 0)$ is decreasing for $t > 0$ and strictly decreasing in a subinterval, then for $0 < \alpha \leq 1$, we still have the results of Theorem 5.5.1.

Proof We find

$$\omega_\alpha(m) = \sum_{n=1}^\infty h(mn) \frac{m^\alpha}{n^{1-\alpha}} < \int_0^\infty h(my) \frac{m^\alpha}{y^{1-\alpha}} dy.$$

Setting $t = my$, we have $\omega_\alpha(m) < k_\alpha$. We still have

$$\omega_\alpha(m) \geq \int_1^\infty h(my) \frac{m^\alpha}{y^{1-\alpha}} dy = \int_m^\infty h(t) t^{\alpha-1} dt = k_\alpha - \int_0^m h(t) t^{\alpha-1} dt = k_\alpha [1 - \theta_\alpha(m)],$$

where, it follows

$$0 < \theta_\alpha(m) := \frac{1}{k_\alpha} \int_0^m h(t) t^{\alpha-1} dt < 1.$$

If $0 < \alpha \leq 1$, $h(t)(> 0)$ is decreasing for $t > 0$ and strictly decreasing in a subinterval, then $h(t)t^{\alpha-1} (> 0)$ still keeps the same decreasing property, and we have the results of Theorem 5.5.1. □

Example 5.5.4 For $0 < \alpha \leq 1$, setting $h(t) = e^{-\beta t}$ $(\beta > 0)$, we find

$$k_\alpha = \int_0^\infty e^{-\beta t} t^{\alpha-1} dt = \frac{1}{\beta^\alpha} \int_0^\infty e^{-v} v^{\alpha-1} dv = \frac{1}{\beta^\alpha} \Gamma(\alpha) > 0,$$

and $0 < \theta_\alpha(n) := \frac{\beta^\alpha}{\Gamma(\alpha)} \int_0^n e^{-\beta u} u^{\alpha-1} du < 1$.

By Corollary 5.5.3, (1) for $p > 1$, $0 < \sum_{n=1}^\infty n^{p(1-\alpha)-1} a_n^p < \infty$ and $0 < \sum_{n=1}^\infty n^{q(1-\alpha)-1} b_n^q < \infty$, we have the following equivalent inequalities:

$$I = \sum_{n=1}^\infty \sum_{m=1}^\infty \frac{a_m b_n}{e^{\beta mn}} < \frac{1}{\beta^\alpha} \Gamma(\alpha)$$

$$\times\{\sum_{n=1}^{\infty} n^{p(1-\alpha)-1} a_n^p\}^{\frac{1}{p}} \{\sum_{n=1}^{\infty} n^{q(1-\alpha)-1} b_n^q\}^{\frac{1}{q}}, \quad (5.5.22)$$

$$J = \sum_{n=1}^{\infty} n^{p\alpha-1} (\sum_{m=1}^{\infty} \frac{a_m}{e^{\beta mn}})^p$$

$$< [\frac{1}{\beta^\alpha}\Gamma(\alpha)]^p \sum_{n=1}^{\infty} n^{p(1-\alpha)-1} a_n^p; \quad (5.5.23)$$

(2) for $0 < p < 1$,

$0 < \sum_{n=1}^{\infty} [1-\theta_\alpha(n)] n^{p(1-\alpha)-1} a_n^p < \infty$ and $0 < \sum_{n=1}^{\infty} n^{q(1-\alpha)-1} b_n^q < \infty$, we have the following reverse equivalent inequalities:

$$I > \frac{1}{\beta^\alpha}\Gamma(\alpha)\{\sum_{n=1}^{\infty} [1-\theta_\alpha(n)] n^{p(1-\alpha)-1} a_n^p\}^{\frac{1}{p}}$$

$$\times\{\sum_{n=1}^{\infty} n^{q(1-\alpha)-1} b_n^q\}^{\frac{1}{q}}, \quad (5.5.24)$$

$$J > [\frac{1}{\beta^\alpha}\Gamma(\alpha)]^p \sum_{n=1}^{\infty} [1-\theta_\alpha(n)] n^{p(1-\alpha)-1} a_n^p, \quad (5.5.25)$$

$$\sum_{m=1}^{\infty} \frac{m^{q\alpha-1}}{[1-\theta_\alpha(m)]^{q-1}} (\sum_{n=1}^{\infty} \frac{b_n}{e^{\beta mn}})^q$$

$$< [\frac{1}{\beta^\alpha}\Gamma(\alpha)]^q \sum_{n=1}^{\infty} n^{q(1-\alpha)-1} b_n^q. \quad (5.5.26)$$

Corollary 5.5.5 If $\lambda \in \mathbf{R}$, $k_\lambda(x,y)$ is a finite homogeneous function of degree $-\lambda$ in $\mathbf{R}_+^2$, such that $k_\lambda(1,t)t^{\frac{\lambda}{2}-1} > 0$ (or $k_\lambda(t,1)t^{\frac{\lambda}{2}-1} > 0$) is decreasing for $t > 0$ and strictly decreasing in a subinterval, then we have the results of Theorem 5.5.1 with the kernel

$$h(mn) = k_\lambda(1,mn) \text{ (or } k_\lambda(mn,1) \text{)};$$

if $k_\lambda(1,t) > 0$ (or $k_\lambda(t,1) > 0$) is decreasing for $t > 0$ and strictly decreasing in a subinterval, then for $0 < \lambda \le 2$, we have the results of Theorem 5.5.1 with the kernel $k_\lambda(1,mn)$ (or $k_\lambda(mn,1)$).

Note 5.5.6 We still can't prove that the best possible property of the constant factors in the above new inequalities.

In the following, assuming that $p > 1, \frac{1}{p}+\frac{1}{q}=1$, $\alpha \in \mathbf{R}$, $\tilde{\phi}(x) = x^{p(1-\alpha)-1}$, $\tilde{\psi}(x) = x^{q(1-\alpha)-1}$, $x \in (0,\infty)$ and $[\tilde{\psi}(x)]^{1-p} = x^{p\alpha-1}$, we define the following real sequences space:

$$l_{\tilde{\phi}}^p = \{a = \{a_n\}_{n=1}^\infty; \|a\|_{p,\tilde{\phi}}$$

$$= \{\sum_{n=1}^\infty \tilde{\phi}(n)|a_n|^p\}^{\frac{1}{p}} < \infty\}.$$

By the same way, we may define the spaces $l_{\tilde{\psi}}^q$ and $l_{\tilde{\psi}^{1-p}}^p$.

For $a = \{a_m\}_{m=1}^\infty \in l_{\tilde{\phi}}^p$, if $c_n := \sum_{m=1}^\infty h(mn)a_m$ $(n \in \mathbf{N})$, then by (5.5.4), we have $c = \{c_n\}_{n=1}^\infty \in l_{\tilde{\psi}^{1-p}}^p$. Define a linear operator $T: l_{\tilde{\phi}}^p \to l_{\tilde{\psi}^{1-p}}^p$ as: for any $a = \{a_m\}_{m=1}^\infty \in l_{\tilde{\phi}}^p$,

$$(Ta)(n) = c_n = \sum_{m=1}^\infty h(mn)a_m, \ n \in \mathbf{N}. \quad (5.5.27)$$

The norm of Ta is expressing as

$$\|Ta\|_{p,\tilde{\psi}^{1-p}} = \{\sum_{n=1}^\infty n^{p\alpha-1}(\sum_{m=1}^\infty h(mn)a_m)^p\}^{\frac{1}{p}}. \quad (5.5.28)$$

For $b = \{b_n\}_{n=1}^\infty \in l_{\tilde{\psi}}^q$, define the following formal inner product of Ta and b:

$$(Ta,b) := \sum_{n=1}^\infty \sum_{m=1}^\infty h(mn)a_m b_n. \quad (5.5.29)$$

If $\|a\|_{p,\tilde{\phi}}, \|b\|_{q,\tilde{\psi}} > 0$, then we may express (5.5.3) and (5.5.4) as the following equivalent forms :

$$(Ta,b) < k_\alpha \|a\|_{p,\tilde{\phi}} \|b\|_{q,\tilde{\psi}}; \quad (5.5.30)$$

$$\|Ta\|_{p,\tilde{\psi}^{1-p}} < k_\alpha \|a\|_{p,\tilde{\phi}}. \quad (5.5.31)$$

In view of (5.5.18), it follows that the linear operator T defined by (5.5.27) is bounded and the norm satisfies $\|T\|_{p,\tilde{\psi}^{1-p}} \le k_\alpha$.

5.5.2 SOME EXAMPLES FOR APPLYING COROLLARY 5.5.5

In the following examples, suppose that $p > 0$ $(p \ne 1)$, $\frac{1}{p}+\frac{1}{q}=1$, $\lambda \in \mathbf{R}$, $\tilde{\phi}(x) = x^{p(1-\frac{\lambda}{2})-1}$, $\tilde{\psi}(x) = x^{q(1-\frac{\lambda}{2})-1}$, $a_n, b_n \ge 0$ and for $p > 1$,

$$0 < \|a\|_{p,\tilde{\phi}} = \{\sum_{n=1}^\infty n^{p(1-\frac{\lambda}{2})-1} a_n^p\}^{\frac{1}{p}} < \infty,$$

$$0 < \|b\|_{q,\tilde{\psi}} = \{\sum_{n=1}^\infty n^{q(1-\frac{\lambda}{2})-1} b_n^q\}^{\frac{1}{q}} < \infty;$$

for $0 < p < 1, 0 < \|b\|_{q,\tilde{\psi}} < \infty$ and

$$0 < \{\sum_{n=1}^\infty [1-\theta_{\frac{\lambda}{2}}(n)] n^{p(1-\frac{\lambda}{2})-1} a_n^p\}^{\frac{1}{p}} < \infty..$$

Example 5.5.7 If $\alpha>0, 0<\lambda\le 2$, $k_\lambda(1,t) = \frac{1}{(1+t^\alpha)^{\lambda/\alpha}}$, then we find $k_{\lambda/2}=\frac{1}{\alpha}B(\frac{\lambda}{2\alpha},\frac{\lambda}{2\alpha})$.

By Corollary 5.5.5, (1) for $p>1$, we have the following equivalent inequalities:

$$\sum_{n=1}^\infty\sum_{m=1}^\infty \frac{a_m b_n}{(1+m^\alpha n^\alpha)^{\lambda/\alpha}} < \frac{1}{\alpha}B(\frac{\lambda}{2\alpha},\frac{\lambda}{2\alpha})\|a\|_{p,\tilde{\phi}}\|b\|_{q,\tilde{\psi}}, \quad (5.5.32)$$

$$\sum_{n=1}^\infty n^{\frac{p\lambda}{2}-1}[\sum_{m=1}^\infty \frac{a_m}{(1+m^\alpha n^\alpha)^{\lambda/\alpha}}]^p < [\frac{1}{\alpha}B(\frac{\lambda}{2\alpha},\frac{\lambda}{2\alpha})]^p\|a\|_{p,\tilde{\phi}}^p; \quad (5.5.33)$$

(1) for $0<p<1$, setting

$$\theta_{\lambda/2}(n):=\frac{\alpha}{B(\frac{\lambda}{2\alpha},\frac{\lambda}{2\alpha})}\int_0^n \frac{1}{(1+t^\alpha)^{\lambda/\alpha}}t^{\frac{\lambda}{2}-1}dt,$$

we have the following reverse equivalent inequalities:

$$\sum_{n=1}^\infty\sum_{m=1}^\infty \frac{a_m b_n}{(1+m^\alpha n^\alpha)^{\lambda/\alpha}} > \frac{1}{\alpha}B(\frac{\lambda}{2\alpha},\frac{\lambda}{2\alpha})$$

$$\times\{\sum_{n=1}^\infty[1-\theta_{\lambda/2}(n)]n^{p(1-\frac{\lambda}{2})-1}a_n^p\}^{\frac{1}{p}}\|b\|_{q,\tilde{\psi}}, \quad (5.5.34)$$

$$\sum_{n=1}^\infty n^{\frac{p\lambda}{2}-1}[\sum_{m=1}^\infty \frac{a_m}{(1+m^\alpha n^\alpha)^{\lambda/\alpha}}]^p > [\frac{1}{\alpha}B(\frac{\lambda}{2\alpha},\frac{\lambda}{2\alpha})]^p$$

$$\times\sum_{n=1}^\infty[1-\theta_{\lambda/2}(n)]n^{p(1-\frac{\lambda}{2})-1}a_n^p, \quad (5.5.35)$$

$$\sum_{m=1}^\infty \frac{m^{(q\lambda/2)-1}}{[1-\theta_{\lambda/2}(m)]^{q-1}}(\sum_{n=1}^\infty \frac{b_n}{(1+m^\alpha n^\alpha)^{\lambda/\alpha}})^q < [\frac{1}{\alpha}B(\frac{\lambda}{2\alpha},\frac{\lambda}{2\alpha})]^q\|b\|_{q,\tilde{\psi}}^q. \quad (5.5.36)$$

In particular, (1) for $0<\alpha=\lambda\le 2$, (i) if $p>1$, then we have the following equivalent inequalities:

$$\sum_{n=1}^\infty\sum_{m=1}^\infty \frac{a_m b_n}{1+m^\lambda n^\lambda} < \frac{\pi}{\lambda}\|a\|_{p,\tilde{\phi}}\|b\|_{q,\tilde{\psi}}, \quad (5.5.37)$$

$$\sum_{n=1}^\infty n^{\frac{p\lambda}{2}-1}(\sum_{m=1}^\infty \frac{a_m}{1+m^\lambda n^\lambda})^p < (\frac{\pi}{\lambda})^p\|a\|_{p,\tilde{\phi}}^p; \quad (5.5.38)$$

(ii) if $0<p<1$, setting $\theta'_{\lambda/2}(n):=\frac{\lambda}{\pi}\int_0^n \frac{1}{1+t^\lambda}t^{\frac{\lambda}{2}-1}dt$,

then we have the following reverse equivalent inequalities:

$$\sum_{n=1}^\infty\sum_{m=1}^\infty \frac{a_m b_n}{1+m^\lambda n^\lambda} > \frac{\pi}{\lambda}$$

$$\times\{\sum_{n=1}^\infty[1-\theta'_{\lambda/2}(n)]n^{p(1-\frac{\lambda}{2})-1}a_n^p\}^{\frac{1}{p}}\|b\|_{q,\tilde{\psi}}, \quad (5.5.39)$$

$$\sum_{n=1}^\infty n^{\frac{p\lambda}{2}-1}(\sum_{m=1}^\infty \frac{a_m}{1+m^\lambda n^\lambda})^p > (\frac{\pi}{\lambda})^p$$

$$\times\sum_{n=1}^\infty[1-\theta'_{\lambda/2}(n)]n^{p(1-\frac{\lambda}{2})-1}a_n^p, \quad (5.5.40)$$

$$\sum_{m=1}^\infty \frac{m^{(q\lambda/2)-1}}{[1-\theta'_{\lambda/2}(m)]^{q-1}}(\sum_{n=1}^\infty \frac{b_n}{1+(mn)^\lambda})^q < (\frac{\pi}{\lambda})^q\|b\|_{q,\tilde{\psi}}^q; \quad (5.5.41)$$

(2) for $\alpha=1, 0<\lambda\le 2$, (i) if $p>1$, then we have the following equivalent inequalities:

$$\sum_{n=1}^\infty\sum_{m=1}^\infty \frac{a_m b_n}{(1+mn)^\lambda} < B(\frac{\lambda}{2},\frac{\lambda}{2})\|a\|_{p,\tilde{\phi}}\|b\|_{q,\tilde{\psi}}, \quad (5.5.42)$$

$$\sum_{n=1}^\infty n^{\frac{p\lambda}{2}-1}[\sum_{m=1}^\infty \frac{a_m}{(1+mn)^\lambda}]^p < [B(\frac{\lambda}{2},\frac{\lambda}{2})]^p\|a\|_{p,\tilde{\phi}}^p; \quad (5.5.43)$$

(ii) if $0<p<1$, setting

$$\tilde{\theta}_{\lambda/2}(n):=\frac{1}{B(\frac{\lambda}{2},\frac{\lambda}{2})}\int_0^n \frac{1}{(1+t)^\lambda}t^{\frac{\lambda}{2}-1}dt,$$

then we have the following reverse equivalent inequalities:

$$\sum_{n=1}^\infty\sum_{m=1}^\infty \frac{a_m b_n}{(1+mn)^\lambda} > B(\frac{\lambda}{2},\frac{\lambda}{2})$$

$$\times\{\sum_{n=1}^\infty[1-\tilde{\theta}_{\lambda/2}(n)]n^{p(1-\frac{\lambda}{2})-1}a_n^p\}^{\frac{1}{p}}\|b\|_{q,\tilde{\psi}}, \quad (5.5.44)$$

$$\sum_{n=1}^\infty n^{\frac{p\lambda}{2}-1}[\sum_{m=1}^\infty \frac{a_m}{(1+mn)^\lambda}]^p > [B(\frac{\lambda}{2},\frac{\lambda}{2})]^p$$

$$\times\sum_{n=1}^\infty[1-\tilde{\theta}_{\lambda/2}(n)]n^{p(1-\frac{\lambda}{2})-1}a_n^p, \quad (5.5.45)$$

$$\sum_{m=1}^\infty \frac{m^{\frac{q\lambda}{2}-1}}{[1-\tilde{\theta}_{\lambda/2}(m)]^{q-1}}[\sum_{n=1}^\infty \frac{b_n}{(1+mn)^\lambda}]^q < [B(\frac{\lambda}{2},\frac{\lambda}{2})]^q\|b\|_{q,\tilde{\psi}}^q. \quad (5.5.46)$$

Example 5.5.8 If $0<\alpha\le\lambda\le 2$ (or $0=\alpha<\lambda\le 2$), $k_\lambda(x,y)=\frac{1}{(x+y)^{\lambda-\alpha}(\max\{x,y\})^\alpha}$,

$$k_\lambda(1,t)=\frac{1}{(1+t)^{\lambda-\alpha}(\max\{1,t\})^\alpha},$$

then we find

$$k_{\lambda/2}:=k_\lambda(\alpha)=\int_0^\infty \frac{u^{\frac{\lambda}{2}-1}}{(1+t)^{\lambda-\alpha}(\max\{1,t\})^\alpha}du$$

$$=\sum_{k=0}^\infty \binom{\alpha-\lambda}{k}\frac{4}{\lambda+2k}. \quad (5.5.47)$$

By Corollary 5.5.5, (1) for $p>1$, we have the following equivalent inequalities:

$$\sum_{n=1}^{\infty}\sum_{m=1}^{\infty}\frac{a_m b_n}{(1+mn)^{\lambda-\alpha}(mn)^{\alpha}}$$

$$< k_{\lambda}(\alpha)\|a\|_{p,\tilde{\phi}}\|b\|_{q,\tilde{\psi}}, \qquad (5.5.48)$$

$$\sum_{n=1}^{\infty}n^{\frac{p\lambda}{2}-1}[\sum_{m=1}^{\infty}\frac{a_m}{(1+mn)^{\lambda-\alpha}(mn)^{\alpha}}]^p$$

$$< [k_{\lambda}(\alpha)]^p\|a\|_{p,\tilde{\phi}}^p; \qquad (5.5.49)$$

(2) for $0 < p < 1$, setting

$$\theta_{\lambda/2}(n) := \frac{1}{k_{\lambda}(\alpha)}\int_0^n\frac{1}{(1+t)^{\lambda-\alpha}(\max\{1,t\})^{\alpha}}t^{\frac{\lambda}{2}-1}dt,$$

we have the following reverse equivalent inequalities:

$$\sum_{n=1}^{\infty}\sum_{m=1}^{\infty}\frac{a_m b_n}{(1+mn)^{\lambda-\alpha}(mn)^{\alpha}} > k_{\lambda}(\alpha)$$

$$\times\{\sum_{n=1}^{\infty}[1-\theta_{\lambda/2}(n)]n^{p(1-\frac{\lambda}{2})-1}a_n^p\}^{\frac{1}{p}}\|b\|_{q,\tilde{\psi}}, \quad (5.5.50)$$

$$\sum_{n=1}^{\infty}n^{\frac{p\lambda}{2}-1}[\sum_{m=1}^{\infty}\frac{a_m}{(1+mn)^{\lambda-\alpha}(mn)^{\alpha}}]^p > [k_{\lambda}(\alpha)]^p$$

$$\times\sum_{n=1}^{\infty}[1-\theta_{\lambda/2}(n)]n^{p(1-\frac{\lambda}{2})-1}a_n^p, \quad (5.5.51)$$

$$\sum_{m=1}^{\infty}\frac{m^{\frac{q\lambda}{2}-1}}{[1-\theta_{\lambda/2}(m)]^{q-1}}[\sum_{n=1}^{\infty}\frac{b_n}{(1+mn)^{\lambda-\alpha}(mn)^{\alpha}}]^q$$

$$< [k_{\lambda}(\alpha)]^q\|b\|_{q,\tilde{\psi}}^q. \qquad (5.5.52)$$

In particular, (1) for $0 < \alpha = \lambda \le 2$, $k_{\lambda}(\lambda) = \frac{4}{\lambda}$,

(i) if $p > 1$, then we have the following equivalent inequalities:

$$\sum_{n=1}^{\infty}\sum_{m=1}^{\infty}\frac{a_m b_n}{(mn)^{\lambda}} < \frac{4}{\lambda}\|a\|_{p,\tilde{\phi}}\|b\|_{q,\tilde{\psi}}, \qquad (5.5.53)$$

$$\sum_{n=1}^{\infty}\frac{1}{n^{\frac{p\lambda}{2}+1}}(\sum_{m=1}^{\infty}\frac{a_m}{m^{\lambda}})^p < (\frac{4}{\lambda})^p\|a\|_{p,\tilde{\phi}}^p; \quad (5.5.54)$$

(ii) if $0 < p < 1$, setting

$$\tilde{\theta}_{\lambda/2}(n) := \frac{\lambda}{4}\int_0^n\frac{1}{(\max\{1,t\})^{\lambda}}t^{\frac{\lambda}{2}-1}dt = 1-\frac{1}{2n^{\lambda/2}},$$

then we have the following reverse equivalent inequalities:

$$\sum_{n=1}^{\infty}\sum_{m=1}^{\infty}\frac{a_m b_n}{(mn)^{\lambda}} > \frac{2}{\lambda}\{\sum_{n=1}^{\infty}n^{p(1-\frac{\lambda}{2})-(1+\frac{\lambda}{2})}a_n^p\}^{\frac{1}{p}}\|b\|_{q,\tilde{\psi}},$$

$$(5.5.55)$$

$$\sum_{n=1}^{\infty}\frac{1}{n^{\frac{p\lambda}{2}+1}}(\sum_{m=1}^{\infty}\frac{a_m}{m^{\lambda}})^p > \frac{1}{2}(\frac{4}{\lambda})^p\sum_{n=1}^{\infty}n^{p(1-\frac{\lambda}{2})-(1+\frac{\lambda}{2})}a_n^p,$$

$$(5.5.56)$$

$$\sum_{m=1}^{\infty}\frac{1}{m^{(\lambda/2)+1}}(\sum_{n=1}^{\infty}\frac{b_n}{n^{\lambda}})^q < 2(\frac{2}{\lambda})^q\|b\|_{q,\tilde{\psi}}^q. \quad (5.5.57)$$

(2) For $\alpha = \frac{1}{2}, \lambda = 1$, we find

$$k_1(\tfrac{1}{2}) = 4\int_0^1\frac{1}{(1+x^2)^{1/2}}dx$$

$$= 4\ln|x+(1+x^2)^{\frac{1}{2}}|\,|_0^1 = 4\ln(1+\sqrt{2}).$$

(i) If $p > 1$, then we have the following equivalent inequalities:

$$\sum_{n=1}^{\infty}\sum_{m=1}^{\infty}\frac{a_m b_n}{\sqrt{(1+mn)mn}} < 4\ln(1+\sqrt{2})$$

$$\times\{\sum_{n=1}^{\infty}n^{\frac{p}{2}-1}a_n^p\}^{\frac{1}{p}}\{\sum_{n=1}^{\infty}n^{\frac{q}{2}-1}b_n^q\}^{\frac{1}{q}}, \qquad (5.5.58)$$

$$\sum_{n=1}^{\infty}\frac{1}{n}[\sum_{m=1}^{\infty}\frac{a_m}{\sqrt{(1+mn)m}}]^p$$

$$< [4\ln(1+\sqrt{2})]^p\sum_{n=1}^{\infty}n^{\frac{p}{2}-1}a_n^p; \qquad (5.5.59)$$

(ii) if $0 < p < 1$, setting

$$\tilde{\theta}_{1/2}(n) := \frac{1}{4\ln(1+\sqrt{2})}\int_0^n\frac{1}{(1+t)^{1/2}(\max\{1,t\})^{1/2}}t^{\frac{-1}{2}}dt$$

$$= \frac{1}{4\ln(1+\sqrt{2})}[\int_0^1\frac{1}{(1+t)^{1/2}}t^{\frac{-1}{2}}dt + \int_1^n\frac{1}{(1+t)^{1/2}t}dt]$$

$$= 1-\frac{1}{2\ln(1+\sqrt{2})}\ln\frac{\sqrt{n+1}+1}{\sqrt{n}},$$

then we have the following reverse equivalent inequalities:

$$\sum_{n=1}^{\infty}\sum_{m=1}^{\infty}\frac{a_m b_n}{\sqrt{(1+mn)mn}} > 2\{\sum_{n=1}^{\infty}(\ln\frac{\sqrt{n+1}+1}{\sqrt{n}})n^{\frac{p}{2}-1}a_n^p\}^{\frac{1}{p}}$$

$$\times\{\sum_{n=1}^{\infty}n^{\frac{q}{2}-1}b_n^q\}^{\frac{1}{q}}, \qquad (5.5.60)$$

$$\sum_{n=1}^{\infty}\frac{1}{n}(\sum_{m=1}^{\infty}\frac{a_m}{\sqrt{(1+mn)m}})^p > 2^{2p-1}[\ln(1+\sqrt{2})]^{p-1}$$

$$\times\sum_{n=1}^{\infty}(\ln\frac{\sqrt{n+1}+1}{\sqrt{n}})n^{\frac{p}{2}-1}a_n^p, \qquad (5.5.61)$$

$$\sum_{m=1}^{\infty}\frac{1}{m[\ln\frac{\sqrt{m+1}+1}{\sqrt{m}}]^{q-1}}(\sum_{n=1}^{\infty}\frac{b_n}{\sqrt{(1+mn)n}})^q$$

$$> 2^{q+1}\ln(1+\sqrt{2})\{\sum_{n=1}^{\infty}n^{\frac{q}{2}-1}b_n^q\}^{\frac{1}{q}}. \qquad (5.5.62)$$

Example 5.5.9 If $0 \le \alpha < \lambda \le 2$,

$$k_{\lambda}(x,y) = \frac{1}{(x^{\lambda-\alpha}+y^{\lambda-\alpha})(\max\{x,y\})^{\alpha}},$$

$$k_{\lambda}(1,t) = \frac{1}{(1+t^{\lambda-\alpha})(\max\{1,t\})^{\alpha}}, \text{ then we find}$$

$$k_{\lambda/2} = \tilde{k}_{\lambda}(\alpha) := \int_0^{\infty}\frac{t^{\frac{\lambda}{2}-1}}{(1+t^{\lambda-\alpha})(\max\{1,t\})^{\alpha}}dt$$

$$=4\sum_{k=0}^{\infty}\frac{(-1)^k}{(2k+1)\lambda-2k\alpha}.$$

By Corollary 5.5.5, (1) for $p>1$, we have the following equivalent inequalities:

$$\sum_{n=1}^{\infty}\sum_{m=1}^{\infty}\frac{a_m b_n}{(1+m^{\lambda-\alpha}n^{\lambda-\alpha})(mn)^{\alpha}}$$

$$<\tilde{k}_\lambda(\alpha)\|a\|_{p,\tilde{\phi}}\|b\|_{q,\tilde{\psi}}, \quad (5.5.63)$$

$$\sum_{n=1}^{\infty}n^{p(\frac{\lambda}{2}-\alpha)-1}[\sum_{m=1}^{\infty}\frac{a_m}{(1+m^{\lambda-\alpha}n^{\lambda-\alpha})m^{\alpha}}]^p$$

$$<[\tilde{k}_\lambda(\alpha)]^p\|a\|_{p,\tilde{\phi}}^p; \quad (5.5.64)$$

(2) for $0<p<1$, setting

$$\tilde{\theta}_{\lambda/2}(n):=\frac{1}{\tilde{k}_\lambda(\alpha)}\int_0^n\frac{1}{(1+t^{\lambda-\alpha})(\max\{1,t\})^{\alpha}}t^{\frac{\lambda}{2}-1}dt,$$

we have the following reverse equivalent inequalities:

$$\sum_{n=1}^{\infty}\sum_{m=1}^{\infty}\frac{a_m b_n}{(1+m^{\lambda-\alpha}n^{\lambda-\alpha})(mn)^{\alpha}}>\tilde{k}_\lambda(\alpha)$$

$$\times\{\sum_{n=1}^{\infty}[1-\tilde{\theta}_{\lambda/2}(n)]n^{p(1-\frac{\lambda}{2})-1}a_n^p\}^{\frac{1}{p}}\|b\|_{q,\tilde{\psi}}, \quad (5.5.65)$$

$$\sum_{n=1}^{\infty}n^{p(\frac{\lambda}{2}-\alpha)-1}[\sum_{m=1}^{\infty}\frac{a_m}{(1+m^{\lambda-\alpha}n^{\lambda-\alpha})m^{\alpha}}]^p>[\tilde{k}_\lambda(\alpha)]^p$$

$$\times\sum_{n=1}^{\infty}[1-\tilde{\theta}_{\lambda/2}(n)]n^{p(1-\frac{\lambda}{2})-1}a_n^p, \quad (5.5.66)$$

$$\sum_{m=1}^{\infty}\frac{m^{q(\frac{\lambda}{2}-\alpha)-1}}{[1-\tilde{\theta}_{\lambda/2}(m)]^{q-1}}[\sum_{n=1}^{\infty}\frac{b_n}{(1+m^{\lambda-\alpha}n^{\lambda-\alpha})n^{\alpha}}]^q$$

$$<[\tilde{k}_\lambda(\alpha)]^q\|b\|_{q,\tilde{\psi}}^q. \quad (5.5.67)$$

In particular, for $\alpha=\frac{1}{2},\lambda=1$,

$$\tilde{k}_1(\tfrac{1}{2})=4\sum_{k=0}^{\infty}\frac{(-1)^k}{1+k}=4\ln 2,$$

(1) if $p>1$, then we have the following equivalent inequalities:

$$\sum_{n=1}^{\infty}\sum_{m=1}^{\infty}\frac{a_m b_n}{(1+\sqrt{mn})\sqrt{mn}}<4\ln 2$$

$$\times\{\sum_{n=1}^{\infty}n^{\frac{p}{2}-1}a_n^p\}^{\frac{1}{p}}\{\sum_{n=1}^{\infty}n^{\frac{q}{2}-1}b_n^q\}^{\frac{1}{q}}, \quad (5.5.68)$$

$$\sum_{n=1}^{\infty}\frac{1}{n}(\sum_{m=1}^{\infty}\frac{a_m}{(1+\sqrt{mn})\sqrt{m}})^p$$

$$<(4\ln 2)^p\sum_{n=1}^{\infty}n^{\frac{p}{2}-1}a_n^p; \quad (5.5.69)$$

(2) if $0<p<1$, setting

$$\tilde{\theta}_{1/2}(n):=\frac{1}{4\ln 2}\int_0^n\frac{1}{(1+t^{1/2})(\max\{1,t\})^{1/2}}t^{\frac{-1}{2}}dt$$

$$=\frac{1}{4\ln 2}[\int_0^1\frac{1}{1+t^{1/2}}t^{\frac{-1}{2}}dt+\int_1^n\frac{1}{(1+t^{1/2})t}dt]$$

$$=\frac{1}{2\ln 2}[2\ln 2-\ln(1+\tfrac{1}{\sqrt{n}})],$$

then we have the following reverse equivalent inequalities:

$$\sum_{n=1}^{\infty}\sum_{m=1}^{\infty}\frac{a_m b_n}{(1+\sqrt{mn})\sqrt{mn}}>2\{\sum_{n=1}^{\infty}\ln(1+\tfrac{1}{\sqrt{n}})n^{\frac{p}{2}-1}a_n^p\}^{\frac{1}{p}}$$

$$\times\{\sum_{n=1}^{\infty}n^{\frac{q}{2}-1}b_n^q\}^{\frac{1}{q}}, \quad (5.5.70)$$

$$\sum_{n=1}^{\infty}\frac{1}{n}(\sum_{m=1}^{\infty}\frac{a_m}{(1+\sqrt{mn})\sqrt{m}})^p>2(4\ln 2)^{p-1}$$

$$\times\sum_{n=1}^{\infty}\ln(1+\tfrac{1}{\sqrt{n}})n^{\frac{p}{2}-1}a_n^p, \quad (5.5.71)$$

$$\sum_{m=1}^{\infty}\frac{1}{m[\ln(1+\frac{1}{\sqrt{m}})]^{q-1}}(\sum_{n=1}^{\infty}\frac{b_n}{(1+\sqrt{mn})\sqrt{n}})^q$$

$$<2^{q+1}\ln 2\{\sum_{n=1}^{\infty}n^{\frac{q}{2}-1}b_n^q\}^{\frac{1}{q}}. \quad (5.5.72)$$

Example 5.5.10 If $0<\alpha\le\lambda\le 2$,

$$k_\lambda(x,y)=\frac{\ln(x/y)}{(x^{\alpha}-y^{\alpha})(\max\{x,y\})^{\lambda-\alpha}},$$

$$k_\lambda(1,t)=\frac{\ln t}{(t^{\alpha}-1)(\max\{1,t\})^{\lambda-\alpha}},$$

then we find

$$k_{\lambda/2}=k_\lambda(\alpha):=\int_0^{\infty}\frac{t^{\frac{\lambda}{2}-1}\ln t}{(t^{\alpha}-1)(\max\{1,t\})^{\lambda-\alpha}}dt$$

$$=2\int_0^1\frac{-\ln t}{1-t^{\alpha}}t^{\frac{\lambda}{2}-1}dt=2\int_0^1(-\ln t)\sum_{k=0}^{\infty}t^{k\alpha+\frac{\lambda}{2}-1}dt$$

$$=\sum_{k=0}^{\infty}\frac{2}{k\alpha+\frac{\lambda}{2}}\int_0^1(-\ln t)dt^{k\alpha+\frac{\lambda}{2}}=\sum_{k=0}^{\infty}\frac{8}{(2k\alpha+\lambda)^2}. \quad (5.5.73)$$

By Corollary 5.5.5, (1) for $p>1$, we have the following equivalent inequalities:

$$\sum_{n=1}^{\infty}\sum_{m=1}^{\infty}\frac{\ln(mn)a_m b_n}{(m^{\alpha}n^{\alpha}-1)(mn)^{\lambda-\alpha}}$$

$$<k_\lambda(\alpha)\|a\|_{p,\tilde{\phi}}\|b\|_{q,\tilde{\psi}}, \quad (5.5.74)$$

$$\sum_{n=1}^{\infty}\frac{1}{n^{p(\frac{\lambda}{2}-\alpha)+1}}[\sum_{m=1}^{\infty}\frac{\ln(mn)a_m}{(m^{\alpha}n^{\alpha}-1)m^{\lambda-\alpha}}]^p$$

$$<[k_\lambda(\alpha)]^p\|a\|_{p,\tilde{\phi}}^p; \quad (5.5.75)$$

(3) for $0<p<1$, setting

$$\tilde{\theta}_{\lambda/2}(n):=\frac{1}{\tilde{k}_\lambda(\alpha)}\int_0^n\frac{\ln t}{(t^{\alpha}-1)(\max\{1,t\})^{\lambda-\alpha}}t^{\frac{\lambda}{2}-1}dt,$$

we have the following reverse equivalent inequalities:

$$\sum_{n=1}^{\infty}\sum_{m=1}^{\infty}\frac{\ln(mn)a_m b_n}{(m^\alpha n^\alpha -1)(mn)^{\lambda-\alpha}} > k_\lambda(\alpha)$$

$$\times\{\sum_{n=1}^{\infty}[1-\tilde{\theta}_{\lambda/2}(n)]n^{p(1-\frac{\lambda}{2})-1}\}^{\frac{1}{p}} \| b \|_{q,\tilde{\psi}}, \quad (5.5.76)$$

$$\sum_{n=1}^{\infty}\frac{1}{n^{p(\frac{\lambda}{2}-\alpha)+1}}[\sum_{m=1}^{\infty}\frac{\ln(mn)}{(m^\alpha n^\alpha -1)m^{\lambda-\alpha}}a_m]^p$$

$$> [k_\lambda(\alpha)]^p \sum_{n=1}^{\infty}[1-\tilde{\theta}_{\lambda/2}(n)]n^{p(1-\frac{\lambda}{2})-1}, \quad (5.5.77)$$

$$\sum_{m=1}^{\infty}\frac{m^{q(\alpha-\frac{\lambda}{2})-1}}{(1-\tilde{\theta}_{\lambda/2}(m))^{q-1}}[\sum_{n=1}^{\infty}\frac{\ln(mn)}{(m^\alpha n^\alpha -1)n^{\lambda-\alpha}}b_n]^q$$

$$< [k_\lambda(\alpha)]^q \| b \|_{q,\tilde{\psi}}^q. \quad (5.5.78)$$

In particular, (1) for $0<\alpha=\lambda\le 2$, $k_\lambda(\lambda)=(\frac{\pi}{\lambda})^2$, (i) if $p>1$, then we have the following equivalent inequalities:

$$\sum_{n=1}^{\infty}\sum_{m=1}^{\infty}\frac{\ln(mn)a_m b_n}{m^\lambda n^\lambda -1}$$

$$< (\tfrac{\pi}{\lambda})^2 \| a \|_{p,\tilde{\phi}} \| b \|_{q,\tilde{\psi}}, \quad (5.5.79)$$

$$\sum_{n=1}^{\infty} n^{\frac{p\lambda}{2}-1}[\sum_{m=1}^{\infty}\frac{\ln(mn)a_m}{m^\lambda n^\lambda -1}]^p$$

$$< (\tfrac{\pi}{\lambda})^{2p} \| a \|_{p,\tilde{\phi}}^p; \quad (5.5.80)$$

(ii) if $0<p<1$, setting

$$\tilde{\theta}'_{\lambda/2}(n) := (\tfrac{\lambda}{\pi})^2 \int_0^n \frac{\ln t}{t^\lambda -1} t^{\frac{\lambda}{2}-1} dt,$$

then we have the following reverse equivalent inequalities:

$$\sum_{n=1}^{\infty}\sum_{m=1}^{\infty}\frac{\ln(mn)a_m b_n}{m^\lambda n^\lambda -1} > (\tfrac{\pi}{\lambda})^2$$

$$\times\{\sum_{n=1}^{\infty}[1-\tilde{\theta}'_{\lambda/2}(n)]n^{p(1-\frac{\lambda}{2})-1}a_n^p\}^{\frac{1}{p}} \| b \|_{q,\tilde{\psi}}, \quad (5.5.81)$$

$$\sum_{n=1}^{\infty} n^{\frac{p\lambda}{2}-1}[\sum_{m=1}^{\infty}\frac{\ln(mn)a_m}{m^\lambda n^\lambda -1}]^p > (\tfrac{\pi}{\lambda})^{2p}$$

$$\times\sum_{n=1}^{\infty}[1-\tilde{\theta}'_{\lambda/2}(n)]n^{p(1-\frac{\lambda}{2})-1}a_n^p, \quad (5.5.82)$$

$$\sum_{m=1}^{\infty}\frac{m^{\frac{q\lambda}{2}-1}}{[1-\tilde{\theta}'_{\lambda/2}(m)]^{q-1}}[\sum_{n=1}^{\infty}\frac{\ln(mn)b_n}{m^\lambda n^\lambda -1}]^q$$

$$> (\tfrac{\pi}{\lambda})^{2q} \| b \|_{q,\tilde{\psi}}^q; \quad (5.5.83)$$

(2) for $\alpha=\frac{1}{2}, \lambda=1, k_1(\frac{1}{2})=8\sum_{k=0}^{\infty}\frac{1}{(1+k)^2}=\frac{4\pi^2}{3}$,

(i) if $p>1$, then we have the following equivalent inequalities:

$$\sum_{n=1}^{\infty}\sum_{m=1}^{\infty}\frac{\ln(mn)a_m b_n}{(\sqrt{mn}-1)\sqrt{mn}}$$

$$< \tfrac{4\pi^2}{3}\{\sum_{n=1}^{\infty} n^{\frac{p}{2}-1}a_n^p\}^{\frac{1}{p}}\{\sum_{n=1}^{\infty} n^{\frac{q}{2}-1}b_n^q\}^{\frac{1}{q}}, \quad (5.5.84)$$

$$\sum_{n=1}^{\infty}\tfrac{1}{n}(\sum_{m=1}^{\infty}\frac{\ln(mn)a_m}{(\sqrt{mn}-1)\sqrt{m}})^p < (\tfrac{4\pi^2}{3})^p \sum_{n=1}^{\infty} n^{\frac{p}{2}-1}a_n^p; \quad (5.5.85)$$

(ii) for $0<p<1$, setting

$$\theta(n) := \frac{3}{4\pi^2}\int_0^n \frac{\ln t}{(t^{1/2}-1)(\max\{1,t\})^{1/2}} t^{\frac{-1}{2}} dt,$$

we have the following reverse equivalent inequalities:

$$\sum_{n=1}^{\infty}\sum_{m=1}^{\infty}\frac{\ln(mn)a_m b_n}{(\sqrt{mn}-1)\sqrt{mn}} > \frac{4\pi^2}{3}$$

$$\times\{\sum_{n=1}^{\infty}[1-\theta(n)]n^{\frac{p}{2}-1}a_n^p\}^{\frac{1}{p}}\{\sum_{n=1}^{\infty} n^{\frac{q}{2}-1}b_n^q\}^{\frac{1}{q}}, \quad (5.5.86)$$

$$\sum_{n=1}^{\infty}\tfrac{1}{n}(\sum_{m=1}^{\infty}\frac{\ln(mn)a_m}{(\sqrt{mn}-1)\sqrt{m}})^p$$

$$> (\tfrac{4\pi^2}{3})^p \sum_{n=1}^{\infty}[1-\theta(n)]n^{\frac{p}{2}-1}a_n^p, \quad (5.5.87)$$

$$\sum_{m=1}^{\infty}\frac{1}{m[1-\theta(m)]^{q-1}}(\sum_{n=1}^{\infty}\frac{\ln(mn)b_n}{(\sqrt{mn}-1)\sqrt{n}})^q$$

$$< (\tfrac{4\pi^2}{3})^q \sum_{n=1}^{\infty} n^{\frac{q}{2}-1}b_n^q. \quad (5.5.88)$$

Example 5.5.11 If $0<\lambda\le 2, \frac{\lambda}{2}<\alpha\le\lambda$,

$$k_\lambda(x,y) = \frac{\ln(x/y)}{(x^\alpha - y^\alpha)(\min\{x,y\})^{\lambda-\alpha}},$$

$k_\lambda(1,t) = \frac{\ln t}{(t^\alpha -1)(\min\{1,t\})^{\lambda-\alpha}}$, then we find

$$k_{\lambda/2} = \tilde{k}_{\lambda,\alpha} := \int_0^\infty \frac{t^{\frac{\lambda}{2}-1}\ln t}{(t^\alpha -1)(\min\{1,t\})^{\lambda-\alpha}} dt$$

$$= 2\int_0^1(-\ln t)\sum_{k=0}^{\infty} t^{(k+1)\alpha-\frac{\lambda}{2}-1} dt$$

$$= 2\sum_{k=0}^{\infty}\frac{1}{(k+1)\alpha-\frac{\lambda}{2}}\int_0^1(-\ln t)dt^{(k+1)\alpha-\frac{\lambda}{2}}$$

$$= \sum_{k=0}^{\infty}\frac{8}{[2(k+1)\alpha-\lambda]^2}.$$

By Corollary 5.5.5, (1) for $p>1$, we have the following equivalent inequalities:

$$\sum_{n=1}^{\infty}\sum_{m=1}^{\infty}\frac{\ln(mn)}{m^\alpha n^\alpha -1}a_m b_n < \tilde{k}_{\lambda,\alpha} \| a \|_{p,\tilde{\phi}} \| b \|_{q,\tilde{\psi}}, \quad (5.5.89)$$

$$\sum_{n=1}^{\infty} n^{\frac{p\lambda}{2}-1}[\sum_{m=1}^{\infty}\frac{\ln(mn)}{m^\alpha n^\alpha -1}a_m]^p < \tilde{k}_{\lambda,\alpha}^p \| a \|_{p,\tilde{\phi}}^p; \quad (5.5.90)$$

(1) for $0<p<1$, setting

$$\theta_{\lambda/2}(n) := \frac{1}{\tilde{k}_{\lambda,\alpha}} \int_0^n \frac{\ln t}{(t^\alpha - 1)(\min\{1,t\})^{\lambda-\alpha}} t^{\frac{\lambda}{2}-1} dt,$$

we have the following reverse equivalent inequalities:

$$\sum_{n=1}^\infty \sum_{m=1}^\infty \frac{\ln(mn) a_m b_n}{m^\alpha n^\alpha - 1} > \tilde{k}_{\lambda,\alpha}$$

$$\times \{\sum_{n=1}^\infty [1-\theta_{\lambda/2}(n)] n^{p(1-\frac{\lambda}{2})-1} a_n^p\}^{\frac{1}{p}} \| b \|_{q,\tilde{\psi}}, \quad (5.5.91)$$

$$\sum_{n=1}^\infty n^{\frac{p\lambda}{2}-1} [\sum_{m=1}^\infty \frac{\ln(mn)}{m^\alpha n^\alpha - 1} a_m]^p > \tilde{k}_{\lambda,\alpha}^p$$

$$\times \sum_{n=1}^\infty [1-\theta_{\lambda/2}(n)] n^{p(1-\frac{\lambda}{2})-1} a_n^p. \quad (5.5.92)$$

$$\sum_{m=1}^\infty \frac{m^{\frac{q\lambda}{2}-1}}{[1-\theta_{\lambda/2}(m)]^{q-1}} [\sum_{n=1}^\infty \frac{\ln(mn)}{m^\alpha n^\alpha - 1} b_n]^q < \tilde{k}_{\lambda,\alpha}^q \| b \|_{q,\tilde{\psi}}^q. \quad (5.5.93)$$

Example 5.5.12 If $0 \le 2\alpha < \lambda \le 2$,

$$k_\lambda(x,y) = \frac{1}{(x+y)^{\lambda-\alpha} (\min\{x,y\})^\alpha},$$

$$k_\lambda(1,t) = \frac{1}{(1+t)^{\lambda-\alpha} (\min\{1,t\})^\alpha},$$

then we find

$$k_{\lambda/2} = K_\lambda(\alpha) := \int_0^\infty \frac{t^{\frac{\lambda}{2}-1}}{(1+t)^{\lambda-\alpha} (\min\{1,t\})^\alpha} dt$$

$$= \sum_{k=0}^\infty \binom{\alpha-\lambda}{k} \frac{4}{\lambda - 2\alpha + 2k}. \quad (5.5.94)$$

By Corollary 5.5.5, (1) for $p > 1$, we have the following equivalent inequalities:

$$\sum_{n=1}^\infty \sum_{m=1}^\infty \frac{a_m b_n}{(1+mn)^{\lambda-\alpha}} < K_\lambda(\alpha) \| a \|_{p,\tilde{\phi}} \| b \|_{q,\tilde{\psi}}, \quad (5.5.95)$$

$$\sum_{n=1}^\infty n^{\frac{p\lambda}{2}-1} [\sum_{m=1}^\infty \frac{a_m}{(1+mn)^{\lambda-\alpha}}]^p < [K_\lambda(\alpha)]^p \| a \|_{p,\tilde{\phi}}^p; \quad (5.5.96)$$

(2) for $0 < p < 1$, setting

$$\theta_{\lambda/2}(n) := \frac{1}{K_\lambda(\alpha)} \int_0^n \frac{1}{(1+t)^{\lambda-\alpha} (\min\{1,t\})^\alpha} t^{\frac{\lambda}{2}-1} dt,$$

we have the following reverse equivalent inequalities:

$$\sum_{n=1}^\infty \sum_{m=1}^\infty \frac{a_m b_n}{(1+mn)^{\lambda-\alpha}} > K_\lambda(\alpha)$$

$$\times \{\sum_{n=1}^\infty [1-\theta_{\lambda/2}(n)] n^{p(1-\frac{\lambda}{2})-1} a_n^p\}^{\frac{1}{p}} \| b \|_{q,\tilde{\psi}}, \quad (5.5.97)$$

$$\sum_{n=1}^\infty n^{\frac{p\lambda}{2}-1} [\sum_{m=1}^\infty \frac{a_m}{(1+mn)^{\lambda-\alpha}}]^p > [k_\lambda(\alpha)]^p$$

$$\times \sum_{n=1}^\infty [1-\theta_{\lambda/2}(n)] n^{p(1-\frac{\lambda}{2})-1} a_n^p, \quad (5.5.98)$$

$$\sum_{m=1}^\infty \frac{m^{\frac{q\lambda}{2}-1}}{[1-\theta_{\lambda/2}(m)]^{q-1}} [\sum_{n=1}^\infty \frac{b_n}{(1+mn)^{\lambda-\alpha}}]^q < [K_\lambda(\alpha)]^q \| b \|_{q,\tilde{\psi}}^q. \quad (5.5.99)$$

Example 5.5.13 If $0 \le 2\alpha < \lambda \le 2$,

$$k_\lambda(x,y) = \frac{1}{(x^{\lambda-\alpha} + y^{\lambda-\alpha})(\min\{x,y\})^\alpha},$$

$$k_\lambda(1,t) = \frac{1}{(1+t^{\lambda-\alpha})(\min\{1,t\})^\alpha},$$

then we find

$$k_{\lambda/2} = \tilde{K}_\lambda(\alpha) := \int_0^\infty \frac{t^{\frac{\lambda}{2}-1} dt}{(1+t^{\lambda-\alpha})(\min\{1,t\})^\alpha}$$

$$= \sum_{k=0}^\infty \frac{2(-1)^k}{[(\frac{1}{2}+k)\lambda - (k+1)\alpha]}. \quad (5.5.100)$$

By Corollary 5.5.5, (1) for $p > 1$, we have the following equivalent inequalities:

$$\sum_{n=1}^\infty \sum_{m=1}^\infty \frac{a_m b_n}{1+m^{\lambda-\alpha} n^{\lambda-\alpha}} < \tilde{K}_\lambda(\alpha) \| a \|_{p,\tilde{\phi}} \| b \|_{q,\tilde{\psi}}, \quad (5.5.101)$$

$$\sum_{n=1}^\infty n^{\frac{p\lambda}{2}-1} (\sum_{m=1}^\infty \frac{a_m}{1+m^{\lambda-\alpha} n^{\lambda-\alpha}})^p < [\tilde{K}_\lambda(\alpha)]^p \| a \|_{p,\tilde{\phi}}^p; \quad (5.5.102)$$

(2) for $0 < p < 1$, setting

$$\theta_{\lambda/2}(n) := \frac{1}{\tilde{K}_\lambda(\alpha)} \int_0^n \frac{1}{(1+t^{\lambda-\alpha})(\min\{1,t\})^\alpha} t^{\frac{\lambda}{2}-1} dt,$$

we have the following reverse equivalent inequalities:

$$\sum_{n=1}^\infty \sum_{m=1}^\infty \frac{a_m b_n}{1+m^{\lambda-\alpha} n^{\lambda-\alpha}} > \tilde{K}_\lambda(\alpha)$$

$$\times \{\sum_{n=1}^\infty [1-\theta_{\lambda/2}(n)] n^{p(1-\frac{\lambda}{2})-1} a_n^p\}^{\frac{1}{p}} \| b \|_{q,\tilde{\psi}}, \quad (5.5.103)$$

$$\sum_{n=1}^\infty n^{\frac{p\lambda}{2}-1} (\sum_{m=1}^\infty \frac{a_m}{1+m^{\lambda-\alpha} n^{\lambda-\alpha}})^p > [\tilde{K}_\lambda(\alpha)]^p \sum_{n=1}^\infty [1-\theta_{\frac{\lambda}{2}}(n)] n^{p(1-\frac{\lambda}{2})-1} a_n^p, \quad (5.5.104)$$

$$\sum_{m=1}^\infty \frac{m^{\frac{q\lambda}{2}-1}}{[1-\theta_{\lambda/2}(m)]^{q-1}} (\sum_{n=1}^\infty \frac{b_n}{1+m^{\lambda-\alpha} n^{\lambda-\alpha}})^q < [\tilde{K}_\lambda(\alpha)]^q \| b \|_{q,\tilde{\psi}}^q. \quad (5.5.105)$$

Example 5.5.14 If $0 < \lambda \le 1, b \ge 0$,

$$k_{2\lambda}(x,y) = \frac{1}{x^{2\lambda} + 2b x^\lambda y^\lambda + y^{2\lambda}},$$

$$k_{2\lambda}(1,t) = \frac{1}{1 + 2b t^\lambda + t^{2\lambda}},$$

then we find

$$k_\lambda = k_\lambda(b) := \int_0^\infty k_{2\lambda}(1,t) t^{-1+\frac{2\lambda}{2}} du$$
$$= \int_0^\infty \frac{1}{1+2bt^\lambda + t^{2\lambda}} t^{-1+\lambda} dt$$
$$= \frac{1}{\lambda}\int_0^\infty \frac{dv}{v^2+2bv^2+1}$$
$$= \begin{cases} \frac{1}{\lambda\sqrt{b^2-1}}\ln(b+\sqrt{b^2-1}) \ , b>1, \\ \frac{1}{\lambda}, \quad b=1, \\ \frac{1}{\lambda\sqrt{1-b^2}}(\frac{\pi}{2} - \arctan\frac{b}{\sqrt{1-b^2}}), \ 0 \le b < 1. \end{cases} \quad (5.5.106)$$

By Corollary 5.5.5, (1) for $p>1$, we have the following equivalent inequalities:

$$\sum_{n=1}^\infty \sum_{m=1}^\infty \frac{a_m b_n}{1+2bm^\lambda n^\lambda + m^{2\lambda} n^{2\lambda}} < k_\lambda(b)$$
$$\times \{\sum_{n=1}^\infty n^{p(1-\lambda)-1} a_n^p\}^{\frac{1}{p}} \{\sum_{n=1}^\infty n^{q(1-\lambda)-1} b_n^q\}^{\frac{1}{q}}, \quad (5.5.107)$$
$$\sum_{n=1}^\infty n^{p\lambda-1} (\sum_{m=1}^\infty \frac{a_m}{1+2bm^\lambda n^\lambda + m^{2\lambda} n^{2\lambda}})^p$$
$$< k_\lambda^p(b) \sum_{n=1}^\infty n^{p(1-\lambda)-1} a_n^p ; \quad (5.5.108)$$

(2) for $0<p<1$, setting

$$\theta_\lambda(n) := \frac{1}{k_{2\lambda}(b)} \int_0^n \frac{1}{1+bt^\lambda + t^{2\lambda}} t^{\lambda-1} dt,$$

we have the following reverse equivalent inequalities:

$$\sum_{n=1}^\infty \sum_{m=1}^\infty \frac{a_m b_n}{1+2bm^\lambda n^\lambda + m^{2\lambda} n^{2\lambda}} > k_\lambda(b)$$
$$\times \{\sum_{n=1}^\infty [1-\theta_\lambda(n)] n^{p(1-\lambda)-1} a_n^p\}^{\frac{1}{p}} \{\sum_{n=1}^\infty n^{q(1-\lambda)-1} b_n^q\}^{\frac{1}{q}}, \quad (5.5.109)$$
$$\sum_{n=1}^\infty n^{p\lambda-1} (\sum_{m=1}^\infty \frac{a_m}{1+2bm^\lambda n^\lambda + m^{2\lambda} n^{2\lambda}})^p$$
$$> k_\lambda^p(b) \sum_{n=1}^\infty [1-\theta_\lambda(n)] n^{p(1-\lambda)-1} a_n^p, \quad (5.5.110)$$
$$\sum_{m=1}^\infty \frac{m^{q\lambda-1}}{[1-\theta_\lambda(m)]^{q-1}} (\sum_{n=1}^\infty \frac{b_n}{1+2bm^\lambda n^\lambda + m^{2\lambda} n^{2\lambda}})^q$$
$$< k_\lambda^q(b) \sum_{n=1}^\infty n^{q(1-\lambda)-1} b_n^q . \quad (5.5.111)$$

5.6 SOME HILBERT- TYPE INEQUALITIES WITH THE GENERAL KERNEL CONTAINING VARIABLES

5.6.1 MAIN RESULTS

Assuming that $p>0\ (p\neq 1), \frac{1}{p}+\frac{1}{q}=1, \alpha \in$ **R**, $h(t)(\geq 0)$ is a measurable function in $\mathbf{R}_+$, such that $0<k_\alpha = \int_0^\infty h(t) t^{\alpha-1} dt < \infty$, $n_0 \in \mathbf{N}_0$, $u(x)$ and $v(x)$ are strictly decreasing functions in $[n_0,\infty)$, satisfying $u(n_0),\ v(n_0)>0,\ u(\infty)=v(\infty)=\infty$ and $u'(x)>0$, $v'(x)>0$, we define the weight coefficients $\tilde{w}_\alpha(m)$ and $\tilde{\vartheta}_\alpha(n)$ as follows:

$$\tilde{w}_\alpha(m) := \sum_{n=n_0}^\infty h(u(m)v(n)) \frac{[u(m)]^\alpha}{[v(n)]^{1-\alpha}} v'(n),$$
$$\tilde{\vartheta}_\alpha(n) := \sum_{m=n_0}^\infty h(u(m)v(n)) \frac{[v(n)]^\alpha}{[u(m)]^{1-\alpha}} u'(m),$$
$$m,n \geq n_0 . \quad (5.6.1)$$

Setting $\phi(x) = [u(x)]^{p(1-\alpha)-1}[u'(x)]^{1-p}$,

$$\psi(x) = [v(x)]^{q(1-\alpha)-1}[v'(x)]^{1-q} \ (x\in[n_0,\infty)),$$

and for $a=\{a_m\}_{m=n_0}^\infty, b=\{b_n\}_{n=n_0}^\infty, a_m, b_n \geq 0$,

$$\| a \|_{p,\phi} = \{\sum_{n=n_0}^\infty [u(n)]^{p(1-\alpha)-1}[u'(n)]^{1-p} a_n^p\}^{\frac{1}{p}},$$
$$\| b \|_{q,\psi} = \{\sum_{n=n_0}^\infty [v(n)]^{q(1-\alpha)-1}[v'(n)]^{1-q} b_n^q\}^{\frac{1}{q}},$$

in view of the above assumption, we have the following theorems:

Theorem 5.6.1 If $p>1, \alpha\in\mathbf{R}$,

$$0<k_\alpha = \int_0^\infty h(t) t^{\alpha-1} dt < \infty,$$

such that

$$\tilde{w}_\alpha(m) < k_\alpha, \tilde{\vartheta}_\alpha(n) < k_\alpha, \ m,n\geq n_0, \quad (5.6.2)$$

then for

$$0<\| a \|_{p,\phi} < \infty, 0 < \| b \|_{q,\psi} < \infty,$$

we have the following equivalent inequalities:

$$\tilde{I}_\alpha := \sum_{n=n_0}^\infty \sum_{m=n_0}^\infty h(u(m)v(n)) a_m b_n$$
$$< k_\alpha \| a \|_{p,\phi} \| b \|_{q,\psi}, \quad (5.6.3)$$
$$\tilde{J}_\alpha := \sum_{n=n_0}^\infty [v(n)]^{p\alpha-1} v'(n) [\sum_{m=n_0}^\infty h(u(m)v(n)) a_m]^p$$
$$< k_\alpha^p \| a \|_{p,\phi}^p . \quad (5.6.4)$$

Proof By Hölder's inequality and (5.6.1), it follows

$$\tilde{I}_\alpha = \sum_{n=n_0}^{\infty}\sum_{m=n_0}^{\infty} h(u(m)v(n))$$

$$\times[\frac{(u(m))^{(1-\alpha)/q}(v'(n))^{1/p}}{(v(n))^{(1-\alpha)/p}(u'(m))^{1/q}}a_m][\frac{(v(n))^{(1-\alpha)/p}(u'(m))^{1/q}}{(u(m))^{(1-\alpha)/q}(v'(n))^{1/p}}b_n]$$

$$\leq \{\sum_{m=n_0}^{\infty}\tilde{w}_\alpha(m)\phi(m)a_m^p\}^{\frac{1}{p}}\{\sum_{n=n_0}^{\infty}\tilde{\vartheta}_\alpha(n)\psi(n)b_n^q\}^{\frac{1}{q}} \quad (5.6.5)$$

By (5.6.2), we have (5.6.3).

If $\tilde{J}_\alpha = 0$, then (5.6.4) is naturally valid; if $\tilde{J}_\alpha > 0$, then there exists a $N \in \mathbf{N}, N > n_0$, such that $\sum_{n=n_0}^{N}[u(n)]^{p(1-\alpha)-1}[u'(n)]^{1-p}a_n^p > 0$ and

$$\tilde{J}_\alpha(N) := \sum_{n=n_0}^{N}[v(n)]^{p\alpha-1}v'(n)$$

$$\times[\sum_{m=n_0}^{N}h(u(m)v(n))a_m]^p > 0.$$

Setting

$$\tilde{I}_\alpha(N) := \sum_{n=n_0}^{N}\sum_{m=n_0}^{N}h(u(m)v(n))a_m b_n,$$

$$b_n(N) = \frac{v'(n)}{[v(n)]^{1-p\alpha}}[\sum_{m=n_0}^{N}h(u(m)v(n))a_m]^{p-1},$$

$$n \in \mathbf{N}, n_0 \leq n \leq N,$$

by (5.6.3), we have

$$0 < \sum_{n=n_0}^{N}[v(n)]^{q(1-\alpha)-1}[v'(n)]^{1-q}b_n^q(N)$$

$$= \tilde{J}_\alpha(N) = \tilde{I}_\alpha(N)$$

$$< k_\alpha\{\sum_{n=n_0}^{N}[u(n)]^{p(1-\alpha)-1}[u'(n)]^{1-p}a_n^p\}^{\frac{1}{p}}$$

$$\times\{\sum_{n=n_0}^{N}[v(n)]^{q(1-\alpha)-1}[v'(n)]^{1-q}b_n^q(N)\}^{\frac{1}{q}};$$

$$\tilde{J}_\alpha(N) = \sum_{n=n_0}^{N}[v(n)]^{q(1-\alpha)-1}[v'(n)]^{1-q}b_n^q(N)$$

$$< k_\alpha^p\sum_{n=n_0}^{\infty}[u(n)]^{p(1-\alpha)-1}[u'(n)]^{1-p}a_n^p.$$

It follows that

$$0 < \sum_{n=n_0}^{\infty}[v(n)]^{q(1-\alpha)-1}[v'(n)]^{1-q}b_n^q(\infty) < \infty$$

and for $N \to \infty$, using (5.6.3), both the above inequalities still keep the forms of strict-sign inequalities. Hence we have (5.6.4).

On the other-hand, suppose that (5.6.4) is valid. By Hölder's inequality, we find

$$\tilde{I}_\alpha = \sum_{n=n_0}^{\infty}\{\frac{[v'(n)]^{\frac{1}{p}}}{[v(n)]^{\frac{1}{p}-\alpha}}\sum_{m=n_0}^{\infty}h(u(m)v(n))a_m\}$$

$$\times\{\frac{[v(n)]^{\frac{1}{p}-\alpha}}{[v'(n)]^{1/p}}b_n\} \leq \tilde{J}_\alpha^{\frac{1}{p}}\|b\|_{q,\psi}. \quad (5.6.6)$$

By (5.6.4), we have (5.6.3), which is equivalent to (5.6.4). □

Theorem 5.6.2 If $0 < p < 1$,

$$0 < k_\alpha = \int_0^\infty h(t)t^{\alpha-1}dt < \infty,$$

$0 < \tilde{\theta}_\alpha(m) < 1$, such that

$$k_\alpha[1-\tilde{\theta}_\alpha(m)] \leq \tilde{w}_\alpha(m),$$

$$\tilde{\vartheta}_\alpha(n) < k_\alpha, \ m, n \geq n_0, \quad (5.6.7)$$

then for

$$0 < \sum_{n=n_0}^{\infty}[1-\tilde{\theta}_\alpha(n)][u(n)]^{p(1-\alpha)-1}$$

$$\times[u'(n)]^{1-p}a_n^p < \infty$$

and $0 < \|b\|_{q,\psi} < \infty$, we have the following reverse equivalent inequalities:

$$\sum_{n=n_0}^{\infty}\sum_{m=n_0}^{\infty}h(u(m)v(n))a_m b_n$$

$$> k_\alpha\{\sum_{n=n_0}^{\infty}[1-\tilde{\theta}_\alpha(n)]\phi(n)a_n^p\}^{\frac{1}{p}}\|b\|_{q,\psi}, \quad (5.6.8)$$

$$\sum_{n=n_0}^{\infty}[v(n)]^{p\alpha-1}v'(n)[\sum_{m=n_0}^{\infty}h(u(m)v(n))a_m]^p$$

$$> k_\alpha^p\sum_{n=n_0}^{\infty}[1-\tilde{\theta}_\alpha(n)]\phi(n)a_n^p, \quad (5.6.9)$$

$$\sum_{m=n_0}^{\infty}\frac{[u(m)]^{q\alpha-1}u'(m)}{[1-\tilde{\theta}_\alpha(m)]^{q-1}}[\sum_{n=n_0}^{\infty}h(u(m)v(n))b_n]^q$$

$$< k_\alpha^q\|b\|_{q,\psi}^q. \quad (5.6.10)$$

We leave the proof of this theorem to the readers.

Note 5.6.3 (1) For $u(x) = v(x) = x, n_0 = 1$ in Theorem 5.6.1-Theorem 5.6.2, we have Theorem 5.5.1. (2) For $\lambda \in \mathbf{R}$, $k_\lambda(x,y)(\geq 0)$ is a finite homogeneous function of degree $-\lambda$, setting $\alpha = \frac{\lambda}{2}$, $h(t) = k_\lambda(1,t)$ (or $k_\lambda(t,1)$), in Theorem 5.6.1-Theorem 5.6.2, if the assumptions of Theorem 5.6.1-Theorem 5.6.2 are fulfilled, then we still have all the results with the kernel $k_\lambda(1,mn)$ (or $k_\lambda(mn,1)$).

Corollary 5.6.4 For $0<\lambda\le 2$, $\alpha=\frac{\lambda}{2}$, $h(t)=k_\lambda(1,t)$ (or $k_\lambda(t,1)$) in Theorem 5.6.1 - Theorem 5.6.2, if $k_\lambda(1,t)>0$ (or $k_\lambda(t,1)>0$) is decreasing for $u>0$ and strictly decreasing in a subinterval, $u(n_0-1)\ge 0$, $v(n_0-1)\ge 0$, $u'(x)$ and $v'(x)$ are decreasing in $\mathbf{R}_+$, then (1) for $p>1$, we have the equivalent inequalities (5.6.3) and (5.6.4) with the kernel $h(mn)=k_\lambda(1,mn)$ (or $k_\lambda(mn,1)$); (2) for $0<p<1$, we have the reverse equivalent inequalities (5.6.8)-(5.6.10) with the kernel $k_\lambda(1,mn)$ (or $k_\lambda(mn,1)$).

Proof Setting $t=u(m)v(y)$, we find

$$\tilde{w}_{\lambda/2}(m)=\sum_{n=n_0}^{\infty}k_\lambda(1,u(m)v(n))\frac{[u(m)]^{\lambda/2}}{[v(n)]^{1-\lambda/2}}v'(n)$$

$$<\int_{n_0-1}^{\infty}k_\lambda(1,u(m)v(y))\frac{[u(m)]^{\lambda/2}}{[v(y)]^{1-\lambda/2}}v'(y)dy$$

$$=\int_{u(m)v(n_0-1)}^{\infty}k_\lambda(1,t)t^{\frac{\lambda}{2}-1}dt$$

$$\le\int_0^{\infty}k_\lambda(1,t)t^{\frac{\lambda}{2}-1}dt=k_{\lambda/2};$$

$$\tilde{w}_{\lambda/2}(m)$$

$$\ge\int_{n_0}^{\infty}k_\lambda(1,u(m)v(y))\frac{[u(m)]^{\lambda/2}}{[v(y)]^{1-\lambda/2}}v'(y)dy$$

$$=\int_{u(m)v(n_0)}^{\infty}k_\lambda(1,t)t^{\frac{\lambda}{2}-1}dt=k_{\lambda/2}[1-\tilde{\theta}_{\lambda/2}(m)],$$

where, $\tilde{\theta}_{\lambda/2}(m)$ is defined by

$$\tilde{\theta}_{\lambda/2}(m):=\frac{1}{k_{\lambda/2}}\int_0^{u(m)v(n_0)}k_\lambda(1,t)t^{\frac{\lambda}{2}-1}dt,$$

and $0<\tilde{\theta}_{\lambda/2}(m)<1$. By the same way, we have $\tilde{\vartheta}_{\lambda/2}(n)<k_{\lambda/2}$. Hence we have the results of Theorem 5.6.1- Theorem 5.6.2. □

Example 5.6.5 Putting $h(t)=e^{-t}$,

$$u(x)=\ln(x+1),v(x)=x,n_0=1,$$

for $\alpha>0$, we find

$$k_\alpha=\int_0^{\infty}e^{-t}t^{\alpha-1}dt=\Gamma(\alpha),$$

where $\Gamma(\cdot)$ is the gamma function (Wang SC 1979)[7]. Setting $t=y\ln(1+m)$, we obtain

$$\tilde{w}_\alpha(m)=\sum_{n=1}^{\infty}\frac{1}{(m+1)^n}\frac{[\ln(m+1)]^\alpha}{n^{1-\alpha}}$$

$$<\int_0^{\infty}\frac{1}{(m+1)^y}\frac{[\ln(m+1)]^\alpha}{y^{1-\alpha}}dy=k_\alpha,$$

$$\tilde{w}_\alpha(m)>\int_1^{\infty}\frac{1}{(m+1)^y}\frac{[\ln(m+1)]^\alpha}{y^{1-\alpha}}dy$$

$$=\int_{\ln(m+1)}^{\infty}e^{-t}t^{\alpha-1}dt=\Gamma(\alpha)[1-\tilde{\theta}_\alpha(m)],$$

where, it follows

$$0<\tilde{\theta}_\alpha(m)=1-\frac{1}{\Gamma(\alpha)}\int_0^{\ln(1+m)}e^{-t}t^{\alpha-1}dt<1.$$

By the same way, we find $\tilde{\vartheta}_\alpha(n)<k_\alpha$.

By Theorem 5.6.1 and Theorem 5.6.2, (1) for $p>1$, $0<\sum_{n=1}^{\infty}\frac{[\ln(1+n)]^{p(1-\alpha)-1}}{(n+1)^{1-p}}a_n^p<\infty$ and

$$0<\sum_{n=1}^{\infty}n^{q(1-\alpha)-1}b_n^q<\infty,$$

we have the following equivalent inequalities:

$$\sum_{n=1}^{\infty}\sum_{m=1}^{\infty}\frac{a_m b_n}{(m+1)^n}<\Gamma(\alpha)\{\sum_{n=1}^{\infty}\frac{[\ln(1+n)]^{p(1-\alpha)-1}}{(n+1)^{1-p}}a_n^p\}^{\frac{1}{p}}$$

$$\times\{\sum_{n=1}^{\infty}n^{q(1-\alpha)-1}b_n^q\}^{\frac{1}{q}},\qquad(5.6.11)$$

$$\sum_{n=1}^{\infty}n^{p\alpha-1}[\sum_{m=1}^{\infty}\frac{1}{(m+1)^n}a_m]^p$$

$$<\Gamma^p(\alpha)\sum_{n=1}^{\infty}\frac{[\ln(1+n)]^{p(1-\alpha)-1}}{(n+1)^{1-p}}a_n^p;\qquad(5.6.12)$$

(2) for $0<p<1$,

$$0<\sum_{n=1}^{\infty}[1-\tilde{\theta}_\alpha(n)]\frac{[\ln(1+n)]^{p(1-\alpha)-1}}{(n+1)^{1-p}}a_n^p<\infty$$

and $0<\sum_{n=1}^{\infty}n^{q(1-\alpha)-1}b_n^q<\infty$, we have the following reverse equivalent inequalities:

$$\sum_{n=1}^{\infty}\sum_{m=1}^{\infty}\frac{a_m b_n}{(m+1)^n}>\Gamma(\alpha)$$

$$\times\{\sum_{n=1}^{\infty}[1-\tilde{\theta}_\alpha(n)]\frac{[\ln(1+n)]^{p(1-\alpha)-1}}{(n+1)^{1-p}}a_n^p\}^{\frac{1}{p}}$$

$$\times\{\sum_{n=1}^{\infty}n^{q(1-\alpha)-1}b_n^q\}^{\frac{1}{q}},\qquad(5.6.13)$$

$$\sum_{n=1}^{\infty}n^{p\alpha-1}[\sum_{m=1}^{\infty}\frac{1}{(m+1)^n}a_m]^p$$

$$>\Gamma^p(\alpha)\sum_{n=1}^{\infty}[1-\tilde{\theta}_\alpha(n)]\frac{[\ln(1+n)]^{p(1-\alpha)-1}}{(n+1)^{1-p}}a_n^p;\qquad(5.6.14)$$

$$\sum_{m=1}^{\infty} \frac{[\ln(1+m)]^{q\alpha-1}}{(1+m)[1-\tilde{\theta}_\alpha(m)]^{q-1}} [\sum_{n=1}^{\infty} \frac{1}{(1+m)^n} b_n]^q$$

$$< \Gamma^q(\alpha) \sum_{n=1}^{\infty} n^{q(1-\alpha)-1} b_n^q. \quad (5.6.15)$$

5.6.2 SOME EXAMPLES FOR APPLYING COROLLARY 5.6.4

In the following examples, suppose that $u(x)$ and $v(x)$ satisfy the assumptions of Corollary 5.6.4, and $a = \{a_m\}_{m=n_0}^{\infty}$, $b = \{b_n\}_{n=n_0}^{\infty}$, $a_m, b_n \geq 0$,

$$\| a \|_{p,\phi} = \{\sum_{n=n_0}^{\infty} [u(n)]^{p(1-\frac{\lambda}{2})-1} [u'(n)]^{1-p} a_n^p\}^{\frac{1}{p}},$$

$$\| b \|_{q,\psi} = \{\sum_{n=n_0}^{\infty} [v(n)]^{q(1-\frac{\lambda}{2})-1} [v'(n)]^{1-q} b_n^q\}^{\frac{1}{q}}.$$

Example 5.6.6 Setting $k_\lambda(x,y) = \frac{1}{(x+y)^\lambda}$, we find

$$k_{\lambda/2} = B(\tfrac{\lambda}{2}, \tfrac{\lambda}{2}), 0 < \lambda \leq 2.$$

By Corollary 5.6.4, (1) for $p > 1$, $0 < \| a \|_{p,\phi} < \infty$ and $0 < \| b \|_{q,\psi} < \infty$, we have the following equivalent inequalities:

$$\sum_{n=n_0}^{\infty} \sum_{m=n_0}^{\infty} \frac{a_m b_n}{(1+u(m)v(n))^\lambda}$$

$$< B(\tfrac{\lambda}{2}, \tfrac{\lambda}{2}) \| a \|_{p,\phi} \| b \|_{q,\psi}, \quad (5.6.16)$$

$$\sum_{n=n_0}^{\infty} [v(n)]^{\frac{p\lambda}{2}-1} v'(n) [\sum_{m=n_0}^{\infty} \frac{a_m}{(1+u(m)v(n))^\lambda}]^p$$

$$< [B(\tfrac{\lambda}{2}, \tfrac{\lambda}{2})]^p \| a \|_{p,\phi}^p; \quad (5.6.17)$$

(2) for $0 < p < 1$, setting

$$\tilde{\theta}_{\lambda/2}(m) := \frac{1}{B(\frac{\lambda}{2}, \frac{\lambda}{2})} \int_0^{u(m)v(n_0)} \frac{1}{(1+t)^\lambda} t^{\frac{\lambda}{2}-1} dt,$$

such that $0 < \sum_{n=n_0}^{\infty} [1 - \tilde{\theta}_{\lambda/2}(n)] \phi(n) a_n^p < \infty$ and $0 < \| b \|_{q,\psi} < \infty$, we have the following reverse equivalent inequalities:

$$\sum_{n=n_0}^{\infty} \sum_{m=n_0}^{\infty} \frac{a_m b_n}{(1+u(m)v(n))^\lambda} > B(\tfrac{\lambda}{2}, \tfrac{\lambda}{2})$$

$$\times \{\sum_{n=n_0}^{\infty} [1 - \tilde{\theta}_{\lambda/2}(n)] \phi(n) a_n^p\}^{\frac{1}{p}} \| b \|_{q,\psi}, \quad (5.6.18)$$

$$\sum_{n=n_0}^{\infty} [v(n)]^{\frac{p\lambda}{2}-1} v'(n) [\sum_{m=n_0}^{\infty} \frac{a_m}{(1+u(m)v(n))^\lambda}]^p$$

$$> [B(\tfrac{\lambda}{2}, \tfrac{\lambda}{2})]^p \sum_{n=n_0}^{\infty} [1 - \tilde{\theta}_{\lambda/2}(n)] \phi(n) a_n^p, \quad (5.6.19)$$

$$\sum_{m=n_0}^{\infty} \frac{[u(m)]^{\frac{q\lambda}{2}-1} u'(m)}{[1-\tilde{\theta}_{\lambda/2}(m)]^{q-1}} [\sum_{n=n_0}^{\infty} \frac{b_n}{(1+u(m)v(n))^\lambda}]^q$$

$$< [B(\tfrac{\lambda}{2}, \tfrac{\lambda}{2})]^q \| b \|_{q,\psi}^q. \quad (5.6.20)$$

In particular, setting $u(x) = v(x) = \ln x, n_0 = 2$, (1) for $p > 1$, we have the following equivalent inequalities:

$$\sum_{n=2}^{\infty} \sum_{m=2}^{\infty} \frac{a_m b_n}{(1+\ln m \ln n)^\lambda}$$

$$< B(\tfrac{\lambda}{2}, \tfrac{\lambda}{2}) \{\sum_{n=2}^{\infty} n^{p-1} (\ln n)^{p(1-\frac{\lambda}{2})-1} a_n^p\}^{\frac{1}{p}}$$

$$\times \{\sum_{n=2}^{\infty} n^{q-1} (\ln n)^{q(1-\frac{\lambda}{2})-1} b_n^q\}^{\frac{1}{q}}, \quad (5.6.21)$$

$$\sum_{n=2}^{\infty} \frac{1}{n} [\ln n]^{\frac{p\lambda}{2}-1} [\sum_{m=2}^{\infty} \frac{a_m}{(1+\ln m \ln n)^\lambda}]^p$$

$$< [B(\tfrac{\lambda}{2}, \tfrac{\lambda}{2})]^p \sum_{n=2}^{\infty} n^{p-1} (\ln n)^{p(1-\frac{\lambda}{2})-1} a_n^p; \quad (5.6.22)$$

(2) for $0 < p < 1$, setting

$$\tilde{\theta}_{\lambda/2}(m) := \frac{1}{B(\frac{\lambda}{2}, \frac{\lambda}{2})} \int_0^{\ln m \ln 2} \frac{1}{(1+t)^\lambda} t^{\frac{\lambda}{2}-1} dt,$$

we have the following reverse equivalent inequalities:

$$\sum_{n=2}^{\infty} \sum_{m=2}^{\infty} \frac{a_m b_n}{(1+\ln m \ln n)^\lambda} > B(\tfrac{\lambda}{2}, \tfrac{\lambda}{2})$$

$$\times \{\sum_{n=2}^{\infty} [1 - \tilde{\theta}_{\lambda/2}(n)] n^{p-1} (\ln n)^{p(1-\frac{\lambda}{2})-1} a_n^p\}^{\frac{1}{p}}$$

$$\times \{\sum_{n=2}^{\infty} n^{q-1} (\ln n)^{q(1-\frac{\lambda}{2})-1} b_n^q\}^{\frac{1}{q}}, \quad (5.6.23)$$

$$\sum_{n=2}^{\infty} \frac{1}{n} (\ln n)^{\frac{p\lambda}{2}-1} [\sum_{m=2}^{\infty} \frac{a_m}{(1+\ln m \ln n)^\lambda}]^p > [B(\tfrac{\lambda}{2}, \tfrac{\lambda}{2})]^p$$

$$\times \sum_{n=2}^{\infty} [1 - \tilde{\theta}_{\lambda/2}(n)] n^{p-1} (\ln n)^{p(1-\frac{\lambda}{2})-1} a_n^p, \quad (5.6.24)$$

$$\sum_{m=2}^{\infty} \frac{(\ln m)^{\frac{q\lambda}{2}-1}}{m[1-\tilde{\theta}_{\lambda/2}(m)]^{q-1}} [\sum_{n=2}^{\infty} \frac{b_n}{(1+\ln m \ln n)^\lambda}]^q$$

$$< [B(\tfrac{\lambda}{2}, \tfrac{\lambda}{2})]^q \sum_{n=2}^{\infty} n^{q-1} (\ln n)^{q(1-\frac{\lambda}{2})-1} b_n^q. \quad (5.6.25)$$

Example 5.6.7 Setting

$$k_\lambda(x,y) = \frac{1}{(\max\{x,y\})^\lambda},$$

we find $k_{\lambda/2}=\frac{4}{\lambda}$, $0<\lambda\leq 2$. By Corollary 5.6.4, (1) for $p>1$, $0<\|a\|_{p,\phi}<\infty$ and $0<\|b\|_{q,\psi}<\infty$, we have the following equivalent inequalities:

$$\sum_{n=n_0}^{\infty}\sum_{m=n_0}^{\infty}\frac{a_m b_n}{(\max\{1,u(m)v(n)\})^{\lambda}}$$

$$<\frac{4}{\lambda}\|a\|_{p,\phi}\|b\|_{q,\psi}, \qquad (5.6.26)$$

$$\sum_{n=n_0}^{\infty}[v(n)]^{\frac{p\lambda}{2}-1}v'(n)[\sum_{m=n_0}^{\infty}\frac{a_m}{(\max\{1,n(m)v(n)\})^{\lambda}}]^p$$

$$<(\frac{4}{\lambda})^p\|a\|_{p,\phi}^p; \qquad (5.6.27)$$

(2) for $0<p<1$, setting

$$\tilde{\theta}_{\lambda/2}(m):=\frac{\lambda}{4}\int_0^{u(m)v(n_0)}\frac{1}{(\max\{1,t\})^{\lambda}}t^{\frac{\lambda}{2}-1}dt$$

$$=1-\frac{1}{2[v(n_0)u(m)]^{\lambda/2}},$$

such that $0<\sum_{n=n_0}^{\infty}[1-\tilde{\theta}_{\lambda/2}(n)]\phi(n)a_n^p<\infty$ and $0<\|b\|_{q,\psi}<\infty$, we have the following reverse equivalent inequalities:

$$\sum_{n=n_0}^{\infty}\sum_{m=n_0}^{\infty}\frac{a_m b_n}{(\max\{1,u(m)v(n)\})^{\lambda}}>\frac{2}{\lambda[v(n_0)]^{\lambda/(2p)}}$$

$$\times\{\sum_{n=n_0}^{\infty}\frac{1}{[u(n)]^{\lambda/2}}\phi(n)a_n^p\}^{\frac{1}{p}}\|b\|_{q,\psi}, \qquad (5.6.28)$$

$$\sum_{n=n_0}^{\infty}[v(n)]^{\frac{p\lambda}{2}-1}v'(n)[\sum_{m=n_0}^{\infty}\frac{a_m}{(\max\{1,u(m)v(n)\})^{\lambda}}]^p$$

$$>\frac{1}{2[v(n_0)]^{\lambda/2}}(\frac{4}{\lambda})^p\sum_{n=n_0}^{\infty}\frac{1}{[u(n)]^{\lambda/2}}\phi(n)a_n^p, \qquad (5.6.29)$$

$$\sum_{m=n_0}^{\infty}[u(m)]^{\lambda(q-\frac{1}{2})-1}u'(m)[\sum_{n=n_0}^{\infty}\frac{b_n}{(\max\{1,u(m)v(n)\})^{\lambda}}]^q$$

$$<\frac{1}{2^{q-1}[v(n_0)]^{\frac{\lambda}{2}(q-1)}}(\frac{4}{\lambda})^q\|b\|_{q,\psi}^q. \qquad (5.6.30)$$

Example 5.6.8 Setting $k_{\lambda}(x,y)=\frac{\ln(x/y)}{x^{\lambda}-y^{\lambda}}$,

we find $k_{\lambda/2}=(\frac{\pi}{\lambda})^2$, $0<\lambda\leq 2$. By Corollary 5.6.4, (1) for $p>1$, $0<\|a\|_{p,\phi}<\infty$ and $0<\|b\|_{q,\psi}<\infty$, we have the following equivalent inequalities:

$$\sum_{n=n_0}^{\infty}\sum_{m=n_0}^{\infty}\frac{\ln[(u(m)v(n)]a_m b_n}{[(u(m)v(n)]^{\lambda}-1}$$

$$<(\frac{\pi}{\lambda})^2\|a\|_{p,\phi}\|b\|_{q,\psi}, \qquad (5.6.31)$$

$$\sum_{n=n_0}^{\infty}[v(n)]^{\frac{p\lambda}{2}-1}v'(n)[\sum_{m=n_0}^{\infty}\frac{\ln[u(m)v(n)]a_m}{[u(m)v(n)]^{\lambda}-1}]^p$$

$$<(\frac{\pi}{\lambda})^{2p}\|a\|_{p,\phi}^p; \qquad (5.6.32)$$

(2) for $0<p<1$, setting

$$\tilde{\theta}_{\lambda/2}(m):=(\frac{\lambda}{\pi})^2\int_0^{u(m)v(n_0)}\frac{\ln t}{t^{\lambda}-1}t^{\frac{\lambda}{2}-1}dt,$$

such that $0<\sum_{n=n_0}^{\infty}[1-\tilde{\theta}_{\lambda/2}(n)]\phi(n)a_n^p<\infty$ and $0<\|b\|_{q,\psi}<\infty$, we have the following reverse equivalent inequalities:

$$\sum_{n=n_0}^{\infty}\sum_{m=n_0}^{\infty}\frac{\ln[u(m)v(n)]a_m b_n}{[u(m)v(n)]^{\lambda}-1}>(\frac{\pi}{\lambda})^2$$

$$\times\{\sum_{n=n_0}^{\infty}[1-\tilde{\theta}_{\lambda/2}(n)]\phi(n)a_n^p\}^{\frac{1}{p}}\|b\|_{q,\psi}, \qquad (5.6.33)$$

$$\sum_{n=n_0}^{\infty}[v(n)]^{\frac{p\lambda}{2}-1}v'(n)[\sum_{m=n_0}^{\infty}\frac{\ln[u(m)v(n)]a_m}{[u(m)v(n)]^{\lambda}-1}]^p$$

$$>(\frac{\pi}{\lambda})^{2p}\sum_{n=n_0}^{\infty}[1-\tilde{\theta}_{\lambda/2}(n)]\phi(n)a_n^p, \qquad (5.6.34)$$

$$\sum_{m=n_0}^{\infty}\frac{[u(m)]^{\frac{q\lambda}{2}-1}u'(m)}{[1-\tilde{\theta}_{\lambda/2}(m)]^{q-1}}[\sum_{n=n_0}^{\infty}\frac{\ln[u(m)v(n)]b_n}{[u(m)v(n)]^{\lambda}-1}]^q$$

$$<(\frac{\pi}{\lambda})^{2q}\|b\|_{q,\psi}^q. \qquad (5.6.35)$$

5.7 REFERENCES

1. Yang BC. A Hilbert-type inequality with the homogeneous kernel of real number-degree. Journal of Southwest China Normal University (Natural Science Edition), 2010,35(1):40-44.
2. Yang BC. The norm of operator and Hilbert-type inequalities. Science Press, 2009.
3. Zhong YQ. Complex functions. Higher Education Press, 2003.
4. Yang BC. On a decomposition of Hilbert's inequality. Journal of Inequalities in Pure and Applied Mathematics, 1009,10(1), Art. 25, 8 pp.
5. Kuang JC. Introduction to real analysis. Changsha: Hunan Education Press, 1996.
6. Kuang JC. Applied inequalities. Jinan: Shangdong Science Technology Press, 2004.
7. Wang ZQ, Guo DR. Introduction to special functions. Beijing: Science Press, 1979.

CHAPTER 6

Multiple Hilbert-Type Inequalities

Abstract: In this chapter, by using the technique of real analysis and the way of weight coefficients, we establish some lemmas and obtain two equivalent multiple Hilbert-type inequalities and the reverses with the general homogeneous kernel of real number-degree and some variables, which are the best extensions of the corresponding results in Chapter 4. Some particular examples with the best constant factors are considered.

6.1. TWO MULTIPLE EQUIVALENT INEQUALITIES AND THE REVERSES

6.1.1. SOME LEMMAS

Lemma 6.1.1 If $n \in \mathbf{N} \setminus \{1\}$, $\lambda_i \in \mathbf{R}$, $u(m_i) > 0$ $(m_i \in \mathbf{N}_0)$, $p_i \in \mathbf{R} \setminus \{0,1\}$ $(i = 1,2,\cdots,n)$, $\sum_{i=1}^{n} \frac{1}{p_i} = 1$, then we have

$$A := \prod_{i=1}^{n} \{ \frac{[u(m_i)]^{(\lambda_i-1)(1-p_i)}}{[u'(m_i)]^{p_i-1}} \times \prod_{\substack{j=1 \\ (j \neq i)}}^{n} [u(m_j)]^{\lambda_j-1} u'(m_j) \}^{\frac{1}{p_i}} = 1. \tag{6.1.1}$$

Proof We find

$$A = \prod_{i=1}^{n} \{ \frac{[u(m_i)]^{(\lambda_i-1)(1-p_i)-(\lambda_i-1)}}{[u'(m_i)]^{p_i}} \times \prod_{j=1}^{n} [u(m_j)]^{\lambda_j-1} u'(m_j) \}^{\frac{1}{p_i}}$$

$$= \prod_{i=1}^{n} \{ \frac{[u(m_i)]^{(\lambda_i-1)(-p_i)}}{[u'(m_i)]^{p_i}} \}^{\frac{1}{p_i}} \times \{ \prod_{j=1}^{n} [u(m_j)]^{\lambda_j-1} u'(m_j) \}^{\sum_{i=1}^{n} \frac{1}{p_i}}$$

$$= \prod_{i=1}^{n} \frac{[u(m_i)]^{1-\lambda_i}}{u'(m_i)} \prod_{j=1}^{n} [u(m_j)]^{\lambda_j-1} u'(m_j) = 1.$$

Hence (6.1.1) is valid. □

Lemma 6.1.2 If $u(x)$ is a strictly increasing derivable function in $[t_0, \infty)$, $n_0 \in \mathbf{N}_0$ ($n_0 > t_0$), $u(n_0) > 0, u(\infty) = \infty$, and $u'(x)$ is a positive decreasing function in (t_0, ∞), then we have the following inequalities:

$$\sum_{m=n_0}^{\infty} \frac{u'(m)}{[u(m)]^{1+\varepsilon}} \geq \frac{1}{\varepsilon} O_1(1) \quad (\varepsilon > 0), \tag{6.1.2}$$

$$\sum_{m=n_0}^{\infty} \frac{u'(m)}{[u(m)]^{1+\delta}} \leq \frac{1}{\delta} O_2(1) \quad (\delta > 0). \tag{6.1.3}$$

Proof Since $\frac{u'(m)}{[u(m)]^{1+\varepsilon}} (> 0)$ is decreasing, then we find

$$\sum_{m=n_0}^{\infty} \frac{u'(m)}{[u(m)]^{1+\varepsilon}} \geq \int_{n_0}^{\infty} \frac{u'(x)dx}{[u(x)]^{1+\varepsilon}} = \frac{1}{\varepsilon [u(n_0)]^{\varepsilon}},$$

$$\sum_{m=n_0}^{\infty} \frac{u'(m)}{[u(m)]^{1+\delta}} = \frac{u'(n_0)}{[u(n_0)]^{1+\delta}} + \sum_{m=n_0+1}^{\infty} \frac{u'(m)}{[u(m)]^{1+\delta}}$$

$$\leq \frac{u'(n_0)}{[u(n_0)]^{1+\delta}} + \int_{n_0}^{\infty} \frac{u'(x)}{[u(x)]^{1+\delta}} dx$$

$$= \frac{1}{\delta} \{ \frac{\delta u'(n_0)}{[u(n_0)]^{1+\delta}} + \frac{1}{[u(n_0)]^{\delta}} \},$$

and then (6.1.2) and (6.1.3) are valid. □

6.1.2. TWO EQUIVALENT INEQUALITIES

Definition 6.1.3 If $n \in \mathbf{N}$,

$$\mathbf{R}_+^n = \{(x_1, \cdots, x_n) \mid x_i > 0 \ (i = 1, \cdots, n)\},$$

$\lambda \in \mathbf{R}$, $k_\lambda(x_1, \cdots, x_n)$ is a measurable function in $\mathbf{R}_+^n$, such that for any $t > 0, (x_1, \cdots, x_n) \in \mathbf{R}_+^n$,

$$k_\lambda(tx_1, \cdots, tx_n) = t^{-\lambda} k_\lambda(x_1, \cdots, x_n),$$

then we call $k_\lambda(x_1, \cdots, x_n)$ the homogeneous function of degree $-\lambda$.

Definition 6.1.4 Assuming that $n \in \mathbf{N} \setminus \{1\}$, $\lambda_i, \lambda \in \mathbf{R}$, $(i = 1, \cdots, n)$, $\sum_{i=1}^{n} \lambda_i = \lambda$, $k_\lambda(x_1, \cdots, x_n) (\geq 0)$ is a finite measurable function in $\mathbf{R}_+^n, n_0 \in \mathbf{N}_0$, $u(x)$ is a strictly increasing derivable function in $[n_0, \infty)$, $u(n_0) > 0$, $u(\infty) = \infty$, and $u'(x)$ is a positive decreasing function, setting

$$\tilde{k}(m_1, \cdots, m_n) := k_\lambda(u(m_1), \cdots, u(m_n)),$$

for $m_i \in \mathbf{N}_0$, define the weight coefficient $\omega_i(m_i) = \omega_i(m_i; \lambda_1, \cdots, \lambda_n)$ as follows:

Bicheng Yang

$$\omega_i(m_i) := [u(m_i)]^{\lambda_i} \sum_{m_n=n_0}^{\infty}\cdots\sum_{m_{i+1}=n_0}^{\infty}\sum_{m_{i-1}=n_0}^{\infty}\cdots\sum_{m_1=n_0}^{\infty}$$

$$\times\tilde{k}(m_1,\cdots,m_n)\prod_{\substack{j=1\\(j\neq i)}}^{n}[u(m_j)]^{\lambda_j-1}u'(m_j)$$

$$(m_i \ge n_0; i=1,2,\cdots,n). \tag{6.1.4}$$

Then we have the following theorem:

Theorem 6.1.5 Let the assumptions of Definition 6.1.4 be fulfilled and additionally, there exists a constant $k_\lambda > 0$, satisfying

$$\omega_i(m_i) < k_\lambda \quad (m_i \ge n_0; i=1,\cdots,n). \tag{6.1.5}$$

If $p_i > 1 \quad (i=1,\cdots,n),\ \sum_{i=1}^{n}\frac{1}{p_i}=1,\ \frac{1}{q_n}=1-\frac{1}{p_n}$, and $a_{m_i}^{(i)} \ge 0$, such that

$$0<\sum_{m_i=1}^{\infty}[u(m_i)]^{p_i(1-\lambda_i)-1}[u'(m_i)]^{1-p_i}(a_{m_i}^{(i)})^{p_i}<\infty,$$

$$i=1,\cdots,n,$$

then we have the following equivalent inequalities:

$$I := \sum_{m_n=n_0}^{\infty}\cdots\sum_{m_1=n_0}^{\infty}\tilde{k}(m_1,\cdots,m_n)\prod_{i=1}^{n}a_{m_i}^{(i)}$$

$$<k_\lambda\prod_{i=1}^{n}\left\{\sum_{m_i=n_0}^{\infty}\frac{[u(m_i)]^{p_i(1-\lambda_i)-1}}{[u'(m_i)]^{p_i-1}}(a_{m_i}^{(i)})^{p_i}\right\}^{\frac{1}{p_i}}, \tag{6.1.6}$$

$$J_n := \left\{\sum_{m_n=n_0}^{\infty}\frac{u'(m_n)}{[u(m_n)]^{1-q_n\lambda_n}}\right.$$

$$\left.\times\left[\sum_{m_{n-1}=n_0}^{\infty}\cdots\sum_{m_1=n_0}^{\infty}\tilde{k}(m_1,\cdots,m_n)\prod_{i=1}^{n-1}a_{m_i}^{(i)}\right]^{q_n}\right\}^{\frac{1}{q_n}}$$

$$<k_\lambda\prod_{i=1}^{n-1}\left\{\sum_{m_i=n_0}^{\infty}\frac{[u(m_i)]^{p_i(1-\lambda_i)-1}}{[u'(m_i)]^{p_i-1}}(a_{m_i}^{(i)})^{p_i}\right\}^{\frac{1}{p_i}}. \tag{6.1.7}$$

Proof By (6.1.5), in view of $\frac{1}{q_n}+\frac{1}{p_n}=1$, applying Hölder's inequality (Kuang SSTP 2004)[1], we have

$$\left[\sum_{m_{n-1}=n_0}^{\infty}\cdots\sum_{m_1=n_0}^{\infty}\tilde{k}(m_1,\cdots,m_n)\prod_{i=1}^{n-1}a_{m_i}^{(i)}\right]^{q_n}$$

$$=\left\{\sum_{m_{n-1}=n_0}^{\infty}\cdots\sum_{m_1=n_0}^{\infty}\tilde{k}(m_1,\cdots,m_n)\right.$$

$$\times\prod_{i=1}^{n-1}\left[\frac{[u(m_i)]^{(\lambda_i-1)(1-p_i)}}{[u'(m_i)]^{p_i-1}}\prod_{\substack{j=1\\(j\neq i)}}^{n}[u(m_j)]^{\lambda_j-1}u'(m_j)\right]^{\frac{1}{p_i}}a_{m_i}^{(i)}$$

$$\left.\times\left[\frac{[u(m_n)]^{(\lambda_n-1)(1-p_n)}}{[u'(m_n)]^{p_n-1}}\prod_{j=1}^{n-1}[u(m_j)]^{\lambda_j-1}u'(m_j)\right]^{\frac{1}{p_n}}\right\}^{q_n}$$

$$\le\sum_{m_{n-1}=n_0}^{\infty}\cdots\sum_{m_1=n_0}^{\infty}\tilde{k}(m_1,\cdots,m_n)\prod_{i=1}^{n-1}$$

$$\times\left[\frac{[u(m_i)]^{(\lambda_i-1)(1-p_i)}}{[u'(m_i)]^{p_i-1}}\prod_{\substack{j=1\\(j\neq i)}}^{n}[u(m_j)]^{\lambda_j-1}u'(m_j)\right]^{\frac{q_n}{p_i}}(a_{m_i}^{(i)})^{q_n}$$

$$\times\left\{\sum_{m_{n-1}=n_0}^{\infty}\cdots\sum_{m_1=n_0}^{\infty}\tilde{k}(m_1,\cdots,m_n)\right.$$

$$\left.\times\frac{[u(m_n)]^{(\lambda_n-1)(1-p_n)}}{[u'(m_n)]^{p_n-1}}\prod_{j=1}^{n-1}[u(m_j)]^{\lambda_j-1}u'(m_j)\right\}^{q_n-1}$$

$$=\left\{\frac{[u(m_n)]^{p_n(1-\lambda_n)-1}}{[u'(m_n)]^{p_n-1}}\omega_n(m_n)\right\}^{q_n-1}$$

$$\times\sum_{m_{n-1}=n_0}^{\infty}\cdots\sum_{m_1=n_0}^{\infty}\tilde{k}(m_1,\cdots,m_n)\prod_{i=1}^{n-1}$$

$$\times\left[\frac{[u(m_i)]^{(\lambda_i-1)(1-p_i)}}{[u'(m_i)]^{p_i-1}}\prod_{\substack{j=1\\(j\neq i)}}^{n}[u(m_j)]^{\lambda_j-1}u'(m_j)\right]^{\frac{q_n}{p_i}}(a_{m_i}^{(i)})^{q_n}$$

$$\le k_\lambda^{q_n-1}\frac{[u(m_n)]^{1-q_n\lambda_n}}{u'(m_n)}\sum_{m_{n-1}=n_0}^{\infty}\cdots\sum_{m_1=n_0}^{\infty}\tilde{k}(m_1,\cdots,m_n)$$

$$\times\prod_{i=1}^{n-1}\left[\frac{[u(m_i)]^{(\lambda_i-1)(1-p_i)}}{[u'(m_i)]^{p_i-1}}\right.$$

$$\left.\times\prod_{\substack{j=1\\(j\neq i)}}^{n}[u(m_j)]^{\lambda_j-1}u'(m_j)\right]^{\frac{q_n}{p_i}}(a_{m_i}^{(i)})^{q_n},$$

$$J_n\le\left\{k_\lambda^{q_n-1}\sum_{m_n=n_0}^{\infty}\sum_{m_{n-1}=n_0}^{\infty}\cdots\sum_{m_1=n_0}^{\infty}\tilde{k}(m_1,\cdots,m_n)\right.$$

$$\times\prod_{i=1}^{n-1}\left[\frac{[u(m_i)]^{(\lambda_i-1)(1-p_i)}}{[u'(m_i)]^{p_i-1}}\prod_{\substack{j=1\\(j\neq i)}}^{n}[u(m_j)]^{\lambda_j-1}u'(m_j)\right]^{\frac{q_n}{p_i}}$$

$$\left.\times(a_{m_i}^{(i)})^{q_n}\right\}^{\frac{1}{q_n}}$$

$$=k_\lambda^{1/p_n}\left\{\sum_{m_{n-1}=n_0}^{\infty}\cdots\sum_{m_1=n_0}^{\infty}\left[\sum_{m_n=n_0}^{\infty}\tilde{k}(m_1,\cdots,m_n)\right.\right.$$

$$\times[u(m_n)]^{\lambda_n-1}u'(m_n)\Big]\prod_{i=1}^{n-1}\Big[\frac{[u(m_i)]^{(\lambda_i-1)(1-p_i)}}{[u'(m_i)]^{p_i-1}}$$

$$\times\prod_{\substack{j=1\\(j\neq i)}}^{n-1}[u(m_j)]^{\lambda_j-1}u'(m_j)\Big]^{\frac{q_n}{p_i}}(a_{m_i}^{(i)})^{q_n}\Bigg\}^{\frac{1}{q_n}}. \quad (6.1.8)$$

For $n\geq 3$ in (6.1.8), since $\sum_{i=1}^{n-1}\frac{q_n}{p_i}=1$, still by Hölder's inequality, we find

$$J_n\leq k_\lambda^{1/p_n}\Bigg\{\prod_{i=1}^{n-1}\Bigg[\sum_{m_{n-1}=n_0}^{\infty}\cdots\sum_{m_1=n_0}^{\infty}\sum_{m_n=n_0}^{\infty}\tilde{k}(m_1,\cdots,m_n)$$

$$\times[u(m_n)]^{\lambda_n-1}u'(m_n)\frac{[u(m_i)]^{(\lambda_i-1)(1-p_i)}}{[u'(m_i)]^{p_i-1}}$$

$$\times\prod_{\substack{j=1\\(j\neq i)}}^{n-1}[u(m_j)]^{\lambda_j-1}u'(m_j)(a_{m_i}^{(i)})^{p_i}\Bigg]^{\frac{q_n}{p_i}}\Bigg\}^{\frac{1}{q_n}}$$

$$=k_\lambda^{1/p_n}\prod_{i=1}^{n-1}\Bigg\{\sum_{m_i=n_0}^{\infty}\Bigg[\sum_{m_n=n_0}^{\infty}\cdots\sum_{m_{i+1}=n_0}^{\infty}\sum_{m_{i-1}=n_0}^{\infty}\cdots$$

$$\times\sum_{m_1=n_0}^{\infty}\tilde{k}(m_1,\cdots,m_n)[u(m_i)]^{\lambda_i}$$

$$\times\prod_{\substack{j=1\\(j\neq i)}}^{n}[u(m_j)]^{\lambda_j-1}u'(m_j)\Bigg]\frac{[u(m_i)]^{p_i(1-\lambda_i)-1}}{[u'(m_i)]^{p_i-1}}(a_{m_i}^{(i)})^{p_i}\Bigg\}^{\frac{1}{p_i}}$$

$$=k_\lambda^{1/p_n}\prod_{i=1}^{n-1}\Bigg\{\sum_{m_i=n_0}^{\infty}\omega_i(m_i)\frac{[u(m_i)]^{p_i(1-\frac{\lambda}{r_i})-1}}{[u'(m_i)]^{p_i-1}}(a_{m_i}^{(i)})^{p_i}\Bigg\}^{\frac{1}{p_i}}. \quad (6.1.9)$$

Then by (6.1.5), we have (6.1.7)
(Note. for $n=2$, we do not use Hölder's inequality again).

Since $\frac{1}{q_n}+\frac{1}{p_n}=1$, then by Hölder's inequality, we find

$$I=\sum_{m_n=n_0}^{\infty}\Bigg\{\frac{[u'(m_n)]^{1/q_n}}{[u(m_n)]^{(1/q_n)-\lambda_n}}\sum_{m_{n-1}}^{\infty}\cdots\sum_{m_1=n_0}^{\infty}\tilde{k}(m_1,\cdots,m_n)$$

$$\times\prod_{i=1}^{n-1}a_{m_i}^{(i)}\Bigg\}\Bigg\{\frac{[u(m_n)]^{(1/q_n)-\lambda_n}}{[u'(m_n)]^{1/q_n}}a_{m_n}^{(n)}\Bigg\}$$

$$\leq J_n\Bigg\{\sum_{m_n=n_0}^{\infty}\frac{[u(m_n)]^{p_n(1-\lambda_n)-1}}{[u'(m_n)]^{p_n-1}}(a_{m_n}^{(n)})^{p_n}\Bigg\}^{\frac{1}{p_n}}. \quad (6.1.10)$$

Hence by (6.1.7), we have (6.1.6).

On the other-hand, Supposing that (6.1.6) is valid, setting

$$a_{m_n}^{(n)}:=\frac{u'(m_n)}{[u(m_n)]^{1-q_n\lambda_n}}$$

$$\times\Bigg[\sum_{m_{n-1}=n_0}^{\infty}\cdots\sum_{m_1=n_0}^{\infty}\tilde{k}(m_1,\cdots,m_n)\prod_{i=1}^{n-1}a_{m_i}^{(i)}\Bigg]^{q_n-1},$$

we have

$$\sum_{m_n=n_0}^{\infty}\frac{[u(m_n)]^{p_n(1-\lambda_n)-1}}{[u'(m_n)]^{p_n-1}}(a_{m_n}^{(n)})^{p_n}=J_n^{q_n}=I. \quad (6.1.11)$$

By (6.1.9), it follows that $J_n<\infty$. If $J_n=0$, then (6.1.7) is naturally valid; if $0<J_n<\infty$, then

$$0<J_n^{q_n}=\sum_{m_n=n_0}^{\infty}\frac{[u(m_n)]^{p_n(1-\lambda_n)-1}}{[u'(m_n)]^{p_n-1}}(a_{m_n}^{(n)})^{p_n}<\infty,$$

and then by (6.1.6) and (6.1.11), it follows

$$J_n=\Bigg\{\sum_{m_n=n_0}^{\infty}\frac{[u(m_n)]^{p_n(1-\lambda_n)-1}}{[u'(m_n)]^{p_n-1}}(a_{m_n}^{(n)})^{p_n}\Bigg\}^{1-\frac{1}{p_n}}$$

$$<k_\lambda\prod_{i=1}^{n-1}\Bigg\{\sum_{m_i=n_0}^{\infty}\frac{[u(m_i)]^{p_i(1-\lambda_i)-1}}{[u'(m_i)]^{p_i-1}}(a_{m_i}^{(i)})^{p_i}\Bigg\}^{\frac{1}{p_i}}.$$

Hence (6.1.7) is valid, which is equivalent to (6.1.6). □

Theorem 6.1.6 Let the assumptions of Theorem 6.1.4 be fulfilled and additionally, there exists $\delta,\lambda'>0$, such that for any

$$(\tilde{\lambda}_1,\cdots\tilde{\lambda}_n)\in A_\delta:=(\lambda_1-\delta,\lambda_1]\times\cdots$$
$$\times(\lambda_{n-1}-\delta,\lambda_{n-1}]\times[\lambda_n,\lambda_n+\delta)$$

($\sum_{i=1}^{n}\tilde{\lambda}_i=\lambda$),

$$\tilde{\omega}_n(m_n)\geq(k_\lambda+o(1))(1-O(\tfrac{1}{[u(m_n)]^{\lambda'}}))$$
$$(\tilde{\lambda}_i\to\lambda_i, i=1,\cdots,n), \quad (6.1.12)$$

where $\tilde{\omega}_n(m_n)=\omega_n(m_n;\tilde{\lambda}_1,\cdots,\tilde{\lambda}_n)$. Then the constant factor k_λ in (6.1.6) and (6.1.7) is the best possible.

Proof For $0<\varepsilon<\delta\min_{1\leq i\leq n-1}\{p_i,q_n\}$, setting

$\tilde{\lambda}_i = \lambda_i - \frac{\varepsilon}{p_i}$ $(i = 1, \cdots, n-1)$, $\tilde{\lambda}_n = \lambda_n + \frac{\varepsilon}{q_n}$ ($(\tilde{\lambda}_1, \cdots \tilde{\lambda}_n) \in A_\delta$), $\tilde{a}^{(i)}_{m_i} = [u(m_i)]^{\tilde{\lambda}_i - 1} u'(m_i)$ $(i = 1, \cdots, n-1)$,

$$\tilde{a}^{(n)}_{m_n} = [u(m_n)]^{\tilde{\lambda}_n - 1 - \varepsilon} u'(m_n),$$

by (6.1.12), we find

$$\tilde{I} := \sum_{m_n = n_0}^{\infty} \cdots \sum_{m_1 = n_0}^{\infty} \tilde{k}(m_1, \cdots, m_n) \prod_{i=1}^{n} \tilde{a}^{(i)}_{m_i}$$

$$= \sum_{m_n = n_0}^{\infty} \frac{u'(m_n)}{[u(m_n)]^{1+\varepsilon}} \omega_n(m_n; \tilde{\lambda}_1, \cdots \tilde{\lambda}_n)$$

$$\geq (k_\lambda + o(1)) \sum_{m_n = n_0}^{\infty} \frac{u'(m_n)}{[u(m_n)]^{1+\varepsilon}} (1 - O(\frac{1}{[u(m)]^{\lambda'}}))$$

$$\geq (k_\lambda + o(1))[\sum_{m_n = n_0}^{\infty} \frac{u'(m_n)}{[u(m_n)]^{1+\varepsilon}} - \sum_{m_n = n_0}^{\infty} O(\frac{u'(m_n)}{[u(m_n)]^{1+\lambda'}})]$$

$$= (k_\lambda + o(1)) \sum_{m_n = n_0}^{\infty} \frac{u'(m_n)}{[u(m_n)]^{1+\varepsilon}} \left\{ 1 - \frac{\sum_{m_n = n_0}^{\infty} O(\frac{u'(m_n)}{[u(m_n)]^{1+\lambda'}})}{\sum_{m_n = n_0}^{\infty} \frac{u'(m_n)}{[u(m_n)]^{1+\varepsilon}}} \right\}. \tag{6.1.13}$$

If there exists a positive number $k \leq k_\lambda$, such that (6.1.6) is still valid as we replace k_λ by k, then in particular, we find

$$\tilde{I} < k \prod_{i=1}^{n} \{ \sum_{m_i = n_0}^{\infty} \frac{[u(m_i)]^{p_i(1-\lambda_i)-1}}{[u'(m_i)]^{p_i - 1}} (\tilde{a}^{(i)}_{m_i})^{p_i} \}^{\frac{1}{p_i}}$$

$$= k \sum_{m_n = n_0}^{\infty} \frac{u'(m_n)}{[u(m_n)]^{1+\varepsilon}}. \tag{6.1.14}$$

In view of (6.1.13) and (6.1.14), we have

$$(k_\lambda + o(1)) \left\{ 1 - \frac{\sum_{m_n = n_0}^{\infty} O(\frac{u'(m_n)}{[u(m_n)]^{1+\lambda'}})}{\sum_{m_n = n_0}^{\infty} \frac{u'(m_n)}{[u(m_n)]^{1+\varepsilon}}} \right\} < k,$$

and then for $\varepsilon \to 0^+$, by(6.1.2) and (6.1.3), it follows $k_\lambda \leq k$. Hence $k = k_\lambda$ is the best value of (6.1.6). We conclude that the constant factor k_λ in (6.1.7) is the best possible, otherwise we can get a contradiction by (6.1.10) that the constant factor in (6.1.6) is not the best possible. □

In particular, for $u(x) = x, n_0 = 1$, by Theorem 6.1.5 and Theorem 6.1.6, we have the following corollary:

Corollary 6.1.7 Assuming that $n \in \mathbf{N} \setminus \{1\}$, $\lambda_i, \lambda \in \mathbf{R}$, $\sum_{i=1}^{n} \lambda_i = \lambda$, $k_\lambda(x_1, \cdots, x_n) \geq 0$ is a finite measurable function in $\mathbf{R}^n_+$, for $m_i \in \mathbf{N}$, we define the weight coefficients $\varpi_i(m_i) = \varpi_i(m_i; \lambda_1, \cdots, \lambda_n)$ as:

$$\varpi_i(m_i) := m_i^{\lambda_i} \sum_{m_n = 1}^{\infty} \cdots \sum_{m_{i+1} = 1}^{\infty} \sum_{m_{i-1} = 1}^{\infty} \cdots \sum_{m_1 = 1}^{\infty} k_\lambda(m_1, \cdots, m_n)$$

$$\times \prod_{\substack{j=1 \\ (j \neq i)}}^{n} m_j^{\lambda_j - 1} \quad (i = 1, 2, \cdots, n). \tag{6.1.15}$$

If there exist constants $k_\lambda, \delta, \lambda' > 0$, such that for any $m_i \in \mathbf{N}$,

$$\varpi_i(m_i) < k_\lambda \quad (i = 1, \cdots, n), \tag{6.1.16}$$

and for

$$(\tilde{\lambda}_1, \cdots \tilde{\lambda}_n) \in A_\delta$$
$$= (\lambda_1 - \delta, \lambda_1] \times \cdots \times (\lambda_{n-1} - \delta, \lambda_{n-1}] \times [\lambda_n, \lambda_n + \delta)$$

$$(\sum_{i=1}^{n} \tilde{\lambda}_i = \lambda, \tilde{\varpi}_n(m_n) = \varpi_n(m_n; \tilde{\lambda}_1, \cdots \tilde{\lambda}_n)),$$
$$\tilde{\varpi}_n(m_n) \geq (k_\lambda + o(1))(1 - O(\frac{1}{m_n^{\lambda'}}))$$
$$(\tilde{\lambda}_i \to \lambda_i, i = 1, \cdots, n). \tag{6.1.17}$$

then for $p_i > 1$ $(i = 1, \cdots, n)$, $\sum_{i=1}^{n} \frac{1}{p_i} = 1$, $\frac{1}{q_n} = 1 - \frac{1}{p_n}$, $a^{(i)}_{m_i} \geq 0$, such that

$$0 < \sum_{m_i = 1}^{\infty} m_i^{p_i(1-\lambda_i)-1} (a^{(i)}_{m_i})^{p_i} < \infty, \ i = 1, \cdots, n,$$

we have the following equivalent inequalities:

$$\sum_{m_n = 1}^{\infty} \cdots \sum_{m_1 = 1}^{\infty} k_\lambda(m_1, \cdots, m_n) \prod_{i=1}^{n} a^{(i)}_{m_i}$$

$$< k_\lambda \prod_{i=1}^{n} \left\{ \sum_{m_i = 1}^{\infty} m_i^{p_i(1-\lambda_i)-1} (a^{(i)}_{m_i})^{p_i} \right\}^{\frac{1}{p_i}}, \tag{6.1.18}$$

$$\left\{ \sum_{m_n = 1}^{\infty} m_n^{q_n \lambda_n - 1} \times \left[\sum_{m_{n-1} = 1}^{\infty} \cdots \sum_{m_1 = 1}^{\infty} k_\lambda(m_1, \cdots, m_n) \prod_{i=1}^{n-1} a^{(i)}_{m_i} \right]^{q_n} \right\}^{\frac{1}{q_n}}$$

$$< k_\lambda \prod_{i=1}^{n-1} \left\{ \sum_{m_i = 1}^{\infty} m_i^{p_i(1-\lambda_i)-1} (a^{(i)}_{m_i})^{p_i} \right\}^{\frac{1}{p_i}}, \tag{6.1.19}$$

where the constant factor k_λ is the best possible.

6.1.3. SOME EQUIVALENT REVERSE INEQUALITIES

Lemma 6.1.8 Let the assumptions of Definition 6.1.4 be fulfilled. If $p_i<0,(i=1,\cdots,n-1),\ 0<p_n<1$, $\sum_{i=1}^{n}\frac{1}{p_i}=1,\ \frac{1}{q_i}=1-\frac{1}{p_i}$ and $a_{m_i}^{(i)}\ge 0$, then for any $k=1,\cdots,n$, we have the following equivalent inequalities:

$$\tilde{J}_k:=\left\{\sum_{m_k=n_0}^{\infty}\frac{[\omega_k(m_k)]^{1-q_k}u'(m_k)}{[u(m_k)]^{1-q_k\lambda_k}}\left[\sum_{m_n=n_0}^{\infty}\cdots\sum_{m_{k+1}=n_0}^{\infty}\right.\right.$$

$$\left.\left.\times\sum_{m_{k-1}=n_0}^{\infty}\cdots\sum_{m_1=n_0}^{\infty}\tilde{k}(m_1,\cdots,m_n)\prod_{\substack{i=1\\(i\neq k)}}^{n}a_{m_i}^{(i)}\right]^{q_k}\right\}^{\frac{1}{q_k}}$$

$$\ge\prod_{\substack{i=1\\(i\neq k)}}^{n}\left\{\sum_{m_i=n_0}^{\infty}\omega_i(m_i)\frac{[u(m_i)]^{p_i(1-\lambda_i)-1}}{[u'(m_i)]^{p_i-1}}(a_{m_i}^{(i)})^{p_i}\right\}^{\frac{1}{p_i}},\quad(6.1.20)$$

$$I=\sum_{m_n=n_0}^{\infty}\cdots\sum_{m_1=n_0}^{\infty}\tilde{k}(m_1,\cdots,m_n)\prod_{i=1}^{n}a_{m_i}^{(i)}$$

$$\ge\prod_{i=1}^{n}\left\{\sum_{m_i=n_0}^{\infty}\omega_i(m_i)\frac{[u(m_i)]^{p_i(1-\lambda_i)-1}}{[u'(m_i)]^{p_i-1}}(a_{m_i}^{(i)})^{p_i}\right\}^{\frac{1}{p_i}}.\quad(6.1.21)$$

Proof Since $p_kq_k<0,\frac{1}{p_k}+\frac{1}{q_k}=1$, by (6.1.1) and the reverse Hö lder's inequality, we have

$$\sum_{m_n=n_0}^{\infty}\cdots\sum_{m_{k+1}=n_0}^{\infty}\sum_{m_{k-1}=n_0}^{\infty}\cdots\sum_{m_1=n_0}^{\infty}\tilde{k}(m_1,\cdots,m_n)\prod_{\substack{i=1\\(i\neq k)}}^{n}a_{m_i}^{(i)}$$

$$=\sum_{m_n=n_0}^{\infty}\cdots\sum_{m_{k+1}=n_0}^{\infty}\sum_{m_{k-1}=n_0}^{\infty}\cdots\sum_{m_1=n_0}^{\infty}\tilde{k}(m_1,\cdots,m_n)$$

$$\times\left\{\frac{[u(m_k)]^{(\lambda_k-1)(1-p_k)}}{[u'(m_k)]^{p_k-1}}\prod_{\substack{j=1\\(j\neq k)}}^{n}[u(m_j)]^{\lambda_j-1}u'(m_j)\right\}^{\frac{1}{p_k}}$$

$$\times\prod_{\substack{i=1\\(i\neq k)}}^{n}\left\{\frac{[u(m_i)]^{(\lambda_i-1)(1-p_i)}}{[u'(m_i)]^{p_i-1}}\prod_{\substack{j=1\\(j\neq i)}}^{n}[u(m_j)]^{\lambda_j-1}u'(m_j)\right\}^{\frac{1}{p_i}}a_{m_i}^{(i)}$$

$$\ge\left\{\omega_k(m_k)\frac{[u(m_k)]^{p_k(1-\lambda_k)-1}}{[u'(m_k)]^{p_k-1}}\right\}^{\frac{1}{p_k}}$$

$$\times\left\{\sum_{m_n=n_0}^{\infty}\cdots\sum_{m_{k+1}=n_0}^{\infty}\sum_{m_{k-1}=n_0}^{\infty}\cdots\sum_{m_1=n_0}^{\infty}\tilde{k}(m_1,\cdots,m_n)\right.$$

$$\left.\times\prod_{\substack{i=1\\(i\neq k)}}^{n}\left\{\frac{[u(m_i)]^{(\lambda_i-1)(1-p_i)}}{[u'(m_i)]^{p_i-1}}\prod_{\substack{j=1\\(j\neq i)}}^{n}\frac{u'(m_j)}{[u(m_j)]^{1-\lambda_j}}\right\}^{\frac{q_k}{p_i}}(a_{m_i}^{(i)})^{q_k}\right\}^{\frac{1}{q_k}}.$$

No matter what $q_k<0$ or $0<q_k<1$, we have

$$\tilde{J}_k\ge\left\{\sum_{m_k=n_0}^{\infty}\sum_{m_n=n_0}^{\infty}\cdots\sum_{m_{k+1}=n_0}^{\infty}\sum_{m_{k-1}=n_0}^{\infty}\cdots\sum_{m_1=n_0}^{\infty}\tilde{k}(m_1,\cdots,m_n)\right.$$

$$\left.\times\prod_{\substack{i=1\\(i\neq k)}}^{n}\left\{\frac{[u(m_i)]^{(\lambda_i-1)(1-p_i)}}{[u'(m_i)]^{p_i-1}}\prod_{\substack{j=1\\(j\neq i)}}^{n}\frac{u'(m_j)}{[u(m_j)]^{1-\lambda_j}}\right\}^{\frac{q_k}{p_i}}(a_{m_i}^{(i)})^{q_k}\right\}^{\frac{1}{q_k}}$$

$$=\left\{\sum_{m_n=n_0}^{\infty}\cdots\sum_{m_{k+1}=n_0}^{\infty}\sum_{m_{k-1}=n_0}^{\infty}\cdots\sum_{m_1=n_0}^{\infty}\left[\sum_{m_k=n_0}^{\infty}\tilde{k}(m_1,\cdots,m_n)\right.\right.$$

$$\times u(m_k)]^{\lambda_k-1}u'(m_k)]\prod_{\substack{i=1\\(i\neq k)}}^{n}\left[\frac{[u(m_i)]^{(\lambda_i-1)(1-p_i)}}{[u'(m_i)]^{p_i-1}}\right.$$

$$\left.\left.\times\prod_{\substack{j=1\\(j\neq i,k)}}^{n}\frac{u'(m_j)}{[u(m_j)]^{1-\lambda_j}}\right]^{\frac{q_k}{p_i}}(a_{m_i}^{(i)})^{q_k}\right\}^{\frac{1}{q_k}}.\quad(6.1.22)$$

For $n\ge 3$, if $k=n$, then $q_k<0,\frac{q_k}{p_i}>0$ $(i=1,\cdots,n-1),\sum_{i=1}^{n-1}\frac{q_k}{p_i}=1$, we may use Hö lder's inequality in (6.1.22); if $1\le k<n$,

$$q_k>0,\frac{q_k}{p_i}<0\ (i=1,\cdots,n-1),$$

$\sum_{i=1}^{n-1}\frac{q_k}{p_i}=1$, then we may use the reverse Hö lder's inequality in (6.1.22). It follows

$$\tilde{J}_k\ge\prod_{i=1(i\neq k)}^{n}\left\{\sum_{m_k=n_0}^{\infty}\sum_{m_n=n_0}^{\infty}\cdots\sum_{m_{k+1}=n_0}^{\infty}\sum_{m_{k-1}=n_0}^{\infty}\cdots\sum_{m_1=n_0}^{\infty}\right.$$

$$\times[\tilde{k}(m_1,\cdots,m_n)\frac{u'(m_k)}{[u(m_k)]^{1-\lambda_k}}]\left[\frac{[u(m_i)]^{(\lambda_i-1)(1-p_i)}}{[u'(m_i)]^{p_i-1}}\right.$$

$$\left.\left.\times\prod_{\substack{j=1\\(j\neq i,k)}}^{n}\frac{u'(m_j)}{[u(m_j)]^{1-\lambda_j}}\right](a_{m_i}^{(i)})^{q_k}\right\}^{\frac{1}{q_k}}$$

$$=\prod_{\substack{i=1\\(i\neq k)}}^{n}\left\{\sum_{m_i=n_0}^{\infty}\omega_i(m_i)\frac{[u(m_i)]^{p_i(1-\frac{\lambda}{r_i})-1}}{[u'(m_i)]^{p_i-1}}(a_{m_i}^{(i)})^{p_i}\right\}^{\frac{1}{p_i}},$$

(6.1.23)

and then (6.1.20) is valid (Note. For $n=2$, we don't use Hö lder's inequality in (6.1.22) again).

Since $p_kq_k<0,\frac{1}{q_k}+\frac{1}{p_k}=1$, by the reverse Hö lder's inequality, we have

$$I=\sum_{m_k=n_0}^{\infty}\left[\frac{[\omega_k(m_k)]^{\frac{-1}{p_k}}[u'(m_k)]^{1/q_k}}{[u(m_k)]^{\frac{1}{q_k}-\lambda_k}}\sum_{m_n=n_0}^{\infty}\cdots\sum_{m_{k+1}=n_0}^{\infty}\right.$$

$$\times \sum_{m_{k-1}=n_0}^{\infty} \cdots \sum_{m_1=n_0}^{\infty} \tilde{k}(m_1,\cdots,m_n) \prod_{i=1(i\neq k)}^{n} a_{m_i}^{(i)}\Bigg]$$

$$\times\left\{[\omega_k(m_k)]^{\frac{1}{p_k}} \frac{[u(m_k)]^{\frac{1}{q_k}-\lambda_k}}{[u'(m_k)]^{1/q_k}} a_{m_k}^{(k)}\right\}$$

$$\leq \tilde{J}_k \left\{\sum_{m_k=n_0}^{\infty} \omega_k(m_k) \frac{[u(m_k)]^{p_k(1-\lambda_k)-1}}{[u'(m_k)]^{p_k-1}} (a_{m_k}^{(k)})^{p_k}\right\}^{\frac{1}{p_k}}. \tag{6.1.24}$$

Hence by (6.1.20), we have (6.1.21).

On the other-hand, supposing that (6.1.21) is valid, setting

$$a_{m_k}^{(k)} = \frac{u'(m_k)}{[u(m_k)]^{1-q_k\lambda_k}}\left[\sum_{m_n=n_0}^{\infty} \cdots \sum_{m_{k+1}=n_0}^{\infty}\right.$$

$$\left.\times \sum_{m_{k-1}=n_0}^{\infty} \cdots \sum_{m_1=n_0}^{\infty} \tilde{k}(m_1,\cdots,m_n)\prod_{\substack{i=1\\(i\neq k)}}^{n} a_{m_i}^{(i)}\right]^{q_k-1},$$

then by (6.1.21), we find

$$\sum_{m_k=n_0}^{\infty} \omega_k(m_k) \frac{[u(m_k)]^{p_k(1-\lambda_k)-1}}{[u'(m_k)]^{p_k-1}} (a_{m_k}^{(k)})^{p_k} = \tilde{J}_k^{q_k} = I$$

$$\geq \prod_{i=1}^{n}\left\{\sum_{m_i=n_0}^{\infty} \omega_i(m_i) \frac{[u(m_i)]^{p_i(1-\lambda_i)-1}}{[u'(m_i)]^{p_i-1}} (a_{m_i}^{(i)})^{p_i}\right\}^{\frac{1}{p_i}}. \tag{6.1.25}$$

By (6.1.24), if $\tilde{J}_k = \infty$ or $\tilde{J}_k = 0$, then (6.1.20) is naturally valid; if $0 < \tilde{J}_k < \infty$, then

$$0 < \sum_{m_k=n_0}^{\infty} \omega_k(m_k) \frac{[u(m_k)]^{p_k(1-\lambda_k)-1}}{[u'(m_k)]^{p_k-1}} (a_{m_k}^{(k)})^{p_k}$$

$$= \tilde{J}_k^{q_k} < \infty,$$

by (6.1.25), we find

$$\left\{\sum_{m_k=n_0}^{\infty} \omega_k(m_k) \frac{[u(m_k)]^{p_k(1-\lambda_k)-1}}{[u'(m_k)]^{p_k-1}} (a_{m_k}^{(k)})^{p_k}\right\}^{1-\frac{1}{p_k}} = \tilde{J}_k$$

$$\geq \prod_{\substack{i=1\\(i\neq k)}}^{n}\left\{\sum_{m_i=n_0}^{\infty} \omega_i(m_i) \frac{[u(m_i)]^{p_i(1-\lambda_i)-1}}{[u'(m_i)]^{p_i-1}} (a_{m_i}^{(i)})^{p_i}\right\}^{\frac{1}{p_i}}.$$

Hence we have (6.1.20), which is equivalent to (6.1.21). □

Theorem 6.1.9 Let the assumptions of Lemma 6.1.8 be fulfilled and additionally, there exists a constant $k_\lambda > 0$, satisfying

$$\omega_i(m_i) < k_\lambda (m_i \geq n_0; i = 1,\cdots,n-1), \tag{6.1.26}$$

$$\omega_n(m_n) \geq k_\lambda(1-\theta_\lambda(m_n)) \geq c_\lambda > 0 (m_n \geq n_0). \tag{6.1.27}$$

If $a_{m_i}^{(i)} \geq 0$, such that

$$0 < \sum_{m_i=n_0}^{\infty} \frac{[u(m_i)]^{p_i(1-\lambda_i)-1}}{[u'(m_i)]^{p_i-1}} (a_{m_i}^{(i)})^{p_i} < \infty,$$

$$i = 1,2,\cdots,n,$$

then we have the following equivalent inequalities:

$$I = \sum_{m_n=n_0}^{\infty} \cdots \sum_{m_1=n_0}^{\infty} \tilde{k}(m_1,\cdots,m_n)\prod_{i=1}^{n} a_{m_i}^{(i)}$$

$$> k_\lambda \prod_{i=1}^{n-1}\left\{\sum_{m_i=n_0}^{\infty} \frac{[u(m_i)]^{p_i(1-\lambda_i)-1}}{[u'(m_i)]^{p_i-1}} (a_{m_i}^{(i)})^{p_i}\right\}^{\frac{1}{p_i}}$$

$$\times\left\{\sum_{m_n=n_0}^{\infty} [1-\theta_\lambda(m_n)] \frac{[u(m_n)]^{p_n(1-\lambda_n)-1}}{[u'(m_n)]^{p_n-1}} (a_{m_n}^{(n)})^{p_n}\right\}^{\frac{1}{p_n}}, \tag{6.1.28}$$

$$\widehat{J}_n := \left\{\sum_{m_n=n_0}^{\infty} \frac{[1-\theta_\lambda(m_n)]^{1-q_n} u'(m_n)}{[u(m_n)]^{1-q_n\lambda_n}}\right.$$

$$\left.\times\left[\sum_{m_{n-1}=n_0}^{\infty} \cdots \sum_{m_1=n_0}^{\infty} \tilde{k}(m_1,\cdots,m_n)\prod_{i=1}^{n-1} a_{m_i}^{(i)}\right]^{q_n}\right\}^{\frac{1}{q_n}}$$

$$< k_\lambda \prod_{i=1}^{n-1}\left\{\sum_{m_i=n_0}^{\infty} \frac{[u(m_i)]^{p_i(1-\lambda_i)-1}}{[u'(m_i)]^{p_i-1}} (a_{m_i}^{(i)})^{p_i}\right\}^{\frac{1}{p_i}}, \tag{6.1.29}$$

$$\widehat{J}_k := \left\{\sum_{m_k=n_0}^{\infty} \frac{u'(m_k)}{[u(m_k)]^{1-q_k\lambda_k}}\left[\sum_{m_n=n_0}^{\infty} \cdots \sum_{m_{k+1}=n_0}^{\infty}\right.\right.$$

$$\left.\left.\times \sum_{m_{k-1}=n_0}^{\infty} \cdots \sum_{m_1=n_0}^{\infty} \tilde{k}(m_1,\cdots,m_n)\prod_{\substack{i=1\\(i\neq k)}}^{n} a_{m_i}^{(i)}\right]^{q_k}\right\}^{\frac{1}{q_k}}$$

$$> k_\lambda \prod_{i=1(i\neq k)}^{n-1}\left\{\sum_{m_i=n_0}^{\infty} \frac{[u(m_i)]^{p_i(1-\lambda_i)-1}}{[u'(m_i)]^{p_i-1}} (a_{m_i}^{(i)})^{p_i}\right\}^{\frac{1}{p_i}}$$

$$\times\left\{\sum_{m_n=n_0}^{\infty} [1-\theta_\lambda(m_n)] \frac{[u(m_n)]^{p_n(1-\lambda_n)-1}}{[u'(m_n)]^{p_n-1}} (a_{m_n}^{(n)})^{p_n}\right\}^{\frac{1}{p_n}}$$

$$(1 \leq k \leq n-1). \tag{6.1.30}$$

Proof Substitution of (6.1.26) and (6.1.27) in (6.1.21) and (6.1.20), we have (6.1.28), (6.1.29) and (6.1.30). Since by Lemma 6.1.7, it is obvious that (6.1.29) is equivalent to (6.1.28), and (6.1.28) is equivalent to (6.1.30), then inequalities (6.1.28), (6.1.29) and (6.1.30) are all equivalent. □

Theorem 6.1.10 Let the assumptions of Theorem 6.1.9 be fulfilled and additionally,

$$\theta_\lambda(m_n) = O(\frac{1}{u(m_n)^{\lambda'}})(\lambda' > 0; m_n \to \infty),$$

there exists a constant $\delta > 0$, such that for any

$$(\tilde{\lambda}_1, \cdots \tilde{\lambda}_n) \in \tilde{A}_\delta := [\lambda_1, \lambda_1 + \delta) \times \cdots \times [\lambda_{n-1}, \lambda_{n-1} + \delta) \times (\lambda_n - \delta, \lambda_n] \quad (\sum_{i=1}^{n} \tilde{\lambda}_i = \lambda),$$

$$\tilde{\omega}_n(m_n) \le k_\lambda + o(1)(\tilde{\lambda}_i \to \lambda_i; i = 1, \cdots, n), \tag{6.1.31}$$

where $\tilde{\omega}_n(m_n) = \omega_n(m_n; \tilde{\lambda}_1, \cdots, \tilde{\lambda}_n)$. Then the constant factor k_λ in (6.1.28), (6.1.29) and (6.1.30) is the best possible.

Proof For $0 < \varepsilon < \delta \min_{1 \le i \le n-1} \{-p_i, -q_n\}$, setting

$$\tilde{\lambda}_i = \lambda_i - \frac{\varepsilon}{p_i} (i = 1, \cdots, n-1), \tilde{\lambda}_n = \lambda_n + \frac{\varepsilon}{q_n}$$

$$((\tilde{\lambda}_1, \cdots \tilde{\lambda}_n) \in A_\delta),\ \tilde{a}_{m_i}^{(i)} = [u(m_i)]^{\tilde{\lambda}_i - 1} u'(m_i)$$

$(i = 1, \cdots, n-1)$,

$$\tilde{a}_{m_n}^{(n)} = [u(m_n)]^{\tilde{\lambda}_n - 1 - \varepsilon} u'(m_n),$$

by (6.1.31), we find

$$\tilde{I} := \sum_{m_n = n_0}^{\infty} \cdots \sum_{m_1 = n_0}^{\infty} \tilde{k}(m_1, \cdots, m_n) \prod_{i=1}^{n} \tilde{a}_{m_i}^{(i)}$$

$$= \sum_{m_n = n_0}^{\infty} \frac{u'(m_n)}{[u(m_n)]^{1+\varepsilon}} \omega_n(m_n; \tilde{\lambda}_1, \cdots \tilde{\lambda}_n)$$

$$\le (k_\lambda + o(1)) \sum_{m_n = n_0}^{\infty} \frac{u'(m_n)}{[u(m_n)]^{1+\varepsilon}}. \tag{6.1.32}$$

If there exists a positive constant $K \ge k_\lambda$, such that (6.1.28) is still valid as we replace k_λ by K, then in particular, we find

$$\tilde{I} > K \prod_{i=1}^{n-1} \left\{ \sum_{m_i = n_0}^{\infty} \frac{[u(m_i)]^{p_i(1-\lambda_i)-1}}{[u'(m_i)]^{p_i - 1}} (\tilde{a}_{m_i}^{(i)})^{p_i} \right\}^{\frac{1}{p_i}}$$

$$\times \left\{ \sum_{m_n = n_0}^{\infty} [1 - O(\frac{1}{u(m_n)^{\lambda'}})] \frac{[u(m_n)]^{p_n(1-\lambda_n)-1}}{[u'(m_n)]^{p_n - 1}} (\tilde{a}_{m_n}^{(n)})^{p_n} \right\}^{\frac{1}{p_n}}$$

$$= K \sum_{m_n = n_0}^{\infty} \frac{u'(m_n)}{[u(m_n)]^{1+\varepsilon}} \left\{ 1 - \frac{\sum_{m_n = n_0}^{\infty} O(\frac{1}{u(m_n)^{1+\lambda'+\varepsilon}})}{\sum_{m_n = n_0}^{\infty} \frac{u'(m_n)}{[u(m_n)]^{1+\varepsilon}}} \right\}^{\frac{1}{p_n}}. \tag{6.1.33}$$

In view of (6.1.32) and (6.1.33), we have

$$K \left\{ 1 - \frac{\sum_{m_n = n_0}^{\infty} O(\frac{u'(m_n)}{[u(m_n)]^{1+\lambda'+\varepsilon}})}{\sum_{m_n = n_0}^{\infty} \frac{u'(m_n)}{[u(m_n)]^{1+\varepsilon}}} \right\} < k_\lambda + o(1),$$

and then for $\varepsilon \to 0^+$, by (6.1.2) and (6.1.3), it follows $K \le k_\lambda$. Hence $K = k_\lambda$ is the best value of (6.1.28).

We conclude that the constant factor k_λ in (6.1.29) and (6.1.30) is the best possible, otherwise we can get a contradiction by (6.1.24) that the constant factor in (6.1.28) is not the best possible. □

In particular, for $u(x) = x, n_0 = 1$, by Theorem 6.1.9 and Theorem 6.1.10, we have the following corollary:

Corollary 6.1.11 Let the assumptions of Corollary 6.1.7 be fulfilled. If there exist constants $k_\lambda, \delta > 0$, such that for any $m_i \in \mathbf{N}$,

$$\varpi_i(m_i) < k_\lambda \quad (i = 1, \cdots, n-1), \tag{6.1.34}$$

$$\varpi_n(m_n) \ge k_\lambda(1 - \tilde{\theta}_\lambda(m_n)) \ge \tilde{c}_\lambda > 0, \tag{6.1.35}$$

where

$$\tilde{\theta}_\lambda(m_n) = O(\frac{1}{m_n^{\lambda'}})(\lambda' > 0; m_n \to \infty),$$

and for

$$(\tilde{\lambda}_1, \cdots \tilde{\lambda}_n) \in \tilde{A}_\delta = [\lambda_1, \lambda_1 + \delta) \times \cdots \times [\lambda_{n-1}, \lambda_{n-1} + \delta) \times (\lambda_n - \delta, \lambda_n]$$

$(\sum_{i=1}^{n} \tilde{\lambda}_i = \lambda$, $\tilde{\varpi}_n(m_n) = \varpi_n(m_n; \tilde{\lambda}_1, \cdots \tilde{\lambda}_n))$,

$$\tilde{\varpi}_n(m_n) \le k_\lambda + o(1)(\tilde{\lambda}_i \to \lambda_i; i = 1, \cdots, n). \tag{6.1.36}$$

then for $0 < p_n < 1, p_i < 0 \quad (i = 1, \cdots, n-1)$, $\sum_{i=1}^{n} \frac{1}{p_i} = 1$, $\frac{1}{q_i} = 1 - \frac{1}{p_i}$, $a_{m_i}^{(i)} \ge 0$, such that

$$0 < \sum_{m_i = 1}^{\infty} m_i^{p_i(1-\lambda_i)-1} (a_{m_i}^{(i)})^{p_i} < \infty,$$

$$i = 1, \cdots, n,$$

we have the following equivalent inequalities:

$$\sum_{m_n = 1}^{\infty} \cdots \sum_{m_1 = 1}^{\infty} k_\lambda(m_1, \cdots, m_n) \prod_{i=1}^{n} a_{m_i}^{(i)}$$

$$> k_\lambda \prod_{i=1}^{n-1} \left\{ \sum_{m_i = 1}^{\infty} m_i^{p_i(1-\lambda_i)-1} (a_{m_i}^{(i)})^{p_i} \right\}^{\frac{1}{p_i}}$$

$$\times \left\{ \sum_{m_n = 1}^{\infty} [1 - O(\frac{1}{m_n^{\lambda'}})] m_n^{p_n(1-\lambda_n)-1} (a_{m_n}^{(n)})^{p_n} \right\}^{\frac{1}{p_n}}, \tag{6.1.37}$$

$$\left\{\sum_{m_n=1}^{\infty}\frac{m_n^{q_n\lambda_n-1}}{[1-O(\frac{1}{m_n^{\lambda'}})]^{q_n-1}}\right.$$

$$\left.\times[\sum_{m_{n-1}=1}^{\infty}\cdots\sum_{m_1=1}^{\infty}k_\lambda(m_1,\cdots,m_n)\prod_{i=1}^{n-1}a_{m_i}^{(i)}]^{q_n}\right\}^{\frac{1}{q_n}}$$

$$>k_\lambda\prod_{i=1}^{n-1}\{\sum_{m_i=1}^{\infty}m_i^{p_i(1-\lambda_i)-1}(a_{m_i}^{(i)})^{p_i}\}^{\frac{1}{p_i}}, \qquad (6.1.38)$$

$$\left\{\sum_{m_k=1}^{\infty}m_k^{q_k\lambda_k-1}\left[\sum_{m_n=1}^{\infty}\cdots\sum_{m_{k+1}=1}^{\infty}\right.\right.$$

$$\left.\left.\times\sum_{m_{k-1}=1}^{\infty}\cdots\sum_{m_1=1}^{\infty}k_\lambda(m_1,\cdots,m_n)\prod_{\substack{i=1\\(i\neq k)}}^{n}a_{m_i}^{(i)}\right]^{q_k}\right\}^{\frac{1}{q_k}}$$

$$>k_\lambda\prod_{i=1(i\neq k)}^{n-1}\left\{\sum_{m_i=1}^{\infty}m_i^{p_i(1-\lambda_i)-1}(a_{m_i}^{(i)})^{p_i}\right\}^{\frac{1}{p_i}}$$

$$\times\left\{\sum_{m_n=1}^{\infty}[1-O(\frac{1}{m_n^{\lambda'}})]m_n^{p_n(1-\lambda_n)-1}(a_{m_n}^{(n)})^{p_n}\right\}^{\frac{1}{p_n}}$$

$$(1\le k\le n-1). \qquad (6.1.39)$$

where the constant factor k_λ is the best possible.

6.2. SOME PARTICULAR CASES

6.2.1. SOME LEMMAS

For giving some multiple Hilbert-type inequalities with the particular kernels, we introduce some lemmas in the following.

Lemma 6.2.1 Assuming that $n\in\mathbf{N}\setminus\{1\}$, $\lambda_i,\lambda\in\mathbf{R}$, $(i=1,\cdots,n)$, $\sum_{i=1}^{n}\lambda_i=\lambda$, $k_\lambda(x_1,\cdots,x_n)(\ge 0)$ is a finite measurable function in $\mathbf{R}_+^n$, if k_λ is indicated by the following convergent multiple integral:

$$k_\lambda=k(\lambda_1,\cdots,\lambda_{n-1}):=\int_{R_+^{n-1}}k_\lambda(v_1,\cdots,v_{n-1},1)$$

$$\times\prod_{j=1}^{n-1}v_j^{\lambda_j-1}dv_1\cdots dv_{n-1}\in\mathbf{R}, \qquad (6.2.1)$$

then for any $i=1,\cdots,n$,

$$\vartheta_i(t_i):=t_i^{\lambda_i}\int_{R_+^{n-1}}k_\lambda(t_1,\cdots,t_n)$$

$$\times\prod_{\substack{j=1\\(j\neq i)}}^{n}t_j^{\lambda_j-1}dt_1\cdots dt_{i-1}dt_{i+1}\cdots dt_n=k_\lambda. \qquad (6.2.2)$$

Proof Setting $u_j=\frac{t_j}{t_i}(j\neq i)$ in the integral (6.2.2), we find

$$\vartheta_i(t_i)=t_i^{\lambda_i}\int_{R_+^{n-1}}k_\lambda(t_iu_1,\cdots,t_iu_{i-1},t_i,t_iu_{i+1},$$

$$\cdots,t_iu_n)\prod_{\substack{j=1\\(j\neq i)}}^{n}u_j^{\lambda_j-1}t_i^{\sum_{j=1(j\neq i)}^{n}(\lambda_j-1)}t_i^{n-1}$$

$$\times du_1\cdots du_{i-1}du_{i+1}\cdots du_n$$

$$=\int_{R_+^{n-1}}k_\lambda(u_1,\cdots,u_{i-1},1,u_{i+1},\cdots,u_n)$$

$$\times\prod_{\substack{j=1\\(j\neq i)}}^{n}u_j^{\lambda_j-1}du_1\cdots du_{i-1}du_{i+1}\cdots du_n. \qquad (6.2.3)$$

Setting $v_j=\frac{u_j}{u_n}(j\neq i,n)$ in (6.2.3), we obtain the following expression:

$$\vartheta_i(t_i)=\int_{R_+^{n-1}}k_\lambda(u_nv_1,\cdots,u_nv_{i-1},1,u_nv_{i+1},$$

$$\cdots,u_nv_{n-1},u_n)\prod_{\substack{j=1\\(j\neq i)}}^{n-1}v_j^{\lambda_j-1}u_n^{\sum_{j=1(j\neq i)}^{n}(\lambda_j-1)}u_n^{n-2}$$

$$\times dv_1\cdots dv_{i-1}dv_{i+1}\cdots dv_{n-1}du_n$$

$$=\int_{R_+^{n-1}}k_\lambda(v_1,\cdots,v_{i-1},u_n^{-1},v_{i+1},\cdots v_{n-1},1)$$

$$\times\prod_{\substack{j=1\\(j\neq i)}}^{n-1}v_j^{\lambda_j-1}u_n^{-1-\lambda_i}dv_1\cdots dv_{i-1}dv_{i+1}\cdots dv_{n-1}du_n. \qquad (6.2.4)$$

Setting $v_i=u_n^{-1}$ in (6.2.4), we have

$$\vartheta_i(t_i)=k(\lambda_1,\cdots,\lambda_{n-1})=k_\lambda.$$

□

Lemma 6.2.2 Assuming that $n\in\mathbf{N}\setminus\{1\}$, $\lambda_i,\lambda\in\mathbf{R}$, $(i=1,\cdots,n)$, $\sum_{i=1}^{n}\lambda_i=\lambda$, $k_\lambda(x_1,\cdots,x_n)(\ge 0)$ is a finite measurable function in $\mathbf{R}_+^n$, if $k(\lambda_1,\cdots,\lambda_{n-1})$ is defined by (6.2.1), then for any $i=1,\cdots,n-1$, $k(\lambda_1,\cdots,\lambda_{n-1})$ is left continuous at λ_i if there exists a constant $\delta>0$, such that for any $\tilde{\lambda}_i\in(\lambda_i-\delta,\lambda_i]$,

$$k(\lambda_1,\cdots\lambda_{i-1}\tilde{\lambda}_i,\lambda_{i+1},\cdots,\lambda_{n-1})\in\mathbf{R}.$$

Proof For any $\lambda_i^{(k)}\in(\lambda_i-\frac{\delta}{2},\lambda_i](k=1,2,\cdots)$, $\lambda_i^{(k)}\to\lambda_i(k\to\infty)$, we find

$$k_\lambda(v_1,\cdots,v_{n-1},1)\prod_{j=1(j\neq i)}^{n-1} v_j^{\lambda_j-1}v_i^{\lambda_i^{(k)}-1}$$
$$\le k_\lambda(v_1,\cdots,v_{n-1},1)\prod_{\substack{j=1\\(j\neq i)}}^{n-1} v_j^{\lambda_j-1}v_i^{\lambda_i-\frac{\delta}{2}-1}(v_i\in(0,1]);$$
$$k_\lambda(v_1,\cdots,v_{n-1},1)\prod_{\substack{j=1\\(j\neq i)}}^{n-1} v_j^{\lambda_j-1}v_i^{\lambda_i^{(k)}-1}$$
$$\le k_\lambda(v_1,\cdots,v_{n-1},1)\prod_{\substack{j=1\\(j\neq i)}}^{n-1} v_j^{\lambda_j-1}v_i^{\lambda_i-1}(v_i\in(1,\infty)).$$

Since it follows

$$0\le\int_{R_+^{n-1}}\int_0^1 k_\lambda(v_1,\cdots,v_{n-1},1)v_i^{\lambda_i-\frac{\delta}{2}-1}dv_i \times\prod_{j=1(j\neq i)}^{n-1} v_j^{\lambda_j-1}dv_1\cdots dv_{i-1}dv_{i+1}\cdots dv_{n-1}$$
$$\le k(\lambda_1,\cdots,\lambda_{i-1},\lambda_i-\tfrac{\delta}{2},\lambda_{i+1},\cdots,\lambda_{n-1})<\infty;$$
$$0\le\int_{R_+^{n-1}}\int_1^\infty k_\lambda(v_1,\cdots,v_{n-1},1)v_i^{\lambda_i-1}dv_i \times\prod_{j=1(j\neq i)}^{n-1} v_j^{\lambda_j-1}dv_1\cdots dv_{i-1}dv_{i+1}\cdots dv_{n-1}$$
$$\le k(\lambda_1,\cdots,\lambda_{i-1},\lambda_i,\lambda_{i+1},\cdots,\lambda_{n-1})<\infty,$$

then by Lebesgue control convergent theorem (Kuang HEP 1996) [2], we have

$$\lim_{k\to\infty}k(\lambda_1,\cdots\lambda_{i-1},\lambda_i^{(k)},\lambda_{i+1},\cdots,\lambda_{n-1})$$
$$=\int_{R_+^{n-1}}\lim_{k\to\infty}k_\lambda(v_1,\cdots,v_{n-1},1)\times\prod_{\substack{j=1\\(j\neq i)}}^{n-1} v_j^{\lambda_j-1}v_i^{\lambda_i^{(k)}-1}dv_1\cdots dv_{n-1}$$
$$=k(\lambda_1,\cdots,\lambda_{i-1},\lambda_i,\lambda_{i+1},\cdots,\lambda_{n-1}),$$

and then $k(\lambda_1,\cdots,\lambda_{n-1})$ is left continuous at λ_i. □

In the same way, we have the following lemmas:

Lemma 6.2.3 Assuming that $n\in\mathbf{N}\setminus\{1\}$, $\lambda_i,\lambda\in\mathbf{R}$, $(i=1,\cdots,n)$, $\sum_{i=1}^n\lambda_i=\lambda$, $k_\lambda(x_1,\cdots,x_n)\ge 0$ is a finite measurable function in $\mathbf{R}_+^n$, if $k(\lambda_1,\cdots,\lambda_{n-1})$ is defined by (6.2.1), then for any $i=1,\cdots,n-1$, $k(\lambda_1,\cdots,\lambda_{n-1})$ is right continuous at λ_i if there exists a constant $\delta>0$, such that for any $\tilde\lambda_i\in[\lambda_i,\lambda_i+\delta)$,

$$k(\lambda_1,\cdots\lambda_{i-1}\tilde\lambda_i,\lambda_{i+1},\cdots,\lambda_{n-1})\in\mathbf{R}.$$

Lemma 6.2.4 Assuming that $n\in\mathbf{N}\setminus\{1\}$, $\lambda_i,\lambda\in\mathbf{R}$, $(i=1,\cdots,n)$, $\sum_{i=1}^n\lambda_i=\lambda$, $k_\lambda(x_1,\cdots,x_n)(\ge 0)$ is a finite measurable function in $\mathbf{R}_+^n$, if $k(\lambda_1,\cdots,\lambda_{n-1})$ is defined by (6.2.1), then for any $i=1,\cdots,n-1$, $k(\lambda_1,\cdots,\lambda_{n-1})$ is continuous at λ_i if there exists $\delta>0$, such that for any $\tilde\lambda_i\in(\lambda_i-\delta,\lambda_i+\delta)$,

$$k(\lambda_1,\cdots\lambda_{i-1}\tilde\lambda_i,\lambda_{i+1},\cdots,\lambda_{n-1})\in\mathbf{R}.$$

Lemma 6.2.5 Let the assumptions of Definition 6.1.4 be fulfilled and additionally, for $p_i>1$ $\sum_{i=1}^n\frac{1}{p_i}=1$, $\frac{1}{q_n}=1-\frac{1}{p_n}$, there exists a $\delta>0$, such that for any

$$(\tilde\lambda_1,\cdots,\tilde\lambda_{n-1})\in B_\delta: =(\lambda_1-\delta,\lambda_1]\times\cdots\times(\lambda_{n-1}-\delta,\lambda_{n-1}],$$

$\tilde k_\lambda=k(\tilde\lambda_1,\cdots,\tilde\lambda_{n-1})\in\mathbf{R}$. If

$$0<\varepsilon<\delta\min_{1\le i\le n-1}\{p_i\},\ \tilde\lambda_i=\lambda_i-\frac{\varepsilon}{p_i}$$

$(i=1,\cdots,n-1)$, then we have

$$I_\varepsilon:=\varepsilon\int_{n_0}^\infty\frac{u'(x_n)}{[u(x_n)]^{1+\varepsilon}}\Big[\int_{\frac{u(n_0)}{u(x_n)}}^\infty\cdots\int_{\frac{u(n_0)}{u(x_n)}}^\infty k_\lambda(t_1,\cdots,t_{n-1},1)\times\prod_{j=1}^{n-1}t_j^{\tilde\lambda_j-1}dt_1\cdots dt_{n-1}\Big]dx_n\ge k_\lambda+o(1)(\varepsilon\to 0^+).\tag{6.2.5}$$

Proof Setting

$$D_j:=\Big\{(t_1,\cdots,t_{n-1})\Big|t_j\in(0,\tfrac{u(n_0)}{u(x_n)}),\ t_k\in(0,\infty),k\neq j\Big\}$$

and

$$A_j(x_n)=\int_{D_j}k_\lambda(t_1,\cdots,t_{n-1},1)\prod_{j=1}^{n-1}t_j^{\tilde\lambda_j-1}dt_1\cdots dt_{n-1},\tag{6.2.6}$$

We have

$$I_\varepsilon\ge\varepsilon\int_{n_0}^\infty\frac{u'(x_n)dx_n}{[u(x_n)]^{1+\varepsilon}}\tilde k_\lambda-\varepsilon\sum_{j=1}^{n-1}\int_{n_0}^\infty\frac{u'(x_n)A_j(x_n)}{[u(x_n)]^{1+\varepsilon}}dx_n$$
$$=\frac{\tilde k_\lambda}{[u(n_0)]^\varepsilon}-\varepsilon\sum_{j=1}^{n-1}\int_{n_0}^\infty\frac{u'(x_n)}{[u(x_n)]^{1+\varepsilon}}A_j(x_n)dx_n.\tag{6.2.7}$$

Without loses of generality, we show the following estimation:

$$\int_{n_0}^\infty\frac{u'(x_n)A_{n-1}(x_n)}{[u(x_n)]^{1+\varepsilon}}dx_n=O_{n-1}(1)(\varepsilon\to 0^+).\tag{6.2.8}$$

Hence we can reduce (6.2.7) as follows

$$I_\varepsilon \geq \frac{\tilde{k}_\lambda}{[u(n_0)]^\varepsilon} - \varepsilon \sum_{j=1}^{n-1} O_j(1)(\varepsilon \to 0^+). \quad (6.2.9)$$

In fact, setting a $\alpha > 0$, such that $\frac{\varepsilon}{p_{n-1}} + \alpha < \delta$, and then $(\tilde{\lambda}_1, \cdots \tilde{\lambda}_{n-2}, \tilde{\lambda}_{n-1} - \alpha) \in B_\delta$. Since

$$-t_{n-1}^\alpha \ln u_{n-1} \to 0(u_{n-1} \to 0^+),$$

there exists a constant $M > 0$, satisfying

$$-t_{n-1}^\alpha \ln u_{n-1} \leq M \quad (u_{n-1} \in (0,1]),$$

and then by Fubini theorem, it follows that

$$0 \leq \int_{n_0}^\infty \frac{u'(x_n)}{[u(x_n)]^{1+\varepsilon}} A_{n-1}(x_n) dx_n$$

$$\leq \int_{n_0}^\infty \frac{u'(x_n)}{u(x_n)} [\int_{R_+^{n-2}} \int_0^{\frac{u(n_0)}{u(x_n)}} k_\lambda(t_1, \cdots, t_{n-1}, 1)$$

$$\times \prod_{j=1}^{n-1} t_j^{\tilde{\lambda}_j - 1} dt_{n-1} dt_1 \cdots dt_{n-2}] dx_n$$

$$= \int_0^1 \int_{R_+^{n-2}} k_\lambda(t_1, \cdots, t_{n-1}, 1)$$

$$\times \prod_{j=1}^{n-1} t_j^{\tilde{\lambda}_j - 1} (\int_{n_0}^{u^{-1}(\frac{u(n_0)}{t_{n-1}})} \frac{u'(x_n)}{u(x_n)} dx_n) dt_1 \cdots dt_{n-1}$$

$$= \int_0^1 \int_{R_+^{n-2}} k_\lambda(t_1, \cdots, t_{n-1}, 1)$$

$$\times \prod_{j=1}^{n-1} t_j^{\tilde{\lambda}_j - 1} (-\ln t_{n-1}) dt_1 \cdots dt_{n-1}$$

$$\leq \int_0^1 \int_{R_+^{n-2}} k_\lambda(t_1, \cdots, t_{n-1}, 1)$$

$$\times \prod_{j=1}^{n-2} t_j^{\tilde{\lambda}_j - 1} t_{n-1}^{\tilde{\lambda}_{n-1} - \alpha - 1} dt_1 \cdots dt_{n-1}$$

$$= M \cdot k(\tilde{\lambda}_1, \cdots \tilde{\lambda}_{n-2}, \tilde{\lambda}_{n-1} - \alpha) < \infty.$$

Hence we have (6.2.9). In view of Lemma 6.2.2 , it follows $\lim_{\varepsilon \to 0^+} \tilde{k}_\lambda = k_\lambda$ and then by (6.2.9), we have $\lim_{\varepsilon \to 0^+} I_\varepsilon \geq k_\lambda$, and (6.2.5) is valid. □

6.2.2. MULTIPLE HILBERT-TYPE INEQUALITIES AND THE REVERSES WITH THE PARTICULAR KERNELS

Theorem 6.2.6 Let the assumptions of Definition 6.1.4 be fulfilled and additionally,(1) $n_0 - 1 \geq t_0$, $u(n_0 - 1) \geq 0$; (2) there exists constant $\delta > 0$, such that for any

$$(\tilde{\lambda}_1, \cdots \tilde{\lambda}_n) \in A_\delta = (\lambda_1 - \delta, \lambda_1] \times \cdots$$
$$\times (\lambda_{n-1} - \delta, \lambda_{n-1}] \times [\lambda_n, \lambda_n + \delta),$$

$\tilde{k}_\lambda = k(\tilde{\lambda}_1, \cdots, \tilde{\lambda}_{n-1}) > 0$, and for any $i = 1, \cdots, n$,

$$\tilde{f}_i(x_1, \cdots, x_n) := \tilde{k}(x_1, \cdots, x_n)$$

$$\times \prod_{\substack{j=1 \\ (j \neq i)}}^n [u(x_j)]^{\tilde{\lambda}_j - 1} u'(x_j) \quad (6.2.10)$$

is decreasing and strictly decreasing in a subinterval for any variable $x_j > n_0 - 1 (j \neq i)$. Then for $p_i > 1$ $(i = 1, \cdots, n)$, $\sum_{i=1}^n \frac{1}{p_i} = 1$, $\frac{1}{q_n} = 1 - \frac{1}{p_n}$, $a_{m_i}^{(i)} \geq 0$, satisfying

$$0 < \sum_{m_i = n_0}^\infty \frac{[u(m_i)]^{p_i(1-\lambda_i)-1}}{[u'(m_i]^{p_i - 1}}) (a_{m_i}^{(i)})^{p_i} < \infty,$$

$$i = 1, \cdots, n,$$

we have equivalent inequalities (6.1.6) and (6.1.7) with the best constant factor k_λ.

Proof For any $i = 1, \cdots, n$, by (6.1.4), in view of Conditions (1) and (2) (for $\tilde{\lambda}_j = \lambda_j, j = 1, \cdots, n$), we find

$$\omega_i(m_i) = [u(m_i)]^{\lambda_i} \sum_{m_n = n_0}^\infty \cdots \sum_{m_{i+1} = n_0}^\infty \sum_{m_{i-1} = n_0}^\infty \cdots \sum_{m_1 = n_0}^\infty$$

$$\times \tilde{f}_i(m_1, \cdots, m_n) < [u(m_i)]^{\lambda_i} \int_{n_0 - 1}^\infty \cdots \int_{n_0 - 1}^\infty \int_{n_0 - 1}^\infty \cdots$$

$$\times \int_{n_0 - 1}^\infty \tilde{f}_i(x_1, \cdots, x_{i-1}, m_i, x_{i+1}, \cdots, x_n)$$

$$\times dx_1 \cdots dx_{i-1} dx_i \cdots dx_n. \quad (6.2.11)$$

Setting

$$t_j = \frac{u(x_j)}{u(m_i)} (j = 1, \cdots, i-1, i+1, \cdots, n)$$

in (6.2.11), by (6.2.2), we have

$$\omega_i(m_i) \leq \int_{\frac{u(n_0 - 1)}{u(m_i)}}^\infty \cdots \int_{\frac{u(n_0 - 1)}{u(m_i)}}^\infty \int_{\frac{u(n_0 - 1)}{u(m_i)}}^\infty \cdots \int_{\frac{u(n_0 - 1)}{u(m_i)}}^\infty$$

$$\times k_\lambda(t_1, \cdots t_{i-1}, 1, t_{i+1}, \cdots, t_n)$$

$$\times \prod_{\substack{j=1 \\ (j \neq i)}}^n t_j^{\lambda_j - 1} dt_1 \cdots dt_{i-1} dt_{i+1} \cdots dt_n$$

$$\leq \int_0^\infty \cdots \int_0^\infty \int_0^\infty \cdots \int_0^\infty k_\lambda(t_1, \cdots t_{i-1}, 1, t_{i+1}, \cdots, t_n)$$

$$\times \prod_{\substack{j=1 \\ (j \neq i)}}^n t_j^{\lambda_j - 1} dt_1 \cdots dt_{i-1} dt_{i+1} \cdots dt_n = \vartheta_i(1) = k_\lambda.$$

(6.2.12)

Then by Theorem 6.1.5, we have equivalent inequalities (6.1.6) and (6.1.7).

For $0 < \varepsilon < \delta \min_{1 \leq i \leq n-1} \{p_i, q_n\}$, setting $\tilde{\lambda}_i = \lambda_i - \frac{\varepsilon}{p_i}$, $\tilde{\lambda}_n = \lambda_n + \frac{\varepsilon}{q_n}$,

$\tilde{a}_{m_i}^{(i)} = [u(m_i)]^{\tilde{\lambda}_i - 1} u'(m_i)$ $(i = 1, \cdots, n-1)$, $\tilde{a}_{m_n}^{(n)} = [u(m_n)]^{\tilde{\lambda}_n - 1 - \varepsilon} u'(m_n)$, then by Condition (2), we find

$$\tilde{I} := \sum_{m_n = n_0}^{\infty} \cdots \sum_{m_1 = n_0}^{\infty} \tilde{k}(m_1, \cdots, m_n) \prod_{i=1}^{n} \tilde{a}_{m_i}^{(i)}$$

$$= \sum_{m_n = n_0}^{\infty} \frac{u'(m_n)}{[u(m_n)]^{1+\varepsilon}} \omega_n(m_n; \tilde{\lambda}_1, \cdots \tilde{\lambda}_n)$$

$$\geq \int_{n_0}^{\infty} \frac{u'(x_n)}{[u(x_n)]^{1+\varepsilon}} [u(x_n)]^{\tilde{\lambda}_n}$$

$$\times [\int_{n_0}^{\infty} \cdots \int_{n_0}^{\infty} \tilde{k}(x_1, \cdots, x_{n-1}, x_n)$$

$$\times \prod_{j=1}^{n-1} [u(x_j)]^{\tilde{\lambda}_j - 1} u'(x_j) dx_1 \cdots dx_{n-1}] dx_n. \tag{6.2.13}$$

Setting

$$t_j = \frac{u(x_j)}{u(x_n)} (j = 1, \cdots, n-1)$$

in (6.2.12), we have

$$\varepsilon \tilde{I} \geq \varepsilon \int_{n_0}^{\infty} \frac{u'(x_n)}{[u(x_n)]^{1+\varepsilon}} [\int_{\frac{u(n_0)}{u(x_n)}}^{\infty} \cdots \int_{\frac{u(n_0)}{u(x_n)}}^{\infty} k_\lambda(t_1, \cdots, t_{n-1}, 1)$$

$$\times \prod_{j=1}^{n-1} t_j^{\tilde{\lambda}_j - 1} dt_1 \cdots dt_{n-1}] dx_n = I_\varepsilon. \tag{6.2.14}$$

If there exists a positive number $k \leq k_\lambda$, such that (6.1.6) is valid as we replace k_λ by k, then in particular, by (6.2.14), it follows

$$I_\varepsilon \leq \varepsilon \tilde{I} < \varepsilon k \prod_{i=1}^{n} \{ \sum_{m_i = n_0}^{\infty} \frac{[u(m_i)]^{p_i(1-\tilde{\lambda}_i)-1}}{[u'(m_i)]^{p_i - 1}} (\tilde{a}_{m_i}^{(i)})^{p_i} \}^{\frac{1}{p_i}}$$

$$= \varepsilon k \sum_{m=n_0}^{\infty} \frac{u'(m)}{[u(m)]^{1+\varepsilon}}$$

$$\leq \varepsilon k [\frac{u'(n_0)}{[u(n_0)]^{1+\varepsilon}} + \int_{n_0}^{\infty} \frac{u'(x) dx}{[u(x)]^{1+\varepsilon}}]$$

$$= \frac{\varepsilon k \cdot u'(n_0)}{[u(n_0)]^{1+\varepsilon}} + \frac{k}{[u(n_0)]^{\varepsilon}}. \tag{6.2.15}$$

For $\varepsilon \to 0^+$ in (6.2.15), in view of (6.2.5), we have $k_\lambda \leq k$. Hence $k = k_\lambda$ is the best value of (6.1.6). Since (6.1.7) is equivalent to (6.1.6), we can conclude that the constant factor in (6.1.6) is still the best possible. □

Theorem 6.2.7 Let the assumptions of Lemma 6.1.8 be fulfilled and additionally, (1) $n_0 - 1 \geq t_0$, $u(n_0 - 1) \geq 0$; (2) there exists constant $\delta > 0$, such that for any

$$(\tilde{\lambda}_1, \cdots \tilde{\lambda}_n) \in \tilde{A}_\delta = [\lambda_1, \lambda_1 + \delta) \times \cdots$$
$$\times [\lambda_{n-1}, \lambda_{n-1} + \delta) \times (\lambda_n - \delta, \lambda_n],$$

$\tilde{k}_\lambda = k(\tilde{\lambda}_1, \cdots, \tilde{\lambda}_{n-1}) > 0$, and for any $i = 1, \cdots, n$, $\tilde{f}_i(x_1, \cdots, x_n)$ defined by (6.2.10) is decreasing and strictly decreasing in a subinterval for any variable $x_j > n_0 - 1 (j \neq i)$; (3)

$$\omega_n(m_n) \geq k_\lambda (1 - \theta_\lambda(m_n))$$
$$\geq c_\lambda > 0 (m_n \geq n_0), \tag{6.2.16}$$

where

$$\theta_\lambda(m_n) = O(\frac{1}{u(m_n)^{\lambda'}}) (\lambda' > 0; m_n \to \infty).$$

Then for $p_i < 0$ $(i = 1, \cdots, n-1)$, $0 < p_n < 1$, $\sum_{i=1}^{n} \frac{1}{p_i} = 1$, $\frac{1}{q_i} = 1 - \frac{1}{p_i}$, $a_{m_i}^{(i)} \geq 0$, such that

$$0 < \sum_{m_i = n_0}^{\infty} \frac{[u(m_i)]^{p_i(1-\lambda_i)-1}}{[u'(m_i]^{p_i - 1}}) (a_{m_i}^{(i)})^{p_i} < \infty,$$
$$i = 1, \cdots, n,$$

we have the equivalent inequalities (6.1.28), (6.1.29) and(6.1.30) with the best constant factor k_λ.

Proof By Conditions (1) and (2), in the same way of Theorem 6.2.6, we can prove that (6.1.26) and (6.1.31) are valid. Since we have (6.1.27), then by Theorem 6.1.9 and Theorem 6.1.10, we obtain all results of this theorem. □

Corollary 6.2.8 Let the assumptions of Lemma 6.1.8 be fulfilled and additionally, (1) $n_0 - \gamma \geq t_0$, $u(n_0 - \gamma) \geq 0 (0 < \gamma \leq 1)$; (2) there exists constant $\delta > 0$, such that for any

$$(\tilde{\lambda}_1, \cdots \tilde{\lambda}_n) \in A_\delta = (\lambda_1 - \delta, \lambda_1 + \delta) \times \cdots$$
$$\times (\lambda_n - \delta, \lambda_n + \delta),$$

$\tilde{k}_\lambda = k(\tilde{\lambda}_1, \cdots, \tilde{\lambda}_{n-1}) > 0$, and for any $i = 1, \cdots, n$, $\tilde{f}_i(x_1, \cdots, x_n)$ defined by (6.2.10) is decreasing and strictly decreasing in a subinterval for any variable $x_j > n_0 - \gamma (j \neq i)$; (3)

$$\omega_i(m_i) < k_\lambda (m_i \geq n_0; i = 1, \cdots, n),$$
$$\omega_n(m_n) \geq k_\lambda (1 - \theta_\lambda(m_n))$$
$$\geq c_\lambda > 0 (m_n \geq n_0), \tag{6.2.17}$$

where

$$\theta_\lambda(m_n) = O(\frac{1}{u(m_n)^{\lambda'}}) (\lambda' > 0; m_n \to \infty).$$

Then for $p_i > 1$ $(i = 1, \cdots, n)$, $\sum_{i=1}^{n} \frac{1}{p_i} = 1$, $\frac{1}{q_n} = 1 - \frac{1}{p_n}$, $a_{m_i}^{(i)} \geq 0$, satisfying

$$0<\sum_{m_i=n_0}^{\infty}\frac{[u(m_i)]^{p_i(1-\lambda_i)-1}}{[u'(m_i)]^{p_i-1}})(a_{m_i}^{(i)})^{p_i}<\infty,$$

$$i=1,\cdots,n,$$

we have equivalent inequalities (6.1.6) and (6.1.7) with the best constant factor k_λ; for $p_i<0$ $(i=1,\cdots,n-1)$, $0<p_n<1$, we have the equivalent inequalities (6.1.28), (6.1.29) and (6.1.30) with the best constant factor k_λ.

Theorem 6.2.9 Let the assumptions of Definition 6.1.4 be fulfilled and additionally, (1) $n_0-\frac{1}{2}\ge t_0$, $u(n_0-\frac{1}{2})\ge 0$; (2) there exists a constant $\delta>0$, such that for any

$$(\tilde{\lambda}_1,\cdots\tilde{\lambda}_n)\in A_\delta=(\lambda_1-\delta,\lambda_1]\times\cdots$$
$$\times(\lambda_{n-1}-\delta,\lambda_{n-1}]\times[\lambda_n,\lambda_n+\delta),$$
$$\tilde{k}_\lambda=k(\tilde{\lambda}_1,\cdots,\tilde{\lambda}_{n-1})>0,$$

and for any $i=1,\cdots,n$, $\tilde{f}_i(x_1,\cdots,x_n)$ defined by (6.2.10) satisfies

$$\frac{\partial\tilde{f}_i}{\partial x_j}<0,\ \frac{\partial^2\tilde{f}_i}{\partial x_j^2}>0,$$

for any variable $x_j>n_0-\frac{1}{2}(j\ne i)$. Then for $p_i>1$ $(i=1,\cdots,n)$, $\sum_{i=1}^{n}\frac{1}{p_i}=1$, $\frac{1}{q_n}=1-\frac{1}{p_n}$, and $a_{m_i}^{(i)}\ge 0$, such that

$$0<\sum_{m_i=n_0}^{\infty}\frac{[u(m_i)]^{p_i(1-\lambda_i)-1}}{[u'(m_i)]^{p_i-1}})(a_{m_i}^{(i)})^{p_i}<\infty,$$

$$i=1,\cdots,n,$$

we have equivalent inequalities (6.1.6) and (6.1.7) with the best constant factor k_λ.

Proof For any $i=1,\cdots,n$, by (6.1.4), in view of the Conditions (1) and (2), by Hadamard's inequality (Yang JGEI 2010) [3], we find

$$\omega_i(m_i)=[u(m_i)]^{\lambda_i}\sum_{m_n=n_0}^{\infty}\cdots\sum_{m_{i+1}=n_0}^{\infty}\sum_{m_{i-1}=n_0}^{\infty}\cdots\sum_{m_1=n_0}^{\infty}$$
$$\times\tilde{f}_i(m_1,\cdots,m_n)<[u(m_i)]^{\lambda_i}\int_{n_0-\frac{1}{2}}^{\infty}\cdots\int_{n_0-\frac{1}{2}}^{\infty}\int_{n_0-\frac{1}{2}}^{\infty}\cdots$$
$$\times\int_{n_0-\frac{1}{2}}^{\infty}\tilde{f}_i(x_1,\cdots,x_{i-1},m_i,x_{i+1},\cdots,x_n)$$
$$\times dx_1\cdots dx_{i-1}dx_i\cdots dx_n. \qquad (6.2.18)$$

Setting

$$t_j=\frac{u(x_j)}{u(m_i)}(j=1,\cdots,i-1,i+1,\cdots,n)$$

in (6.2.16), by (6.2.2), we have

$$\omega_i(m_i)\le\int_{\frac{u(n_0-\frac{1}{2})}{u(m_i)}}^{\infty}\cdots\int_{\frac{u(n_0-\frac{1}{2})}{u(m_i)}}^{\infty}\int_{\frac{u(n_0-\frac{1}{2})}{u(m_i)}}^{\infty}\cdots\int_{\frac{u(n_0-\frac{1}{2})}{u(m_i)}}^{\infty}$$
$$\times k_\lambda(t_1,\cdots t_{i-1},1,t_{i+1},\cdots,t_n)$$
$$\times\prod_{\substack{j=1\\(j\ne i)}}^{n}t_j^{\lambda_j-1}dt_1\cdots dt_{i-1}dt_{i+1}\cdots dt_n$$
$$\le\int_0^\infty\cdots\int_0^\infty\int_0^\infty\cdots\int_0^\infty k_\lambda(t_1,\cdots t_{i-1},1,t_{i+1},\cdots,t_n)$$
$$\times\prod_{\substack{j=1\\(j\ne i)}}^{n}t_j^{\lambda_j-1}dt_1\cdots dt_{i-1}dt_{i+1}\cdots dt_n=\vartheta_i(1)=k_\lambda.$$
$$(6.2.19)$$

Then by Theorem 6.1.5, we obtain equivalent inequalities (6.1.6) and (6.1.7). The other part of this proof is as Theorem 6.2.6. □

In the same way, we still have the following theorem:

Theorem 6.2.10 Let the assumptions of Lemma 6.1.8 be fulfilled and additionally, (1) $n_0-\frac{1}{2}\ge t_0$, $u(n_0-\frac{1}{2})\ge 0$; (2) there exists a constant $\delta>0$, such that for any

$$(\tilde{\lambda}_1,\cdots\tilde{\lambda}_n)\in\tilde{A}_\delta=[\lambda_1,\lambda_1+\delta)\times\cdots$$
$$\times[\lambda_{n-1},\lambda_{n-1}+\delta)\times(\lambda_n-\delta,\lambda_n],$$
$$\tilde{k}_\lambda=k(\tilde{\lambda}_1,\cdots,\tilde{\lambda}_{n-1})>0,$$

and for $i=1,\cdots,n$, $\tilde{f}_i(x_1,\cdots,x_n)$ defined by (6.2.10) satisfies

$$\frac{\partial\tilde{f}_i}{\partial x_j}<0,\ \frac{\partial^2\tilde{f}_i}{\partial x_j^2}>0,$$

for any variable $x_j>n_0-\frac{1}{2}$ $(j=1,\cdots,n)$; (3) there exists a positive constant c_λ satisfying

$$\omega_n(m_n)\ge k_\lambda(1-\theta_\lambda(m_n))$$
$$\ge c_\lambda>0(m_n\ge n_0), \qquad (6.2.20)$$

where

$$\theta_\lambda(m_n)=O(\frac{1}{u(m_n)^{\lambda'}})(\lambda'>0;m_n\to\infty).$$

Then for $p_i<0$ $(i=1,\cdots,n-1)$, $0<p_n<1$, $\sum_{i=1}^{n}\frac{1}{p_i}=1$, $\frac{1}{q_i}=1-\frac{1}{p_i}$, $a_{m_i}^{(i)}\ge 0$, such that

$$0<\sum_{m_i=n_0}^{\infty}\frac{[u(m_i)]^{p_i(1-\lambda_i)-1}}{[u'(m_i)]^{1-p_i}})(a_{m_i}^{(i)})^{p_i}<\infty,$$

$$i=1,\cdots,n,$$

we have the equivalent inequalities (6.1.28), (6.1.29) and (6.1.30) with the best constant factor k_λ.

6.2.3. SOME COROLLARIES

Corollary 6.2.11 Let the assumptions of Definition 6.1.4 be fulfilled and additionally, (1) $n_0 - 1 \geq t_0$, $u(n_0-1) \geq 0$; (2) $r_i > 1, \lambda_i = \frac{\lambda}{r_i} < 1 \quad (i = 1, \cdots, n)$, $\sum_{i=1}^{n} \frac{1}{r_i} = 1$,

$$0 < k_\lambda = k(\lambda_1, \cdots, \lambda_{n-1}) < \infty,$$

and for any $i = 1, \cdots, n$,

$$\tilde{f}_i(x_1, \cdots, x_n) := k_\lambda(u(x_1), \cdots, u(x_n)) \times \prod_{\substack{j=1 \\ (j \neq i)}}^{n} [u(x_j)]^{\frac{\lambda}{r_j}-1} u'(x_j) \tag{6.2.21}$$

is decreasing and strictly decreasing in a subinterval for any variable $x_j > n_0 - 1 (j \neq i)$; (3)

$$c_\lambda := \int_1^\infty \cdots \int_1^\infty k_\lambda(t_1, \cdots, t_{n-1}, 1) \times \prod_{j=1}^{n-1} t_j^{\frac{\lambda}{r_j}-1} dt_1 \cdots dt_{n-1} > 0,$$

and for $n \geq 3, 0 < t_j \leq 1 (j = 1, \cdots, n-1)$,

$$\varphi(t_j) := \int_{R_+^{n-2}} k_\lambda(t_1, \cdots, t_{n-1}, 1) \prod_{\substack{i=1 \\ (i \neq j)}}^{n-1} t_i^{\frac{\lambda}{r_i}-1} \times dt_1 \cdots dt_{j-1} dt_{j+1} \cdots dt_{n-1} \leq M_j < \infty. \tag{6.2.22}$$

Then for $p_n > 0 (\neq 1)$, $\sum_{i=1}^{n} \frac{1}{p_i} = 1$, $\frac{1}{q_i} = 1 - \frac{1}{p_i}$, and $a_{m_i}^{(i)} \geq 0$, satisfying

$$0 < \sum_{m_i=n_0}^{\infty} \frac{[u(m_i)]^{p_i(1-\frac{\lambda}{r_i})-1}}{[u'(m_i)]^{p_i-1}} (a_{m_i}^{(i)})^{p_i} < \infty, \quad i = 1, \cdots, n, \tag{6.2.23}$$

(a) if $p_i > 1 (i = 1, \cdots, n)$, then we have equivalent inequalities (6.1.6) and (6.1.7) with the best constant factor k_λ; (b) if $p_i < 0 \quad (i = 1, \cdots, n-1)$, $0 < p_n < 1$, then we have the equivalent inequalities (6.1.28), (6.1.29) and (6.1.30) with the best constant factor k_λ, and

$$\theta_\lambda(m_n) = O\left(\frac{1}{[u(m_n)]^{\lambda'}}\right)$$

$$(\lambda' = \lambda \min_{1 \leq j \leq n-1} \{\tfrac{1}{r_j}\} > 0; m_n \to \infty).$$

Proof We only prove that (6.2.16) is value. By Condition (2), setting

$$t_j = \frac{u(x_j)}{u(m_n)} \quad (j = 1, \cdots, n-1),$$

we find

$$\omega_n(m_n) = [u(m_n)]^{\frac{\lambda}{r_n}} \sum_{m_{n-1}=n_0}^{\infty} \cdots \sum_{m_1=n_0}^{\infty} \times \tilde{k}(m_1, \cdots, m_n) \prod_{j=1}^{n-1} [u(m_j)]^{\frac{\lambda}{r_j}-1} u'(m_j)$$

$$\geq [u(m_n)]^{\frac{\lambda}{r_n}} \int_{n_0}^{\infty} \cdots \int_{n_0}^{\infty} \tilde{k}(x_1, \cdots, x_{n-1}, m_n) \times \prod_{j=1}^{n-1} [u(x_j)]^{\frac{\lambda}{r_j}-1} u'(x_j) dx_1 \cdots dx_{n-1}$$

$$= \vartheta_\lambda(m_n) := \int_{\frac{u(n_0)}{u(m_n)}}^{\infty} \cdots \int_{\frac{u(n_0)}{u(m_n)}}^{\infty} k_\lambda(t_1, \cdots, t_{n-1}, 1) \times \prod_{j=1}^{n-1} t_j^{\frac{\lambda}{r_j}-1} dt_1 \cdots dt_{n-1}. \tag{6.2.24}$$

Hence we find

$$\omega_n(m_n) \geq \vartheta_\lambda(m_n) \geq c_\lambda = \int_1^\infty \cdots \int_1^\infty k_\lambda(t_1, \cdots, t_{n-1}, 1) \times \prod_{j=1}^{n-1} t_j^{\frac{\lambda}{r_j}-1} dt_1 \cdots dt_{n-1} > 0. \tag{6.2.25}$$

In view of (6.2.24), setting

$$\theta_\lambda(m_n) := 1 - \frac{1}{k_\lambda} \vartheta_\lambda(m_n),$$

we have

$$\vartheta_\lambda(m_n) = k_\lambda(1 - \theta_\lambda(m_n)). \tag{6.2.26}$$

Setting

$$D_j := \{(t_1, \cdots, t_{n-1}) \big| t_j \in (0, \tfrac{u(n_0)}{u(m_n)}), t_k \in (0, \infty)(k \neq j)\}$$

and

$$A_j(m_n) := \int_{D_j} k_\lambda(t_1, \cdots, t_{n-1}, 1) \times \prod_{j=1}^{n-1} t_j^{\frac{\lambda}{r_j}-1} dt_1 \cdots dt_{n-1},$$

By (6.2.24) and (6.2.26), we find

$$k_\lambda - k_\lambda \theta_\lambda(m_n) = \vartheta_\lambda(m_n) \geq k_\lambda - \sum_{j=1}^{n-1} A_j(m_n),$$

and

$$0 < \theta_\lambda(m_n) \leq \frac{1}{k_\lambda} \sum_{j=1}^{n-1} A_j(m_n). \tag{6.2.27}$$

For $n = 2$, it is obvious that (6.2.16) is valid; for $n \geq 3$, without loses of generality, we estimate $A_j(m_n)$ for $j = n-1$. By Condition (3), setting $\lambda' = \lambda \min_{1 \leq j \leq n-1} \{\frac{1}{r_j}\}$, we find

$$0 \leq A_{n-1}(m_n) = \int_0^{\frac{u(n_0)}{u(m_n)}} [\int_{R_+^{n-2}} k_\lambda(t_1, \cdots, t_{n-1}, 1) \times \prod_{j=1}^{n-1} t_j^{\frac{\lambda}{r_j}-1} dt_1 \cdots dt_{n-2}] dt_{n-1}$$

$$\leq M_{n-1}\int_0^{\frac{u(n_0)}{u(m_n)}} t_{n-1}^{\frac{\lambda}{r_{n-1}}-1} dt_{n-1}$$

$$= M_{n-1}\frac{r_{n-1}}{\lambda}[\frac{u(n_0)}{u(m_n)}]^{\frac{\lambda}{r_{n-1}}}$$

$$\leq M_{n-2}\frac{r_{n-1}}{\lambda}[u(n_0)]^{\frac{\lambda}{r_{n-1}}}[\frac{1}{u(m_n)}]^{\lambda'}.$$

Hence by (6.2.27), we have

$$\theta_\lambda(m_n) = O(\frac{1}{[u(m_n)]^{\lambda'}})(m_n \to \infty).$$

In view of (6.2.25) and (6.2.26), we have (6.2.16).□

Corollary 6.2.12 Let the assumptions of Definition 6.1.4 be fulfilled and additionally, (1) $n_0 - 1 \geq t_0$, $u(n_0-1)\geq 0$; (2) $r_i > 1, \lambda_i = \frac{-\lambda}{r_i} > -1$ $(i=1,\cdots,n)$, $\sum_{i=1}^n \frac{1}{r_i} = 1$,

$$0 < k_{-\lambda} = k(\lambda_1,\cdots,\lambda_{n-1}) < \infty,$$

and for $i=1,\cdots,n$,

$$\tilde{f}_i(x_1,\cdots,x_n) := k_{-\lambda}(u(x_1),\cdots,u(x_n)) \times\prod_{\substack{j=1\\(j\neq i)}}^{n}[u(x_j)]^{\frac{-\lambda}{r_j}-1}u'(x_j) \quad (6.2.28)$$

is decreasing and strictly decreasing in a subinterval for any variable $x_j > n_0 - 1 (j \neq i)$; (3)

$$c_{-\lambda} := \int_1^\infty \cdots \int_1^\infty k_{-\lambda}(t_1,\cdots,t_{n-1},1) \times\prod_{j=1}^{n-1} t_j^{\frac{-\lambda}{r_j}-1} dt_1\cdots dt_{n-1} > 0,$$

and for $n\geq 3$, $0 < t_j \leq 1$ ($j=1,\cdots,n-1$), there exists $M_j > 0$, such that

$$\tilde{\varphi}(t_j) := \int_{R_+^{n-2}} k_{-\lambda}(t_1,\cdots,t_{n-1},1)\prod_{\substack{i=1\\(i\neq j)}}^{n-1} t_i^{\frac{-\lambda}{r_i}-1} \times dt_1\cdots dt_{j-1}dt_{j+1}\cdots dt_{n-1} \leq M_j\cdot t_j^{\frac{\lambda}{r_j}}. \quad (6.2.29)$$

Then for $p_n > 0(\neq 1)$, $\sum_{i=1}^n \frac{1}{p_i} = 1$, $\frac{1}{q_i} = 1 - \frac{1}{p_i}$, $a_{m_i}^{(i)} \geq 0$, satisfying

$$0 < \sum_{m_i=n_0}^\infty \frac{[u(m_i)]^{p_i(1-\frac{\lambda}{r_i})-1}}{[u'(m_i)]^{p_i-1}})(a_{m_i}^{(i)})^{p_i} < \infty,$$

$$i=1,\cdots,n, \quad (6.2.30)$$

(a) if $p_i > 1$ $(i=1,\cdots,n)$, then we have equivalent inequalities (6.1.6) and (6.1.7) with the best constant factor $k_{-\lambda}$; (b) if $p_i < 0$ $(i=1,\cdots,n-1)$, $0 < p_n < 1$, then we have the equivalent inequalities (6.1.28), (6.1.29) and (6.1.30) with the best constant factor $k_{-\lambda}$ and

$$\theta_{-\lambda}(m_n) = O(\frac{1}{[u(m_n)]^{\lambda'}})$$

$$(\lambda' = \lambda \min_{1\leq j\leq n-1}\{1-\frac{1}{r_j}\} > 0; m_n \to \infty).$$

Proof We only prove that (6.2.16) is value. By Condition (2), setting

$$t_j = \frac{u(x_j)}{u(m_n)}(j=1,\cdots,n-1),$$

we find

$$\omega_n(m_n) = [u(m_n)]^{\frac{-\lambda}{r_n}}\sum_{m_{n-1}=n_0}^\infty\cdots\sum_{m_1=n_0}^\infty \times\tilde{k}(m_1,\cdots,m_n)\prod_{j=1}^{n-1}[u(m_j)]^{\frac{-\lambda}{r_j}-1}u'(m_j)$$

$$\geq [u(m_n)]^{\frac{-\lambda}{r_n}}\int_{n_0}^\infty\cdots\int_{n_0}^\infty \tilde{k}(x_1,\cdots,x_{n-1},m_n) \times\prod_{j=1}^{n-1}[u(x_j)]^{\frac{-\lambda}{r_j}-1}u'(x_j)dx_1\cdots dx_{n-1}$$

$$= \vartheta_{-\lambda}(m_n) := \int_{\frac{u(n_0)}{u(m_n)}}^\infty\cdots\int_{\frac{u(n_0)}{u(m_n)}}^\infty k_{-\lambda}(t_1,\cdots,t_{n-1},1) \times\prod_{j=1}^{n-1} t_j^{\frac{-\lambda}{r_j}-1}dt_1\cdots dt_{n-1}. \quad (6.2.31)$$

Hence we obtain

$$\omega_n(m_n) \geq \vartheta_{-\lambda}(m_n) \geq c_{-\lambda} = \int_1^\infty\cdots\int_1^\infty k_{-\lambda}(t_1,\cdots,t_{n-1},1) \times\prod_{j=1}^{n-1} t_j^{\frac{-\lambda}{r_j}-1}dt_1\cdots dt_{n-1} > 0. \quad (6.2.32)$$

Setting

$$\theta_{-\lambda}(m_n) := 1 - \frac{1}{k_{-\lambda}}\vartheta_{-\lambda}(m_n),$$

we have

$$\vartheta_{-\lambda}(m_n) = k_{-\lambda}(1-\theta_{-\lambda}(m_n)). \quad (6.2.33)$$

Setting

$$D_j := \left\{(t_1,\cdots,t_{n-1})\middle| t_j \in (0,\frac{u(n_0)}{u(m_n)}), t_k \in (0,\infty)(k\neq j)\right\}$$

and

$$\tilde{A}_j(m_n) := \int_{D_j} k_\lambda(t_1,\cdots,t_{n-1},1) \times\prod_{j=1}^{n-1} t_j^{\frac{-\lambda}{r_j}-1}dt_1\cdots dt_{n-1},$$

by (6.2.32) and (6.2.33), we find

$$k_{-\lambda} - k_{-\lambda}\theta_{-\lambda}(m_n)$$

$$= \vartheta_{-\lambda}(m_n) \geq k_{-\lambda} - \sum_{j=1}^{n-1} \tilde{A}_j(m_n),$$

and

$$0 < \theta_{-\lambda}(m_n) \leq \frac{1}{k_{-\lambda}} \sum_{j=1}^{n-1} \tilde{A}_j(m_n). \quad (6.2.34)$$

For $n = 2$, by the conditions, we have (6.2.16); for $n \geq 3$, without loses of generality, we estimate $\tilde{A}_j(m_n)$ for $j = n-1$. Since by Condition (3), for

$$\lambda' = \lambda \min_{1\leq j\leq n-1} \{1 - \tfrac{1}{r_j}\} > 0,$$

we have

$$0 \leq \tilde{A}_{n-1}(m_n) = \int_0^{\frac{u(n_0)}{u(m_n)}} [\int_{R_+^{n-2}} k_{-\lambda}(t_1, \cdots, t_{n-1}, 1)$$

$$\times \prod_{j=1}^{n-1} t_j^{\frac{-\lambda}{r_j}-1} dt_1 \cdots dt_{n-2}] dt_{n-1}$$

$$\leq M_{n-1} \int_0^{\frac{u(n_0)}{u(m_n)}} t_{n-1}^{\lambda - \frac{\lambda}{r_{n-1}} - 1} dt_{n-1}$$

$$= M_{n-1} \frac{r_{n-1}}{\lambda(r_{n-1}-1)} [\frac{u(n_0)}{u(m_n)}]^{\lambda - \frac{\lambda}{r_{n-1}}}$$

$$\leq \frac{M_{n-1} r_{n-1}}{\lambda(r_{n-1}-1)} [u(n_0)]^{\lambda - \frac{\lambda}{r_{n-1}}} [\frac{1}{u(m_n)}]^{\lambda'}.$$

Hence by (6.2.34), we have

$$\theta_\lambda(m_n) = O(\frac{1}{[u(m_n)]^{\lambda'}})(m_n \to \infty).$$

In view of (6.2.32) and (6.2.33), we have (6.2.16). □

In the same way, we still have the following corollary:

Corollary 6.2.13 Let the assumptions of Definition 6.1.4 be fulfilled and additionally, (1) $n_0 - \frac{1}{2} \geq t_0$, $u(n_0 - \frac{1}{2}) \geq 0$; (2) $r_i > 1, \lambda_i = \frac{\lambda}{r_i} > 1$ $(i = 1, \cdots, n)$, $\sum_{i=1}^{n} \frac{1}{r_i} = 1$, $0 < k_\lambda = k(\lambda_1, \cdots, \lambda_{n-1}) < \infty$, and for any $i = 1, \cdots, n$, the function $\tilde{f}_i(x_1, \cdots, x_n)$ defined by (6.2.20) satisfies $\frac{\partial \tilde{f}_i}{\partial x_j} < 0, \frac{\partial^2 \tilde{f}_i}{\partial x_j^2} > 0$, for any variable $x_j > n_0 - \frac{1}{2}$ $(j \neq i)$; (3)

$$c_\lambda := \int_1^\infty \cdots \int_1^\infty k_\lambda(t_1, \cdots, t_{n-1}, 1)$$

$$\times \prod_{j=1}^{n-1} t_j^{\frac{\lambda}{r_j}-1} dt_1 \cdots dt_{n-1} > 0,$$

and for any $0 < t_j \leq 1$ $(j = 1, \cdots, n-1)$,

$$\varphi(t_j) = \int_{R_+^{n-2}} k_\lambda(t_1, \cdots, t_{n-1}, 1) \prod_{\substack{i=1 \\ (i\neq j)}}^{n-1} t_i^{\frac{\lambda}{r_i}-1}$$

$$\times dt_1 \cdots dt_{j-1} dt_{j+1} \cdots dt_{n-1} \leq M_j < \infty.$$

Then for $p_n > 0(\neq 1)$, $\sum_{i=1}^{n} \frac{1}{p_i} = 1$, $\frac{1}{q_i} = 1 - \frac{1}{p_i}$, $a_{m_i}^{(i)} \geq 0$, satisfying(6.2.22), (a) if $p_i > 1$ $(i = 1, \cdots, n)$, then we have equivalent inequalities (6.1.6) and (6.1.7) with the best constant factor k_λ; (b) if $p_i < 0$ $(i = 1, \cdots, n-1)$, $0 < p_n < 1$, then we have the equivalent inequalities (6.1.28), (6.1.29) and (6.1.30) with the best constant factor k_λ, and

$$\theta_\lambda(m_n) = O(\frac{1}{[u(m_n)]^{\lambda'}})$$

$$(\lambda' = \lambda \min_{1\leq j\leq n-1} \{\tfrac{1}{r_j}\} > 0; m_n \to \infty).$$

Corollary 6.2.14 Let the assumptions of Definition 6.1.4 be fulfilled and additionally, (1) $n_0 - \gamma \geq t_0$, $u(n_0 - \gamma) \geq 0 (0 < \gamma \leq 1)$;(2) $r_i > 1, \lambda_i = \frac{\lambda}{r_i} > 1$ $(i = 1, \cdots, n)$, $\sum_{i=1}^{n} \frac{1}{r_i} = 1$,

$$0 < k_\lambda = k(\lambda_1, \cdots, \lambda_{n-1}) < \infty,$$

and for any $i = 1, \cdots, n$, $\tilde{f}_i(x_1, \cdots, x_n)$ defined by (6.2.20) is decreasing and strictly decreasing in a subinterval for any variable $x_j > n_0 - \gamma (j \neq i)$; (3)

$$\omega_i(m_i) < k_\lambda (m_i \geq n_0; i = 1, \cdots, n),$$

$$c_\lambda := \int_1^\infty \cdots \int_1^\infty k_\lambda(t_1, \cdots, t_{n-1}, 1)$$

$$\times \prod_{j=1}^{n-1} t_j^{\frac{\lambda}{r_j}-1} dt_1 \cdots dt_{n-1} > 0,$$

and for any $0 < t_j \leq 1$ $(j = 1, \cdots, n-1)$,

$$\varphi(t_j) = \int_{R_+^{n-2}} k_\lambda(t_1, \cdots, t_{n-1}, 1) \prod_{\substack{i=1 \\ (i\neq j)}}^{n-1} t_i^{\frac{\lambda}{r_i}-1}$$

$$\times dt_1 \cdots dt_{j-1} dt_{j+1} \cdots dt_{n-1} \leq M_j < \infty. \quad (6.2.35)$$

Then for $p_n > 0(\neq 1)$, $\sum_{i=1}^{n} \frac{1}{p_i} = 1$, $\frac{1}{q_i} = 1 - \frac{1}{p_i}$, $a_{m_i}^{(i)} \geq 0$, satisfying (6.2.22), (a) if $p_i > 1$ $(i = 1, \cdots, n)$, then we have equivalent inequalities (6.1.6) and (6.1.7) with the best constant factor k_λ; (b) if $p_i < 0$ $(i = 1, \cdots, n-1)$, $0 < p_n < 1$, then we have the equivalent inequalities (6.1.28), (6.1.29) and (6.1.30) with the best constant factor k_λ, and

$$\theta_\lambda(m_n) = O(\frac{1}{[u(m_n)]^{\lambda'}})$$

$$(\lambda' = \lambda \min_{1\leq j\leq n-1} \{\tfrac{1}{r_j}\} > 0; m_n \to \infty).$$

Corollary 6.2.15 Let the assumptions of Definition 6.1.4 be fulfilled and additionally, (1) $n_0 - \frac{1}{2} \geq t_0$, $u(n_0 - \frac{1}{2}) \geq 0$; (2) for $r_i > 1, \lambda_i = \frac{-\lambda}{r_i} > -1$ $(i = 1, \cdots, n)$, $\sum_{i=1}^{n} \frac{1}{r_i} = 1$,

$$0 < k_{-\lambda} = k(\lambda_1, \cdots, \lambda_{n-1}) < \infty,$$

and for $i = 1, \cdots, n$, $\tilde{f}_i(x_1, \cdots, x_n)$ defined by (6.2.27) satisfies $\frac{\partial \tilde{f}_i}{\partial x_j} < 0, \frac{\partial^2 \tilde{f}_i}{\partial x_j^2} > 0$, for any variable $x_j > n_0 - \frac{1}{2} (j \neq i)$; (3)

$$c_{-\lambda} := \int_1^{\infty} \cdots \int_1^{\infty} k_{-\lambda}(t_1, \cdots, t_{n-1}, 1) \times \prod_{j=1}^{n-1} t_j^{\frac{-\lambda}{r_j} - 1} dt_1 \cdots dt_{n-1} > 0,$$

and for $n \geq 3$, $0 < t_j \leq 1$ ($j = 1, \cdots, n-1$), there exists a constant $\tilde{M}_j > 0$, such that

$$\tilde{\varphi}(t_j) = \int_{R_+^{n-2}} k_{-\lambda}(t_1, \cdots, t_{n-1}, 1) \prod_{\substack{i=1 \\ (i \neq j)}}^{n-1} t_i^{\frac{-\lambda}{r_i} - 1} \times dt_1 \cdots dt_{j-1} dt_{j+1} \cdots dt_{n-1} \leq \tilde{M}_j \cdot t_j^{\lambda}.$$

Then for $p_n > 0 (\neq 1)$, $\sum_{i=1}^{n} \frac{1}{p_i} = 1$, $\frac{1}{q_i} = 1 - \frac{1}{p_i}$, $a_{m_i}^{(i)} \geq 0$, satisfying (6.2.29), (a) if $p_i > 1$ $(i = 1, \cdots, n)$, then we have equivalent inequalities (6.1.6) and (6.1.7) with the best constant factor $k_{-\lambda}$; (b) if $p_i < 0$ $(i = 1, \cdots, n-1)$, $0 < p_n < 1$, then we have the equivalent inequalities (6.1.28), (6.1.29) and (6.1.30) with the best constant factor $k_{-\lambda}$ and

$$\theta_{-\lambda}(m_n) = O(\frac{1}{[u(m_n)]^{\lambda'}})$$
$$(\lambda' = \lambda \min_{1 \leq j \leq n-1} \{1 - \frac{1}{r_j}\} > 0; m_n \to \infty).$$

Corollary 6.2.16 Let the assumptions of Definition 6.1.4 be fulfilled and additionally, (1) $n_0 - \gamma \geq t_0$, $u(n_0 - \gamma) \geq 0 (0 < \gamma \leq 1)$; (2) for $r_i > 1$, $\lambda_i = \frac{-\lambda}{r_i} > -1$ $(i = 1, \cdots, n)$, $\sum_{i=1}^{n} \frac{1}{r_i} = 1$,

$$0 < k_{-\lambda} = k(\lambda_1, \cdots, \lambda_{n-1}) < \infty,$$

and for $i = 1, \cdots, n$, $\tilde{f}_i(x_1, \cdots, x_n)$ defined by (6.2.28) is decreasing and strictly decreasing in a subinterval for any variable $x_j > n_0 - \gamma (j \neq i)$; (3)

$$\omega_i(m_i) < k_{\lambda}(m_i \geq n_0; i = 1, \cdots, n),$$

$$c_{-\lambda} := \int_1^{\infty} \cdots \int_1^{\infty} k_{-\lambda}(t_1, \cdots, t_{n-1}, 1) \times \prod_{j=1}^{n-1} t_j^{\frac{-\lambda}{r_j} - 1} dt_1 \cdots dt_{n-1} > 0,$$

and for $n \geq 3$, $0 < t_j \leq 1$ ($j = 1, \cdots, n-1$), there exists a constant $\tilde{M}_j > 0$, such that

$$\tilde{\varphi}(t_j) = \int_{R_+^{n-2}} k_{-\lambda}(t_1, \cdots, t_{n-1}, 1) \prod_{\substack{i=1 \\ (i \neq j)}}^{n-1} t_i^{\frac{-\lambda}{r_i} - 1} \times dt_1 \cdots dt_{j-1} dt_{j+1} \cdots dt_{n-1} \leq \tilde{M}_j \cdot t_j^{\lambda}.$$

Then for $p_n > 0 (\neq 1)$, $\sum_{i=1}^{n} \frac{1}{p_i} = 1$, $\frac{1}{q_i} = 1 - \frac{1}{p_i}$, $a_{m_i}^{(i)} \geq 0$, satisfying(6.2.30), (a) if $p_i > 1$ $(i = 1, \cdots, n)$, then we have equivalent inequalities (6.1.6) and (6.1.7) with the best constant factor $k_{-\lambda}$; (b) if $p_i < 0$ $(i = 1, \cdots, n-1)$, $0 < p_n < 1$, then we have the equivalent inequalities (6.1.28), (6.1.29) and (6.1.30) with the best constant factor $k_{-\lambda}$ and

$$\theta_{-\lambda}(m_n) = O(\frac{1}{[u(m_n)]^{\lambda'}})$$
$$(\lambda' = \lambda \min_{1 \leq j \leq n-1} \{1 - \frac{1}{r_j}\} > 0; m_n \to \infty).$$

Note 6.2.17 If for $i = 1, \cdots, n$, $\tilde{f}_i(x_1, \cdots, x_n)$ defined by (6.2.20) and (6.2.27) is convex decreasing and strictly decreasing in a subinterval for any variable $x_j > n_0 - \frac{1}{2} (j = 1, \cdots, n)$, then we still have Corollary 6.2.13 and Corollary 6.2.15.

6.3. SOME EXAMPLES

6.3.1. EXAMPLES FOR $k_{\lambda}(t_1, \cdots, t_{n-1}, 1) = \frac{1}{(t_1 + \cdots + t_{n-1} + 1)^{\lambda}}$

Lemma 6.3.1 For $n \geq 2$, $\lambda > 0, r_i > 1, \lambda_i = \frac{\lambda}{r_i} < 1$ $(i = 1, \cdots, n)$, $\sum_{i=1}^{n} \frac{1}{r_i} = 1$, we have

$$k_{\lambda} = \int_{R_+^{n-1}} \frac{1}{(t_1 + \cdots + t_{n-1} + 1)^{\lambda}} \prod_{i=1}^{n-1} t_i^{\frac{\lambda}{r_i} - 1} dt_1 \cdots dt_{n-1}$$
$$= \frac{1}{\Gamma(\lambda)} \prod_{i=1}^{n} \Gamma(\frac{\lambda}{r_i}). \quad (6.3.1)$$

Proof We prove (6.3.1) by mathematical induction. For $n = 2, r_1 > 1, \frac{1}{r_1} + \frac{1}{r_2} = 1$,

$$k_{\lambda} = \int_{R_+} \frac{1}{(t_1 + 1)^{\lambda}} t_1^{\frac{\lambda}{r_1} - 1} dt_1 = \frac{\Gamma(\frac{\lambda}{r_1}) \Gamma(\frac{\lambda}{r_2})}{\Gamma(\lambda)}$$

(Wang SP 1979) [4]. Assuming that for $n(\geq 2)$, (6.3.1) is valid, then for $n+1$, we have

$$k_\lambda = \int_{R_+^n} \frac{1}{(t_1+\cdots+t_n+1)^\lambda} \prod_{i=1}^{n} t_i^{\frac{\lambda}{r_i}-1} dt_1 \cdots dt_n$$

$$= \int_{R_+^{n-1}} \prod_{i=2}^{n} t_i^{\frac{\lambda}{r_i}-1} [\int_0^\infty \frac{t_1^{\frac{\lambda}{r_1}-1} dt_1}{[t_1+(t_2\cdots+t_n+1)]^\lambda}] dt_2 \cdots dt_n$$

$$\overset{s_1=\frac{t_1}{t_2+\cdots t_n+1}}{=} \int_0^\infty \frac{1}{(s_1+1)^\lambda} s_1^{\frac{\lambda}{r_1}-1} ds_1$$

$$\times \int_{R_+^{n-1}} \frac{1}{(t_2+\cdots+t_n+1)^{\lambda(1-\frac{1}{r_1})}} \prod_{i=2}^{n} t_i^{\frac{\lambda}{r_i}-1} dt_2 \cdots dt_n$$

$$= \frac{1}{\Gamma(\lambda)} \Gamma(\frac{\lambda}{r_1}) \Gamma(\lambda(1-\frac{1}{r_1}))$$

$$\times \frac{1}{\Gamma(\lambda(1-\frac{1}{r_1}))} \prod_{i=2}^{n+1} \Gamma(\frac{\lambda}{r_i})$$

$$= \frac{1}{\Gamma(\lambda)} \prod_{i=1}^{n+1} \Gamma(\frac{\lambda}{r_i}).$$

In view of mathematical induction, we have (6.3.1). □

It is obvious that

$$c_\lambda = \int_1^\infty \cdots \int_1^\infty \frac{1}{(t_1+\cdots+t_{n-1}+1)^\lambda}$$

$$\times \prod_{j=1}^{n-1} t_j^{\frac{\lambda}{r_j}-1} dt_1 \cdots dt_{n-1} > 0.$$

Without loses of generality, we show that for $n \geq 3$, $j = n-1, 0 < t_{n-1} \leq 1$,

$$\varphi(t_{n-1}) := \int_{R_+^{n-2}} \frac{1}{(t_1+\cdots+t_{n-1}+1)^\lambda} \prod_{i=1}^{n-2} t_i^{\frac{\lambda}{r_i}-1}$$

$$\times dt_1 \cdots dt_{n-2} \leq M_{n-1} < \infty. \quad (6.3.2)$$

In fact,

$$\varphi(t_{n-1}) \leq \int_{R_+^{n-2}} \frac{1}{(t_1+\cdots+t_{n-2}+1)^\lambda} \prod_{i=1}^{n-2} t_i^{\frac{\lambda}{r_i}-1} dt_1 \cdots dt_{n-2}$$

$$= \frac{1}{\Gamma(\lambda)} \prod_{i=1}^{n-2} \Gamma(\frac{\lambda}{r_i}) \Gamma(\lambda(\frac{1}{r_{n-1}}+\frac{1}{r_n})) < \infty. \quad (6.3.3)$$

For $(-1)^i u^{(i)}(t) < 0 (i=1,2,3; t > t_0)$, $n_0 - \frac{1}{2} \geq t_0$, $u(n_0-\frac{1}{2}) \geq 0$, $i = 1,\cdots,n$, since

$$0 < \lambda < \min_{1\leq i\leq n}\{r_i\},$$

it is obvious that the function

$$\tilde{f}_i(x_1,\cdots,x_n)$$

$$= \frac{1}{(u(x_1)+\cdots+u(x_n))^\lambda} \prod_{\substack{j=1\\(j\neq i)}}^{n} [u(x_j)]^{\frac{\lambda}{r_j}-1} u'(x_j)$$

satisfies $\frac{\partial \tilde{f}_i}{\partial x_j} < 0, \frac{\partial^2 \tilde{f}_i}{\partial x_j^2} > 0$, for any variable $x_j > n_0 - \frac{1}{2}$ $(j \neq i)$.

By Corollary 6.2.13, (i) setting $u(x) = (x+\beta)^\alpha$ $(\beta \geq \frac{-1}{2}, 0 < \alpha \leq 1)$, $n_0 = 1$, we have the following theorem:

Theorem 6.3.2 (cf. Huang JIA 2010) [5] Assuming that $n \geq 2$, $\beta \geq \frac{-1}{2}$, $0 < \alpha \leq 1$, $p_n > 0 (\neq 1)$, $\frac{1}{q_i} = 1 - \frac{1}{p_i}$, $r_i > 1$, $0 < \lambda < \min_{1\leq i\leq n}\{r_i\}$, $\sum_{i=1}^{n} \frac{1}{r_i} = \sum_{i=1}^{n} \frac{1}{p_i} = 1$, $a_{m_i}^{(i)} \geq 0$, such that

$$0 < \sum_{m_i=1}^{\infty} (m_i+\beta)^{p_i(1-\frac{\lambda\alpha}{r_i})-1} (a_{m_i}^{(i)})^{p_i} < \infty,$$

$$i = 1,\cdots,n,$$

(a) if $p_i > 1$ $(i=1,\cdots,n)$, then we have the following equivalent inequalities with the best constant factor $\frac{\alpha^{1-n}}{\Gamma(\lambda)} \prod_{i=1}^{n} \Gamma(\frac{\lambda}{r_i})$:

$$\sum_{m_n=1}^{\infty} \cdots \sum_{m_1=1}^{\infty} \frac{1}{[\sum_{j=1}^{n} (m_j+\beta)^\alpha]^\lambda} \prod_{i=1}^{n} a_{m_i}^{(i)}$$

$$< \frac{\alpha^{1-n}}{\Gamma(\lambda)} \prod_{i=1}^{n} \Gamma(\frac{\lambda}{r_i}) \left\{ \sum_{m_i=1}^{\infty} (m_i+\beta)^{p_i(1-\frac{\lambda\alpha}{r_i})-1} (a_{m_i}^{(i)})^{p_i} \right\}^{\frac{1}{p_i}}, \quad (6.3.4)$$

$$\left\{ \sum_{m_n=1}^{\infty} (m_n+\beta)^{\frac{q_n\alpha\lambda}{r_n}-1} \times \left[\sum_{m_{n-1}=1}^{\infty} \cdots \sum_{m_1=1}^{\infty} \frac{1}{[\sum_{j=1}^{n} (m_j+\beta)^\alpha]^\lambda} \prod_{i=1}^{n-1} a_{m_i}^{(i)} \right]^{q_n} \right\}^{\frac{1}{q_n}}$$

$$< \frac{\alpha^{1-n}}{\Gamma(\lambda)} \Gamma(\frac{\lambda}{r_n}) \prod_{i=1}^{n-1} \Gamma(\frac{\lambda}{r_i})$$

$$\times \left\{ \sum_{m_i=1}^{\infty} (m_i+\beta)^{p_i(1-\frac{\lambda\alpha}{r_i})-1} (a_{m_i}^{(i)})^{p_i} \right\}^{\frac{1}{p_i}}; \quad (6.3.5)$$

(b) if $0 < p_n < 1, p_i < 0$ $(i=1,\cdots,n-1)$, then we have the following equivalent inequalities with the best constant factor $\frac{\alpha^{1-n}}{\Gamma(\lambda)} \prod_{i=1}^{n} \Gamma(\frac{\lambda}{r_i})$:

$$\sum_{m_n=1}^{\infty} \cdots \sum_{m_1=1}^{\infty} \frac{1}{[\sum_{j=1}^{n} (m_j+\beta)^\alpha]^\lambda} \prod_{i=1}^{n} a_{m_i}^{(i)}$$

$$> \frac{\alpha^{1-n}}{\Gamma(\lambda)} \Gamma(\frac{\lambda}{r_n}) \prod_{i=1}^{n-1} \Gamma(\frac{\lambda}{r_i})$$

$$\times\left\{\sum_{m_i=1}^{\infty}(m_i+\beta)^{p_i(1-\frac{\lambda\alpha}{r_i})-1}(a_{m_i}^{(i)})^{p_i}\right\}^{\frac{1}{p_i}}$$

$$\times\left\{\sum_{m_n=1}^{\infty}[1-\theta_\lambda(m_n)](m_n+\beta)^{p_n(1-\frac{\lambda\alpha}{r_n})-1}(a_{m_n}^{(n)})^{p_n}\right\}^{\frac{1}{p_n}}, \tag{6.3.6}$$

$$\left\{\sum_{m_n=1}^{\infty}\frac{(m_n+\beta)^{\frac{q_n\alpha\lambda}{r_n}-1}}{[1-\theta_\lambda(m_n)]^{q_n-1}}\right.$$

$$\left.\times\left[\sum_{m_{n-1}=1}^{\infty}\cdots\sum_{m_1=1}^{\infty}\frac{1}{[\sum_{j=1}^{n}(m_j+\beta)^{\alpha}]^{\lambda}}\prod_{i=1}^{n-1}a_{m_i}^{(i)}\right]^{q_n}\right\}^{\frac{1}{q_n}}$$

$$<\frac{\alpha^{1-n}}{\Gamma(\lambda)}\Gamma(\frac{\lambda}{r_n})\prod_{i=1}^{n-1}\Gamma(\frac{\lambda}{r_i})$$

$$\times\left\{\sum_{m_i=1}^{\infty}(m_i+\beta)^{p_i(1-\frac{\lambda\alpha}{r_i})-1}(a_{m_i}^{(i)})^{p_i}\right\}^{\frac{1}{p_i}}, \tag{6.3.7}$$

$$\left\{\sum_{m_k=1}^{\infty}(m_k+\beta)^{\frac{q_k\alpha\lambda}{r_k}-1}\left[\sum_{m_n=1}^{\infty}\cdots\sum_{m_{k+1}=1}^{\infty}\right.\right.$$

$$\left.\left.\times\sum_{m_{k-1}=1}^{\infty}\cdots\sum_{m_1=1}^{\infty}\frac{1}{[\sum_{j=1}^{n}(m_j+\beta)^{\alpha}]^{\lambda}}\prod_{\substack{i=1\\(i\neq k)}}^{n}a_{m_i}^{(i)}\right]^{q_k}\right\}^{\frac{1}{q_k}}$$

$$>\frac{\alpha^{1-n}}{\Gamma(\lambda)}\Gamma(\frac{\lambda}{r_n})\Gamma(\frac{\lambda}{r_k})\prod_{i=1(i\neq k)}^{n-1}\Gamma(\frac{\lambda}{r_i})$$

$$\times\left\{\sum_{m_i=1}^{\infty}(m_i+\beta)^{p_i(1-\frac{\lambda\alpha}{r_i})-1}(a_{m_i}^{(i)})^{p_i}\right\}^{\frac{1}{p_i}}$$

$$\times\left\{\sum_{m_n=1}^{\infty}[1-\theta_\lambda(m_n)](m_n+\beta)^{p_n(1-\frac{\lambda\alpha}{r_n})-1}(a_{m_n}^{(n)})^{p_n}\right\}^{\frac{1}{p_n}}$$

$$(1\le k\le n-1), \tag{6.3.8}$$

where

$$\theta_\lambda(m_n)=O(\frac{1}{(m_n+\beta)^{\lambda'}})\in(0,c_0)$$

$$(0<c_0<1, 0<\lambda'<\frac{\lambda}{\alpha}\min_{1\le i\le n}\{\frac{1}{r_i}\}; m_n\to\infty).$$

In particular, for $\beta=0$, (a) if $p_i>1$ $(i=1,\cdots,n)$, then we have the following equivalent inequalities with the best constant factor $\frac{\alpha^{1-n}}{\Gamma(\lambda)}\prod_{i=1}^{n}\Gamma(\frac{\lambda}{r_i})$:

$$\sum_{m_n=1}^{\infty}\cdots\sum_{m_1=1}^{\infty}\frac{1}{(\sum_{j=1}^{n}m_j^{\alpha})^{\lambda}}\prod_{i=1}^{n}a_{m_i}^{(i)}$$

$$<\frac{\alpha^{1-n}}{\Gamma(\lambda)}\prod_{i=1}^{n}\Gamma(\frac{\lambda}{r_i})\left\{\sum_{m_i=1}^{\infty}m_i^{p_i(1-\frac{\lambda\alpha}{r_i})-1}(a_{m_i}^{(i)})^{p_i}\right\}^{\frac{1}{p_i}}, \tag{6.3.9}$$

$$\left\{\sum_{m_n=1}^{\infty}m_n^{\frac{q_n\alpha\lambda}{r_n}-1}\right.$$

$$\left.\times\left[\sum_{m_{n-1}=1}^{\infty}\cdots\sum_{m_1=1}^{\infty}\frac{1}{(\sum_{j=1}^{n}m_j^{\alpha})^{\lambda}}\prod_{i=1}^{n-1}a_{m_i}^{(i)}\right]^{q_n}\right\}^{\frac{1}{q_n}}$$

$$<\frac{\alpha^{1-n}}{\Gamma(\lambda)}\Gamma(\frac{\lambda}{r_n})\prod_{i=1}^{n-1}\Gamma(\frac{\lambda}{r_i})$$

$$\times\left\{\sum_{m_i=1}^{\infty}m_i^{p_i(1-\frac{\lambda\alpha}{r_i})-1}(a_{m_i}^{(i)})^{p_i}\right\}^{\frac{1}{p_i}}; \tag{6.3.10}$$

(b) if $0<p_n<1, p_i<0$ $(i=1,\cdots,n-1)$, then we have the following equivalent inequalities with the best constant factor $\frac{\alpha^{1-n}}{\Gamma(\lambda)}\prod_{i=1}^{n}\Gamma(\frac{\lambda}{r_i})$:

$$\sum_{m_n=1}^{\infty}\cdots\sum_{m_1=1}^{\infty}\frac{1}{(\sum_{j=1}^{n}m_j^{\alpha})^{\lambda}}\prod_{i=1}^{n}a_{m_i}^{(i)}$$

$$>\frac{\alpha^{1-n}}{\Gamma(\lambda)}\Gamma(\frac{\lambda}{r_n})\prod_{i=1}^{n-1}\Gamma(\frac{\lambda}{r_i})$$

$$\times\left\{\sum_{m_i=1}^{\infty}m_i^{p_i(1-\frac{\lambda\alpha}{r_i})-1}(a_{m_i}^{(i)})^{p_i}\right\}^{\frac{1}{p_i}}$$

$$\times\left\{\sum_{m_n=1}^{\infty}[1-\theta_\lambda(m_n)]m_n^{p_n(1-\frac{\lambda\alpha}{r_n})-1}(a_{m_n}^{(n)})^{p_n}\right\}^{\frac{1}{p_n}}, \tag{6.3.11}$$

$$\left\{\sum_{m_n=1}^{\infty}\frac{m_n^{\frac{q_n\alpha\lambda}{r_n}-1}}{[1-\theta_\lambda(m_n)]^{q_n-1}}\right.$$

$$\left.\times\left[\sum_{m_{n-1}=1}^{\infty}\cdots\sum_{m_1=1}^{\infty}\frac{1}{(\sum_{j=1}^{n}m_j^{\alpha})^{\lambda}}\prod_{i=1}^{n-1}a_{m_i}^{(i)}\right]^{q_n}\right\}^{\frac{1}{q_n}}$$

$$<\frac{\alpha^{1-n}}{\Gamma(\lambda)}\Gamma(\frac{\lambda}{r_n})\prod_{i=1}^{n-1}\Gamma(\frac{\lambda}{r_i})$$

$$\times\left\{\sum_{m_i=1}^{\infty}m_i^{p_i(1-\frac{\lambda\alpha}{r_i})-1}(a_{m_i}^{(i)})^{p_i}\right\}^{\frac{1}{p_i}}, \tag{6.3.12}$$

$$\left\{\sum_{m_k=1}^{\infty}m_k^{\frac{q_k\alpha\lambda}{r_k}-1}\left[\sum_{m_n=1}^{\infty}\cdots\sum_{m_{k+1}=1}^{\infty}\right.\right.$$

$$\times\sum_{m_{k-1}=1}^{\infty}\cdots\sum_{m_1=1}^{\infty}\frac{1}{(\sum_{j=1}^{n}m_j^{\alpha})^{\lambda}}\prod_{\substack{i=1\\(i\neq k)}}^{n}a_{m_i}^{(i)}\Bigg]^{q_k}\Bigg\}^{\frac{1}{q_k}}$$

$$>\frac{\alpha^{1-n}}{\Gamma(\lambda)}\Gamma(\tfrac{\lambda}{r_n})\Gamma(\tfrac{\lambda}{r_k})\prod_{i=1(i\neq k)}^{n-1}\Gamma(\tfrac{\lambda}{r_i})$$

$$\times\left\{\sum_{m_i=1}^{\infty}m_i^{p_i(1-\frac{\lambda\alpha}{r_i})-1}(a_{m_i}^{(i)})^{p_i}\right\}^{\frac{1}{p_i}}$$

$$\times\left\{\sum_{m_n=1}^{\infty}[1-\theta_{\lambda}(m_n)]m_n^{p_n(1-\frac{\lambda\alpha}{r_n})-1}(a_{m_n}^{(n)})^{p_n}\right\}$$

$$(1\le k\le n-1),\qquad(6.3.13)$$

where

$$\theta_{\lambda}(m_n)=O(\tfrac{1}{m_n^{\lambda'}})\in(0,c_0)$$

$$(0<c_0<1,0<\lambda'<\tfrac{\lambda}{\alpha}\min_{1\le i\le n}\{\tfrac{1}{r_i}\};m_n\to\infty).$$

(ii) Setting $u(x)=\ln^{\alpha}(x+\beta)$ $(\beta\ge-\frac{1}{2}$, $0<\alpha\le1)$, $n_0=2$, we have the following theorem:

Theorem 6.3.3 Assuming that $n\ge2$, $\beta\ge-\frac{1}{2}$, $0<\alpha\le1$, $p_n>0(\neq1)$, $\frac{1}{q_i}=1-\frac{1}{p_i}$, $r_i>1$, $0<\lambda<\min_{1\le i\le n}\{r_i\}$, $\sum_{i=1}^{n}\frac{1}{r_i}=\sum_{i=1}^{n}\frac{1}{p_i}=1$, and $a_{m_i}^{(i)}\ge0$, such that

$$0<\sum_{m_i=2}^{\infty}\frac{(m_i+\beta)^{p_i-1}}{[\ln(m_i+\beta)]^{p_i(\frac{\lambda\alpha}{r_i}-1)-1}}(a_{m_i}^{(i)})^{p_i}<\infty,$$

$$i=1,\cdots,n,$$

(a) if $p_i>1$ $(i=1,\cdots,n)$, then we have the following equivalent inequalities with the best constant factor $\frac{\alpha^{1-n}}{\Gamma(\lambda)}\prod_{i=1}^{n}\Gamma(\frac{\lambda}{r_i})$:

$$\sum_{m_n=2}^{\infty}\cdots\sum_{m_1=2}^{\infty}\frac{1}{[\sum_{j=1}^{n}\ln^{\alpha}(m_j+\beta)]^{\lambda}}\prod_{i=1}^{n}a_{m_i}^{(i)}$$

$$<\frac{\alpha^{1-n}}{\Gamma(\lambda)}\prod_{i=1}^{n}\Gamma(\tfrac{\lambda}{r_i})\left\{\sum_{m_i=2}^{\infty}\frac{(m_i+\beta)^{p_i-1}(a_{m_i}^{(i)})^{p_i}}{[\ln(m_i+\beta)]^{p_i(\frac{\lambda\alpha}{r_i}-1)-1}}\right\}^{\frac{1}{p_i}},\qquad(6.3.14)$$

$$\Bigg\{\sum_{m_n=2}^{\infty}\frac{1}{m_n+\beta}[\ln(m_n+\beta)]^{\frac{q_n\alpha\lambda}{r_n}-1}$$

$$\times\left[\sum_{m_{n-1}=2}^{\infty}\cdots\sum_{m_1=2}^{\infty}\frac{1}{[\sum_{j=1}^{n}\ln^{\alpha}(m_j+\beta)]^{\lambda}}\prod_{i=1}^{n-1}a_{m_i}^{(i)}\right]^{q_n}\Bigg\}^{\frac{1}{q_n}}$$

$$<\frac{\alpha^{1-n}}{\Gamma(\lambda)}\Gamma(\tfrac{\lambda}{r_n})\prod_{i=1}^{n-1}\Gamma(\tfrac{\lambda}{r_i})$$

$$\times\left\{\sum_{m_i=2}^{\infty}\frac{(m_i+\beta)^{p_i-1}(a_{m_i}^{(i)})^{p_i}}{[\ln(m_i+\beta)]^{p_i(\frac{\lambda\alpha}{r_i}-1)-1}}\right\}^{\frac{1}{p_i}};\qquad(6.3.15)$$

(b) if $0<p_n<1, p_i<0$ $(i=1,\cdots,n-1)$, then we have the following equivalent inequalities with the best constant factor $\frac{\alpha^{1-n}}{\Gamma(\lambda)}\prod_{i=1}^{n}\Gamma(\frac{\lambda}{r_i})$:

$$\sum_{m_n=2}^{\infty}\cdots\sum_{m_1=2}^{\infty}\frac{1}{[\sum_{j=1}^{n}\ln^{\alpha}(m_j+\beta)]^{\lambda}}\prod_{i=1}^{n}a_{m_i}^{(i)}$$

$$>\frac{\alpha^{1-n}}{\Gamma(\lambda)}\Gamma(\tfrac{\lambda}{r_n})\prod_{i=1}^{n-1}\Gamma(\tfrac{\lambda}{r_i})$$

$$\times\left\{\sum_{m_i=2}^{\infty}\frac{(m_i+\beta)^{p_i-1}(a_{m_i}^{(i)})^{p_i}}{[\ln(m_i+\beta)]^{p_i(\frac{\lambda\alpha}{r_i}-1)-1}}\right\}^{\frac{1}{p_i}}$$

$$\times\left\{\sum_{m_n=2}^{\infty}[1-\theta_{\lambda}(m_n)]\frac{(m_n+\beta)^{p_n-1}(a_{m_n}^{(n)})^{p_n}}{[\ln(m_n+\beta)]^{p_n(\frac{\lambda\alpha}{r_n}-1)-1}}\right\}^{\frac{1}{p_n}},\qquad(6.3.16)$$

$$\Bigg\{\sum_{m_n=2}^{\infty}\frac{[\ln(m_n+\beta)]^{\frac{q_n\alpha\lambda}{r_n}-1}}{(m_n+\beta)[1-\theta_{\lambda}(m_n)]^{q_n-1}}$$

$$\times\left[\sum_{m_{n-1}=2}^{\infty}\cdots\sum_{m_1=2}^{\infty}\frac{1}{[\sum_{j=1}^{n}\ln^{\alpha}(m_j+\beta)]^{\lambda}}\prod_{i=1}^{n-1}a_{m_i}^{(i)}\right]^{q_n}\Bigg\}^{\frac{1}{q_n}}$$

$$<\frac{\alpha^{1-n}}{\Gamma(\lambda)}\Gamma(\tfrac{\lambda}{r_n})\prod_{i=1}^{n-1}\Gamma(\tfrac{\lambda}{r_i})$$

$$\times\left\{\sum_{m_i=2}^{\infty}\frac{(m_i+\beta)^{p_i-1}(a_{m_i}^{(i)})^{p_i}}{[\ln(m_i+\beta)]^{p_i(\frac{\lambda\alpha}{r_i}-1)-1}}\right\}^{\frac{1}{p_i}},\qquad(6.3.17)$$

$$\Bigg\{\sum_{m_k=2}^{\infty}\frac{[\ln(m_k+\beta)]^{\frac{q_k\alpha\lambda}{r_k}-1}}{m_k+\beta}\Bigg[\sum_{m_n=2}^{\infty}\cdots\sum_{m_{k+1}=2}^{\infty}$$

$$\times\sum_{m_{k-1}=2}^{\infty}\cdots\sum_{m_1=2}^{\infty}\frac{1}{[\sum_{j=1}^{n}\ln^{\alpha}(m_j+\beta)]^{\lambda}}\prod_{\substack{i=1\\(i\neq k)}}^{n}a_{m_i}^{(i)}\Bigg]^{q_k}\Bigg\}^{\frac{1}{q_k}}$$

$$>\frac{\alpha^{1-n}}{\Gamma(\lambda)}\Gamma(\tfrac{\lambda}{r_n})\prod_{i=1(i\neq k)}^{n-1}\Gamma(\tfrac{\lambda}{r_i})$$

$$\times\left\{\sum_{m_i=2}^{\infty}\frac{(m_i+\beta)^{p_i-1}(a_{m_i}^{(i)})^{p_i}}{[\ln(m_i+\beta)]^{p_i(\frac{\lambda\alpha}{r_i}-1)-1}}\right\}^{\frac{1}{p_i}}$$

$$\times\left\{\sum_{m_n=2}^{\infty}[1-\theta_\lambda(m_n)]\frac{(m_n+\beta)^{p_n-1}(a_{m_n}^{(n)})^{p_n}}{[\ln(m_n+\beta)]^{p_n(\frac{\lambda\alpha}{r_n}-1)-1}}\right\}^{\frac{1}{p_n}}$$

$$(1\le k\le n-1), \quad (6.3.18)$$

where

$$\theta_\lambda(m_n)=O\left(\frac{1}{\ln^{\lambda'}(m_n+\beta)}\right)\in(0,c_0)$$

$$(0<c_0<1, 0<\lambda'<\frac{\lambda}{\alpha}\min_{1\le i\le n}\{\frac{1}{r_i}\}; m_n\to\infty).$$

In particular, for $\beta=0$, (a) if $p_i>1$ $(i=1,\cdots,n)$, then we have the following equivalent inequalities with the best constant factor $\frac{\alpha^{1-n}}{\Gamma(\lambda)}\prod_{i=1}^{n}\Gamma(\frac{\lambda}{r_i})$:

$$\sum_{m_n=2}^{\infty}\cdots\sum_{m_1=2}^{\infty}\frac{1}{(\sum_{j=1}^{n}\ln^\alpha m_j)^\lambda}\prod_{i=1}^{n}a_{m_i}^{(i)}$$

$$<\frac{\alpha^{1-n}}{\Gamma(\lambda)}\prod_{i=1}^{n}\Gamma(\frac{\lambda}{r_i})\left\{\sum_{m_i=2}^{\infty}\frac{m_i^{p_i-1}(a_{m_i}^{(i)})^{p_i}}{(\ln m_i)^{p_i(\frac{\lambda\alpha}{r_i}-1)-1}}\right\}^{\frac{1}{p_i}}, \quad (6.3.19)$$

$$\left\{\sum_{m_n=2}^{\infty}\frac{1}{m_n}(\ln m_n)^{\frac{q_n\alpha\lambda}{r_n}-1}\right.$$

$$\left.\times\left[\sum_{m_{n-1}=2}^{\infty}\cdots\sum_{m_1=2}^{\infty}\frac{1}{(\sum_{j=1}^{n}\ln^\alpha m_j)^\lambda}\prod_{i=1}^{n-1}a_{m_i}^{(i)}\right]^{q_n}\right\}^{\frac{1}{q_n}}$$

$$<\frac{\alpha^{1-n}}{\Gamma(\lambda)}\Gamma(\frac{\lambda}{r_n})\prod_{i=1}^{n-1}\Gamma(\frac{\lambda}{r_i})$$

$$\times\left\{\sum_{m_i=2}^{\infty}\frac{m_i^{p_i-1}(a_{m_i}^{(i)})^{p_i}}{(\ln m_i)^{p_i(\frac{\lambda\alpha}{r_i}-1)-1}}\right\}^{\frac{1}{p_i}}; \quad (6.3.20)$$

(b) if $0<p_n<1, p_i<0$ $(i=1,\cdots,n-1)$, then we have the following equivalent inequalities with the best constant factor $\frac{\alpha^{1-n}}{\Gamma(\lambda)}\prod_{i=1}^{n}\Gamma(\frac{\lambda}{r_i})$:

$$\sum_{m_n=2}^{\infty}\cdots\sum_{m_1=2}^{\infty}\frac{1}{(\sum_{j=1}^{n}\ln^\alpha m_j)^\lambda}\prod_{i=1}^{n}a_{m_i}^{(i)}$$

$$>\frac{\alpha^{1-n}}{\Gamma(\lambda)}\Gamma(\frac{\lambda}{r_n})\prod_{i=1}^{n-1}\Gamma(\frac{\lambda}{r_i})$$

$$\times\left\{\sum_{m_i=2}^{\infty}\frac{m_i^{p_i-1}(a_{m_i}^{(i)})^{p_i}}{(\ln m_i)^{p_i(\frac{\lambda\alpha}{r_i}-1)-1}}\right\}^{\frac{1}{p_i}}$$

$$\times\left\{\sum_{m_n=2}^{\infty}[1-\theta_\lambda(m_n)]\frac{m_n^{p_n-1}(a_{m_n}^{(n)})^{p_n}}{(\ln m_n)^{p_n(\frac{\lambda\alpha}{r_n}-1)-1}}\right\}^{\frac{1}{p_n}},$$

$$(6.3.21)$$

$$\left\{\sum_{m_n=2}^{\infty}\frac{(\ln m_n)^{\frac{q_n\alpha\lambda}{r_n}-1}}{m_n[1-\theta_\lambda(m_n)]^{q_n-1}}\right.$$

$$\left.\times\left[\sum_{m_{n-1}=2}^{\infty}\cdots\sum_{m_1=2}^{\infty}\frac{1}{(\sum_{j=1}^{n}\ln^\alpha m_j)^\lambda}\prod_{i=1}^{n-1}a_{m_i}^{(i)}\right]^{q_n}\right\}^{\frac{1}{q_n}}$$

$$<\frac{\alpha^{1-n}}{\Gamma(\lambda)}\Gamma(\frac{\lambda}{r_n})\prod_{i=1}^{n-1}\Gamma(\frac{\lambda}{r_i})$$

$$\times\left\{\sum_{m_i=2}^{\infty}\frac{m_i^{p_i-1}(a_{m_i}^{(i)})^{p_i}}{(\ln m_i)^{p_i(\frac{\lambda\alpha}{r_i}-1)-1}}\right\}^{\frac{1}{p_i}}, \quad (6.3.22)$$

$$\left\{\sum_{m_k=2}^{\infty}\frac{(\ln m_k)^{\frac{q_k\alpha\lambda}{r_k}-1}}{m_k}\left[\sum_{m_n=2}^{\infty}\cdots\sum_{m_{k+1}=2}^{\infty}\right.\right.$$

$$\left.\left.\times\sum_{m_{k-1}=2}^{\infty}\cdots\sum_{m_1=2}^{\infty}\frac{1}{(\sum_{j=1}^{n}\ln^\alpha m_j)^\lambda}\prod_{\substack{i=1\\(i\ne k)}}^{n}a_{m_i}^{(i)}\right]^{q_k}\right\}^{\frac{1}{q_k}}$$

$$>\frac{\alpha^{1-n}}{\Gamma(\lambda)}\Gamma(\frac{\lambda}{r_n})\prod_{i=1(i\ne k)}^{n-1}\Gamma(\frac{\lambda}{r_i})$$

$$\times\left\{\sum_{m_i=2}^{\infty}\frac{m_i^{p_i-1}(a_{m_i}^{(i)})^{p_i}}{(\ln m_i)^{p_i(\frac{\lambda\alpha}{r_i}-1)-1}}\right\}^{\frac{1}{p_i}}$$

$$\times\left\{\sum_{m_n=2}^{\infty}[1-\theta_\lambda(m_n)]\frac{m_n^{p_n-1}(a_{m_n}^{(n)})^{p_n}}{(\ln m_n)^{p_n(\frac{\lambda\alpha}{r_n}-1)-1}}\right\}^{\frac{1}{p_n}}$$

$$(1\le k\le n-1), \quad (6.3.23)$$

where

$$\theta_\lambda(m_n)=O\left(\frac{1}{\ln^{\lambda'}m_n}\right)\in(0,c_0)$$

$$(0<c_0<1, 0<\lambda'<\frac{\lambda}{\alpha}\min_{1\le i\le n}\{\frac{1}{r_i}\}; m_n\to\infty).$$

(iii) Setting $u(x)=\ln^\alpha\beta x$ $(\beta\ge 2, 0<\alpha\le 1)$, $n_0=1$, we have the following theorem:

Theorem 6.3.4 Assuming that $n\ge 2$, $\beta\ge 2$, $0<\alpha\le 1$, $p_n>0(\ne 1)$, $\frac{1}{q_i}=1-\frac{1}{p_i}$, $r_i>1$,

$$0<\lambda<\min_{1\le i\le n}\{r_i\},$$

$\sum_{i=1}^{n}\frac{1}{r_i}=\sum_{i=1}^{n}\frac{1}{p_i}=1$, $a_{m_i}^{(i)}\ge 0$, such that

$$0<\sum_{m_i=1}^{\infty}\frac{m_i^{p_i-1}(a_{m_i}^{(i)})^{p_i}}{(\ln\beta m_i)^{p_i(\frac{\lambda\alpha}{r_i}-1)-1}}<\infty,$$

$$i=1,\cdots,n,$$

(a) if $p_i>1$ $(i=1,\cdots,n)$, then we have the following equivalent inequalities with the best constant factor $\frac{\alpha^{1-n}}{\Gamma(\lambda)}\prod_{i=1}^{n}\Gamma(\frac{\lambda}{r_i})$:

$$\sum_{m_n=1}^{\infty}\cdots\sum_{m_1=1}^{\infty}\frac{1}{(\sum_{j=1}^{n}\ln^{\alpha}\beta m_j)^{\lambda}}\prod_{i=1}^{n}a_{m_i}^{(i)}$$

$$<\frac{\alpha^{1-n}}{\Gamma(\lambda)}\prod_{i=1}^{n}\Gamma(\frac{\lambda}{r_i})\left\{\sum_{m_i=1}^{\infty}\frac{m_i^{p_i-1}(a_{m_i}^{(i)})^{p_i}}{(\ln\beta m_i)^{p_i(\frac{\lambda\alpha}{r_i}-1)-1}}\right\}^{\frac{1}{p_i}}, \tag{6.3.24}$$

$$\left\{\sum_{m_n=1}^{\infty}\frac{1}{m_n}(\ln\beta m_n)^{\frac{q_n\alpha\lambda}{r_n}-1}\right.$$

$$\left.\times\left[\sum_{m_{n-1}=1}^{\infty}\cdots\sum_{m_1=1}^{\infty}\frac{1}{(\sum_{j=1}^{n}\ln^{\alpha}\beta m_j)^{\lambda}}\prod_{i=1}^{n-1}a_{m_i}^{(i)}\right]^{q_n}\right\}^{\frac{1}{q_n}}$$

$$<\frac{\alpha^{1-n}}{\Gamma(\lambda)}\Gamma(\frac{\lambda}{r_n})\prod_{i=1}^{n-1}\Gamma(\frac{\lambda}{r_i})$$

$$\times\left\{\sum_{m_i=1}^{\infty}\frac{m_i^{p_i-1}(a_{m_i}^{(i)})^{p_i}}{(\ln\beta m_i)^{p_i(\frac{\lambda\alpha}{r_i}-1)-1}}\right\}^{\frac{1}{p_i}}; \tag{6.3.25}$$

(b) if $0<p_n<1, p_i<0$ $(i=1,\cdots,n-1)$, then we have the following equivalent inequalities with the best constant factor $\frac{\alpha^{1-n}}{\Gamma(\lambda)}\prod_{i=1}^{n}\Gamma(\frac{\lambda}{r_i})$:

$$\sum_{m_n=1}^{\infty}\cdots\sum_{m_1=1}^{\infty}\frac{1}{(\sum_{j=1}^{n}\ln^{\alpha}\beta m_j)^{\lambda}}\prod_{i=1}^{n}a_{m_i}^{(i)}$$

$$>\frac{\alpha^{1-n}}{\Gamma(\lambda)}\Gamma(\frac{\lambda}{r_n})\prod_{i=1}^{n-1}\Gamma(\frac{\lambda}{r_i})$$

$$\times\left\{\sum_{m_i=1}^{\infty}\frac{m_i^{p_i-1}(a_{m_i}^{(i)})^{p_i}}{(\ln\beta m_i)^{p_i(\frac{\lambda\alpha}{r_i}-1)-1}}\right\}^{\frac{1}{p_i}}$$

$$\times\left\{\sum_{m_n=1}^{\infty}[1-\theta_{\lambda}(m_n)]\frac{m_n^{p_n-1}(a_{m_n}^{(n)})^{p_n}}{(\ln\beta m_n)^{p_n(\frac{\lambda\alpha}{r_n}-1)-1}}\right\}^{\frac{1}{p_n}}, \tag{6.3.26}$$

$$\left\{\sum_{m_n=1}^{\infty}\frac{(\ln\beta m_n)^{\frac{q_n\alpha\lambda}{r_n}-1}}{m_n[1-\theta_{\lambda}(m_n)]^{q_n-1}}\right.$$

$$\left.\times\left[\sum_{m_{n-1}=1}^{\infty}\cdots\sum_{m_1=1}^{\infty}\frac{1}{(\sum_{j=1}^{n}\ln^{\alpha}\beta m_j)^{\lambda}}\prod_{i=1}^{n-1}a_{m_i}^{(i)}\right]^{q_n}\right\}^{\frac{1}{q_n}}$$

$$<\frac{\alpha^{1-n}}{\Gamma(\lambda)}\Gamma(\frac{\lambda}{r_n})\prod_{i=1}^{n-1}\Gamma(\frac{\lambda}{r_i})$$

$$\times\left\{\sum_{m_i=1}^{\infty}\frac{m_i^{p_i-1}(a_{m_i}^{(i)})^{p_i}}{(\ln\beta m_i)^{p_i(\frac{\lambda\alpha}{r_i}-1)-1}}\right\}^{\frac{1}{p_i}}, \tag{6.3.27}$$

$$\left\{\sum_{m_k=1}^{\infty}\frac{(\ln\beta m_k)^{\frac{q_k\alpha\lambda}{r_k}-1}}{m_k}\left[\sum_{m_n=1}^{\infty}\cdots\sum_{m_{k+1}=1}^{\infty}\right.\right.$$

$$\left.\left.\times\sum_{m_{k-1}=1}^{\infty}\cdots\sum_{m_1=1}^{\infty}\frac{1}{(\sum_{j=1}^{n}\ln^{\alpha}\beta m_j)^{\lambda}}\prod_{\substack{i=1\\(i\neq k)}}^{n}a_{m_i}^{(i)}\right]^{q_k}\right\}^{\frac{1}{q_k}}$$

$$>\frac{\alpha^{1-n}}{\Gamma(\lambda)}\Gamma(\frac{\lambda}{r_n})\prod_{i=1(i\neq k)}^{n-1}\Gamma(\frac{\lambda}{r_i})$$

$$\times\left\{\sum_{m_i=1}^{\infty}\frac{m_i^{p_i-1}(a_{m_i}^{(i)})^{p_i}}{(\ln\beta m_i)^{p_i(\frac{\lambda\alpha}{r_i}-1)-1}}\right\}^{\frac{1}{p_i}}$$

$$\times\left\{\sum_{m_n=1}^{\infty}[1-\theta_{\lambda}(m_n)]\frac{m_n^{p_n-1}(a_{m_n}^{(n)})^{p_n}}{(\ln\beta m_n)^{p_n(\frac{\lambda\alpha}{r_n}-1)-1}}\right\}^{\frac{1}{p_n}}$$

$$(1\le k\le n-1), \tag{6.3.28}$$

where

$$\theta_{\lambda}(m_n)=O(\frac{1}{\ln^{\lambda'}\beta m_n})\in(0,c_0)$$

$$(0<c_0<1, 0<\lambda'<\frac{\lambda}{\alpha}\min_{1\le i\le n}\{\frac{1}{r_i}\}; m_n\to\infty).$$

6.3.2. EXAMPLES FOR $k_{\lambda}(t_1,\cdots,t_{n-1},1)=\frac{1}{(\max\{t_1,\cdots,t_{n-1},1\})^{\lambda}}$

Lemma 6.3.5 For $n\ge 2$, $\lambda>0, r_i>1, \lambda_i=\frac{\lambda}{r_i}<1$ $(i=1,\cdots,n), \sum_{i=1}^{n}\frac{1}{r_i}=1$, we have

$$k_{\lambda}=\int_{R_+^{n-1}}\frac{1}{(\max\{t_1,\cdots,t_{n-1},1\})^{\lambda}}\prod_{i=1}^{n-1}t_i^{\frac{\lambda}{r_i}-1}dt_1\cdots dt_{n-1}$$

$$=\frac{1}{\lambda^{n-1}}\prod_{i=1}^{n}r_i. \tag{6.3.29}$$

Proof We prove (6.3.29) by mathematical induction. For $n=2$, we find

$$k_{\lambda}=\int_{R_+}\frac{1}{(\max\{t_1,1\})^{\lambda}}t_1^{\frac{\lambda}{r_1}-1}dt_1$$

$$=\int_0^1 t_1^{\frac{\lambda}{r_1}-1}dt_1+\int_1^{\infty}\frac{1}{t_1^{\lambda}}t_1^{\frac{\lambda}{r_1}-1}dt_1$$

$$=\frac{r_1}{\lambda}+\frac{r_2}{\lambda}=\frac{1}{\lambda}r_1r_2.$$

Assuming that for $n(\ge 2)$, (6.3.29) is valid, then for $n+1$, we have

$$k_{\lambda}=\int_{R_+^{n}}\frac{1}{(\max\{t_1,\cdots,t_n,1\})^{\lambda}}\prod_{i=1}^{n}t_i^{\frac{\lambda}{r_i}-1}dt_1\cdots dt_n$$

$$=\int_{R_+^{n-1}}\prod_{i=2}^{n}t_i^{\frac{\lambda}{r_i}-1}\left[\int_0^{\infty}\frac{t_1^{\frac{\lambda}{r_1}-1}dt_1}{(\max\{t_1,t_2,\cdots,t_n,1\})^{\lambda}}\right]dt_2\cdots dt_n$$

$$=\int_{R_+^{n-1}}\prod_{i=2}^{n}t_i^{\frac{\lambda}{r_i}-1}$$

$$\times\Big[\int_0^{\max\{t_2,\cdots,t_n,1\}}\frac{t_1^{\frac{\lambda}{r_1}-1}dt_1}{(\max\{t_2,\cdots,t_n,1\})^{\lambda}}\Big]dt_2\cdots dt_n$$

$$+\int_{R_+^{n-1}}\prod_{i=2}^{n}t_i^{\frac{\lambda}{r_i}-1}\Big[\int_{\max\{t_2,\cdots,t_n,1\}}^{\infty}\frac{t_1^{\frac{\lambda}{r_1}-1}dt_1}{t_1^{\lambda}}\Big]dt_2\cdots dt_n$$

$$=\frac{r_1}{\lambda}\int_{R_+^{n-1}}\frac{1}{(\max\{t_2,\cdots,t_n,1\})^{\lambda(1-\frac{1}{r_1})}}\prod_{i=2}^{n}t_i^{\frac{\lambda}{r_i}-1}dt_2\cdots dt_n$$

$$+\frac{1}{\lambda(1-\frac{1}{r_1})}\int_{R_+^{n-1}}\frac{1}{(\max\{t_2,\cdots,t_n,1\})^{\lambda(1-\frac{1}{r_1})}}$$

$$\times\prod_{i=2}^{n}t_i^{\frac{\lambda}{r_i}-1}dt_2\cdots dt_n$$

$$=\frac{r_1}{\lambda}\frac{1}{(1-\frac{1}{r_1})}\int_{R_+^{n-1}}\frac{1}{(\max\{t_2,\cdots,t_n,1\})^{\lambda(1-\frac{1}{r_1})}}$$

$$\times\prod_{i=2}^{n}t_i^{\frac{\lambda(1-\frac{1}{r_1})}{r_i(1-\frac{1}{r_1})}-1}dt_2\cdots dt_n$$

$$=\frac{r_1}{\lambda}\frac{1}{(1-\frac{1}{r_1})}\frac{1}{\lambda^{n-1}(1-\frac{1}{r_1})^{n-1}}\prod_{i=2}^{n+1}r_i(1-\frac{1}{r_1})$$

$$=\frac{1}{\lambda^n}\prod_{i=1}^{n+1}r_i\,.$$

In view of mathematical induction, we have (6.3.29). □

It is obvious that

$$c_\lambda=\int_1^{\infty}\cdots\int_1^{\infty}\frac{1}{(\max\{t_1,\cdots,t_{n-1},1\})^{\lambda}}$$

$$\times\prod_{j=1}^{n-1}t_j^{\frac{\lambda}{r_j}-1}dt_1\cdots dt_{n-1}>0\,.$$

Without loses of generality, we show that for $n\ge 3$, $j=n-1, 0<t_{n-1}\le 1$,

$$\varphi(t_{n-1}):=\int_{R_+^{n-2}}\frac{1}{(\max\{t_1,\cdots,t_{n-1},1\})^{\lambda}}\prod_{i=1}^{n-2}t_i^{\frac{\lambda}{r_i}-1}$$

$$\times dt_1\cdots dt_{n-2}\le M_{n-1}<\infty\,. \qquad (6.3.30)$$

In fact, we find

$$\varphi(t_{n-1})=\int_{R_+^{n-2}}\frac{1}{(\max\{t_1,\cdots,t_{n-2},1\})^{\lambda}}$$

$$\times\prod_{i=1}^{n-2}t_i^{\frac{\lambda}{r_i}-1}dt_1\cdots dt_{n-2}$$

$$\le\int_{R_+^{n-2}}\frac{1}{(\max\{t_1,\cdots,t_{n-2},1\})^{\lambda(1-\frac{1}{r_n})}}$$

$$\times\prod_{i=1}^{n-2}t_i^{\frac{\lambda(1-\frac{1}{r_n})}{r_i(1-\frac{1}{r_n})}-1}dt_1\cdots dt_{n-2}$$

$$=\frac{1}{\lambda^{n-2}(1-\frac{1}{r_n})^{n-2}}\prod_{i=1}^{n-1}r_i(1-\frac{1}{r_n})=\frac{1-\frac{1}{r_n}}{\lambda^{n-2}}\prod_{i=1}^{n-1}r_i<\infty\,.$$

(6.3.31)

For $n_0-1\ge t_0$, $u(n_0-1)\ge 0$, $i=1,\cdots,n$, since $0<\lambda<\min_{1\le i\le n}\{r_i\}$, it is obvious that the function

$$\tilde{f}_i(x_1,\cdots,x_n)$$

$$=\frac{1}{(\max\{u(x_1),\cdots,u(x_n)\})^{\lambda}}\prod_{\substack{j=1\\(j\ne i)}}^{n}[u(x_j)]^{\frac{\lambda}{r_j}-1}u'(x_j)$$

is strictly decreasing for any variable $x_j>n_0-1$ $(j\ne i)$.

By Corollary 6.2.11, (i) setting $u(x)=x^{\alpha}$ $(0<\alpha\le 1), n_0=1$, we have the following theorem:

Theorem 6.3.6 Assuming that $n\ge 2$, $0<\alpha\le 1$, $p_n>0(\ne 1), \frac{1}{q_i}=1-\frac{1}{p_i}, r_i>1$,

$$0<\lambda<r_i\ (i=1,\cdots,n),$$

$\sum_{i=1}^{n}\frac{1}{r_i}=\sum_{i=1}^{n}\frac{1}{p_i}=1$, and $a_{m_i}^{(i)}\ge 0$, such that

$$0<\sum_{m_i=1}^{\infty}m_i^{p_i(1-\frac{\lambda\alpha}{r_i})-1}(a_{m_i}^{(i)})^{p_i}<\infty,$$

$$i=1,\cdots,n,$$

(a) if $p_i>1$ $(i=1,\cdots,n)$, then we have the following equivalent inequalities with the best constant factor $\frac{1}{(\alpha\lambda)^{n-1}}\prod_{i=1}^{n}r_i$:

$$\sum_{m_n=1}^{\infty}\cdots\sum_{m_1=1}^{\infty}\frac{1}{(\max_{1\le j\le n}\{m_j^{\alpha}\})^{\lambda}}\prod_{i=1}^{n}a_{m_i}^{(i)}$$

$$<\frac{1}{(\alpha\lambda)^{n-1}}\prod_{i=1}^{n}r_i\left\{\sum_{m_i=1}^{\infty}m_i^{p_i(1-\frac{\lambda\alpha}{r_i})-1}(a_{m_i}^{(i)})^{p_i}\right\}^{\frac{1}{p_i}},$$

(6.3.32)

$$\left\{\sum_{m_n=1}^{\infty}m_n^{\frac{q_n\alpha\lambda}{r_n}-1}\left[\sum_{m_{n-1}=1}^{\infty}\cdots\sum_{m_1=1}^{\infty}\frac{\prod_{i=1}^{n-1}a_{m_i}^{(i)}}{(\max_{1\le j\le n}\{m_j^{\alpha}\})^{\lambda}}\right]^{q_n}\right\}^{\frac{1}{q_n}}$$

$$<\frac{r_n}{(\alpha\lambda)^{n-1}}\prod_{i=1}^{n-1}r_i\left\{\sum_{m_i=1}^{\infty}m_i^{p_i(1-\frac{\lambda\alpha}{r_i})-1}(a_{m_i}^{(i)})^{p_i}\right\}^{\frac{1}{p_i}};$$

(6.3.33)

(b) if $0<p_n<1, p_i<0$ $(i=1,\cdots,n-1)$, then we have the following equivalent inequalities with the best constant factor $\frac{1}{(\alpha\lambda)^{n-1}}\prod_{i=1}^{n}r_i$:

$$\sum_{m_n=1}^{\infty}\cdots\sum_{m_1=1}^{\infty}\frac{1}{(\max_{1\le j\le n}\{m_j^{\alpha}\})^{\lambda}}\prod_{i=1}^{n}a_{m_i}^{(i)}$$

$$> \frac{r_n}{(\alpha\lambda)^{n-1}} \prod_{i=1}^{n-1} r_i \left\{ \sum_{m_i=1}^{\infty} m_i^{p_i(1-\frac{\lambda\alpha}{r_i})-1} (a_{m_i}^{(i)})^{p_i} \right\}^{\frac{1}{p_i}}$$

$$\times \left\{ \sum_{m_n=1}^{\infty} [1-\theta_\lambda(m_n)] m_n^{p_n(1-\frac{\lambda\alpha}{r_n})-1} (a_{m_n}^{(n)})^{p_n} \right\}^{\frac{1}{p_n}}, \tag{6.3.34}$$

$$\left\{ \sum_{m_n=1}^{\infty} \frac{m_n^{\frac{q_n\alpha\lambda}{r_n}-1}}{[1-\theta_\lambda(m_n)]^{q_n-1}} \right.$$

$$\left. \times \left[\sum_{m_{n-1}=1}^{\infty} \cdots \sum_{m_1=1}^{\infty} \frac{1}{(\max_{1\le j\le n}\{m_j^\alpha\})^\lambda} \prod_{i=1}^{n-1} a_{m_i}^{(i)} \right]^{q_n} \right\}^{\frac{1}{q_n}}$$

$$< \frac{r_n}{(\alpha\lambda)^{n-1}} \prod_{i=1}^{n-1} r_i \left\{ \sum_{m_i=1}^{\infty} m_i^{p_i(1-\frac{\lambda\alpha}{r_i})-1} (a_{m_i}^{(i)})^{p_i} \right\}^{\frac{1}{p_i}}, \tag{6.3.35}$$

$$\left\{ \sum_{m_k=1}^{\infty} m_k^{\frac{q_k\alpha\lambda}{r_k}-1} \left[\sum_{m_n=1}^{\infty} \cdots \sum_{m_{k+1}=1}^{\infty} \right. \right.$$

$$\left. \left. \times \sum_{m_{k-1}=1}^{\infty} \cdots \sum_{m_1=1}^{\infty} \frac{1}{(\max_{1\le j\le n}\{m_j^\alpha\})^\lambda} \prod_{\substack{i=1\\(i\ne k)}}^{n} a_{m_i}^{(i)} \right]^{q_k} \right\}^{\frac{1}{q_k}}$$

$$> \frac{r_n r_k}{(\alpha\lambda)^{n-1}} \prod_{\substack{i=1\\(i\ne k)}}^{n-1} r_i \left\{ \sum_{m_i=1}^{\infty} m_i^{p_i(1-\frac{\lambda\alpha}{r_i})-1} (a_{m_i}^{(i)})^{p_i} \right\}^{\frac{1}{p_i}}$$

$$\times \left\{ \sum_{m_n=1}^{\infty} [1-\theta_\lambda(m_n)] m_n^{p_n(1-\frac{\lambda\alpha}{r_n})-1} (a_{m_n}^{(n)})^{p_n} \right\}^{\frac{1}{p_n}}$$

$$(1 \le k \le n-1), \tag{6.3.36}$$

where

$$\theta_\lambda(m_n) = O(\frac{1}{m_n^{\lambda'}}) \in (0, c_0)$$

$$(0 < c_0 < 1,\ 0 < \lambda' < \frac{\lambda}{\alpha} \min_{1\le i\le n}\{\frac{1}{r_i}\}; m_n \to \infty).$$

(ii) Setting $u(x) = \ln^\alpha x (0 < \alpha \le 1)$, $n_0 = 2$, we have the following theorem:

Theorem 6.3.7 Assuming that $n \ge 2$, $0 < \alpha \le 1$, $p_n > 0 (\ne 1)$, $\frac{1}{q_i} = 1 - \frac{1}{p_i}$, $r_i > 1$,

$$0 < \lambda < \min_{1\le i\le n}\{r_i\},$$

$\sum_{i=1}^{n} \frac{1}{r_i} = \sum_{i=1}^{n} \frac{1}{p_i} = 1$, and $a_{m_i}^{(i)} \ge 0$, such that

$$0 < \sum_{m_i=2}^{\infty} \frac{m_i^{p_i-1}}{(\ln m_i)^{p_i(\frac{\lambda\alpha}{r_i}-1)-1}} (a_{m_i}^{(i)})^{p_i} < \infty,$$

$$i = 1, \cdots, n,$$

(a) if $p_i > 1$ $(i = 1, \cdots, n)$, then we have the following equivalent inequalities with the best constant factor $\frac{1}{\lambda^{n-1}} \prod_{i=1}^{n} r_i$:

$$\sum_{m_n=2}^{\infty} \cdots \sum_{m_1=2}^{\infty} \frac{1}{(\max_{1\le j\le n}\{\ln^\alpha m_j\})^\lambda} \prod_{i=1}^{n} a_{m_i}^{(i)}$$

$$< \frac{1}{(\alpha\lambda)^{n-1}} \prod_{i=1}^{n} r_i \left\{ \sum_{m_i=2}^{\infty} \frac{m_i^{p_i-1}(a_{m_i}^{(i)})^{p_i}}{(\ln m_i)^{p_i(\frac{\lambda\alpha}{r_i}-1)-1}} \right\}^{\frac{1}{p_i}}, \tag{6.3.37}$$

$$\left\{ \sum_{m_n=2}^{\infty} \frac{1}{m_n} (\ln m_n)^{\frac{q_n\alpha\lambda}{r_n}-1} \right.$$

$$\left. \times \left[\sum_{m_{n-1}=2}^{\infty} \cdots \sum_{m_1=2}^{\infty} \frac{1}{(\max_{1\le j\le n}\{\ln^\alpha m_j\})^\lambda} \prod_{i=1}^{n-1} a_{m_i}^{(i)} \right]^{q_n} \right\}^{\frac{1}{q_n}}$$

$$< \frac{r_n}{(\alpha\lambda)^{n-1}} \prod_{i=1}^{n-1} r_i \left\{ \sum_{m_i=2}^{\infty} \frac{m_i^{p_i-1}(a_{m_i}^{(i)})^{p_i}}{(\ln m_i)^{p_i(\frac{\lambda\alpha}{r_i}-1)-1}} \right\}^{\frac{1}{p_i}}; \tag{6.3.38}$$

(b) if $0 < p_n < 1$, $p_i < 0$ $(i = 1, \cdots, n-1)$, then we have the following equivalent inequalities with the best constant factor $\frac{1}{\lambda^{n-1}} \prod_{i=1}^{n} r_i$:

$$\sum_{m_n=2}^{\infty} \cdots \sum_{m_1=2}^{\infty} \frac{1}{(\max_{1\le j\le n}\{\ln^\alpha m_j\})^\lambda} \prod_{i=1}^{n} a_{m_i}^{(i)}$$

$$> \frac{r_n}{(\alpha\lambda)^{n-1}} \prod_{i=1}^{n-1} r_i \left\{ \sum_{m_i=2}^{\infty} \frac{m_i^{p_i-1}(a_{m_i}^{(i)})^{p_i}}{(\ln m_i)^{p_i(\frac{\lambda\alpha}{r_i}-1)-1}} \right\}^{\frac{1}{p_i}}$$

$$\times \left\{ \sum_{m_n=2}^{\infty} [1-\theta_\lambda(m_n)] \frac{m_n^{p_n-1}(a_{m_n}^{(n)})^{p_n}}{(\ln m_n)^{p_n(\frac{\lambda\alpha}{r_n}-1)-1}} \right\}^{\frac{1}{p_n}}, \tag{6.3.39}$$

$$\left\{ \sum_{m_n=2}^{\infty} \frac{(\ln m_n)^{\frac{q_n\alpha\lambda}{r_n}-1}}{m_n[1-\theta_\lambda(m_n)]^{q_n-1}} \right.$$

$$\left. \times \left[\sum_{m_{n-1}=2}^{\infty} \cdots \sum_{m_1=2}^{\infty} \frac{1}{(\max_{1\le j\le n}\{\ln^\alpha m_j\})^\lambda} \prod_{i=1}^{n-1} a_{m_i}^{(i)} \right]^{q_n} \right\}^{\frac{1}{q_n}}$$

$$< \frac{r_n}{(\alpha\lambda)^{n-1}} \prod_{i=1}^{n-1} r_i \left\{ \sum_{m_i=2}^{\infty} \frac{m_i^{p_i-1}(a_{m_i}^{(i)})^{p_i}}{(\ln m_i)^{p_i(\frac{\lambda\alpha}{r_i}-1)-1}} \right\}^{\frac{1}{p_i}}, \tag{6.3.40}$$

$$\left\{ \sum_{m_k=2}^{\infty} \frac{(\ln m_k)^{\frac{q_k\alpha\lambda}{r_k}-1}}{m_k} \left[\sum_{m_n=2}^{\infty} \cdots \sum_{m_{k+1}=2}^{\infty} \right. \right.$$

$$\times \sum_{m_{k-1}=2}^{\infty} \cdots \sum_{m_1=2}^{\infty} \frac{1}{(\max_{1\le j\le n}\{\ln^{\alpha} m_j\})^{\lambda}} \prod_{\substack{i=1\\(i\neq k)}}^{n} a_{m_i}^{(i)} \Bigg]^{q_k} \Bigg\}^{\frac{1}{q_k}}$$

$$> \frac{r_n r_k}{(\alpha\lambda)^{n-1}} \prod_{\substack{i=1\\(i\neq k)}}^{n-1} r_i \left\{ \sum_{m_i=2}^{\infty} \frac{m_i^{p_i-1}(a_{m_i}^{(i)})^{p_i}}{(\ln m_i)^{p_i(\frac{\lambda\alpha}{r_i}-1)-1}} \right\}^{\frac{1}{p_i}}$$

$$\times \left\{ \sum_{m_n=2}^{\infty} [1-\theta_{\lambda}(m_n)] \frac{m_n^{p_n-1}(a_{m_n}^{(n)})^{p_n}}{(\ln m_n)^{p_n(\frac{\lambda\alpha}{r_n}-1)-1}} \right\}^{\frac{1}{p_n}}$$

$$(1 \le k \le n-1), \qquad (6.3.41)$$

where

$$\theta_{\lambda}(m_n) = O(\tfrac{1}{\ln^{\lambda'} m_n}) \in (0, c_0)$$

$$(0 < c_0 < 1,\ 0 < \lambda' < \tfrac{\lambda}{\alpha} \min_{1\le i\le n}\{\tfrac{1}{r_i}\}; m_n \to \infty).$$

For giving Theorem 6.3.10, we consider the following two lemmas.

Lemma 6.3.8 If $r > 1, \frac{1}{r}+\frac{1}{s}=1$, $\beta \ge -\frac{1}{4}$, $0<\lambda < \min\{r,s\}$, then for any $k \in \mathbf{N}$ and $y_0 = k+\beta$, we have the following inequality

$$\sum_{m=1}^{\infty} \frac{y_0^{\lambda/s}(m+\beta)^{(\lambda/r)-1}}{(\max\{y_0, m+\beta\})^{\lambda}} < \frac{rs}{\lambda}. \qquad (6.3.42)$$

Proof Setting $f(x)$ as follows:

$$f(x) := \frac{y_0^{\lambda/s}(x+\beta)^{(\lambda/r)-1}}{(\max\{y_0, x+\beta\})^{\lambda}},$$

then we have

$$f(x) = \begin{cases} f_1(x), & -\beta < x \le y_0 - \beta = k; \\ f_2(x), & x > y_0 - \beta, \end{cases}$$

where,

$$f_1(x) := \frac{(x+\beta)^{(\lambda/r)-1}}{y_0^{\lambda/r}},\ f_2(x) := \frac{y_0^{\lambda/s}}{(x+\beta)^{1+(\lambda/s)}}.$$

Since $f_1(k) = f_2(k)$, by (3.1.49) and (3.1.50), we find

$$\sum_{m=1}^{\infty} f(m) = \sum_{m=1}^{k} f_1(m) + \sum_{m=k}^{\infty} f_2(m) - f_2(k)$$

$$< \int_1^k f_1(x)dx + \tfrac{1}{2}[f_1(1)+f_1(k)] + \tfrac{1}{12} f_1'(x)|_1^k$$

$$+ \int_k^{\infty} f_2(x)dx + \tfrac{1}{2} f_2(k) - \tfrac{1}{12} f_2'(k) - f_2(k)$$

$$= \int_1^{\infty} f(x)dx + \tfrac{f_1(1)}{2} + \tfrac{1}{12}[f_1'(x)|_1^k - f_2'(k)]. \qquad (6.3.43)$$

We find

$$f_1'(1) = (\tfrac{\lambda}{r}-1)\tfrac{1}{y_0^{\lambda/r}}(1+\beta)^{\frac{\lambda}{r}-2},$$

$$f_1'(k) = (\tfrac{\lambda}{r}-1)\tfrac{1}{y_0^2},\ f_2'(k) = -(\tfrac{\lambda}{s}+1)\tfrac{1}{y_0^2},$$

$$\int_1^{\infty} f(x)dx = \int_1^k f_1(x)dx + \int_k^{\infty} f_2(x)dx$$

$$= \tfrac{rs}{\lambda} - \tfrac{r}{\lambda}\tfrac{(1+\beta)^{\lambda/r}}{y_0^{\lambda/r}}.$$

By (6.3.3) and (6.3.43), it follows

$$\sum_{m=1}^{\infty} \frac{y_0^{\lambda/s}(m+\beta)^{(\lambda/r)-1}}{(\max\{y_0, m+\beta\})^{\lambda}} = \sum_{m=1}^{\infty} f(m)$$

$$< \tfrac{rs}{\lambda} - \tfrac{r}{\lambda}\tfrac{(1+\beta)^{\lambda/r}}{y_0^{\lambda/r}} + \tfrac{1}{2}\tfrac{(1+\beta)^{(\lambda/r)-1}}{y_0^{\lambda/r}}$$

$$+ \tfrac{1}{12}\Big\{(\tfrac{\lambda}{r}-1)[\tfrac{1}{y_0^2} - \tfrac{1}{y_0^{\lambda/r}}(1+\beta)^{\frac{\lambda}{r}-2}]$$

$$+(\tfrac{\lambda}{s}+1)\tfrac{1}{y_0^2}\Big\}$$

$$= \tfrac{rs}{\lambda} - \tfrac{(1+\beta)^{(\lambda/r)-2}}{y_0^{\lambda/r}}\rho, \qquad (6.3.44)$$

where,

$$\rho := \tfrac{r}{\lambda}(1+\beta)^2 - \tfrac{1+\beta}{2}$$

$$+\tfrac{1}{12}(\tfrac{\lambda}{r}-1) - \tfrac{\lambda}{12}(\tfrac{n+\beta}{1+\beta})^{\frac{\lambda}{r}-2}.$$

Since $\frac{\lambda}{r}-2<0, 1+\beta \ge \frac{3}{4}, \lambda < \min\{r,s\}$, then we find

$$R > \tfrac{r}{\lambda}(1+\beta)^2 - \tfrac{1+\beta}{2} + \tfrac{1}{12}(\tfrac{\lambda}{r}-1) - \tfrac{\lambda}{12}$$

$$= \tfrac{r}{\lambda}(1+\beta)^2 - \tfrac{1}{2}(1+\beta) - \tfrac{1}{12}(\tfrac{\lambda}{s}+1)$$

$$> (1+\beta)^2 - \tfrac{1}{2}(1+\beta) - \tfrac{1}{6}$$

$$= (1+\beta)(\tfrac{1}{2}+\beta) - \tfrac{1}{6} \ge \tfrac{3}{4}\times\tfrac{1}{4} - \tfrac{1}{6} > 0.$$

Hence by (6.3.44), we have (6.3.42). □

Lemma 6.3.9 If $n \ge 2, \beta \ge -\frac{1}{4}, r_j > 1$, $j = 1,\cdots,n$, $\sum_{j=1}^{n} \frac{1}{r_j} = 1, 0 < \lambda < \min_{1\le j\le n}\{r_j\}$, define the weight coefficients $\omega_i(m_i) = \omega(m_i; r_1,\cdots,r_n)$ as

$$\omega_i(m_i) := (m_i+\beta)^{\frac{\lambda}{r_i}} \sum_{m_n=1}^{\infty} \cdots \sum_{m_{i+1}=1}^{\infty} \sum_{m_{i-1}=1}^{\infty} \cdots \sum_{m_1=1}^{\infty}$$

$$\times \frac{\prod_{j=1(j\neq i)}^{n} (m_j+\beta)^{(\lambda/r_j)-1}}{[\max_{1\le j\le n}\{m_j+\beta\}]^{\lambda}}\ (i = 1,\cdots,n), \qquad (6.3.45)$$

then we have

$$\omega_i(m_i) < \tfrac{1}{\lambda^{n-1}} \prod_{j=1}^{n} r_j\,(i=1,\cdots,n). \qquad (6.3.46)$$

Proof We first prove that the following inequality:

$$\omega_n(m_n) = \omega(m_n; r_1,\cdots,r_n) < \tfrac{1}{\lambda^{n-1}} \prod_{j=1}^{n} r_j \qquad (6.3.47)$$

is valid for $n \ge 2$ by mathematics induction.

For n=2, we set $r=r_1, s=r_2, m=m_1$ and $y_0=m_2+\beta$, by (6.3.42), we have the following inequality:

$$\omega_2(m_2)=\sum_{m=1}^{\infty}\frac{y_0^{\lambda/s}(m+\beta)^{(\lambda/r)-1}}{(\max\{y_0,m+\beta\})^{\lambda}}<\frac{rs}{\lambda}=\frac{r_1r_2}{\lambda}. \quad (6.3.48)$$

Assuming that for $n(\geq 2)$, (6.3.47) is valid, setting $s_1=(1-\frac{1}{r_1})^{-1}, \tilde{\lambda}=\frac{\lambda}{s_1}, \tilde{r}_j=\frac{r_{j+1}}{s_1}$ and $\tilde{m}_j=m_{j+1}$ $(j=1,\cdots,n)$, we find $\sum_{j=1}^{n}\frac{1}{\tilde{r}_j}=(1-\frac{1}{r_1})s_1=1$ and $\tilde{\lambda}<\min_{1\leq j\leq n}\{\tilde{r}_j\}$. By (6.3.48) and the assumption of induction, for

$$y_0=\max_{2\leq j\leq n+1}\{m_j\}+\beta$$
$$=\max_{2\leq j\leq n+1}\{m_j+\beta\},$$

it follows

$$\omega_{n+1}(m_{n+1})=(m_{n+1}+\beta)^{\frac{\lambda}{r_{n+1}}}\sum_{m_n=1}^{\infty}\cdots\sum_{m_1=1}^{\infty}$$
$$\times\frac{\prod_{j=1}^{n}(m_j+\beta)^{(\lambda/r_j)-1}}{[\max_{1\leq j\leq n+1}\{m_j+\beta\}]^{\lambda}}$$
$$=(m_{n+1}+\beta)^{\frac{\lambda}{r_{n+1}}}\sum_{m_n=1}^{\infty}\cdots\sum_{m_2=1}^{\infty}\frac{\prod_{j=2}^{n}(m_j+\beta)^{(\lambda/r_j)-1}}{y_0^{\lambda/s_1}}$$
$$\times\left\{\sum_{m_1=1}^{\infty}\frac{y_0^{\lambda/s_1}(m_1+\beta)^{(\lambda/r_1)-1}}{[\max\{y_0,m_1+\beta\}]^{\lambda}}\right\}$$
$$<\frac{r_1s_1}{\lambda}(\tilde{m}_n+\beta)^{\frac{\tilde{\lambda}}{\tilde{r}_n}}\sum_{\tilde{m}_{n-1}=1}^{\infty}\cdots\sum_{\tilde{m}_1=1}^{\infty}\frac{\prod_{j=1}^{n-1}(\tilde{m}_j+\beta)^{(\tilde{\lambda}/\tilde{r}_j)-1}}{[\max_{1\leq j\leq n}\{\tilde{m}_j+\beta\}]^{\tilde{\lambda}}}$$
$$<\frac{r_1s_1}{\lambda}\frac{1}{\tilde{\lambda}^{n-1}}\prod_{j=1}^{n}\tilde{r}_j=\frac{r_1s_1}{\lambda}\frac{s_1^{n-1}}{\lambda^{n-1}}\prod_{j=2}^{n+1}\frac{r_j}{s_1}=\frac{1}{\lambda^n}\prod_{j=1}^{n+1}r_j.$$

Hence (6.3.47) is valid for any $n\geq 2$.

Setting $\tilde{m}_j=m_j, \tilde{r}_j=r_j(j=1,\cdots,i-1)$, $\tilde{m}_j=m_{j+1}, \tilde{r}_j=r_{j+1}(j=i,\cdots,n-1), \tilde{m}_n=m_i$ and $\tilde{r}_n=r_i$ in (6.3.45) (for i=n), by (6.3.47), we have

$$\omega_i(m_i)=\omega(\tilde{m}_n;\tilde{r}_1,\cdots,\tilde{r}_n)$$
$$<\frac{1}{\lambda^{n-1}}\prod_{j=1}^{n}\tilde{r}_j=\frac{1}{\lambda^{n-1}}\prod_{j=1}^{n}r_j.$$

Hence (6.3.46) is valid for any $n\geq 2$ and $i=1,\cdots,n$. □

By Corollary 6.2.14 and (6.3.30), for $c_\lambda>0$, we have

Theorem 6.3.10 Assuming that $n\geq 2$, $\beta\geq -\frac{1}{4}$, $p_n>0(\neq 1)$, $\frac{1}{q_i}=1-\frac{1}{p_i}$, $r_i>1$ $(i=1,\cdots,n)$, $0<\lambda<\min_{1\leq i\leq n}\{r_i\}$, $\sum_{i=1}^{n}\frac{1}{r_i}=\sum_{i=1}^{n}\frac{1}{p_i}=1$, and $a_{m_i}^{(i)}\geq 0$, such that for any $i=1,\cdots,n$,

$$0<\sum_{m_i=1}^{\infty}(m_i+\beta)^{p_i(1-\frac{\lambda}{r_i})-1}(a_{m_i}^{(i)})^{p_i}<\infty,$$

(a) if $p_i>1$ $(i=1,\cdots,n)$, then we have the following equivalent inequalities with the best constant factor $\frac{1}{\lambda^{n-1}}\prod_{i=1}^{n}r_i$:

$$\sum_{m_n=1}^{\infty}\cdots\sum_{m_1=1}^{\infty}\frac{1}{(\max_{1\leq j\leq n}\{m_j+\beta\})^{\lambda}}\prod_{i=1}^{n}a_{m_i}^{(i)}$$
$$<\frac{1}{\lambda^{n-1}}\prod_{i=1}^{n}r_i\left\{\sum_{m_i=1}^{\infty}(m_i+\beta)^{p_i(1-\frac{\lambda}{r_i})-1}(a_{m_i}^{(i)})^{p_i}\right\}^{\frac{1}{p_i}},$$
(6.3.49)

$$\left\{\sum_{m_n=1}^{\infty}(m_n+\beta)^{\frac{q_n\lambda}{r_n}-1}\right.$$
$$\left.\times\left[\sum_{m_{n-1}=1}^{\infty}\cdots\sum_{m_1=1}^{\infty}\frac{\prod_{i=1}^{n-1}a_{m_i}^{(i)}}{(\max_{1\leq j\leq n}\{m_j+\beta\})^{\lambda}}\right]^{q_n}\right\}^{\frac{1}{q_n}}$$
$$<\frac{r_n}{\lambda^{n-1}}\prod_{i=1}^{n-1}r_i\left\{\sum_{m_i=1}^{\infty}(m_i+\beta)^{p_i(1-\frac{\lambda}{r_i})-1}(a_{m_i}^{(i)})^{p_i}\right\}^{\frac{1}{p_i}};$$
(6.3.50)

(b) if $0<p_n<1, p_i<0$ $(i=1,\cdots,n-1)$, then we have the following equivalent inequalities with the best constant factor $\frac{1}{\lambda^{n-1}}\prod_{i=1}^{n}r_i$:

$$\sum_{m_n=1}^{\infty}\cdots\sum_{m_1=1}^{\infty}\frac{1}{(\max_{1\leq j\leq n}\{m_j+\beta\})^{\lambda}}\prod_{i=1}^{n}a_{m_i}^{(i)}$$
$$>\frac{r_n}{\lambda^{n-1}}\prod_{i=1}^{n-1}r_i\left\{\sum_{m_i=1}^{\infty}(m_i+\beta)^{p_i(1-\frac{\lambda}{r_i})-1}(a_{m_i}^{(i)})^{p_i}\right\}^{\frac{1}{p_i}}$$
$$\times\left\{\sum_{m_n=1}^{\infty}[1-\theta_\lambda(m_n)](m_n+\beta)^{p_n(1-\frac{\lambda}{r_n})-1}(a_{m_n}^{(n)})^{p_n}\right\}^{\frac{1}{p_n}},$$
(6.3.51)

$$\left\{\sum_{m_n=1}^{\infty}\frac{(m_n+\beta)^{\frac{q_n\lambda}{r_n}-1}}{[1-\theta_\lambda(m_n)]^{q_n-1}}\right.$$
$$\left.\times\left[\sum_{m_{n-1}=1}^{\infty}\cdots\sum_{m_1=1}^{\infty}\frac{1}{(\max_{1\leq j\leq n}\{m_j+\beta\})^{\lambda}}\prod_{i=1}^{n-1}a_{m_i}^{(i)}\right]^{q_n}\right\}^{\frac{1}{q_n}}$$

$$< \frac{r_n}{\lambda^{n-1}} \prod_{i=1}^{n-1} r_i \left\{ \sum_{m_i=1}^{\infty} (m_i+\beta)^{p_i(1-\frac{\lambda}{r_i})-1} (a_{m_i}^{(i)})^{p_i} \right\}^{\frac{1}{p_i}},$$

(6.3.52)

$$\left\{ \sum_{m_k=1}^{\infty} (m_k+\beta)^{\frac{q_k\lambda}{r_k}-1} \left[\sum_{m_n=1}^{\infty} \cdots \sum_{m_{k+1}=1}^{\infty} \right.\right.$$

$$\left.\left. \times \sum_{m_{k-1}=1}^{\infty} \cdots \sum_{m_1=1}^{\infty} \frac{1}{(\max\limits_{1\le j\le n}\{m_j+\beta\})^{\lambda}} \prod_{\substack{i=1\\(i\ne k)}}^{n} a_{m_i}^{(i)} \right]^{q_k} \right\}^{\frac{1}{q_k}}$$

$$> \frac{r_n r_k}{\lambda^{n-1}} \prod_{\substack{i=1\\(i\ne k)}}^{n-1} r_i \left\{ \sum_{m_i=1}^{\infty} (m_i+\beta)^{p_i(1-\frac{\lambda}{r_i})-1} (a_{m_i}^{(i)})^{p_i} \right\}^{\frac{1}{p_i}}$$

$$\times \left\{ \sum_{m_n=1}^{\infty} [1-\theta_\lambda(m_n)] \right.$$

$$\left. \times (m_n+\beta)^{p_n(1-\frac{\lambda}{r_n})-1} (a_{m_n}^{(n)})^{p_n} \right\}^{\frac{1}{p_n}}$$

$$(1 \le k \le n-1), \qquad (6.3.53)$$

where $\theta_\lambda(m_n) = O(\frac{1}{(m_n+\beta)^{\lambda'}}) \in (0, c_0)$ $(0 < c_0 < 1,\ 0 < \lambda' < \lambda \min\limits_{1\le i\le n}\{\frac{1}{r_i}\}; m_n \to \infty)$.

6.3.3. EXAMPLE FOR $k_{-\lambda}(t_1,\cdots,t_{n-1},1) = (\min\{t_1,\cdots,t_{n-1},1\})^{\lambda}$

Lemma 6.3.11 For $n \ge 2, \lambda > 0$, $r_i > 1$, $\lambda_i = \frac{-\lambda}{r_i} > -1$ $(i = 1,\cdots,n)$, $\sum_{i=1}^{n} \frac{1}{r_i} = 1$, we have

$$k_{-\lambda} = \int_{R_+^{n-1}} (\min\{t_1,\cdots,t_{n-1},1\})^{\lambda} \prod_{i=1}^{n-1} t_i^{\frac{-\lambda}{r_i}-1}$$

$$\times dt_1 \cdots dt_{n-1} = \frac{1}{\lambda^{n-1}} \prod_{i=1}^{n} r_i. \qquad (6.3.54)$$

Proof We prove (6.3.42) by mathematical induction. For $n = 2$, we find

$$k_{-\lambda} = \int_{R_+} (\min\{t_1,1\})^{\lambda} t_1^{\frac{-\lambda}{r_1}-1} dt_1$$

$$= \int_0^1 t_1^{\lambda} t_1^{\frac{-\lambda}{r_1}-1} dt_1 + \int_1^{\infty} t_1^{\frac{-\lambda}{r_1}-1} dt_1 = \frac{r_2}{\lambda} + \frac{r_1}{\lambda} = \frac{1}{\lambda} r_1 r_2 .$$

Assuming that for $n(\ge 2)$, (6.3.42) is valid, then for $n+1$, we have

$$k_{-\lambda} = \int_{R_+^n} (\min\{t_1,\cdots,t_n,1\})^{\lambda}$$

$$\times \prod_{i=1}^{n} t_i^{\frac{-\lambda}{r_i}-1} dt_1 \cdots dt_n$$

$$= \int_{R_+^{n-1}} \prod_{i=2}^{n} t_i^{\frac{-\lambda}{r_i}-1} [\int_0^{\infty} (\min\{t_1,\cdots,t_n,1\})^{\lambda}$$

$$\times t_1^{\frac{-\lambda}{r_1}-1} dt_1] dt_2 \cdots dt_n$$

$$= \int_{R_+^{n-1}} \prod_{i=2}^{n} t_i^{\frac{-\lambda}{r_i}-1}$$

$$\times [\int_0^{\min\{t_2,\cdots,t_n,1\}} t_1^{\lambda-\frac{\lambda}{r_1}-1} dt_1] dt_2 \cdots dt_n$$

$$+ \int_{R_+^{n-1}} (\min\{t_2,\cdots,t_n,1\})^{\lambda} \prod_{i=2}^{n} t_i^{\frac{-\lambda}{r_i}-1}$$

$$\times [\int_{\min\{t_2,\cdots,t_n,1\}}^{\infty} t_1^{\frac{-\lambda}{r_1}-1} dt] dt_2 \cdots dt_n$$

$$= \frac{r_1}{\lambda(r_1-1)} \int_{R_+^{n-1}} (\min\{t_2,\cdots,t_n,1\})^{\lambda(1-\frac{1}{r_1})}$$

$$\times \prod_{i=2}^{n} t_i^{\frac{-\lambda}{r_i}-1} dt_2 \cdots dt_n$$

$$+ \frac{r_1}{\lambda} \int_{R_+^{n-1}} (\min\{t_2,\cdots,t_n,1\})^{\lambda(1-\frac{\lambda}{r_1})}$$

$$\times \prod_{i=2}^{n} t_i^{\frac{-\lambda}{r_i}-1} dt_2 \cdots dt_n$$

$$= \frac{r_1}{\lambda} \frac{1}{(1-\frac{1}{r_1})} \int_{R_+^{n-1}} (\min\{t_2,\cdots,t_n,1\})^{\lambda(1-\frac{\lambda}{r_1})}$$

$$\times \prod_{i=2}^{n} t_i^{\frac{-\lambda(1-\frac{1}{r_1})}{r_i(1-\frac{1}{r_1})}-1} dt_2 \cdots dt_n$$

$$= \frac{r_1}{\lambda} \frac{1}{(1-\frac{1}{r_1})} \frac{1}{\lambda^{n-1}(1-\frac{1}{r_1})^{n-1}} \prod_{i=2}^{n+1} r_i (1-\frac{1}{r_1}) = \frac{1}{\lambda^n} \prod_{i=1}^{n+1} r_i .$$

In view of mathematical induction, we have (6.3.42). □

It is obvious that

$$c_{-\lambda} = \int_1^{\infty} \cdots \int_1^{\infty} (\min\{t_1,\cdots,t_{n-1},1\})^{\lambda}$$

$$\times \prod_{j=1}^{n-1} t_j^{\frac{-\lambda}{r_j}-1} dt_1 \cdots dt_{n-1} > 0 .$$

Without loses of generality, we show that for $n \ge 3$, $j = n-1, 0 < t_{n-1} \le 1$,

$$\tilde{\varphi}(t_{n-1}) := \int_{R_+^{n-2}} (\min\{t_1,\cdots,t_{n-1},1\})^{\lambda}$$

$$\times \prod_{i=1}^{n-2} t_i^{\frac{-\lambda}{r_i}-1} dt_1 \cdots dt_{n-2} \le M_{n-1} < \infty .$$

In fact, we find

$$\tilde{\varphi}(t_{n-1}) \le \int_{R_+^{n-2}} (\min\{t_1,\cdots,t_{n-1},1\})^{\lambda(1-\frac{1}{r_n})}$$

$$\times\prod_{i=1}^{n-2} t_i^{\frac{-\lambda}{r_i}-1} dt_1\cdots dt_{n-2}$$

$$\leq \int_{R_+^{n-2}} (\min\{t_1,\cdots,t_{n-2},1\})^{\lambda(1-\frac{1}{r_n})}$$

$$\times\prod_{i=1}^{n-2} t_i^{\frac{-\lambda}{r_n}-1} dt_1\cdots dt_{n-2}$$

$$\leq \int_{R_+^{n-2}} (\min\{t_1,\cdots,t_{n-2},1\})^{\lambda(1-\frac{1}{r_n})}$$

$$\times\prod_{i=1}^{n-2} t_i^{\frac{-\lambda(1-\frac{1}{r_n})}{r_i(1-\frac{1}{r_n})}-1} dt_1\cdots dt_{n-2}$$

$$=\frac{1}{\lambda^{n-2}(1-\frac{1}{r_n})^{n-2}}\prod_{i=1}^{n-1} r_i(1-\frac{1}{r_n})$$

$$=\frac{1-\frac{1}{r_n}}{\lambda^{n-2}}\prod_{i=1}^{n-1} r_i<\infty. \qquad (6.3.55)$$

For $n_0-1\geq t_0, u(n_0-1)\geq 0$, $i=1,\cdots,n$, since $0<\lambda<\min_{1\leq i\leq n}\{\frac{r_i}{r_i-1}\}$, it is obvious that the function

$$\tilde{f}_i(x_1,\cdots,x_n)=(\min_{1\leq j\leq n}\{u(x_j)\})^{\lambda}\prod_{\substack{j=1\\(j\neq i)}}^{n}\frac{u'(x_j)}{[u(x_j)]^{\frac{\lambda}{r_j}+1}}$$

$$=\begin{cases}\dfrac{u'(x_k)}{[u(x_k)]^{\lambda(\frac{1}{r_k}-1)+1}}\displaystyle\prod_{\substack{j=1\\(j\neq i,k)}}^{n}\dfrac{u'(x_j)}{[u(x_j)]^{\frac{\lambda}{r_j}+1}}, \\ \qquad u(x_k)\leq \min_{j\neq k}\{u(x_j)\}; \\ (\min_{j\neq k}\{u(x_j)\})^{\lambda}\displaystyle\prod_{\substack{j=1\\(j\neq i)}}^{n}\dfrac{u'(x_j)}{[u(x_j)]^{\frac{\lambda}{r_j}+1}}, \\ \qquad u(x_k)>\min_{j\neq k}\{u(x_j)\}\end{cases}$$

is strictly decreasing for any variable $x_k>n_0-1$ $(k\neq i)$.

By Corollary 6.2.11,(i) setting $u(x)=x^{\alpha}$ $0<\alpha\leq 1$, $n_0=1$, we have the following theorem:

Theorem 6.3.12 Assuming that $n\geq 2$, $0<\alpha\leq 1$, $p_n>0(\neq 1)$, $\frac{1}{q_i}=1-\frac{1}{p_i}$, $r_i>1$ $(i=1,\cdots,n)$,

$$0<\lambda<\min_{1\leq i\leq n}\{\tfrac{r_i}{r_i-1}\},$$

$\sum_{i=1}^{n}\frac{1}{r_i}=\sum_{i=1}^{n}\frac{1}{p_i}=1$, and $a_{m_i}^{(i)}\geq 0$, such that

$$0<\sum_{m_i=1}^{\infty} m_i^{p_i(1+\frac{\lambda\alpha}{r_i})-1}(a_{m_i}^{(i)})^{p_i}<\infty,$$

$$i=1,\cdots,n,$$

(a) if $p_i>1$ $(i=1,\cdots,n)$, then we have the following equivalent inequalities with the best constant factor $\frac{1}{\lambda^{n-1}}\prod_{i=1}^{n} r_i$:

$$\sum_{m_n=1}^{\infty}\cdots\sum_{m_1=1}^{\infty}(\min_{1\leq j\leq n}\{m_j^{\alpha}\})^{\lambda}\prod_{i=1}^{n}a_{m_i}^{(i)}$$

$$<\frac{1}{(\alpha\lambda)^{n-1}}\prod_{i=1}^{n} r_i\left\{\sum_{m_i=1}^{\infty} m_i^{p_i(1+\frac{\lambda\alpha}{r_i})-1}(a_{m_i}^{(i)})^{p_i}\right\}^{\frac{1}{p_i}}, \qquad (6.3.56)$$

$$\left\{\sum_{m_n=1}^{\infty}\frac{1}{m_n^{\frac{q_n\alpha\lambda}{r_n}+1}}\right.$$

$$\left.\times\left[\sum_{m_{n-1}=1}^{\infty}\cdots\sum_{m_1=1}^{\infty}(\min_{1\leq j\leq n}\{m_j^{\alpha}\})^{\lambda}\prod_{i=1}^{n-1}a_{m_i}^{(i)}\right]^{q_n}\right\}^{\frac{1}{q_n}}$$

$$<\frac{r_n}{(\alpha\lambda)^{n-1}}\prod_{i=1}^{n-1} r_i\left\{\sum_{m_i=1}^{\infty} m_i^{p_i(1+\frac{\lambda\alpha}{r_i})-1}(a_{m_i}^{(i)})^{p_i}\right\}^{\frac{1}{p_i}}; \qquad (6.3.57)$$

(b) if $0<p_n<1, p_i<0$ $(i=1,\cdots,n-1)$, then we have the following equivalent inequalities with the best constant factor $\frac{1}{(\alpha\lambda)^{n-1}}\prod_{i=1}^{n} r_i$:

$$\sum_{m_n=1}^{\infty}\cdots\sum_{m_1=1}^{\infty}(\min_{1\leq j\leq n}\{m_j^{\alpha}\})^{\lambda}\prod_{i=1}^{n}a_{m_i}^{(i)}$$

$$>\frac{r_n}{(\alpha\lambda)^{n-1}}\prod_{i=1}^{n-1} r_i\left\{\sum_{m_i=1}^{\infty} m_i^{p_i(1+\frac{\lambda\alpha}{r_i})-1}(a_{m_i}^{(i)})^{p_i}\right\}^{\frac{1}{p_i}}$$

$$\times\left\{\sum_{m_n=1}^{\infty}[1-\theta_{\lambda}(m_n)]m_n^{p_n(1+\frac{\lambda\alpha}{r_n})-1}(a_{m_n}^{(n)})^{p_n}\right\}^{\frac{1}{p_n}}, \qquad (6.3.58)$$

$$\left\{\sum_{m_n=1}^{\infty}\frac{[1-\theta_{\lambda}(m_n)]^{1-q_n}}{m_n^{\frac{q_n\alpha\lambda}{r_n}+1}}\right.$$

$$\left.\times\left[\sum_{m_{n-1}=1}^{\infty}\cdots\sum_{m_1=1}^{\infty}(\min_{1\leq j\leq n}\{m_j^{\alpha}\})^{\lambda}\prod_{i=1}^{n-1}a_{m_i}^{(i)}\right]^{q_n}\right\}^{\frac{1}{q_n}}$$

$$<\frac{r_n}{(\alpha\lambda)^{n-1}}\prod_{i=1}^{n-1} r_i\left\{\sum_{m_i=1}^{\infty} m_i^{p_i(1+\frac{\lambda\alpha}{r_i})-1}(a_{m_i}^{(i)})^{p_i}\right\}^{\frac{1}{p_i}}, \qquad (6.3.59)$$

$$\left\{\sum_{m_k=1}^{\infty}\frac{1}{m_k^{\frac{q_k\alpha\lambda}{r_k}+1}}\left[\sum_{m_n=1}^{\infty}\cdots\sum_{m_{k+1}=1}^{\infty}\right.\right.$$

$$\times \sum_{m_{k-1}=1}^{\infty} \cdots \sum_{m_1=1}^{\infty} (\min_{1\le j\le n}\{m_j^{\alpha}\})^{\lambda} \prod_{\substack{i=1\\(i\neq k)}}^{n} a_{m_i}^{(i)} \Bigg]^{q_k} \Bigg\}^{\frac{1}{q_k}}$$

$$> \frac{r_n r_k}{(\alpha\lambda)^{n-1}} \prod_{\substack{i=1\\(i\neq k)}}^{n-1} r_i \left\{ \sum_{m_i=1}^{\infty} m_i^{p_i(1+\frac{\lambda\alpha}{r_i})-1} (a_{m_i}^{(i)})^{p_i} \right\}^{\frac{1}{p_i}}$$

$$\times \left\{ \sum_{m_n=1}^{\infty} [1-\theta_\lambda(m_n)] m_n^{p_n(1+\frac{\lambda\alpha}{r_n})-1} (a_{m_n}^{(n)})^{p_n} \right\}^{\frac{1}{p_n}}$$

$$(1\le k\le n-1), \quad (6.3.60)$$

where,

$$\theta_\lambda(m_n) = O(\tfrac{1}{m_n^{\lambda'}}) \in (0, c_0)$$

$$(0<c_0<1,\ 0<\lambda'<\tfrac{\lambda}{\alpha}\min_{1\le i\le n}\{\tfrac{1}{r_i}\}; m_n\to\infty).$$

(ii) Setting $u(x)=\ln^{\alpha} x (0<\alpha\le 1)$, $n_0=2$, we have the following theorem:

Theorem 6.3.13 Assuming that $n\ge 2$, $0<\alpha\le 1$, $p_n>0(\neq 1)$, $\frac{1}{q_i}=1-\frac{1}{p_i}$, $r_i>1 (i=1,\cdots,n)$,

$$0<\lambda<\min_{1\le i\le n}\{\tfrac{r_i}{r_i-1}\},$$

$\sum_{i=1}^{n}\frac{1}{r_i}=\sum_{i=1}^{n}\frac{1}{p_i}=1$, and $a_{m_i}^{(i)}\ge 0$, such that

$$0<\sum_{m_i=2}^{\infty} \frac{(\ln m_i)^{p_i(\frac{\lambda\alpha}{r_i}+1)+1}}{m_i^{1-p_i}} (a_{m_i}^{(i)})^{p_i} <\infty,$$

$$i=1,\cdots,n,$$

(a) if $p_i>1$ $(i=1,\cdots,n)$, then we have the following equivalent inequalities with the best constant factor $\frac{1}{\lambda^{n-1}}\prod_{i=1}^{n} r_i$:

$$\sum_{m_n=2}^{\infty}\cdots\sum_{m_1=2}^{\infty} (\min_{1\le j\le n}\{\ln^{\alpha} m_j\})^{\lambda} \prod_{i=1}^{n} a_{m_i}^{(i)}$$

$$< \frac{1}{(\alpha\lambda)^{n-1}} \prod_{i=1}^{n} r_i \left\{ \sum_{m_i=2}^{\infty} \frac{(\ln m_i)^{p_i(\frac{\lambda\alpha}{r_i}+1)+1}}{m_i^{1-p_i}} (a_{m_i}^{(i)})^{p_i} \right\}^{\frac{1}{p_i}}, \quad (6.3.61)$$

$$\left\{ \sum_{m_n=2}^{\infty} \frac{1}{m_n(\ln m_n)^{\frac{q_n\alpha\lambda}{r_n}+1}} \right.$$

$$\left. \times \left[\sum_{m_{n-1}=2}^{\infty}\cdots\sum_{m_1=2}^{\infty} (\min_{1\le j\le n}\{\ln^{\alpha} m_j\})^{\lambda} \prod_{i=1}^{n-1} a_{m_i}^{(i)} \right]^{q_n} \right\}^{\frac{1}{q_n}}$$

$$< \frac{r_n}{(\alpha\lambda)^{n-1}} \prod_{i=1}^{n-1} r_i \left\{ \sum_{m_i=2}^{\infty} \frac{(\ln m_i)^{p_i(\frac{\lambda\alpha}{r_i}+1)+1}}{m_i^{1-p_i}} (a_{m_i}^{(i)})^{p_i} \right\}^{\frac{1}{p_i}}; \quad (6.3.62)$$

(b) if $0<p_n<1, p_i<0$ $(i=1,\cdots,n-1)$, then we have the following equivalent inequalities with the best constant factor $\frac{1}{\lambda^{n-1}}\prod_{i=1}^{n} r_i$:

$$\sum_{m_n=2}^{\infty}\cdots\sum_{m_1=2}^{\infty} (\min_{1\le j\le n}\{\ln^{\alpha} m_j\})^{\lambda} \prod_{i=1}^{n} a_{m_i}^{(i)}$$

$$> \frac{r_n}{(\alpha\lambda)^{n-1}} \prod_{i=1}^{n-1} r_i \left\{ \sum_{m_i=2}^{\infty} \frac{(\ln m_i)^{p_i(\frac{\lambda\alpha}{r_i}+1)+1}}{m_i^{1-p_i}} (a_{m_i}^{(i)})^{p_i} \right\}^{\frac{1}{p_i}}$$

$$\times \left\{ \sum_{m_n=2}^{\infty} [1-\theta_\lambda(m_n)] \frac{(\ln m_n)^{p_n(\frac{\lambda\alpha}{r_n}+1)+1}}{m_i^{1-p_n}} (a_{m_n}^{(n)})^{p_n} \right\}^{\frac{1}{p_n}}, \quad (6.3.63)$$

$$\left\{ \sum_{m_n=2}^{\infty} \frac{[1-\theta_\lambda(m_n)]^{1-q_n}}{m_n(\ln m_n)^{\frac{q_n\alpha\lambda}{r_n}+1}} \right.$$

$$\left. \times \left[\sum_{m_{n-1}=2}^{\infty}\cdots\sum_{m_1=2}^{\infty} (\min_{1\le j\le n}\{\ln^{\alpha} m_j\})^{\lambda} \prod_{i=1}^{n-1} a_{m_i}^{(i)} \right]^{q_n} \right\}^{\frac{1}{q_n}}$$

$$< \frac{r_n}{(\alpha\lambda)^{n-1}} \prod_{i=1}^{n-1} r_i \left\{ \sum_{m_i=2}^{\infty} \frac{(\ln m_i)^{p_i(\frac{\lambda\alpha}{r_i}+1)+1}}{m_i^{1-p_i}} (a_{m_i}^{(i)})^{p_i} \right\}^{\frac{1}{p_i}}, \quad (6.3.64)$$

$$\left\{ \sum_{m_k=2}^{\infty} \frac{1}{m_k(\ln m_k)^{\frac{q_k\alpha\lambda}{r_k}+1}} \left[\sum_{m_n=2}^{\infty}\cdots\sum_{m_{k+1}=2}^{\infty} \right.\right.$$

$$\left.\left. \times \sum_{m_{k-1}=2}^{\infty}\cdots\sum_{m_1=2}^{\infty} (\min_{1\le j\le n}\{\ln^{\alpha} m_j\})^{\lambda} \prod_{\substack{i=1\\(i\neq k)}}^{n} a_{m_i}^{(i)} \right]^{q_k} \right\}^{\frac{1}{q_k}}$$

$$\times \left\{ \sum_{m_n=2}^{\infty} [1-\theta_\lambda(m_n)] \frac{(\ln m_n)^{p_n(\frac{\lambda\alpha}{r_n}+1)+1}}{m_n^{1-p_n}} (a_{m_n}^{(n)})^{p_n} \right\}^{\frac{1}{p_n}}$$

$$> \frac{r_n r_k}{(\alpha\lambda)^{n-1}} \prod_{\substack{i=1\\(i\neq k)}}^{n-1} r_i \left\{ \sum_{m_i=2}^{\infty} \frac{(\ln m_i)^{p_i(\frac{\lambda\alpha}{r_i}+1)+1}}{m_i^{1-p_i}} (a_{m_i}^{(i)})^{p_i} \right\}^{\frac{1}{p_i}}$$

$$(1\le k\le n-1), \quad (6.3.65)$$

where, $\theta_\lambda(m_n)=O(\frac{1}{\ln^{\lambda'} m_n})\in(0,c_0)$

$$(0<c_0<1,\ 0<\lambda'<\tfrac{\lambda}{\alpha}\min_{1\le i\le n}\{\tfrac{1}{r_i}\}; m_n\to\infty).$$

For given Theorem 6.3.16, we introduce the following lemmas.

Lemma 6.3.14 If $r>1,\frac{1}{r}+\frac{1}{s}=1,\ \beta\ge-\frac{1}{4}$,

$$0<\lambda<\min\{r,s\}=\min\{\frac{s}{s-1},\frac{r}{r-1}\},$$

then for any $k\in\mathbf{N}$ and $y_0=k+\beta$, we have

$$\sum_{m=1}^{\infty}\frac{(\min\{y_0,m+\beta\})^{\lambda}}{y_0^{\lambda/s}(m+\beta)^{(\lambda/r)+1}}<\frac{rs}{\lambda}. \quad (6.3.66)$$

Proof Since it follows

$$(\min\{y_0,m+\beta\})^{\lambda}=\frac{y_0^{\lambda}(m+\beta)^{\lambda}}{(\max\{y_0,m+\beta\})^{\lambda}},$$

then by (6.3.42), we have

$$\sum_{m=1}^{\infty}\frac{(\min\{y_0,m+\beta\})^{\lambda}}{y_0^{\lambda/s}(m+\beta)^{(\lambda/r)+1}}=\sum_{m=1}^{\infty}\frac{1}{y_0^{\lambda/s}(m+\beta)^{(\lambda/r)+1}}\frac{y_0^{\lambda}(m+\beta)^{\lambda}}{(\max\{y_0,m+\beta\})^{\lambda}}$$

$$=\sum_{m=1}^{\infty}\frac{y_0^{\lambda/r}(m+\beta)^{(\lambda/s)-1}}{(\max\{y_0,m+\beta\})^{\lambda}}<\frac{rs}{\lambda}.$$

Hence (6.3.66) is valid. □

Lemma 6.3.15 If $n\ge2,r_i>1(i=1,\cdots,n)$, $\sum_{i=1}^{n}\frac{1}{r_i}=1,\ 0<\lambda<\min_{1\le i\le n}\{\frac{r_i}{r_i-1}\},\beta\ge-\frac{1}{4}$, for $i=1,\cdots,n,$ define the weight coefficients $\omega_i(m_i)=\omega(m_i;r_1,\cdots,r_n)$ as follows:

$$\omega_i(m_i):=\frac{1}{(m_i+\beta)^{\lambda/r_i}}\sum_{m_n=1}^{\infty}\cdots\sum_{m_{i+1}=1}^{\infty}\sum_{m_{i-1}=1}^{\infty}\cdots\sum_{m_1=1}^{\infty}$$

$$\times[\min_{1\le j\le n}\{m_j+\beta\}]^{\lambda}\prod_{\substack{j=1\\(j\ne i)}}^{n}\frac{1}{(m_j+\beta)^{(\lambda/r_j)+1}}, \quad (6.3.67)$$

Then we have

$$\omega_i(m_i)<\frac{1}{\lambda^{n-1}}\prod_{j=1}^{n}r_j\ (i=1,\cdots,n). \quad (6.3.68)$$

Proof We first prove that the following inequality

$$\omega_n(m_n)=\omega(m_n;r_1,\cdots,r_n)<\frac{1}{\lambda^{n-1}}\prod_{j=1}^{n}r_j \quad (6.3.69)$$

is valid for $n\ge2$ by mathematics induction.

For n=2, we set $r=r_1,s=r_2,m=m_1$ and $y_0=m_2+\beta$, by (6.3.66), we have

$$\omega_2(m_2)=\sum_{m=1}^{\infty}\frac{(\min\{y_0,m+\beta\})^{\lambda}}{y_0^{\lambda/s}(m+\beta)^{(\lambda/r)+1}}<\frac{rs}{\lambda}=\frac{r_1r_2}{\lambda}. \quad (6.3.70)$$

Assuming that for $n(\ge2)$, (6.3.69) is valid, setting $s_1=(1-\frac{1}{r_1})^{-1},\tilde{\lambda}=\frac{\lambda}{s_1},\tilde{r}_j=\frac{r_{j+1}}{s_1}$ and $\tilde{m}_j=m_{j+1}$ $(j=1,\cdots,n)$, we find $\sum_{j=1}^{n}\frac{1}{\tilde{r}_j}=(1-\frac{1}{r_1})s_1=1$. Since for any $j=1,\cdots,n$,

$$\tilde{\lambda}=\frac{\lambda}{s_1}<\lambda<\frac{r_{j+1}}{r_{j+1}-1}<\frac{r_{j+1}/s_1}{(r_{j+1}/s_1)-1}=\frac{\tilde{r}_j}{\tilde{r}_j-1},$$

then $\tilde{\lambda}<\min_{1\le j\le n}\{\frac{\tilde{r}_j}{\tilde{r}_j-1}\}$. By (6.3.70) and the assumption of induction, for $y_0=\min_{2\le j\le n+1}\{m_j+\beta\}$, it follows

$$\omega_{n+1}(m_{n+1})=\frac{1}{(m_{n+1}+\beta)^{\frac{\lambda}{r_{n+1}}}}\sum_{m_n=1}^{\infty}\cdots\sum_{m_1=1}^{\infty}$$

$$\times[\min_{1\le j\le n+1}\{m_j+\beta\}]^{\lambda}\prod_{j=1}^{n}\frac{1}{(m_j+\beta)^{(\lambda/r_j)+1}}$$

$$=\frac{1}{(m_{n+1}+\beta)^{\frac{\lambda}{r_{n+1}}}}\sum_{m_n=1}^{\infty}\cdots\sum_{m_2=1}^{\infty}\prod_{j=2}^{n}\frac{y_0^{\lambda/s_1}}{(m_j+\beta)^{(\lambda/r_j)+1}}$$

$$\times\left\{\sum_{m_1=1}^{\infty}\frac{[\min\{y_0,m_1+\beta\}]^{\lambda}}{y_0^{\lambda/s_1}(m_1+\beta)^{(\lambda/r_1)+1}}\right\}$$

$$<\frac{r_1s_1}{\lambda}\frac{1}{(\tilde{m}_n+\beta)^{\tilde{\lambda}/\tilde{r}_n}}\sum_{\tilde{m}_{n-1}=1}^{\infty}\cdots\sum_{\tilde{m}_1=1}^{\infty}\prod_{j=1}^{n-1}\frac{[\min_{1\le j\le n}\{\tilde{m}_j+\beta\}]^{\tilde{\lambda}}}{(\tilde{m}_j+\beta)^{(\tilde{\lambda}/\tilde{r}_j)+1}}$$

$$<\frac{r_1s_1}{\lambda}\frac{1}{\tilde{\lambda}^{n-1}}\prod_{j=1}^{n}\tilde{r}_j=\frac{r_1s_1}{\lambda}\frac{s_1^{n-1}}{\lambda^{n-1}}\prod_{j=2}^{n+1}\frac{r_j}{s_1}=\frac{1}{\lambda^{n}}\prod_{j=1}^{n+1}r_j.$$

Hence (6.3.69) is valid for any $n\ge2$.

Setting $\tilde{m}_j=m_j,\tilde{r}_j=r_j(j=1,\cdots,i-1)$, $\tilde{m}_j=m_{j+1},\ \tilde{r}_j=r_{j+1}(j=i,\cdots,n-1),\tilde{m}_n=m_i$ and $\tilde{r}_n=r_i$ in (6.3.67) (for i=n), by (6.3.69), we have

$$\omega_i(m_i)=\omega(\tilde{m}_n;\tilde{r}_1,\cdots,\tilde{r}_n)$$

$$<\frac{1}{\lambda^{n-1}}\prod_{j=1}^{n}\tilde{r}_j=\frac{1}{\lambda^{n-1}}\prod_{j=1}^{n}r_j.$$

Hence (6.3.68) is valid for any $n\ge2$ and $i=1,\cdots,n$. □

By Corollary 6.2.16 and (6.3.33), for $c_{-\lambda}>0$, we have

Theorem 6.3.16 Assuming that $n\ge2$, $\beta\ge-\frac{1}{4}$, $p_n>0(\ne1)$, $\frac{1}{q_i}=1-\frac{1}{p_i}$, $r_i>1$ $(i=1,\cdots,n)$, $0<\lambda<\min_{1\le j\le n}\{\frac{r_j}{r_j-1}\},\sum_{i=1}^{n}\frac{1}{r_i}=\sum_{i=1}^{n}\frac{1}{p_i}=1$, $a_{m_i}^{(i)}\ge0$, such that

$$0<\sum_{m_i=1}^{\infty}(m_i+\beta)^{p_i(1+\frac{\lambda}{r_i})-1}(a_{m_i}^{(i)})^{p_i}<\infty,$$

$$i=1,\cdots,n,$$

(a) if $p_i>1$ $(i=1,\cdots,n)$, then we have the following equivalent inequalities with the best constant factor $\frac{1}{\lambda^{n-1}}\prod_{i=1}^{n}r_i$:

$$\sum_{m_n=1}^{\infty}\cdots\sum_{m_1=1}^{\infty}(\min_{1\le j\le n}\{m_j+\beta\})^{\lambda}\prod_{i=1}^{n}a_{m_i}^{(i)}$$

$$<\frac{1}{\lambda^{n-1}}\prod_{i=1}^{n}r_i\left\{\sum_{m_i=1}^{\infty}(m_i+\beta)^{p_i(1+\frac{\lambda}{r_i})-1}(a_{m_i}^{(i)})^{p_i}\right\}^{\frac{1}{p_i}},\tag{6.3.71}$$

$$\left\{\sum_{m_n=1}^{\infty}\frac{1}{(m_n+\beta)^{\frac{q_n\lambda}{r_n}+1}}\times\left[\sum_{m_{n-1}=1}^{\infty}\cdots\sum_{m_1=1}^{\infty}(\min_{1\le j\le n}\{m_j+\beta\})^{\lambda}\prod_{i=1}^{n-1}a_{m_i}^{(i)}\right]^{q_n}\right\}^{\frac{1}{q_n}}$$

$$<\frac{r_n}{\lambda^{n-1}}\prod_{i=1}^{n-1}r_i\left\{\sum_{m_i=1}^{\infty}(m_i+\beta)^{p_i(1+\frac{\lambda}{r_i})-1}(a_{m_i}^{(i)})^{p_i}\right\}^{\frac{1}{p_i}};\tag{6.3.72}$$

(b) if $0<p_n<1, p_i<0$ $(i=1,\cdots,n-1)$, then we have the following equivalent inequalities with the best constant factor $\frac{1}{\lambda^{n-1}}\prod_{i=1}^{n}r_i$:

$$\sum_{m_n=1}^{\infty}\cdots\sum_{m_1=1}^{\infty}(\min_{1\le j\le n}\{m_j+\beta\})^{\lambda}\prod_{i=1}^{n}a_{m_i}^{(i)}$$

$$>\frac{r_n}{\lambda^{n-1}}\prod_{i=1}^{n-1}r_i\left\{\sum_{m_i=1}^{\infty}(m_i+\beta)^{p_i(1+\frac{\lambda}{r_i})-1}(a_{m_i}^{(i)})^{p_i}\right\}^{\frac{1}{p_i}}$$

$$\times\left\{\sum_{m_n=1}^{\infty}[1-\theta_\lambda(m_n)](m_n+\beta)^{p_n(1+\frac{\lambda}{r_n})-1}(a_{m_n}^{(n)})^{p_n}\right\}^{\frac{1}{p_n}},\tag{6.3.73}$$

$$\left\{\sum_{m_n=1}^{\infty}\frac{[1-\theta_\lambda(m_n)]^{1-q_n}}{(m_n+\beta)^{\frac{q_n\lambda}{r_n}+1}}\times\left[\sum_{m_{n-1}=1}^{\infty}\cdots\sum_{m_1=1}^{\infty}(\min_{1\le j\le n}\{m_j+\beta\})^{\lambda}\prod_{i=1}^{n-1}a_{m_i}^{(i)}\right]^{q_n}\right\}^{\frac{1}{q_n}}$$

$$<\frac{r_n}{\lambda^{n-1}}\prod_{i=1}^{n-1}r_i\left\{\sum_{m_i=1}^{\infty}(m_i+\beta)^{p_i(1+\frac{\lambda}{r_i})-1}(a_{m_i}^{(i)})^{p_i}\right\}^{\frac{1}{p_i}},\tag{6.3.74}$$

$$\left\{\sum_{m_k=1}^{\infty}\frac{1}{(m_k+\beta)^{\frac{q_k\lambda}{r_k}+1}}\left[\sum_{m_n=1}^{\infty}\cdots\sum_{m_{k+1}=1}^{\infty}\times\sum_{m_{k-1}=1}^{\infty}\cdots\sum_{m_1=1}^{\infty}(\min_{1\le j\le n}\{m_j+\beta\})^{\lambda}\prod_{\substack{i=1\\(i\ne k)}}^{n}a_{m_i}^{(i)}\right]^{q_k}\right\}^{\frac{1}{q_k}}$$

$$>\frac{r_nr_k}{\lambda^{n-1}}\prod_{\substack{i=1\\(i\ne k)}}^{n-1}r_i\left\{\sum_{m_i=1}^{\infty}(m_i+\beta)^{p_i(1+\frac{\lambda}{r_i})-1}(a_{m_i}^{(i)})^{p_i}\right\}^{\frac{1}{p_i}}$$

$$\times\left\{\sum_{m_n=1}^{\infty}[1-\theta_\lambda(m_n)](m_n+\beta)^{p_n(1+\frac{\lambda}{r_n})-1}(a_{m_n}^{(n)})^{p_n}\right\}^{\frac{1}{p_n}}$$

$$(1\le k\le n-1),\tag{6.3.75}$$

where,

$$\theta_\lambda(m_n)=O(\frac{1}{(m_n+\beta)^{\lambda'}})\in(0,c_0)$$

$$(0<c_0<1,\ 0<\lambda'<\lambda\min_{1\le i\le n}\{\tfrac{1}{r_i}\};m_n\to\infty).$$

In particular, for $\beta=0,\lambda=1$, (a) if $r_i=p_i>1$ $(i=1,\cdots,n)$, then we have the following equivalent inequalities with the best constant factor $\prod_{i=1}^{n}p_i$:

$$\sum_{m_n=1}^{\infty}\cdots\sum_{m_1=1}^{\infty}\min_{1\le j\le n}\{m_j\}\prod_{i=1}^{n}a_{m_i}^{(i)}$$

$$<\prod_{i=1}^{n}p_i\left\{\sum_{m_i=1}^{\infty}m_i^{p_i}(a_{m_i}^{(i)})^{p_i}\right\}^{\frac{1}{p_i}},\tag{6.3.76}$$

$$\left\{\sum_{m_n=1}^{\infty}\frac{1}{m_n^{q_n+1}}\left[\sum_{m_{n-1}=1}^{\infty}\cdots\sum_{m_1=1}^{\infty}\min_{1\le j\le n}\{m_j\}\prod_{i=1}^{n-1}a_{m_i}^{(i)}\right]^{q_n}\right\}^{\frac{1}{q_n}}$$

$$<p_n\prod_{i=1}^{n-1}p_i\left\{\sum_{m_i=1}^{\infty}m_i^{p_i}(a_{m_i}^{(i)})^{p_i}\right\}^{\frac{1}{p_i}};\tag{6.3.77}$$

(a) if $0<p_n<1, p_i<0$ $(i=1,\cdots,n-1)$, then we have the following equivalent inequalities with the best constant factor $\prod_{i=1}^{n}r_i$:

$$\sum_{m_n=1}^{\infty}\cdots\sum_{m_1=1}^{\infty}\min_{1\le j\le n}\{m_j\}\prod_{i=1}^{n}a_{m_i}^{(i)}$$

$$>r_n\prod_{i=1}^{n-1}r_i\left\{\sum_{m_i=1}^{\infty}m_i^{p_i(1+\frac{1}{r_i})-1}(a_{m_i}^{(i)})^{p_i}\right\}^{\frac{1}{p_i}}$$

$$\times\left\{\sum_{m_n=1}^{\infty}[1-\theta_1(m_n)]m_n^{p_n(1+\frac{1}{r_n})-1}(a_{m_n}^{(n)})^{p_n}\right\}^{\frac{1}{p_n}},$$

(6.3.78)

$$\left\{\sum_{m_n=1}^{\infty}\frac{[1-\theta_1(m_n)]^{1-q_n}}{m_n^{\frac{q_n}{r_n}+1}}\times\left[\sum_{m_{n-1}=1}^{\infty}\cdots\sum_{m_1=1}^{\infty}\min_{1\le j\le n}\{m_j\}\prod_{i=1}^{n-1}a_{m_i}^{(i)}\right]^{q_n}\right\}^{\frac{1}{q_n}}$$

$$< r_n\prod_{i=1}^{n-1}r_i\left\{\sum_{m_i=1}^{\infty}m_i^{p_i(1+\frac{1}{r_i})-1}(a_{m_i}^{(i)})^{p_i}\right\}^{\frac{1}{p_i}}, \quad (6.3.79)$$

$$\left\{\sum_{m_k=1}^{\infty}\frac{1}{m_k^{\frac{q_k}{r_k}+1}}\left[\sum_{m_n=1}^{\infty}\cdots\sum_{m_{k+1}=1}^{\infty}\times\sum_{m_{k-1}=1}^{\infty}\cdots\sum_{m_1=1}^{\infty}\min_{1\le j\le n}\{m_j\}\prod_{\substack{i=1\\(i\ne k)}}^{n}a_{m_i}^{(i)}\right]^{q_k}\right\}^{\frac{1}{q_k}}$$

$$> r_n r_k\prod_{\substack{i=1\\(i\ne k)}}^{n-1}r_i\left\{\sum_{m_i=1}^{\infty}m_i^{p_i(1+\frac{1}{r_i})-1}(a_{m_i}^{(i)})^{p_i}\right\}^{\frac{1}{p_i}}$$

$$\times\left\{\sum_{m_n=1}^{\infty}[1-\theta_1(m_n)]m_n^{p_n(1+\frac{1}{r_n})-1}(a_{m_n}^{(n)})^{p_n}\right\}^{\frac{1}{p_n}}$$

$$(1\le k\le n-1), \quad (6.3.80)$$

where, $\theta_2(m_n)=O(\frac{1}{m_n^{\lambda'}})\in(0,c_0)$

$(0<c_0<1,\ 0<\lambda'<\min_{1\le i\le n}\{\frac{1}{r_i}\};m_n\to\infty)$.

6.4 REFERENCES

1. Kuang JC. Applied inequalities. Jinan: Shangdong Science Technology Press, 2004.
2. Kuang JC. Introduction to real analysis. Changsha: Hunan Education Press, 1996.
3. Yang BC. On a more accurate Mulhulland's inequality. Journal of Guangdong Education Institute, 2010,30(3):5-11.
4. Wang ZQ, Guo DR. Introduction to special functions. Beijing: Science Press, 1979.
5. Qiliang Huang. On a Multiple Hilbert's inequality with parameters. Journal of Inequalities and Applications. Volume 2010, Article ID 309319, 12 pages. Doi:10.1155/2010/ 309319.

Index

I

K

L

M

N

O

P

R

S

www.ingramcontent.com/pod-product-compliance
Lightning Source LLC
Chambersburg PA
CBHW041710210326
41598CB00007B/602
9781608053315